Applied Statistical
Decision Theory

Applied Statistical Decision Theory

HOWARD RAIFFA
ROBERT SCHLAIFER

Wiley Classics Library Edition Published 2000

A Wiley-Interscience Publication
JOHN WILEY & SONS, INC.
New York • Chichester • Weinheim • Brisbane • Singapore • Toronto

This text is printed on acid-free paper. ♾

Published simultaneously in Canada.

Wiley Classics Library Edition published 2000.

Library of Congress Cataloging in Publication Data:

Library of Congress Catalog Card Number: 68-25380
ISBN 0-471-38349-X (Classics Edition)

10 9 8 7 6 5 4 3 2 1

FOREWORD

The central objective of the research program of the Harvard Business School is the search for and development of findings which will enable business administrators to make wiser decisions in some phase of their activities. The broad field of business administration involves so many facets or dimensions that the nature of research findings varies widely in character. Nearly every field of study and investigation by scholars has something to contribute to business management. In recent years the School has been making special efforts to bring to bear on business management problems the skills and approaches of men trained and knowledgeable in some of the more significantly relevant disciplines which underlie business administration. One such special effort is the application of mathematics and statistics to business problems of decision under conditions of uncertainty. The general nature of this approach is set forth in Professor Schlaifer's *Probability and Statistics for Business Decisions* published by the McGraw-Hill Book Company in 1959, and it is expected that work in such problems will be a continuing part of the research effort of the School.

It is seldom possible, however, to take findings, analytical techniques, or conceptual frameworks from a basic discipline and apply them directly to a concrete business problem. Often the findings of another field must be refined and adapted specifically to the particular problem at hand. At times it is discovered that substantial new work is required of a fundamental character before an approach to a particular class of applied business problems is possible. This volume reports the results of research of the latter type. In the field of statistical decision theory Professors Raiffa and Schlaifer have sought to develop new analytical techniques by which the modern theory of utility and subjective probability can actually be applied to the economic analysis of typical sampling problems.

This book, the first in a group entitled Studies in Managerial Economics, is addressed to persons who are interested in using statistics as a tool in practical problems of decision making under conditions of uncertainty and who also have the necessary training in mathematics and statistics to employ these analytical techniques. It is not written for the general businessman, in contrast to most of the publications of the Division of Research in the past. It is the first of a new class of publications of the Division of Research that will report the results of research in the basic disciplines that underlie the field of business administration. These results, however, are expected to be widely usable in later studies dealing with actual business problems.

Financial support for this study came from an allocation by the School of a portion of a generous grant of the Ford Foundation to provide general support for the School's basic and exploratory research program. The School is indebted to the Ford Foundation for the support of this type of research endeavor.

Soldiers Field
Boston, Massachusetts
November 1960

BERTRAND FOX
Director of Research

PREFACE AND INTRODUCTION

This book is an introduction to the mathematical analysis of decision making when the state of the world is uncertain but further information about it can be obtained by experimentation. For our present purpose we take as given that the objective of such analysis is to identify a course of action (which may or may not include experimentation) that is logically consistent with the decision maker's own preferences for consequences, as expressed by numerical utilities, and with the weights he attaches to the possible states of the world, as expressed by numerical probabilities. The logical and philosophical justification for this statement of the problem has been fully developed by Savage in his *Foundations of Statistics*;† the purpose of the present book is not to discuss these basic principles but to contribute to the body of analytical techniques and numerical results that are needed if practical decision problems are to be solved in accordance with them.

We should like at the outset to call the reader's attention to the fact that the so-called "Bayesian" principles underlying the methods of analysis presented in this book are in no sense in *conflict* with the principles underlying the traditional decision theory of Neyman and Pearson. Statisticians of the school of Neyman and Pearson agree with us—although they use different words—that the decision maker who must choose a particular decision rule from within a suitable family of rules should both carefully appraise the possible consequences of the acts to which the rules may lead and carefully consider the relative importance to him of having a rule which behaves well in certain states of nature versus a rule which behaves well in other states. The only real novelty in the Bayesian approach lies in the fact that it provides a *formal mechanism* for taking account of these preferences and weights instead of leaving it to the decision maker's unaided intuition to determine their implications. We believe, however, that without this formalization decisions under uncertainty have been and will remain essentially arbitrary, as evidenced by the fact that, in most statistical practice, consequences and performance characteristics receive mere lip service while decisions are actually made by treating the numbers .05 and .95 with the same supertitious awe that is usually reserved for the number 13. Even further, we believe that formalization of utilities and weights leads to decisions which are not only less arbitrary but actually more *objective*. In most applied decision problems, both the preferences of a responsible decision maker and his judgments about the weights to be attached to the various possible states of nature are based on very substantial objective evidence; and quantification of his preferences and judgments enables him to arrive at a decision which is consistent with this objective evidence.

We should also like to point out that we are in complete agreement with those who assert that it is rarely if ever possible to find the best of all possible courses

† L. J. Savage, *The Foundations of Statistics*, New York, Wiley, 1954.

of action and who argue that reasonable men "satisfice" much more than they "optimize". We most emphatically do not believe that the objective of an optimizing analysis is to find the best of all possible courses of action; such a task is hopeless. As we see it, the first step in the analysis of any decision problem is necessarily a purely intuitive selection by the decision maker of those courses of action which seem to him worthy of further consideration. Only after a set of "reasonable contenders" has thus been defined does it become *possible* to apply formal procedures for choice among them, but even at this stage it will often be preferable to eliminate some of the original contenders by informal rather than formal analysis. In other words, it is our view that formal methods of optimization such as those described in this book should be used in those parts and only those parts of a complete analysis in which the decision maker believes that it will pay him to use them.† "Satisficing" is as good a word as any to denote both the preliminary choice of contenders and the intuitive elimination of some of them; and we are quite ready to admit that in many situations informal analysis will quite properly reduce the field to a single contender and leave no scope for formal analysis at all.

When we started to write this book our intention was merely to produce a short research report in which some of our as yet nebulous ideas would be submitted to the scrutiny of other specialists. Stimulated by each other's enthusiasm, however, we gradually expanded our objective in the direction of a self-contained introduction to what we believe to be a coherent and important group of analytical methods. As the book now stands, Part I (Chapters 1–3) describes the general structure of this group of methods and indicates in principle how they can be applied to a very wide variety of decision problems. Part II (Chapters 4–6) gives specific analytical results for two specialized classes of problems which are of central importance in applied statistics: (1) problems involving choice among two or more processes when utility is linear in the mean of the chosen process, and (2) problems of point estimation when utility depends on the difference between the estimate and the true value of the quantity being estimated. Finally, Part III (Chapters 7–13) is a systematic compendium of the distribution theory required in Parts I and II, containing definitions of distributions, references to published tables, and formulas for moments and other useful integrals. Because of its self-contained nature, we believe that the book should be accessible as well as of interest to a quite heterogeneous audience. We hope that statisticians interested in practical applications will find our techniques of practical use, but at the same time we hope that the way in which we have formulated some of the standard statistical problems will contribute to the unification of statistical theory with managerial economics and the techniques usually categorized under the heading of "operations research".

Anyone who has the equivalent of a good preliminary course in mathematical probability and statistics, is familiar with the advanced calculus and with the rudiments of matrix theory, and has read a little of the contemporary literature on the foundations of statistics will have no difficulty in reading any part of the

† For a slightly more explicit statement of this principle, see Section 1.4.

book; and much of it can easily be read with considerably less background. The required matrix theory is modest and is essentially confined to Chapter 5B and those chapters in Part III which deal with multivariate distributions. Advanced calculus is scarcely used except in the proofs of the distribution-theoretic results catalogued in Part III. The required knowledge of the theories of utility and subjective probability can easily be obtained by a very little reading: a brief axiomatic treatment of both utility and subjective probability can be found in Luce and Raiffa's *Games and Decisions*,† a brief informal discussion in Schlaifer's *Probability and Statistics for Business Decisions*.‡ For the reader who is already familiar with utility but not with subjective probability, we have given in Section 1.5 a very informal proof of the proposition that self-consistent choice among risk functions implies the existence of numerical weights or probabilities. For a complete discussion of the foundations, the reader should of course consult the book by Savage which we have already cited.

As far as we know, parts of this book are original, although as regards any particular formula or concept we would give better than even money that it has already appeared somewhere in print. Since we unfortunately do not have even the vaguest idea where most of these formulas and concepts appeared for the very first time, we leave it to historians to distribute detailed credits where they are due; but we would be inexcusably remiss if we did not acknowledge those who have most contributed to the main currents of our thinking. Anyone who reads this book will recognize our great debt to Neyman, Pearson, Jeffreys, Von Neumann, Wald, Blackwell, Girshick, and Savage. We should also like to record our gratitude for the innumerable corrections and helpful suggestions which we have received from John Bishop, Marshall Freimer, Andrew Kahr, and I. R. Savage, all of whom have given most generously of their time. Finally, we would like to express our appreciation to the Division of Research of the Harvard Graduate School of Business Administration for the very generous financial support which has made this publication possible.

*   *   *   *   *

A somewhat more detailed description of the topics covered in individual chapters may be of use to the reader who wishes to organize his study of the complete book or who wishes to see what we have to say on certain subjects without reading the complete book. We hope that by providing such an outline together with a rather detailed table of contents we may earn the forgiveness of those who are displeased by the lack of an index—a lack that is due purely and simply to our inability to discover how a book of this sort can be usefully indexed.

In Chapter 1 we start by defining the basic data of any decision problem in which experimentation is possible. These are: a listing of the potential terminal *acts* $\{a\}$ which the decision maker wishes to compare, a listing of the *states* of nature $\{\theta\}$ which he believes possible, a listing of the potential *experiments* $\{e\}$

† R. D. Luce and H. Raiffa, *Games and Decisions*, New York, Wiley, 1957; Chapters 2 and 13.
‡ R. Schlaifer, *Probability and Statistics for Business Decisions*, New York, McGraw-Hill, 1959; Chapters 1 and 2.

which he wishes to consider, a listing of the *outcomes* $\{z\}$ of these experiments which he believes possible, a *utility* function which evaluates his preferences for all (e, z, a, θ) combinations, and listing of the weights or *probabilities* which he assigns to all $\{z, \theta\}$ for each potential e. We then describe two basic modes of analysis of these data; in the language of game theory, they are (1) the *normal form*, in which a choice is made among strategies each of which specifies a particular experiment and assigns a terminal act to every possible outcome of that experiment, and (2) the *extensive form*, in which the optimal strategy is "built up" by taking all possible experimental outcomes one at a time, determining for each one separately what terminal act would be optimal, and then using these results to select an optimal experiment.

After proving that the two modes of analysis must ultimately lead to exactly the same course of action, we argue that even though the extensive form has been very little used in statistical analysis, it often possesses great advantages both conceptual and technical; and we use the extensive form exclusively in the remainder of the book. In particular it permits a clear distinction between two completely different statistical problems: choice of a terminal act *after* an experiment has already been performed, which we call *terminal analysis*, and choice of the experiment which *is to be* performed, which we call *preposterior analysis*. In terminal analysis one simply computes the expected utilities of all possible acts with respect to the posterior distribution resulting from a particular experimental outcome z_o and chooses the act whose expected utility is greatest; in the normal form, on the contrary, a person who has already observed z_o cannot choose a terminal act until he has determined what terminal act he would have chosen given every conceivable z which might have occurred. In the extensive form it is only when one is evaluating potential experiments *before* they are actually performed that one needs to consider outcomes which *might* occur. In this case the utility of any one potential experiment is evaluated by first using terminal analysis to determine the utility of the terminal act which will be optimal given each possible experimental outcome; the "preposterior" expected utility of the experiment is then computed by taking the expectation of these posterior expected utilities with respect to the unconditional prior measure over the experimental outcomes.

It is perhaps worth remarking in passing that although the data of most statistical problems occur in such a form that Bayes' theorem is required when they are analyzed in extensive form but not when they are analyzed in normal form, this relation is by no means necessary. As we point out in Sections 1.1.2 and 1.4.3, problems do occur in which the data are such that analysis in extensive form does not require the use of Bayes' theorem whereas the performance characteristics which play a central role in the usual normal-form analysis can only be computed by use of Bayes' theorem. For this reason we believe that it is essentially misleading to characterize as "Bayesian" the approach to decision theory which is represented by this book, even though we are occasionally driven into using the term *faute de mieux*.

In Chapter 2 we define a sufficient statistic as one which leads to the same posterior distribution that would be obtained by use of a "complete" description of the experimental outcome. Although we show that this definition implies and is implied by the classical definition of sufficiency in terms of factorability of the joint likelihood of the sample observations, we prefer our definition because it leads naturally to the concept of a statistic which is "marginally sufficient" for those unknown parameters which affect utility even though it is not sufficient for any nuisance parameters which may be present (Section 2.2). We then use this concept of marginal sufficiency as a basis for discussion of the problems suggested by the words "optional stopping" and show that after an experiment has already been conducted the experimental data can *usually* be "sufficiently" described without reference to the way in which the "size" of the experiment was determined (Section 2.3). It is usually sufficient, for example, to know simply the number r of successes and the number $(n - r)$ of failures which were observed in an experiment on a Bernoulli process; we have no need to know whether the experimenter decided to observe a predetermined number n of trials and count the number r of success, or to count the number n of trials required to produce a predetermined number r of successes, or simply to experiment until he ran out of time or money. For this reason we uniformly refer to r *and* n as the sufficient statistics of such an experiment, in contrast to the classical practice of reserving the term "statistic" to denote those aspects of the experimental outcome which were not determined in advance.

For essentially the same reason it seems to us impossible to distinguish usefully between "fixed" and "sequential" *experiments*. As for terminal analysis, we have already said that the decision maker usually does not care why the experimenter started or stopped experimenting; and although in preposterior analysis he is of course concerned with various ways of determining when experimentation should cease, even here we believe that the only useful distinction is between fixed and sequential *modes of analysis* rather than fixed and sequential experiments as such. By the fixed mode of analysis we mean analysis in which the entire experiment is evaluated by use of the distribution of those sufficient statistics whose values are not determined in advance, and this mode of analysis can be used just as well when a Bernoulli process is to be observed until the occurrence of the rth success as it can when the process is to be observed until the completion of the nth trial. In the sequential mode of analysis, on the contrary, an experiment is regarded as consisting of a potential sequence of subexperiments each of which is *analyzed* as a separate entity: the analysis does not deal directly with the statistics which will describe the ultimate outcome of the *complete* experiment but asks whether subexperiment e_2 should or should not be performed if subexperiment e_1 results in outcome z_1, and so forth. Any experiment can be analyzed in the fixed mode if the sufficient statistics are appropriately defined and the necessary distribution theory is worked out, but for many experiments the sequential mode of analysis is much more convenient. In this book we deal only with experiments for which the fixed mode of analysis is the more convenient.

In Chapter 3 we take up the problem of assessing prior distributions in a form which will express the essentials of the decision maker's judgments about the possible states of nature and which at the same time will be mathematically tractable. We show that whenever (1) any possible experimental outcome can be described by a sufficient statistic of fixed dimensionality (i.e., an s-tuple $(y_1, y_2, \ldots, y_s)$ where s does not depend on the "size" of the experiment), and (2) the likelihood of every outcome is given by a reasonably simple formula with $y_1, y_2, \ldots, y_s$ as its arguments, we can obtain a very tractable family of "conjugate" prior distributions by simply interchanging the roles of variables and parameters in the algebraic expression for the sample likelihood, and the posterior distribution will be a member of the same family as the prior. This procedure leads, for example, to the beta family of distributions in situations where the state is described by the parameter p of a Bernoulli process, to the Normal family of distributions when the state is the mean μ of a Normal population, and so forth; a conspectus of conjugate distributions for some of the likelihood functions most commonly met in practice is given in Section 3.2.5. In Section 3.3 we show that these conjugate families are often rich enough to allow the decision maker to express the most essential features of his basic judgments about the possible states and to do so with only a reasonable amount of introspection, and we then conclude the chapter by showing in Section 3.4 how the use of conjugate prior distributions facilitates preposterior analysis—i.e., extensive-form evaluation of potential experiments. Briefly, the first step in the analysis is to express the posterior expected utility of optimal terminal action as a function of the parameters of the posterior distribution of the state. Any particular experiment e can then be evaluated by first obtaining the prior distribution of the parameters of the posterior distribution which may result from that experiment and then using this distribution to calculate the prior expectation of the posterior expected utility.

A word about notation is in order at this point. We admit that we should have furnished a magnifying glass with each copy to aid the reader in deciphering such symbols as $\tilde{\omega}''_n$; for such typographical hieroglyphics we apologize to the reader—and more especially, to the compositor. Statistical theory has however always been beset with notational problems—those of keeping parameters distinct from statistics, random variables from values assumed by random variables, and so forth. In Bayesian decision theory these minor annoyances develop into serious headaches. The same state variable which appears as a parameter in the conditional distribution of a statistic appears as a random variable in its own right when expected utility is being computed and the distribution of this state variable has parameters of its own. Sample evidence takes us from a prior distribution to a posterior distribution of the state variable, and we must distinguish between the parameters of these two distributions. When an experiment is being considered but has not yet been performed, the parameters of the posterior distribution which will result from the experiment are as yet unknown; accordingly *they* are random variables and have distributions which in turn have still other parameters.

It is only too obvious that on the one hand we must distinguish notationally between random variables and values of random variables while on the other hand we cannot permit ourselves the luxury of reserving capital letters for random variables and small letters for values. It is equally obvious that the notation must have enough mnemonic value to enable the reader to keep track of the logical progression from prior parameters through sample statistics to posterior parameters to the parameters of the prior distribution of the posterior parameters. Briefly, our solution of this problem is the following. A tilde distinguishes a random variable from its generic value; thus if the state is the parameter p of a Bernoulli process, the state considered as an unknown and therefore a random variable is denoted by $\tilde{p}$. If the sufficient statistics of a sample from this process are r and n, the parameters of the conjugate prior distribution of $\tilde{p}$ will be called r' and n' and the parameters of the posterior distribution will be called r'' and n''. If we are considering an experiment in which n trials are to be made, the statistic $\tilde{r}$ will be a random variable and so will the parameter $\tilde{r}''$ of the posterior distribution. In some situations we shall be more directly interested in the mean of the distribution of $\tilde{p}$ than in its parameters; in such situations the mean of the prior distribution of $\tilde{p}$ will be denoted by $\bar{p}'$, that of the posterior by $\bar{p}''$, and $\tilde{\bar{p}}''$ is a random variable until the sample outcome is known. Despite our apologies for typography, we are proud of this notation and believe that it works for us rather than against us; no system of notation can eliminate complexities which are inherent in the problem being treated.

In Part II of the book (Chapters 4–6), we specialize to the extremely common class of problems in which the utility of an entire (e, z, a, θ) combination can be decomposed into two *additive* parts: a "terminal utility" which depends only on the terminal act a and the true state θ, and a "sampling utility" (the negative of the "cost" of sampling) which depends only on the experiment e and (possibly) its outcome z.

In Chapter 4 we point out that this assumption of additivity by no means restricts us to problems in which all consequences are monetary. Problems in which consequences are purely monetary and the utility of money is linear over a suitable wide range do of course constitute a very important subclass of problems in which sampling and terminal utilities are additive, but additive utilities are frequently encountered when consequences are partially or even wholly of a nonmonetary nature. In general, sampling and terminal utilities will be additive whenever consequences can be measured or scaled in terms of *any* common *numéraire* the utility of which is linear over a suitably wide range; and we point out that number of patients cured or number of hours spent on research may well serve as such a *numéraire* in problems where money plays no role at all.

In situations where terminal and sampling utilities are additive, it is usually possible also to decompose terminal utility itself into a sum or difference of economically meaningful parts; and since we believe that the expert as well as the novice has much to gain from the economic heuristics which such decompositions make possible, we conclude Chapter 4 by defining and interrelating a number of new economic concepts. Opportunity loss (or regret) is introduced in Section 4.4 and it is shown that minimization of expected opportunity loss is equivalent to maximization of expected utility; all of Chapter 6 will be based on this result. The value of perfect information, the value of sample information, and the net gain of sampling (the difference between the value and the cost of sample information) are introduced in Section 4.5 and it is shown that maximization of the net gain of sampling is equivalent to maximization of expected utility; all of Chapter 5 will be based on this result. The chapter closes with a conspectus of the definitions of all of the economic concepts used in the entire book and of the relations among them.

In Chapters 5A and 5B we further specialize by assuming that the terminal utility of every possible act is linear in the state θ or some transformation thereof. This assumption greatly simplifies analysis, and despite its apparently very restrictive character we are convinced that it is satisfied either exactly or as a good working approximation in a very great number of decision problems both within and without the field of business. Given linearity, terminal analysis becomes completely trivial because the expected utility of any act depends only on the mean of the distribution of the state variable; and even preposterior analysis becomes very easy when the mean of the posterior distribution is treated as a random variable. Some general theorems concerning this prior distribution of the posterior mean or "preposterous distribution" are proved in Section 5.4.

Aside from these general theorems about preposterous distributions, Chapter 5A is primarily concerned with problems in which the posterior mean is scalar and the act space contains only a finite number of acts. In Section 5.3 we show that under these conditions the expected net gain of many types of experiments can be very easily expressed in terms of what we call a "linear-loss" integral with respect to the preposterous distribution. The preposterous distributions corresponding to a variety of common statistical experiments are indexed in Table 5.2 (page 110); among other things the table gives references to formulas for the linear-loss integral with respect to each distribution indexed.

The remainder of Chapter 5A specializes still further to the case where there are only two possible terminal acts and examines the problem of using the results previously obtained for the net gain of a sample of arbitrary "size" to find the optimal sample size. For the case where sampling is Normal with known variance, we give complete results including charts from which optimal sample size and the expected net gain of an optimal sample can be read directly.† For the case where sampling is binomial, we describe the behavior of the net gain of sampling as the sample size n increases, discuss the problem of finding the optimum by the use of high-speed computers, and show that surprisingly good approximate results can often be obtained with only trivial computations by treating the beta prior and binomial sampling distributions as if they were Normal.

In Chapter 5B we take up the problem of selecting the best of r "processes" when terminal utility is linear in the mean of the chosen process and sample observations can be taken independently on any or all of the means. Terminal analysis—choice of a process on the basis of whatever information already exists—again turns out to be trivial; the chapter is concerned primarily with preposterior analysis, i.e. the problem of deciding how many observations to make on each

† Similar results for the case where the sampling variance is unknown have been obtained by A. Schleifer, Jr., and the requisite tables are given in J. Bracken and A. Schleifer, Jr., *Tables for Normal Sampling with Unknown Variance*, Boston, Division of Research, Harvard Business School, 1964.

process before any terminal decision is reached. The only case discussed in detail is that in which the sample observations on each process are Normally distributed and the variances of these sampling distributions are known up to a constant of proportionality. Expressions are given for the net gain of any set of samples $(n_1, n_2, \ldots, n_r)$ in terms of multivariate Normal or Student integrals, and we show that these integrals can be easily evaluated with the aid of tables when $r = 2$ or 3 while for $r > 3$ we discuss methods of evaluation by use of high-speed computers. As regards the problem of finding the optimal $(n_1, n_2, \ldots, n_r)$, we show that when $r = 2$ the problem can be completely solved by the use of the univariate results of Chapter 5A; for higher values of r we suggest that the surface representing the net gain of sampling as a function of the r variables $(n_1, n_2, \ldots, n_r)$ might be explored by application of standard techniques for the analysis of response surfaces.

In Chapter 6 we turn to another special class of problems, of which by far the most important representative is the problem of point estimation. The basic ideas are introduced by examining a problem which is *not* usually thought of as involving "estimation": *viz.*, the problem of deciding what quantity q of some commodity should be stocked when the demand d is unknown and an opportunity loss will be incurred if d is not equal to q. We then define the problem of point estimation as the problem which arises when the decision maker wishes to use some number $\hat{\theta}$ *as if* it were the true value θ of some unknown quantity, and we argue that this problem is formally identical to the inventory problem because here too the decision maker will suffer an opportunity loss if $\theta \neq \hat{\theta}$. The inventory problem itself can in fact be expressed as one of finding the optimal "estimate" $\hat{d}$ of the demand d on the understanding that the decision maker will treat this estimate as if it were the true d—i.e., that he will stock $q = \hat{d}$.

Generalizing, we prove that whenever terminal utility depends on an unknown quantity ω, the utility-maximizing terminal act can be found indirectly by (1) determining for all $(\hat{\omega}, \omega)$ pairs the opportunity loss which will result if a terminal act is chosen by treating ω as if it were $\hat{\omega}$, (2) finding the $\hat{\omega}$ which minimizes the expectation of these losses, and finally (3) choosing a terminal act as if this optimal estimate of ω were the true value of ω. This indirect procedure for allowing for uncertainty about ω will rarely if ever be of practical use as it stands, since it will usually be just as difficult to determine the "estimation losses" attached to all $(\hat{\omega}, \omega)$ pairs as it would be to make a direct computation of the expected utilities of the various terminal acts under the distribution of $\tilde{\omega}$; but we believe that when it is *not* possible to take full analytical account of uncertainty about ω, then a very good way of proceeding will be to make a *judgmental* evaluation of the losses which will result from misestimation, to select the estimate which minimizes the expected value of these losses, and to choose the terminal act which would be optimal if ω were in fact equal to this estimate. This procedure will be still more useful in situations where it is possible to obtain additional information about ω and the decision maker must decide how much information it is economically worth while to collect. In Sections 6.3 through 6.6 we consider two types of functions which might be used to represent a judgmental assessment of the losses which may result from misestimation—losses linear in $(\hat{\omega} - \omega)$ and losses quadratic in $(\hat{\omega} - \omega)$—and for each type we give formulas for the optimal estimate, the expected loss of the optimal estimate, and the optimal amount of information to collect in one or two types of experiment.

Part III of the book, comprising Chapters 7 to 13, contains the analytical distribution theory required for application to specific problems of the general methods of analysis described in Parts I and II. The heart of Part III is in the last five Chapters (9 through 13) each of which considers a different "process" generating independent, identically distributed random variables and gives the distribution-theoretic results most likely to be required in a Bayesian analysis of any decision problem in which information about the state of the world can be obtained by observing the process in question. The two preceding chapters of Part III (Chapters 7 and 8) are simply a catalog of definitions of various mass and density functions together with formulas for their moments and references to published tables; this material was collected into one place, ahead of the analyses of the various data generating processes, simply because most of the functions in question appear a great many times in the process analyses rather than just once.

In Chapter 7A we take up successively a number of what we call "natural" univariate mass and density functions, belonging to the binomial, beta, Poisson, gamma, and Normal families. In this chapter the same "basic" distribution often appears in many alternative forms derived by a simple change of variable or even by mere reparametrization, but we believe that this apparent duplication will actually lighten the reader's and the user's task. The alternate parametrizations are introduced in order to keep as clear as possible the relations between sample statistics on the one hand and the parameters of prior and posterior conjugate distributions on the other, while new functions are derived by a change of variable whenever Bayesian analysis requires the distribution of the variable in question. The case for thus multiplying the number of distributions is the same as the case for having both a beta and an F or both a gamma and a chi-square distribution in classical statistics, and we know by bitter experience that without such a systematic catalog of transformed distributions the statistician can waste hours trying to look up a probability in a table and then come out with the wrong answer.

In Chapter 7B we turn to what we consider to be "compound" univariate mass and density functions, obtained by integrating out a parameter of one of the "natural" functions. Among these compound functions the negative-binomial, obtained by taking a gamma mixture of Poisson distributions, will be already familiar to most readers. The beta-binomial and beta-Pascal functions are similar, being obtained by taking a beta mixture of binomial or Pascal functions; in Section 7.11.1 we bring out an interesting relation between these two compound distributions and the well known hypergeometric distribution. Finally, it is interesting to observe that in Bayesian analysis the Student distribution appears as a gamma mixture of Normal densities and not as the distribution of the ratio of two random variables.

In Chapter 8 we complete our catalog with the multivariate distributions we require. There are only three in number: the multivariate Normal, concerning which we have nothing new to say; the multivariate Student, a compound distribution obtained as a gamma mixture of multivariate Normals, and the "inverted Student," obtained from the Student by a change of variable.

Throughout Chapters 7 and 8 the reader may feel that there is an overabundance of proofs and he may well be right in so feeling. We do, however, have two things to say in self-defense. First, many of the formulas we give are hard or impossible to find in other books, and being only moderately sure of our own algebra and calculus we wanted to make it as easy as possible for the reader to verify our results; corrections will be greatly appreciated. Second, as regards the results which are perfectly well known, our desire to produce an introduction to Bayesian analysis which would be accessible even to readers who were not professionally trained in classical statistics seemed to us to imply that we should at least give references to proofs; but when we tried to do so, we quickly came to the conclusion that differences both in notation and in point of view (e.g., as regards compound distributions) would make such references virtually incomprehensible except to those readers who had no need of proofs anyway.

Chapters 9–13 constitute, as we have already said, the essentially Bayesian portion of Part III, in which we give the results required for Bayesian analysis of decision problems where information about the state of nature can be obtained by observing a "data-generating process." The five processes studied in these five chapters are defined in Sections 3.2.5 by the densities of the independent random variables which they generate; they are: the binomial, the Poisson, the (univariate) Normal, the Multivariate Normal, and the Normal Regression processes.

The pattern of the analysis is identical in the case of the two single-parameter processes, the Bernoulli and Poisson (Chapters 9 and 10). The steps in the analysis are the following.

1. Define and characterize the process and interpret its parameter.
2. Exhibit the likelihood of a sample consisting of several observations on the process and give the sufficient statistic.
3. Exhibit the family of conjugate prior distributions for the process parameter and give the algebraic mechanism for obtaining the posterior parameter from the prior parameter and the sample statistic.

3′. Repeat the previous step for an alternative parametrization of the process (in terms of $\rho = 1/p$ in the Bernoulli case, in terms of an analogous substitution in the Poisson case).

4. For a given type of experiment—e.g., fix the number of trials n and leave the number of successes $\tilde{r}$ to chance—derive the conditional distribution of the nonpredetermined part of the sufficient statistic for a given value of the process parameter.
5. Derive the unconditional sampling distribution by integrating out the parameter with respect to its prior distribution.
6. Derive the prior distribution of the posterior mean for both parametrizations of the process.
7. Obtain formulas for certain integrals with respect to the distributions obtained in the previous step.

8. Investigate convenient approximations to the results obtained in the two previous steps.

4′-8′. Repeat steps (4) through (8) for an alternate experimental design—e.g., fix r and leave $\tilde{n}$ to chance.

The analysis of the univariate Normal process in Chapter 11 follows essentially this same pattern except that (1) we consider only one kind of experiment, that in which the number n of observations is predetermined, and (2) we must allow for uncertainty about the process mean μ *and/or* the process precision $h = 1/\sigma^2$. We therefore analyze three cases separately: in case A, μ is known; in case B, h is known; in case C, neither parameter is known. In case A, the conjugate distribution of $\tilde{h}$ is gamma; in case B, the conjugate distribution of $\tilde{\mu}$ is Normal; in both these cases the pattern of the analysis is virtually identical to the pattern in Chapters 9 and 10.

In case C, where both parameters are unknown, the conjugate family is what we call Normal-gamma, with a density which can be regarded as either (1) the product of a marginal gamma density of $\tilde{h}$ times a conditional Normal density of $\tilde{\mu}$ given h, or (2) the product of a marginal Student density of $\tilde{\mu}$ times a conditional gamma density of $\tilde{h}$ given μ. The number of degrees of freedom in the marginal posterior distribution of $\tilde{\mu}$ depends on the marginal prior distribution of $\tilde{h}$, so that Bayesian analysis allows the decision maker a way through the classical dilemma between asserting that he knows nothing and asserting that he knows everything about h. We must point out, however, that although the Normal-gamma distribution has four free parameters and is amply flexible enough to represent the decision maker's judgments when his opinions about μ and h are both "loose" or both "tight" or when his opinions about μ are loose while his opinions about h are tight, it cannot give a good representation of tight opinions about μ in the face of loose opinions about h. In this case we really need to assign independent distributions to $\tilde{\mu}$ and $\tilde{h}$, but Jeffreys has already pointed out the distribution-theoretic difficulties which arise if we do.†

As long as we *can* use a Normal-gamma prior, however, even preposterior analysis is easy. The treatment of preposterior analysis in Chapter 11C follows basically the same pattern that is followed in earlier chapters, except that all the distributions (conditional and unconditional sampling distributions and the distribution of the posterior parameters) involve two random variables; we give in each case the joint distribution of the pair of variables and both the conditional and the marginal distributions of the individual variables.

In Chapter 12 the results of the scalar Normal and Student theory of Chapter 11 are generalized to the multivariate Normal and Student cases. We first assume that sampling results in independent identically distributed, random vectors $\tilde{\mathbf{x}}^{(1)}, \tilde{\mathbf{x}}^{(2)}, \ldots$ where each $\tilde{\mathbf{x}}$ has a multivariate Normal distribution with a mean vector $\boldsymbol{\mu}$ and a precision matrix (inverse of the covariance matrix) $\mathbf{h}$. In part A of the chapter we assume $\mathbf{h}$ is known. In part B we write $\mathbf{h} = h\,\boldsymbol{\eta}$ where $|\boldsymbol{\eta}| = 1$

† H. Jeffreys, *Theory of Probability*, 2nd edition, Oxford, Clarendon Press, 1948; pages 123–124.

and assume $\boldsymbol{\eta}$ is known but h is not. In part C we specialize to the case where $\boldsymbol{\eta}$ is diagonal. Our interest in this special case is due primarily to the fact that it depicts situations in which we wish to compare the means of a number of distinct univariate Normal processes each of which can be sampled independently, and for this reason we give separate analyses for the case where all the univariate processes are to be sampled and the case where only some of them are to be sampled. As regards terminal analysis, the two cases are essentially the same; but in preposterior analysis the second case gives rise to degenerate or singular prior distributions of the posterior parameters.

The Normal regression process, considered in Chapter 13, is defined as a process generating scalar random variables $\tilde{y}_1, \tilde{y}_2, \ldots$ each of which is a linear combination of known xs with unknown β weights plus random error; the errors are assumed to be independent and to have identical Normal distributions with precision h. The state variables (which are also the parameters of the conditional sampling distributions) are $\beta_1, \ldots, \beta_r$ and h; the only experimental designs we consider are those which prescribe a fixed coefficient matrix $\mathbf{X}$, but we carry out the analysis both for the case where $\mathbf{X}$ is of full rank and for the case where it is not. The chapter is divided into three parts. In parts A and B we discuss terminal analysis for any $\mathbf{X}$ and preposterior analysis for $\mathbf{X}$ of full rank; part A deals with the case where h is known and the conjugate distribution of $\tilde{\boldsymbol{\beta}}$ is Normal, part B with the case where h is unknown and the conjugate distribution of $\tilde{\boldsymbol{\beta}}$ and $\tilde{h}$ is Normal-gamma. Part C gives the preposterior analysis for h known or unknown when $\mathbf{X}$ is not of full rank.

The distribution-theoretic results obtained in Chapters 12 and 13 are by no means fully exploited in the applications discussed in Part II of the book; the only real use made there of these last two chapters is in the analysis in Chapter 5B of the problem of choice among a number of univariate Normal processes. We should therefore like to call the reader's attention to a few problems of great practical importance in which we hope that the contents of Chapters 12 and 13 may prove useful.

All the analyses in Part II were based on simple random sampling, but the results of Chapter 12 should make it fairly easy to analyze the same economic applications when the sampling is stratified. If utility depends on the mean μ of an entire population and if we denote by μ_i the mean of the ith stratum of this population, then $\mu = \Sigma\, c_i\, \mu_i$. Bayesian terminal analysis would proceed by putting a multivariate prior on $\tilde{\boldsymbol{\mu}} = (\tilde{\mu}_1 \ldots \tilde{\mu}_r)$, revising this prior in accordance with sample information, and then finding the corresponding distribution of $\tilde{\mu} = \Sigma\, c_i\, \tilde{\mu}_i$. Preposterior analysis might for example first obtain the joint prior distribution of the posterior means $\tilde{\bar{\mu}}''_i$ and from this the preposterior distribution of $\tilde{\bar{\mu}}'' = \Sigma\, c_i\, \tilde{\bar{\mu}}''_i$.

The results on the Regression process obtained in Chapter 13 promise to be useful in two quite different respects. First, we would guess that the best way of obtaining "prior" distributions for many economic variables such as demand will often be to start from a regression of demand on economic predictors. The unconditional distribution of the "next" $\tilde{y}$ obtained from the Regression model can then serve as a *prior* distribution in situations where additional information bearing directly on y can be obtained by sampling the population which creates the demand.

Second, and probably much more important, the regression model will clearly be required for Bayesian analysis of experiments in which blocking and similar devices are used. The great obstacle yet to be overcome in the analysis of such problems is the problem of setting prior distributions, but the game is well worth the candle. Only Bayesian analysis can free the statistician from having to assert either that he is sure that there is no row (or interaction, or what have you) effect, or else that he knows nothing whatever about this effect except what he has learned from one particular sample. We at one time thought of including at least a first attack on some or all of these problems in the present book, but after all one must stop somewhere.

CONTENTS

Selected Tables

Charts

(following page 356)

PART ONE

EXPERIMENTATION AND DECISION: GENERAL THEORY

CHAPTER 1

The Problem and the Two Basic Modes of Analysis

1.1. Description of the Decision Problem

In this monograph we shall be concerned with the logical analysis of choice among courses of action when (a) the consequence of any course of action will depend upon the "state of the world", (b) the true state is as yet unknown, but (c) it is possible at a cost to obtain additional information about the state. We assume that the person responsible for the decision has already eliminated a great many possible courses of action as being unworthy of further consideration and thus has reduced his problem to choice within a well-defined set of contenders; and we assume further that he wishes to choose among these contenders in a way which will be logically consistent with (a) his basic preferences concerning consequences, and (b) his basic judgments concerning the true state of the world.

1.1.1. The Basic Data

Formally, we assume that the decision maker can specify the following basic data defining his decision problem.

1. *Space of terminal acts:* $A = \{a\}$.

The decision maker wishes to select a single act a from some domain A of potential acts.

2. *State space:* $\Theta = \{\theta\}$.

The decision maker believes that the consequence of adopting terminal act a depends on some "state of the world" which he cannot predict with certainty. Each potential state will be labelled by a θ with domain Θ.

3. *Family of experiments:* $E = \{e\}$.

To obtain further information on the importance which he should attach to each θ in Θ, the decision maker may select a single experiment e from a family E of potential experiments.

4. *Sample space:* $Z = \{z\}$.

Each potential outcome of a potential experiment e will be labelled by a z with domain Z. We use the nonstandard convention that Z is rich enough to encompass any outcome of *any* e in E, and for this reason the description of z will in part repeat the description of e; the usefulness of this redundancy will appear in the sequel.

5. *Utility Evaluation:* $u(\cdot, \cdot, \cdot, \cdot)$ on $E \times Z \times A \times \Theta$.

The decision maker assigns a utility $u(e, z, a, \theta)$ to performing a particular e, observing a particular z, taking a particular action a, and then finding that a particular θ obtains. The evaluation u takes account of the costs (monetary and other) of experimentation as well as the consequences (monetary and other) of the terminal act; and the notation leaves open the possibility that the cost of experimentation may depend on the particular experimental outcome z as well as on the mode of experimentation e.

6. *Probability Assessment:* $P_{\theta,z}\{\cdot, \cdot|e\}$ on $\Theta \times Z$.
For every e in E the decision maker directly or indirectly assigns a joint probability measure $P_{\theta,z}\{\cdot, \cdot|e\}$ or more briefly $P_{\theta,z|e}$ to the Cartesian product space $\Theta \times Z$, which will be called the *possibility space.* This joint measure determines four other probability measures:

a. The marginal measure $P'_\theta\{\cdot\}$ or P'_θ on the state space Θ. We assume that P'_θ does not depend on e.

b. The conditional measure $P_z\{\cdot|e, \theta\}$ or $P_{z|e,\theta}$ on the sample space Z for given e and θ.

c. The marginal measure $P_z\{\cdot|e\}$ or $P_{z|e}$ on the sample space Z for given e but unspecified θ.

d. The conditional measure $P''_\theta\{\cdot|z\}$ or $P''_{\theta|z}$ on the state space Θ for given e and z; the condition e is suppressed because the relevant aspects of e will be expressed as part of z.

The prime on the measure P'_θ defined in (a) indicates that it is the measure which the decision maker assigns or would assign to Θ *prior* to knowing the outcome z of the experiment e. The double prime on the measure $P''_{\theta|z}$ defined in (d) indicates that it is the measure on Θ which he assigns or would assign *posterior* to knowing the outcome z of the experiment e. Strictly speaking the primes are redundant, but they will be of help when used to distinguish between prior and posterior values of the parameters of families of probability distributions.

Random Variables and Expectations. In many situations the states $\{\theta\}$ and the sample outcomes $\{z\}$ will be described by real numbers or n-tuples of real numbers. In such cases we shall define the random variables $\tilde{\theta}$ and $\tilde{z}$ by

$$\tilde{\theta}(\theta, z) = \theta \ , \qquad \tilde{z}(\theta, z) = z \ ,$$

using the tilde to distinguish the random variable or function from a particular value of the function.

In taking expectations of random variables, the measure with respect to which the expectation is taken will be indicated *either* (1) by subscripts appended to the expectation operator E, *or* (2) by naming the random variable and the conditions in parentheses following the operator. Thus

E'_θ	or	$E'(\tilde{\theta})$	is taken with respect to	P'_θ
$E''_{\theta\|z}$	or	$E''(\tilde{\theta}\|z)$	is taken with respect to	$P''_{\theta\|z}$
$E_{z\|e,\theta}$	or	$E(\tilde{z}\|e, \theta)$	is taken with respect to	$P_{z\|e,\theta}$
$E_{z\|e}$	or	$E(\tilde{z}\|e)$	is taken with respect to	$P_{z\|e}$

When no condition is shown, the condition e is to be understood and the expectation is with respect to the entire joint measure for that e:

$$E \equiv E_{\theta,z|e} \ .$$

1.1.2. Assessment of Probability Measures

For any given e, there are three "basic" methods for assigning the complete set of measures just defined. (1) We have already said that if a joint measure is assigned to $\Theta \times Z$ directly, the marginal and conditional measures on Θ and Z separately can be computed from it. (2) If a marginal measure is assigned to Θ and a conditional measure is assigned to Z for every θ in Θ, the joint measure on $\Theta \times Z$ can be computed from these, after which the marginal measure on Z and the conditional measures on Θ can be computed from the joint measure. (3) The second procedure can be reversed: if a marginal measure is assigned to Z and conditional measures to Θ, the joint measure on $\Theta \times Z$ can be computed and from it the marginal measure on Θ and the conditional measures on Z.

All three methods of determining the required measures are of practical importance. The decision maker will wish to assess the required measures in whatever way allows him to make the most effective use of his previous experience—including but not restricted to his knowledge of historical relative frequencies in $\Theta \times Z$—and therefore he will want to make direct assessments of those measures on which his experience bears most directly and deduce the other measures from these.

In some situations the decision maker will have had extensive experience on which to base a direct assessment of the joint measure itself; the experience may actually consist of so long a record of joint relative frequencies produced under "constant" conditions that a directly assessed joint measure will be "objective" in the sense that given this same evidence any two "reasonable" men would assign virtually the same joint measure to $\Theta \times Z$.

More commonly, the decision maker will be able to make the most effective use of his experience by making direct assessments of the marginal measure on Θ and the conditional measures on Z, the reason being that very often the latter and not infrequently the former of these two measures can be based on very extensive historical frequencies. In many situations the assessments of the conditional measures $P_{z|e,\theta}$ will be made via theoretical models of the behavior of $\tilde{z}$ which rest on an enormous amount of relevant experience with relative frequencies; thus a particular sampling process may be virtually known with certainty to behave according to the Bernoulli model even though it has never actually been applied in the exact circumstances of the particular problem at hand. In some situations, moreover—e.g., in some acceptance sampling problems—the measure on Θ (e.g., lot quality) may be virtually as "objective" as the measures on Z given θ.

Finally, situations are not at all infrequent in which the marginal measure on Z and the conditional measures on Θ have the most extensive foundation in experience. An example of such a situation is discussed in Section 1.4.3 below.

In addition to these "basic" methods of assigning a measure to the possibility space $\Theta \times Z$, there are of course a variety of special methods which can be used to assess the *parameters* of a measure once the *structure* of the measure has been specified. While it would be impossible to list all such methods, we can give one important example.† If the spaces Θ and Z are each the set of real numbers and

† We owe this example to Professor Arthur Schleifer, Jr.

if the joint measure on the product space is bivariate Normal, then provided that $\tilde{\theta}$ and $\tilde{z}$ are not completely independent the joint measure can be uniquely determined by assigning to the *same* space Z both a marginal measure and a measure conditional on each θ in Θ.

1.1.3. Example

Let $A = \{a_1, a_2\}$ where a_1 stands for "accept the lot of 1000 items" and a_2 for "reject the lot of 1000 items". Let $\Theta = \{\theta_0, \theta_1, \cdots, \theta_{1000}\}$ where θ_l is the state in which l items are defective. Let $E = \{e_0, e_1, \cdots, e_{1000}\}$ where e_i is the experiment in which i items are drawn from the lot and inspected. Let $Z = \{(j, i): 0 \leq j \leq i \leq 1000\}$ where (j, i) represents the outcome that j defectives were found in a sample of i observations. [Notice that if e_5, say, is performed, then the outcomes are constrained to lie in the set $\{(0, 5)(1, 5), \cdots (5, 5)\}$.] Then $u[e_i, (j, i), a_k, \theta_l]$ is the utility which the decision maker attaches to drawing a sample of i items, observing that j of these i items are defective, and adopting a_k ("accept" if $k = 1$, "reject" if $k = 2$) when l items in the lot are in fact defective. [Notice that if the inspection is destructive, the cost of sampling includes the manufacturing cost of the $i - j$ good items destroyed in sampling and that the utility assigned to $\{e_i, (j, i), a_k, \theta_l\}$ should reflect this.]

In a problem of this sort the required probability measure on the possibility space $\Theta \times Z$ will ordinarily be assessed via (1) a family of conditional measures on Z, a typical one being $P_z\{\cdot|e_i, \theta_l\}$, which assign conditional probabilities to the various possible outcomes (j, i) given that i items are inspected and given that there are actually l defectives in the lot, and (2) a marginal measure P'_θ which assigns probabilities to the various possible numbers l of defectives in the lot before observing the outcome of any experiment.

1.1.4. The General Decision Problem as a Game

The general decision problem is: *Given E, Z, A, Θ, u, and $P_{\theta,z|e}$, how should the decision maker choose an e and then, having observed z, choose an a, in such a way as to maximize his expected utility?* This problem can usefully be represented as a game between the decision maker and a fictitious character we shall call "chance". The game has four moves: the decision maker chooses e, chance chooses z, the decision maker chooses a, and finally chance chooses θ. The play is then completed and the decision maker gets the "payoff" $u(e, z, a, \theta)$.

Although the decision maker has full control over his choice of e and a, he has neither control over, nor perfect foreknowledge of, the choices of z and θ which will be made by chance. We have assumed, however, that he *is* able in one way or another to assign probability measures over these choices, and the moves in the game proceed in accordance with these measures as follows:

Move 1: The decision maker selects an e in E.
Move 2: Chance selects a z in Z according to the measure $P_{z|e}$.
Move 3: The decision maker selects an a in A.
Move 4: Chance selects a θ in Θ according to the measure $P''_{\theta|z}$.
Payoff: The decision maker receives $u(e, z, a, \theta)$.

The Decision Tree. When the spaces E, Z, A, and Θ are all finite, the flow of the game can in principle be represented by a tree diagram; and although a complete diagram can actually be drawn only if the number of elements involved in E, Z, A, and Θ is very small, even an incomplete representation of the tree can aid our intuition considerably.

A partial tree of this sort is shown in Figure 1.1, where D denotes the decision maker and C denotes chance. At move 1, D chooses some branch e of the tree;

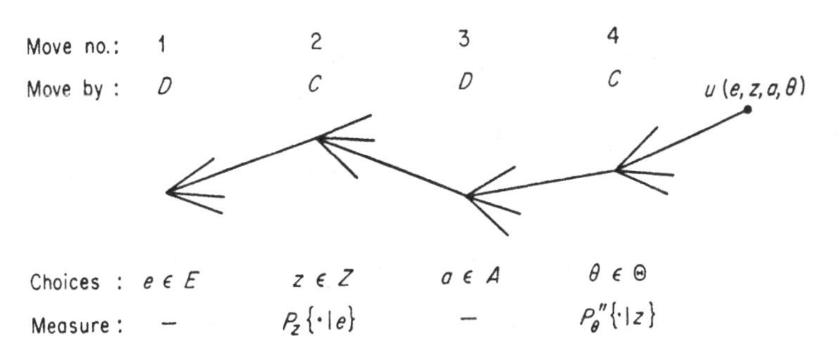

Figure 1.1

One Possible Play of a Game

at move 2, C chooses a branch z; at move 3, D chooses a; at move 4, C chooses θ; and finally, D receives the "payoff" $u(e, z, a, \theta)$. In two examples to follow shortly we shall depict the tree completely and present an analysis of the problem in terms of this representation.

1.2. Analysis in Extensive Form

Once we have at hand all the data of a decision problem as specified in Section 1.1.1, there are two basic modes of analysis which we can use to find the course of action which will maximize the decision maker's expected utility: the *extensive* form of analysis and the *normal* form. Although the two forms are mathematically equivalent and lead to identical results, both will be expounded in this chapter because each has something to contribute to our insight into the decision problem and each has technical advantages in certain situations.

1.2.1. Backwards Induction

The extensive form of analysis proceeds by working backwards from the end of the decision tree (the right side of Figure 1.1) to the initial starting point: instead of starting by asking which experiment e the decision maker should choose at move 1 when he knows neither of the moves which will subsequently be made by chance, we start by asking which terminal act he should choose at move 3 *if* he has already performed a particular experiment e and observed a particular outcome z. Even at this point, with a known history (e, z), the utilities of the various possible terminal acts are uncertain because the θ which will be chosen by chance at move 4 is still unknown; but *this* difficulty is easily resolved by treating the utility of any a for given (e, z) as a random variable $u(e, z, a, \tilde{\theta})$ and applying the operator $E''_{\theta|z}$ which takes the expected value of $u(e, z, a, \tilde{\theta})$ with respect to the *conditional* measure $P''_{\theta|z}$. Symbolically, we can compute for any given history (e, z) and any terminal act a

$$u^*(e, z, a) = E''_{\theta|z}\, u(e, z, a, \tilde{\theta}) \; ; \tag{1-1}$$

this is the utility of being at the juncture (e, z, a) looking forward, before chance has made a choice of θ.

Now since the decision maker's objective is to maximize his expected utility, he will, if faced with a given history (e, z), wish to choose the a (or one of the as, if more than one exist) for which $u^*(e, z, a)$ is greatest;† and since he is free to make this choice as he pleases, we may say that the utility of being at move 3 with history (e, z) and the choice of a still to make is

$$u^*(e, z) \equiv \max_a u^*(e, z, a) \; . \tag{1-2}$$

After we have computed $u^*(e, z)$ in this way for all possible histories (e, z), we are ready to attack the problem of the initial choice of an experiment. At this point, move 1, the utilities of the various possible experiments are uncertain only because the z which will be chosen by chance at move 2 is still unknown, and this difficulty is resolved in exactly the same way that the difficulty in choosing a given (e, z) was resolved: by putting a probability measure over chance's moves and taking expected values. In other words, $u^*(e, \tilde{z})$ is a random variable at move 1 because $\tilde{z}$ is a random variable, and we therefore define for any e

$$u^*(e) \equiv E_{z|e}\, u^*(e, \tilde{z}) \tag{1-3}$$

where $E_{z|e}$ expects with respect to the *marginal* measure $P_{z|e}$.

Now again, the decision maker will wish to choose the e for which $u^*(e)$ is greatest; and therefore we may say that the utility of being at move 1 with the choice of e still to make is

$$u^* \equiv \max_e u^*(e) = \max_e E_{z|e} \max_a E''_{\theta|z}\, u(e, \tilde{z}, a, \tilde{\theta}) \; . \tag{1-4}$$

This procedure of working back from the outermost branches of the decision tree to the base of the trunk is often called "backwards induction." More descriptively it could be called a process of "averaging out and folding back."

1.2.2. Example

Should a certain component be hand-adjusted at extra expense before it is installed in a complicated electronic system? Should the only available test, which is expensive and not infallible, be made on the component before a final decision is reached? The possible choices by the decision maker and chance are listed in Table 1.1; all the possible sequences of choices (e, z, a, θ) and the utility

† In some problems with infinite A it may be impossible to find an a whose utility is equal to the least upper bound of the utilities of all the as. These problems can be handled by considering suprema rather than maxima, but some proofs which are obvious when the maximum exists become complex when the supremum is not attainable and we do not feel that the slight added generality is worth the cost for applications of the sort we are considering in this book.

$u(e, z, a, \theta)$ which the decision maker assigns to each are shown on the decision tree, Figure 1.2.

Table 1.1

Possible Choices

Space	Elements	Interpretation
A	a_1	do not adjust
	a_2	adjust
Θ	θ_1	component does not need adjustment
	θ_2	component needs adjustment
E	e_0	do not experiment
	e_1	experiment
Z	z_0	outcome of e_0 (a dummy)
	z_1	outcome of e_1 which is more favorable to θ_1
	z_2	outcome of e_1 which is more favorable to θ_2

The marginal probability measure $P_{z|e}$ which is shown below each z in Figure 1.2 and the conditional measures $P''_{\theta|z}$ which are shown below each θ are derived from the following measures which we assume to have been *directly* assigned by the decision maker: (1) the conditional measures $P_{z|e,\theta}$ shown for all possible (e, θ) pairs in Table 1.2, and (2) the marginal measure P'_θ shown in Table 1.3. From

Table 1.2

Conditional Measures on Z

	E			
	e_0		e_1	
	θ_1	θ_2	θ_1	θ_2
z_0	1.0	1.0	.0	.0
z_1	.0	.0	.7	.2
z_2	.0	.0	.3	.8
	1.0	1.0	1.0	1.0

Table 1.3

Marginal Measure on Θ

θ	P'_θ
θ_1	.7
θ_2	.3
	1.0

these we first compute the joint measure $P_{\theta,z|e}$ for each of the experiments e_0 and e_1; the results for e_1 are shown in Table 1.4. From this joint measure we then compute the marginal measure $P_{z|e}$ for each e, the results for e_1 being shown in Table 1.4, and the conditional measures $P''_{\theta|z}$ for every z, the results being shown in Table 1.5. (We remind the reader that, as we pointed out in Section 1.1.1 above, the description of z includes everything about e which is relevant to the determination of the conditional measure on Θ.)

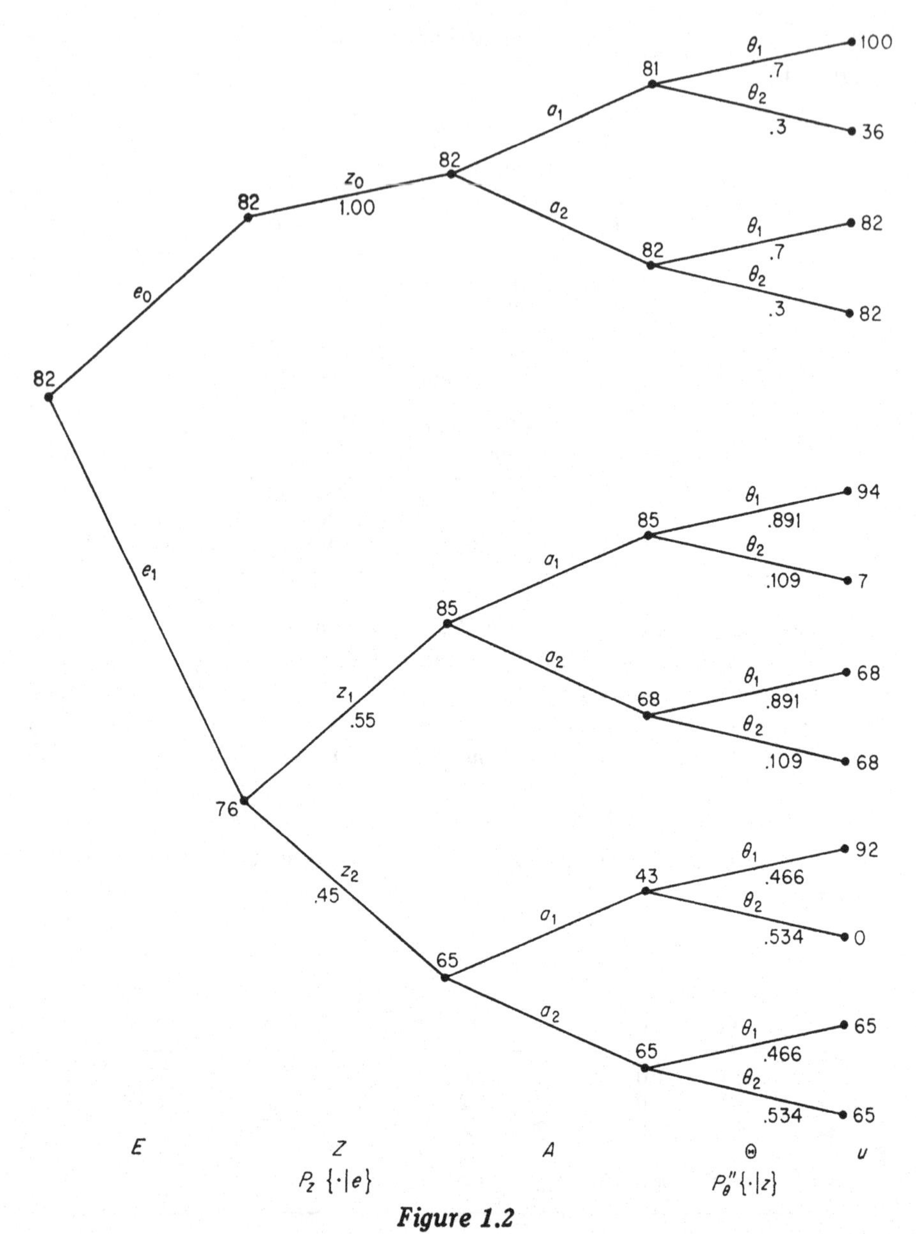

Figure 1.2

Analysis of an "Imperfect Tester"

We are now ready to begin our analysis of the tree Figure 1.2, and our first step is to start from the end (the extreme right) and use the data we find there to evaluate $u^*(e, z, a)$ for all (e, z, a). As a single example of the computations,

$$u^*(e_1, z_1, a_1) = u(e_1, z_1, a_1, \theta_1) \, \mathrm{P}''_\theta\{\theta_1|z_1\} + u(e_1, z_1, a_1, \theta_2) \, \mathrm{P}''_\theta\{\theta_2|z_1\}$$
$$= 94(.891) + 7(.109) = 85$$

Table 1.4

Measures Associated with e_1

Z	Joint Measure on $\Theta \times Z$		Marginal Measure on Z
	θ_1	θ_2	
z_0	.00	.00	.00
z_1	.49	.06	.55
z_2	.21	.24	.45
Marginal Measure on Θ	.70	.30	1.00

as shown on the tree. Having computed $u^*(e, z, a)$ for all (e, z, a) combinations, we are ready to compute $u^*(e, z)$ for all (e, z). As an example,

$$u^*(e_1, z_1) = \max \{u^*(e_1, z_1, a_1), u^*(e_1, z_1, a_2)\}$$
$$= \max \{85, 68\} = 85$$

as shown on the tree. [Notice from the tree that if e_1 is performed, then it is better to take a_1 if z_1 occurs but to take a_2 if z_2 occurs.] Next we compute $u^*(e)$ for all e, i.e., for $e = e_0$ and e_1. For example,

Table 1.5

Conditional Measures on Θ

Z	θ_1	θ_2	Sum
z_0	.700	.300	1.000
z_1	.891	.109	1.000
z_2	.466	.534	1.000

$$u^*(e_1) = u^*(e_1, z_1)\, P_z\{z_1|e_1\} + u^*(e_1, z_2)\, P_z\{z_2|e_1\}$$
$$= 85(.55) + 65(.45) = 76$$

as shown on the tree. Finally, we compute

$$u^* = \max \{u^*(e_0), u^*(e_1)\}$$
$$= \max \{82, 76\} = 82 \;.$$

The optimal course of action is thus to use e_0 and then a_2—i.e., to hand-adjust the component before installing it *without* bothering to test it first. It is better *not* to use the imperfect tester even though if used it would be powerful enough to determine the decision maker's preferred action.

1.3. Analysis in Normal Form

The final product of the extensive form of analysis studied in the previous section can be thought of as the description of an optimal *strategy* consisting of two parts:

1. A prescription of the *experiment* e which should be performed,
2. A *decision rule* prescribing the optimal terminal act a for every possible outcome z of the chosen e.

The whole decision rule for the optimal e can be simply "read off" from the results of that part of the analysis which determined the optimal a for every z in Z; and we may remark incidentally that these same results also enable us to read off the optimal decision rule to accompany any other e in E, even though the e in question is not itself optimal.

The normal form of analysis, which we are now about to examine, also has as its end product the description of an optimal strategy, and it arrives at the same optimal strategy as the extensive form of analysis, but it arrives there by a different route. Instead of first determining the optimal *act* a for *every possible outcome* z, and thus implicitly defining the optimal decision rule for any e, the normal form of analysis starts by explicitly considering *every possible decision rule* for a given e and then choosing the optimal *rule* for that e. After this has been done for all e in E, the optimal e is selected exactly as in the extensive form of analysis.

1.3.1. Decision Rules

Mathematically, a decision rule d for a given experiment e is a mapping which carries z in Z into $d(z)$ in A. In other words, d is a function which assigns a "value" a to each z in Z. Given a particular strategy (e, d) and a particular pair of values (z, θ), the decision maker's act as prescribed by the rule will be $a = d(z)$ and his utility will be $u(e, z, d(z), \theta)$; but before the experiment has been conducted and its outcome observed, $u(e, \tilde{z}, d(\tilde{z}), \tilde{\theta})$ is a random variable because $\tilde{z}$ and $\tilde{\theta}$ are random variables.

The decision maker's objective is therefore to choose the strategy (e, d) which maximizes his *expected* utility

$$u_*(e, d) \equiv \mathrm{E}_{\theta,z|e}\, u(e, \tilde{z}, d(\tilde{z}), \tilde{\theta}) \ .$$

This double expectation will actually be accomplished by iterated expectation, and the iterated expectation can be carried out in either order: we can first expect over $\tilde{\theta}$ holding $\tilde{z}$ fixed and then over $\tilde{z}$, using the same measures $\mathrm{P}''_{\theta|z}$ and $\mathrm{P}_{z|e}$ which were used in the extensive form of analysis; or we can first expect over $\tilde{z}$ holding $\tilde{\theta}$ fixed and then over $\tilde{\theta}$, using the measures $\mathrm{P}_{z|e,\theta}$ and P'_θ. It is traditional and usually more practical to use the measures $\mathrm{P}_{z|e,\theta}$ and P'_θ and we shall therefore proceed here in this manner, reserving for Section 1.4.3 an example of a situation in which the alternative procedure is preferable.

If e and d are given and $\tilde{\theta}$ is held fixed, then by taking the expectation of $u[e, \tilde{z}, d(\tilde{z}), \theta]$ with respect to the *conditional* measure $\mathrm{P}_{z|e,\theta}$ we obtain

$$u_*(e, d, \theta) \equiv \mathrm{E}_{z|e,\theta}\, u[e, \tilde{z}, d(\tilde{z}), \theta] \ , \tag{1-5}$$

which will be called the *conditional* utility of (e, d) *for a given state* θ. Next expecting over $\tilde{\theta}$ with respect to the unconditional measure P'_θ, we obtain

$$u_*(e, d) = \mathrm{E}'_\theta\, u_*(e, d, \tilde{\theta}) \ , \tag{1-6}$$

which will be called the *unconditional* utility of (e,d).

Now given any particular experiment e, the decision maker is free to choose the decision rule d whose expected utility is greatest; and therefore we may say that the utility of any experiment e is

$$u_*(e) \equiv \max_d u_*(e, d) \ . \tag{1-7}$$

After computing the utility of every e in E, the decision maker is free to choose the experiment with the greatest utility, so that we may write finally

$$u_* \equiv \max_e u_*(e) = \max_e \max_d \mathrm{E}'_\theta \, \mathrm{E}_{z|e,\theta} \, u[e, \tilde{z}, d(\tilde{z}), \tilde{\theta}] \ . \tag{1-8}$$

1.3.2. Performance, Error, and Utility Characteristics

For any given strategy (e, d) and state θ, we can compute the probability that the rule d will lead to any specified act a. Formally, we define for any (measurable) subset $A_o \subset A$

$$\mathrm{P}_a\{A_o|e, d, \theta\} \equiv \mathrm{P}_z\{z: d(z) \in A_o|e, \theta\} = \mathrm{P}_z\{d^{-1}(A_o)|e, \theta\} \ ,$$

and we say that the measure $\mathrm{P}_{z|e,\theta}$ on Z induces the measure $\mathrm{P}_{a|e,\theta}$ on A. Although such measures can be defined for any act space A, they are of particular interest when the act space contains only two acts, $A = \{a_1, a_2\}$. In this case $\mathrm{P}_a\{a_1|e, d, \cdot\}$ and $\mathrm{P}_a\{a_2|e, d, \cdot\}$, treated as functions of θ, are called *performance characteristics* of d for e. (When the act a_1 can be identified as "acceptance" and a_2 as "rejection", it is customary to call $\mathrm{P}_a\{a_1|e, d, \cdot\}$ an "operating characteristic" and $\mathrm{P}_a\{a_2|e, d, \cdot\}$ a "power function".)

If the utility measure $u(e, z, a, \theta)$ is sufficiently specified to make it possible to partition Θ into three subsets

$$\begin{aligned} \Theta_0 &\equiv \{\theta : a_1 \text{ and } a_2 \text{ are indifferent}\} \ , \\ \Theta_1 &\equiv \{\theta : a_1 \text{ is preferred to } a_2\} \ , \\ \Theta_2 &\equiv \{\theta : a_2 \text{ is preferred to } a_1\} \ , \end{aligned}$$

we can define another function on Θ which we shall call the *error characteristic* of d; this is the probability that the chosen act will be the one which is *not* preferred for the given θ and is therefore equal to

$$\left.\begin{array}{l} \mathrm{P}_a\{a_2|e, d, \theta\} \\ 0 \\ \mathrm{P}_a\{a_1|e, d, \theta\} \end{array}\right\} \quad \text{if } \theta \in \quad \left\{\begin{array}{l} \Theta_1 \\ \Theta_0 \\ \Theta_2 \ . \end{array}\right.$$

If the utility measure $u(e, z, a, \theta)$ is completely specified, it will be possible to compute $u_*(e, d, \theta)$ for all (e, d, θ), so that for any given (e, d) we can consider $u_*(e, d, \cdot)$, as another function defined on Θ. This function will be called the *utility characteristic*† of (e, d).

Notice that to define the performance characteristics of d all that is required is knowledge of A, Θ, Z, and $\mathrm{P}_{z|e,\theta}$; *neither* P'_θ nor u enters the picture. In order to define the error characteristic of d, we require enough information about u to partition Θ into $\{\Theta_0, \Theta_1, \Theta_2\}$, and in order to define the utility characteristic we

† The negative of what we have called the utility characteristic is sometimes called the "loss" function of (e, d), but because "loss" is also sometimes used to denote the different concept "regret" we prefer to avoid the word altogether for the time being.

need a complete specification of u, but even in these two cases P'_θ does not enter the picture.

1.3.3. Example

As an example for the analysis in normal form we take the same problem which was analyzed in extensive form in Section 1.2.2. We consider four potential strategies:

1: Use e_0 and d_{01} where $d_{01}(z_0) = a_1$.
(Do not experiment and do not adjust.)
2: Use e_0 and d_{02} where $d_{02}(z_0) = a_2$.
(Do not experiment but adjust.)
3: Use e_1 and d_{11} where $d_{11}(z_1) = a_1$ and $d_{11}(z_2) = a_2$.
(Adjust if and only if the experimental outcome is z_2.)
4: Use e_1 and d_{12} where $d_{12}(z_1) = a_2$ and $d_{12}(z_2) = a_1$.
(Adjust if and only if the experimental outcome is z_1.)†

The performance characteristic $P_\theta\{a_2|e, d, \cdot\}$ and the error and utility characteristics of these four possible strategies are shown in Table 1.6; the last column of the table shows the values of $u_*(e, d)$ obtained by applying the prior measure P'_θ to the values in the two preceding columns.

Table 1.6

Strategy	$P\{a_2\|e, d, \theta\}$		Error Characteristic		Utility Characteristic		$u_*(e, d)$
	θ_1	θ_2	θ_1	θ_2	θ_1	θ_2	
1	0	0	0	1.0	100	36	81
2	1.0	1.0	1.0	0	82	82	82
3	.3	.8	.3	.2	85	53	76
4	.7	.2	.7	.8	75	14	57

As an example of the computations underlying this table, consider Strategy 3, which calls for e_1 followed by a_1 if z_1 is observed, a_2 if z_2 is observed.

1. From Table 1.2 we find that the probabilities of z_2 and thus of a_2 are .3 if θ_1 is true and .8 if θ_2 is true.

2. By Table 1.1, a_1 is preferred if θ_1 is true, a_2 if θ_2 is true; therefore the .3 probability of a_2 if θ_1 is true *is* the probability of error if θ_1 is true, while the .8 probability of a_2 if θ_2 is true is the *complement* of the probability of error if θ_2 is true.

3. By Figure 1.2, the utility of $(e_1, z_1, a_1, \theta_1)$ is 94 and the utility of $(e_1, z_2, a_2, \theta_1)$ is 65, so that

† Two more strategies could be defined, *viz.*

5. Experiment but do not adjust regardless of outcome.
6. Experiment but adjust regardless of outcome.

We omit these two strategies from the formal analysis because (5) is obviously inferior to (1) and (6) is obviously inferior to (2).

$$u_*(e_1, d_{11}, \theta_1) = (.7)94 + (.3)65 = 85 \ ;$$

and similarly

$$u_*(e_1, d_{11}, \theta_2) = (.2)7 + (.8)65 = 53 \ .$$

4. Finally, by Table 1.3, $P'_\theta(\theta_1) = .7$ and $P'_\theta(\theta_2) = .3$, so that

$$u_*(e_1, d_{11}) = (.7)85 + (.3)53 = 76 \ .$$

The last column of Table 1.6 shows that the optimal strategy is number 2 (adjust without experimentation) and that its utility is $u_* = 82$. The reader will observe that these results obtained by the normal form of analysis are identical to the results obtained by the extensive form of analysis and shown on the decision tree Figure 1.2. We next proceed to establish that the two forms of analysis are equivalent in complete generality.

1.3.4. Equivalence of the Extensive and Normal Form

The extensive and normal forms of analysis will be equivalent if and only if they assign the same utility to every potential e in E, i.e., if the formula

$$u_*(e) = \max_d E'_\theta \, E_{z|e,\theta} \, u[e, \tilde{z}, d(\tilde{z}), \tilde{\theta}] \tag{1-9}$$

derived as (1-7) by the normal form of analysis agrees for all e with the formula

$$u^*(e) = E_{z|e} \max_a E''_{\theta|z} \, u(e, \tilde{z}, a, \tilde{\theta}) \tag{1-10}$$

derived as (1-3) by the extensive form. We have already pointed out that the operation $E'_\theta \, E_{z|e,\theta}$ in (1-9) is equivalent to expectation over the entire possibility space $\Theta \times Z$ and is therefore equivalent to $E_{z|e} \, E''_{\theta|z}$. It follows that the normal-form result (1-9) can be written

$$u_*(e) = \max_d E_{z|e} \, E''_{\theta|z} \, u[e, \tilde{z}, d(\tilde{z}), \tilde{\theta}] \ ,$$

and it is then obvious that the best d will be the one which for every z maximizes

$$E''_{\theta|z} \, u[e, z, d(z), \tilde{\theta}] \ .$$

This, however, is exactly the same thing as selecting for every z an a_z which satisfies

$$E''_{\theta|z} \, u(e, z, a_z, \tilde{\theta}) = \max_a E''_{\theta|z} \, u(e, z, a, \tilde{\theta})$$

as called for by (1-2) in the extensive form of analysis. Letting $d^*(z)$ denote the optimal decision rule selected by the normal form, we have thus proved that

$$d^*(z) = a_z$$

and that formulas (1-9) for $u_*(e)$ and (1-10) for $u^*(e)$ are equivalent.

Thus we see that if we wish to choose the best e and must therefore evaluate $u^*(e)$ for all e in E, the extensive and normal forms of analysis require exactly the same inputs of information and yield exactly the same results even though the intermediate steps in the analysis are different. If, however, e is fixed and one wishes merely to choose an optimal terminal act a, the extensive form has the merit that one has only to choose an appropriate act for the particular z which actually materializes; there is no need to find the decision rule which selects the best act for every z which might have occurred but in fact did not occur.

1.3.5. Bayesian Decision Theory as a Completion of Classical Theory

In the classical theory of decision based in part on the evidence of a sample, the choice of one among all possible (e, d) pairs is to be made by comparison of their performance characteristics as defined in Section 1.3.2. Such comparisons are inherently extremely difficult because (1) except in two-action problems it will not be obvious which of two (e, d) strategies is superior even for given θ, and (2) even if the comparison *is* easy for any given θ, there will usually be a very great number of (e, d) pairs such that each pair is optimal for *some* θ in Θ but no pair is optimal for *all* θ in Θ.

Difficult as this problem of comparing incomparables may be, however, the decision maker *must* act and therefore he must choose. The normal-form analysis expounded in Section 1.3.1 and illustrated in Section 1.3.3 above amounts to solving the problem of choice by (1) evaluating the utility of every (e, z, a, θ), (2) using this evaluation to translate the performance characteristic of each strategy (e, d) into a utility characteristic, and then (3) computing a weighted average of this utility characteristic with the measure P'_θ used as the weighting factor. It is thus possible to think of the measure P'_θ, not as a measure of "prior probability", but as a mere weighting device required to arrive at a reasoned choice between two utility characteristics one of which is better for some θ in Θ while the other is better for others. It can be shown, moreover, that a few basic principles of logically consistent behavior—principles which are eminently reasonable in our opinion—*compel* one to choose among utility characteristics *as if* one used such a weighting function; a proof is given in Section 1.5 at the end of this chapter.

It is therefore our belief that the Bayesian analysis of decision problems is in no sense in *conflict* with the classical theory of statistical analysis. The classical theory leaves to the decision maker's unaided judgment the task of choosing amongst performance characteristics; Bayesian analysis in the normal form merely formalizes this use of judgment by expressing it in terms of explicitly stated utilities and weights and can thus be thought of as a formal *completion* of the classical theory. From this point of view, the introduction of the weights at the *beginning* of the analysis, as is done in the extensive form of analysis, is justified by the proved equivalence of the two forms; all that is needed to justify the name "probability measure" which we have given to the weighting measure is the fact that *any* normalized system of nonnegative weights obeys the axioms of probability theory.

In this same vein, we view a decision theory which formalizes utilities but not prior probabilities as a partial completion of the classical theory, and in some very simple situations this partial completion may be all that is required for reasoned practical action.

1.3.6. Informal Choice of a Decision Rule

If there are only two possible terminal acts *and* if e is given, the decision maker *may* be able to select a "reasonable" d by an informal, direct comparison of performance characteristics. In making such a comparison he must of course have in mind *some* appraisal of u and P'_θ, but it may not be necessary to formalize these appraisals and express them precisely.

If the utility measure u is completely specified, the decision maker who wishes to choose his d without formal specification of P'_θ will obviously do better to apply his judgment to comparison of the utility characteristics $u_*(e, d, \cdot)$ of the various ds under consideration, since in this way he has only his appraisal of P'_θ to keep in mind informally. This procedure may have special appeal in some group decision problems where there is general consensus on all the data of the problem except P'_θ, since it *may* turn out that the same (e, d) is optimal or near-optimal for all the P'_θs held by the individual members of the group even though these measures are themselves widely divergent.

When e is *not* given and the decision maker must choose an e as well as choosing a d for that e, both these informal methods of choice obviously become much more difficult to apply.

1.4. Combination of Formal and Informal Analysis

1.4.1. Unknown Costs; Cutting the Decision Tree

The general decision model stated in Section 1.1 is often criticized on the grounds that utilities cannot be rationally assigned to the various possible (e, z, a, θ) combinations because the costs, profits, or in general the consequences of these combinations would not be certain even if θ *were* known. In principle, such criticisms represent nothing but an incomplete definition of the state space Θ. If, for example, the decision maker is ignorant not only of the number of defectives but also of the cost per defective, the state space can obviously be made rich enough to include all possible pairs of values of *both* unknowns. The decision maker's uncertainties about these values can then be evaluated together with his other uncertainties in the probability measure which he assigns to Θ, and the analysis of the decision problem can then proceed essentially as before.

Formally, let the state θ be expressible as a doublet $(\theta^{(1)}, \theta^{(2)})$ so that the state space is of the form $\Theta = \Theta^{(1)} \times \Theta^{(2)}$. Thus $\theta^{(1)}$ might be the parameter of a Bernoulli process while $\theta^{(2)}$ might be the cost of accepting a defective item, but observe that the notation does *not* imply that $\theta^{(2)}$ is *necessarily* a single real number: $\theta^{(2)}$ can be a vector quantity representing any number of unknown values.

In terms of our original decision tree Figure 1.1, a play was a 4-tuple (e, z, a, θ) and utilities were assigned directly to each (e, z, a, θ). If θ is split into $(\theta^{(1)}, \theta^{(2)})$, a play is a 5-tuple $(e, z, a, \theta^{(1)}, \theta^{(2)})$ as shown in Figure 1.3; utilities are assigned directly to each $(e, z, a, \theta^{(1)}, \theta^{(2)})$ and the utility of any $(e, z, a, \theta^{(1)})$ is the *expected value* of the random variable $u(e, z, a, \theta^{(1)}, \tilde{\theta}^{(2)})$, the expectation being taken with respect to the *conditional* measure on $\Theta^{(2)}$ given the history $(e, z, a, \theta^{(1)})$.

Use of a Certainty Equivalent for $\theta^{(2)}$. When $\theta = (\theta^{(1)}, \theta^{(2)})$ and $\theta^{(2)}$ stands say for a state (value) of a cost, the probability measure on $\Theta^{(2)}$ will often be *independent* of $\theta^{(1)}$ and z. This will be the case when, for example, $\theta^{(1)}$ is a fraction defective while $\theta^{(2)}$ is the cost of a defective. If in addition $u(e, z, a, \theta^{(1)}, \cdot)$ is a *linear* function of $\theta^{(2)}$, it is easy to see that

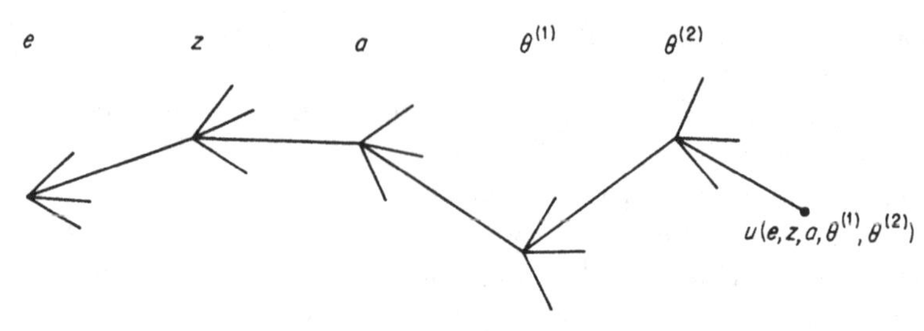

Figure 1.3

One Possible Play when $\boldsymbol{\Theta} = (\boldsymbol{\Theta}^{(1)}, \boldsymbol{\Theta}^{(2)})$

$$u(e, z, a, \theta^{(1)}) = \mathrm{E}\, u(e, z, a, \theta^{(1)}, \tilde{\theta}^{(2)}) = u(e, z, a, \theta^{(1)}, \bar{\theta}^{(2)})$$

where

$$\bar{\theta}^{(2)} \equiv \mathrm{E}(\tilde{\theta}^{(2)})$$

is the *unconditional* expected value of $\tilde{\theta}^{(2)}$. It is therefore legitimate in this case to replace $\tilde{\theta}^{(2)}$ by its expected value $\bar{\theta}^{(2)}$, i.e., to use $\bar{\theta}^{(2)}$ as a *certainty equivalent* of $\theta^{(2)}$.

It is perhaps worth calling attention to one specific type of problem in which application of this principle would, as others have already pointed out, very greatly simplify some of the analyses to be found in the literature. In discussing the economics of acceptance sampling, it is customary to treat each sample as taken from a finite lot, so that the sampling distribution is hypergeometric; and it has several times been proposed to fit the prior distribution of the lot quality by a "mixed binomial", i.e., by the compound distribution created by the assumption that each lot is generated by a Bernoulli process whose parameter p is fixed during the production of any one lot but has a probability distribution from lot to lot. On this assumption we may partition the lot fraction defective θ into two components $(\theta^{(1)}, \theta^{(2)})$ where $\theta^{(1)}$ is the process fraction defective p while $\theta^{(2)}$ is the difference ϵ between the process fraction and the lot fraction defective; but if utility is linear in lot fraction defective, as is usually assumed in problems of this sort, then since the expected value of $\tilde{\epsilon}$ is 0 for any p and z,

$$\mathrm{E}_{\epsilon|p,z} u(e, z, a, p, \tilde{\epsilon}) = u(e, z, a, p, 0)$$

for all p and z. In other words, there is simply no need to look at lot fraction defective if utility is linear in this variable: *we can assign utilities directly to the various possible values of* $\tilde{p}$, assign a prior distribution to $\tilde{p}$, and treat the sampling as being directly from the *process* and therefore binomial rather than from the *lot* and therefore hypergeometric.†

Suppression of $\theta^{(2)}$. When the conditional measure of $\tilde{z}$ given $\theta = (\theta^{(1)}, \theta^{(2)})$ actually depends only on $\theta^{(1)}$, one can formally suppress $\theta^{(2)}$ entirely by assigning utilities directly to the various possible $(e, z, a, \theta^{(1)})$ instead of assigning utilities to all possible $(e, z, a, \theta^{(1)}, \theta^{(2)})$ and then expecting out $\tilde{\theta}^{(2)}$. We remind the reader that *the utilities assigned to* $(e, z, a, \theta^{(1)})$ *are merely expressions of the decision maker's*

† Cf. *Probability and Statistics for Business Decisions*, pages 377–381.

preferences and that in deciding on his preferences he is free to use any level of analysis he likes. While he *may* wish to assign a formal measure to $\tilde{\theta}^{(2)}$ and then average over it in order to decide whether or not he prefers a' to a'' when $\theta_o^{(1)}$ is true, he may prefer to give a direct, intuitive answer to this question keeping his uncertainties about $\theta^{(2)}$ informally in mind.

Actually, of course, a decision tree extending only to $\theta^{(1)}$ is *never* really complete. Life goes on after $(e, z, a, \theta^{(1)})$, and to evaluate $u(e, z, a, \theta^{(1)})$ one could *always* look ahead by adding a $\theta^{(2)}$ component, $\theta^{(2)}$ now standing for "future life", assigning a utility to every $(e, z, a, \theta^{(1)}, \theta^{(2)})$, and then expecting out $\tilde{\theta}^{(2)}$. In practice, however, it will not always be worth the effort of *formally* looking very far ahead.

1.4.2. Incomplete Analysis of the Decision Tree

Besides cutting the decision tree before it is logically complete, the decision maker may rationally decide not to make a complete formal analysis of even the truncated tree which he *has* constructed. Thus if E consists of two experiments e_1 and e_2, he may work out $u^*(e_1)$, say, by formally evaluating

$$u^*(e_1) = \mathrm{E}_{z|e_1} \max_a \mathrm{E}''_{\theta|z}\, u(e_1, \tilde{z}, a, \tilde{\theta}) \;;$$

but after this formal analysis of e_1 is completed he may conclude without any formal analysis at all that e_2 is not so good as e_1 and adopt e_1 without further ado.

That such behavior can be perfectly consistent with the principles of choice expounded earlier in this chapter is easy to see. Before making a formal analysis of e_2, the decision maker can think of the unknown quantity $u^*(e_2)$ as a random variable $\tilde{v}$. If now he were to be told the number v which would result from formal analysis and if this number were *greater* than the known number $u^*(e_1)$, he would adopt e_2 instead of e_1 and we could say that the value of the information that $\tilde{v} = v$ is measured by the difference $v - u^*(e_1)$ in the decision maker's utility which results from this change of choice.† If on the contrary v were *less* than $u^*(e_1)$, the decision maker would adhere to his original choice of e_1 and we could say that the information had been worthless.

In other words, we can meaningfully define the random variable

$$\max\{0, \tilde{v} - u^*(e_1)\} = \textit{value of information} \text{ regarding } e_2 \;,$$

and before "buying" such information at the cost of making a formal analysis of e_2 the decision maker may prefer to compute its *expected* value by assigning a probability measure to $\tilde{v}$ and then expecting with respect to this measure; the geometry of the computation is shown in Figure 1.4. If the expected value is less than the cost, he will quite rationally decide to use e_1 *without* formally evaluating $v = u^*(e_2)$. Operationally one usually does not formally compute either the value or the cost of information on e_2: these are subjectively assessed. The computations *could* be formalized, of course, but *ultimately* direct subjective assessments *must* be used if the decision maker is to avoid an infinite regress. In the last analysis we *must* cut the tree *somewhere.*

† For a more careful discussion of the conditions under which arguments based on utility differences are justified, see Sections 4.4 and 4.5 below.

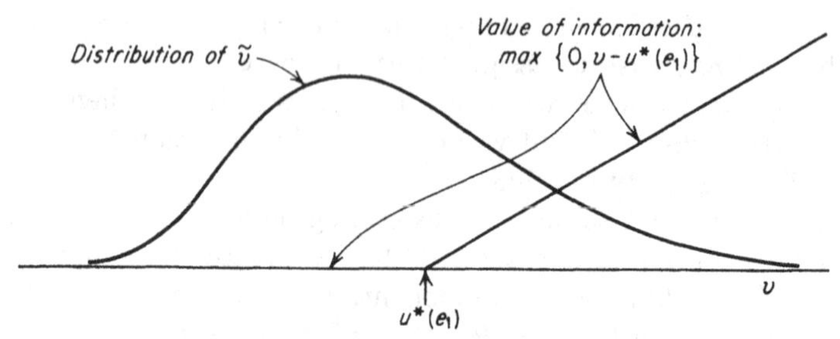

Figure 1.4

Value of Further Analysis of e_2

Before leaving the subject of incomplete analysis we should perhaps remark that completely formal analysis and completely intuitive analysis are not the only possible methods of determining a utility such as $u^*(e_2)$. In many situations it will also be possible to make a *partial* analysis in order to gain some insight but at a less prohibitive cost than a full analysis entails. This operation is very much akin to taking a sample in order to learn more about the state space; and whether or not a sample should be taken and if so of what kind is the main subject of the remainder of this book. Even though we formally take leave of these philosophical considerations at this point, we shall continue to be concerned with close relatives.

1.4.3. Example

An oil wildcatter must choose between drilling a well and selling his rights in a given location. (In a real problem there would of course be many more acts, such as selling partial rights, sharing risks, farmouts, etc.) The desirability of drilling depends on the amount of oil which will be found; we simplify by considering just two states, "oil" and "no oil". Before making his decision the wildcatter can if he wishes obtain more geological and geophysical evidence by means of very

Table 1.7

Possible Choices

Space	Elements	Interpretations
A	a_1	drill, do not sell location
	a_2	do not drill, sell location
Θ	θ_1	no oil
	θ_2	oil
E	e_0	do not take seismic readings
	e_1	take seismic readings
Z	z_0	dummy outcome of e_0
	z_1	e_1 reveals no structure
	z_2	e_1 reveals open structure
	z_3	e_1 reveals closed structure.

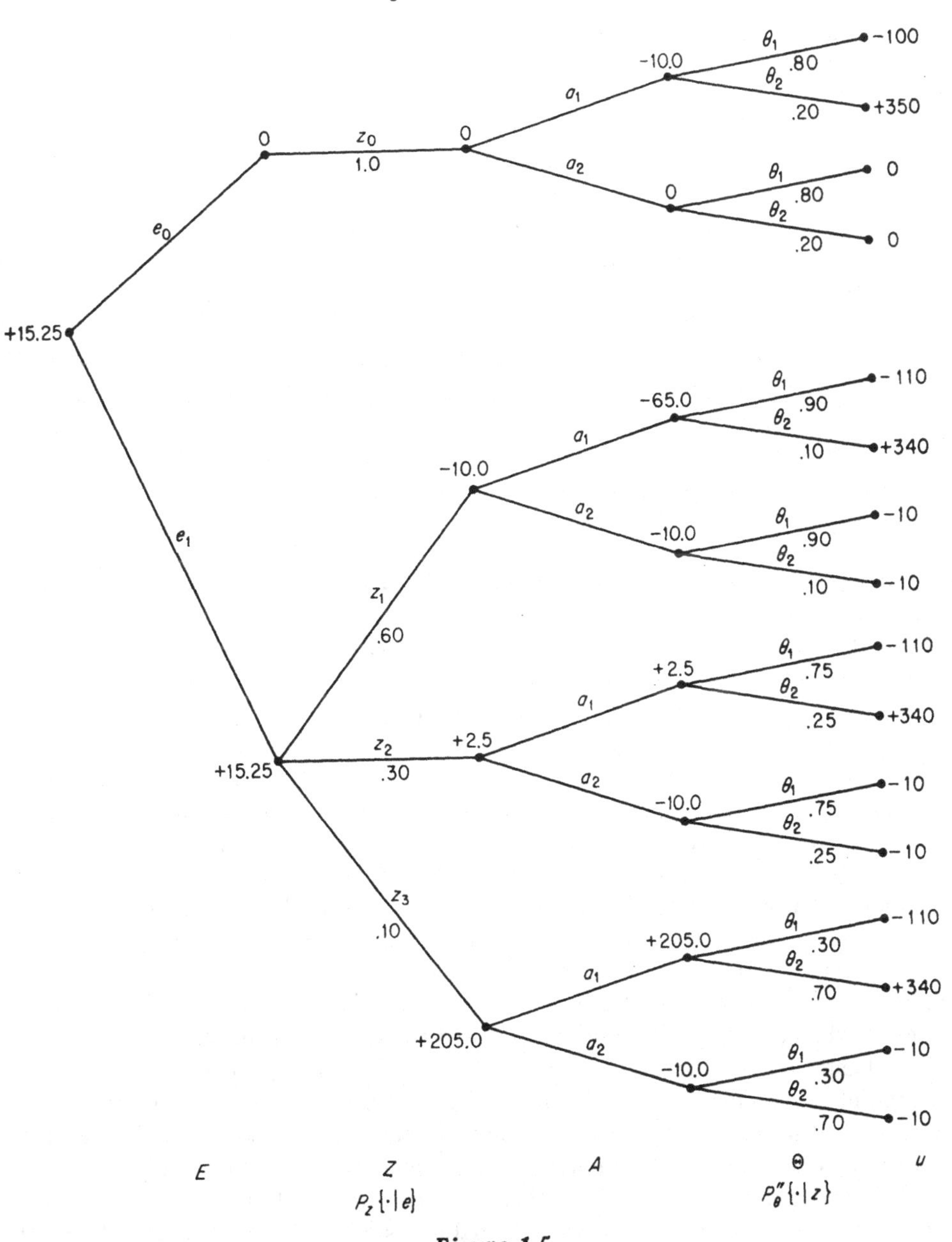

Figure 1.5

Analysis of a Drilling Decision

expensive experiments; we simplify again by allowing for only one form of experiment, seismographic recordings, and by assuming that these recordings, if made, will give completely reliable information that one of three conditions prevails: (1) there is no subsurface structure, (2) there is an open subsurface structure, or (3) there is a closed subsurface structure. The descriptions of the four spaces,

A, Θ, E, and Z, are summarized in Table 1.7, and the possible sequences of choices by the decision maker and chance are shown on the decision tree Figure 1.5. Notice that the three possible "conditions" of the subsurface structure are *not* the *states* $\{\theta\}$ of the problem but correspond to the experimental *outcomes* $\{z\}$.

Assignment of Utilities. The psychological stimulus associated in this problem with an (e, z, a, θ) 4-tuple is highly complicated. Different wells entail different drilling costs, and different strikes produce different quantities and qualities of oil which can be recovered over different periods of time and sold at different prices. Furthermore, each potential consequence of the present drilling venture interacts with future potential drilling ventures. For example, geological information gained in one deal may be crucial for the next, and capital expenditures made in one deal may prohibit the acceptance of the next one however favorable it may then appear.

In other words, there are uncertainties surrounding any particular (e, z, a, θ) complex which must be kept in mind either formally or informally when these complexes are compared. We assume nevertheless that, no matter how intricate these considerations are, *the decision maker is psychologically able to act* and therefore can not only rank complex stimuli of this sort but assign to them utility numbers which reflect his preferences among lotteries having such stimuli as prizes. The kind of problem we are here discussing is simply one specific illustration of the need to cut the decision tree which was discussed in Section 1.4.1; hypothetical utilities are shown at the end of the decision tree Figure 1.5.

Assignment of Probabilities. As regards the assignment of a probability measure to the possibility space $\Theta \times Z$, this problem typifies a class of problems which occur rather frequently in practice but have rarely if ever been recognized in the literature: the available historical evidence bears much more directly on the conditional (or "posterior") measure $P''_{\theta|z}$ and the marginal measure $P_{z|e}$ than on the complementary measures $P_{z|e,\theta}$ and P'_θ. Specifically, previous experience with the amounts of oil found in the three possible types of geologic structure (z_1 = no structure, z_2 = open structure, z_3 = closed structure) may make it possible to assign a nearly if not quite "objective" measure $P''_{\theta|z}$ to the amount of oil which will be found *given* any particular experimental result, whereas it would be much less clear what measure P'_θ should be assigned to the amount of oil in the absence of knowledge of the structure. At the same time it will in general be much more meaningful to a geologist to assign a marginal measure $P_{z|e}$ to the various structures and thus to the sample space Z than it would be to assign conditional measures $P_{z|e,\theta}$ to Z depending on the amount of oil which will be found. Hypothetical measures $P''_{\theta|z}$ and $P_{z|e}$ are shown on those branches of the decision tree Figure 1.5 which emanate from e_1; the "prior" probabilities $P'_\theta\{\theta_1\} = .80$ and $P'_\theta\{\theta_2\} = .20$ shown on the branches emanating from e_0 were computed from them by use of the formula

$$P'_\theta\{\theta_i\} = P''_\theta\{\theta_i|z_1\}\, P_z\{z_1|e_1\} + P''_\theta\{\theta_i|z_2\}\, P_z\{z_2|e_1\} + P''_\theta\{\theta_i|z_3\}\, P_z\{z_3|e_1\} .$$

Analysis. Since all data required for analysis in extensive form appear on the decision tree Figure 1.5, the reader can easily verify that the optimal decision is to pay for seismographic recordings (e_1) and then drill (take act a_1) if and only if the recordings reveal open (z_2) or closed (z_3) structure. It pays to buy the recordings because the expected utility of this decision is 15.25 whereas the expected utility of the optimal act without seismic information is 0.

Observe that in this problem none of the probabilities required for analysis in extensive form has to be computed by the use of Bayes' theorem, but that if the problem is to be analyzed in normal form and if this analysis is to be carried out via a performance characteristic, it *will* be necessary to use Bayes' theorem. Thus, for example, the probability given the state θ_1 (no oil) that the decision rule "drill if and only if z_3 occurs" will lead to drilling can be found only by computing

$$P_z\{z_3|e_1, \theta_1\} = \frac{P''_\theta\{\theta_1|z_3\}\ P_z\{z_3|e_1\}}{P'_\theta\{\theta_1\}} ,$$

where $P'_\theta\{\theta_1\}$ is to be computed from the last previous formula.

It was pointed out in Section 1.3.6 above that when all the elements of a decision problem except P'_θ can be evaluated "objectively", the normal form of analysis has a real advantage in that it permits all the data other than P'_θ to be summarized in the form of a utility characteristic $u_*(e, d, \cdot)$ defined on Θ for each of the strategies (e, d) under consideration. These characteristics can then be compared with the judgmental or "subjective" element P'_θ held informally in mind. In problems of the kind typified by our present example, on the contrary, a utility characteristic $u_*(e, d, \cdot)$ on Θ would be "subjective" because one of the elements involved in its computation is the conditional measure on the sample space and, as we have just seen, this measure has to be derived from the "subjective" marginal measure assigned to the sample space by the geologist.

We have already said, however, that the normal form of analysis does not *require* the use of $P_{z|e,\theta}$ and P'_θ; and when the posterior measure $P''_{\theta|z}$ is "objective" as it is here we can postpone the judgmental calculation to the end by using $P''_{\theta|z}$ to construct a new kind of utility characteristic to which $P_{z|e}$ is then applied as the last step in the analysis. If utilities have been assigned to all (e, z, a, θ), then for any given strategy (e, d) we can use the measure $P''_{\theta|z}$ to compute

$$u^*[e, z, d(z)] \equiv E''_{\theta|z}\, u[e, z, d(z), \tilde\theta]$$

for all z; and we can then make our final evaluation of the (e, d) pair by averaging with respect to $P_{z|e}$:

$$u_*(e, d) = E_{z|e}\, u^*[e, \tilde z, d(\tilde z)] = E_{z|e}\, E''_{\theta|z}\, u[e, \tilde z, d(\tilde z), \tilde\theta] .$$

Considered as a function on Z, the quantity $u^*[e, \cdot, d(\cdot)]$ defined by the former of these two formulas is the new kind of utility characteristic we set out to obtain; and in simple problems the averaging over this characteristic involved in the formula for $u_*(e, d)$ can be intuitive rather than formal.

1.5. Prior Weights and Consistent Behavior

In Section 1.3.5 we remarked that when one is forced to compare utility characteristics because one is forced to act, a few basic principles of logically consistent behavior necessarily lead to the introduction of a *weighting function* over Θ. We then remarked that if this weighting function is normalized it has all the properties of a *probability measure* on Θ and that it can be brought in at the beginning of the analysis just as well as at the end or in the middle. In this section we shall investigate these basic principles *very* informally; complete formal treatments can be found in Blackwell and Girshick, Luce and Raiffa, and Savage.†

We here assume that every conditional measure $P_{z|e,\theta}$ on the sample space Z (the "sampling distribution" of the experimental outcome) is "objective" in the sense that it corresponds to a *known* long-run relative frequency; and we take as given the basic proposition proved by von Neumann and Morgenstern that in situations where all probabilities correspond to known long-run frequencies it is possible to assign utilities to consequences in such a way that choices are logically consistent only if they are such as to maximize expected utility. We shall therefore assume that any strategy (e, d) can be described by a *utility characteristic* $u_*(e, d, \cdot)$ on the state (parameter) space Θ. Readers who do not accept the utility principle even in situations where all probabilities are objective but who believe that within certain limits it is reasonable to maximize objective monetary expectations can substitute "income" for "utility" throughout our argument and find that it applies just as well.

We start by sacrificing a little generality in order to simplify the discussion. (1) We assume that Θ is *finite*, consisting of elements $\{\theta_1, \theta_2, \cdots, \theta_r\}$. The utility characteristic of any strategy (e, d) can then be represented as an r-tuple $[u_*(e, d, \theta_1), u_*(e, d, \theta_2), \cdots, u_*(e, d, \theta_r)]$, and comparison of any two strategies is equivalent to a comparison of r-tuples within a certain interval R in r-space, the boundaries of the region corresponding to the bounds on the utility function $u_*(e, d, \cdot)$—or to the limits within which the decision maker wishes to maximize expected monetary value if he refuses the utility principle. (2) We assume that our task is to find a conceptual principle for ranking *all* r-tuples in R, whether or not they are all actually achievable in any particular situation. Given these two simplifying assumptions, the argument can be visualized in 2-space and we shall give diagrams to aid the reader in this visualization.

We now proceed to make three *basic assumptions concerning logically consistent behavior*. The first of these is the so-called sure-thing or dominance principle:

Assumption 1. Let $\boldsymbol{u} = (u_1, \cdots, u_r)$ be the utility characteristic of strategy (e_1, d_1) and let $\boldsymbol{v} = (v_1, \cdots, v_r)$ be the utility characteristic of strategy (e_2, d_2). If $u_i \geq v_i$ for all i and if $u_i > v_i$ for some i, then (e_1, d_1) is preferred to (e_2, d_2).

In order to express the second basic assumption informally, we shall make use of the notion of indifference surfaces familiar in classical economic theory. In terms of these surfaces we can express what might be called an assumption of continuous substitutability as follows:

Assumption 2. Indifference surfaces extend smoothly from boundary to boundary of the region R in the sense that, if $\boldsymbol{u}$ is a point on any indifference

† D. Blackwell and M. A. Girshick, *Theory of Games and Statistical Decisions*, New York, Wiley, 1954, Chapter 4; R. D. Luce and H. Raiffa, *Games and Decisions*, New York, Wiley, 1957, Chapter 13; L. J. Savage, *The Foundations of Statistics*, New York, Wiley, 1954, Chapters 1–5.

surface and if small changes are made in any $r - 1$ coordinates of $\boldsymbol{u}$, then by making a small compensating change in the remaining coordinate we can obtain a new point on the same indifference surface as $\boldsymbol{u}$.

The meaning of "smooth" can be made more precise by saying that the compensating change in the rth coordinate is a continuous function of the changes in the $r - 1$ coordinates; or we can say that if preferences for utility characteristics $\boldsymbol{u}$ are indexed by a Marshallian index function, this function is continuous on R.

Finally, we make what might be called the assumption that pigs is pigs.

Assumption 3. Let (e_1, d_1), (e_2, d_2) and (e_3, d_3) be three strategies such that (e_1, d_1) and (e_2, d_2) are indifferent. Then given any p such that $0 \leq p \leq 1$, a mixed strategy which selects (e_1, d_1) with "objective" probability p and (e_3, d_3) with probability $1 - p$ is indifferent to a mixed strategy which selects (e_2, d_2) with probability p and (e_3, d_3) with probability $1 - p$.

Before actually following out the implications of these three basic assumptions, we make two observations. (1) *If* the decision maker compares two r-tuples $\boldsymbol{u} = (u_1, \cdots, u_r)$ and $\boldsymbol{v} = (v_1, \cdots, v_r)$ by means of the indices

$$\Sigma_{i=1}^{r} p_i u_i \qquad \text{and} \qquad \Sigma_{i=1}^{r} p_i v_i$$

where $(p_1, \cdots, p_r)$ are preassigned positive weights, *then* clearly the indifference surfaces must constitute a family of parallel hyperplanes whose common normal is the vector $\boldsymbol{p} = (p_1, \cdots, p_r)$. (2) Conversely, *if* the indifference surfaces are parallel hyperplanes with a normal going into the interior of the first orthant, *then* it is not difficult to see that there exist positive weights $(p_1, \cdots, p_r)$ such that r-tuples can be ranked on the basis of an index which associates to each r-tuple the weighted average of its r components and that by our first basic assumption an r-tuple with a greater index must be preferred to an r-tuple with a lesser index.

We shall now prove that, given our three basic assumptions about logically consistent behavior, the decision maker's indifference surfaces *must* be parallel hyperplanes with a common normal going into the interior of the first orthant, from which it follows that all utility characteristics $\boldsymbol{u} = (u_1, \cdots, u_r)$ in R *can in fact* be ranked by an index which applies a predetermined set of weights $\boldsymbol{p} = (p_1, \cdots, p_r)$ to their r components.

We first show that the third basic assumption implies that the indifference surfaces which exist by the second basic assumption must be *hyperplanes*. To do so we assume the contrary and prove a contradiction; the geometry for $r = 2$ is shown in Figure 1.6, where an underscored letter denotes a vector and corresponds to boldface in the text. If an indifference surface is curved, we can always find two points $\boldsymbol{u}$ and $\boldsymbol{v}$ on the surface such that $\boldsymbol{w} = \frac{1}{2}\boldsymbol{u} + \frac{1}{2}\boldsymbol{v}$ is *not* on the surface. Now choose (e_1, d_1) and (e_2, d_2) so that they have utility characteristics $\boldsymbol{u}$ and $\boldsymbol{v}$ respectively. Then by the utility theory we take as given—or by ordinary principles of expected monetary value—the mixed strategy which chooses (e_1, d_1) with "objective" probability $\frac{1}{2}$ and (the same) (e_1, d_1) with probability $\frac{1}{2}$ has the evaluation $\boldsymbol{u}$ while the mixed strategy which chooses (e_1, d_1) with probability $\frac{1}{2}$ and (e_2, d_2) with probability $\frac{1}{2}$ has the evaluation $(\frac{1}{2}\boldsymbol{u} + \frac{1}{2}\boldsymbol{v}) = \boldsymbol{w}$. But by the third basic assump-

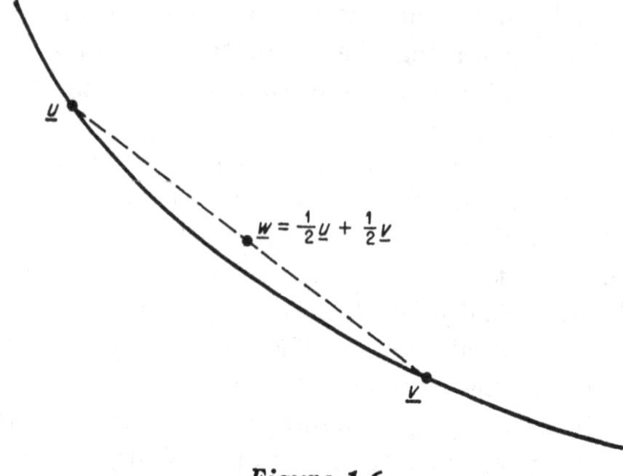

Figure 1.6

Linearity of Indifference Surfaces

tion these two mixed strategies are indifferent, and therefore $\mathbf{w}$ must lie on the same indifference surface as $\mathbf{u}$. This establishes the linearity of the indifference surfaces.

We next show that the indifference hyperplanes must be *parallel.* If the region in r-space in which the comparisons are being made is *all* of r-space, the proof is obvious: indifference surfaces cannot intersect, and the hyperplanes *will* intersect unless they are parallel. If, however, R is *not* all of r-space, an additional argument must be added because hyperplanes which are not parallel need not intersect in a finite region. We shall supply this argument by showing that all indifference hyperplanes must have the same normal; the geometry for $r = 2$ is shown in Figure 1.7.

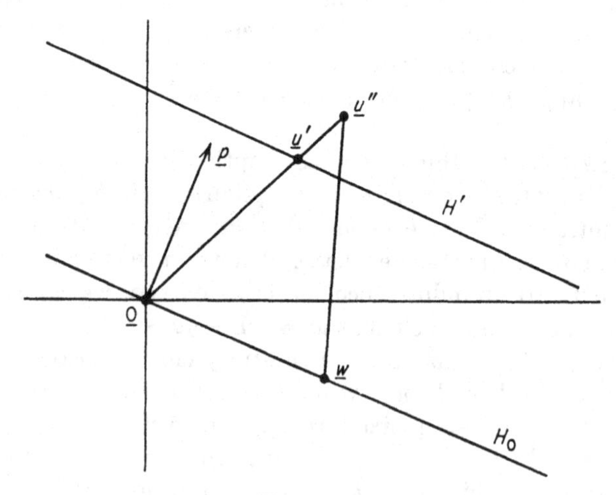

Figure 1.7

Parallelism of Indifference Surfaces

Assuming without loss of generality that the origin $\boldsymbol{0} = (0, 0, \cdots, 0)$ belongs to the interior of R, let H_o be the indifference hyperplane containing $\boldsymbol{0}$ and let $\boldsymbol{p} = (p_1, \cdots, p_r)$ be the normalized normal to H_o (i.e. the normal with $\Sigma\, p_i = 1$), so that H_o consists of all points $\boldsymbol{w} = (w_1, w_2, \cdots, w_r)$ such that

$$\Sigma_{i=1}^{r}\, p_i\, w_i = 0 \ . \tag{1-11}$$

Let $\boldsymbol{u}'$ be a point in R on H', let $\boldsymbol{u}''$ be a point in R on the prolongation of the ray from $\boldsymbol{0}$ through $\boldsymbol{u}'$, and define λ (where $0 < \lambda < 1$) by

$$\boldsymbol{u}' = \lambda\, \boldsymbol{u}'' + (1 - \lambda)\, \boldsymbol{0} = \lambda\, \boldsymbol{u}'' \ .$$

Now $\boldsymbol{0}$ is indifferent to any $\boldsymbol{w}$ in R which lies on H_o, and therefore by the third basic assumption and the ordinary rules of utility or expected monetary value the mixed strategy which selects $\boldsymbol{u}''$ with "objective" probability λ and $\boldsymbol{0}$ with probability $1 - \lambda$ is indifferent to the mixed strategy which selects $\boldsymbol{u}''$ with probability λ and $\boldsymbol{w}$ with probability $(1 - \lambda)$, so that the points

$$\lambda\, \boldsymbol{u}'' + (1 - \lambda)\, \boldsymbol{0} = \boldsymbol{u}' \qquad \text{and} \qquad \lambda\, \boldsymbol{u}'' + (1 - \lambda)\, \boldsymbol{w} = \boldsymbol{u}' + (1 - \lambda)\, \boldsymbol{w}$$

lie on the same hyperplane H'. Now if $\boldsymbol{p}' = (p_1', p_2', \cdots, p_r')$ is the normalized normal to H', the projections onto this normal of any two points of H' must be the same. Consequently

$$\Sigma_{i=1}^{r}\, p_i'\, u_i' = \Sigma_{i=1}^{r}\, p_i'[u_i' + (1 - \lambda)\, w_i]$$

and therefore

$$\Sigma_{i=1}^{r}\, p_i'\, w_i = 0 \ . \tag{1-12}$$

Since (1-11) and (1-12) hold for all $\boldsymbol{w}$ in H_o, both $\boldsymbol{p}$ and $\boldsymbol{p}'$ are orthogonal to H_o; and since in addition both $\boldsymbol{p}$ and $\boldsymbol{p}'$ are normalized by definition, $\boldsymbol{p}' = \boldsymbol{p}$ as was to be shown. That all components of $\boldsymbol{p}$ are positive follows from assumptions 1 and 2.

CHAPTER 2

Sufficient Statistics and Noninformative Stopping

2.1. Introduction

Of the various methods of assigning a probability measure to the possibility space $\Theta \times Z$ which were discussed in Section 1.1.2, the most *frequently* useful is the method which starts by assigning a marginal measure P'_θ to Θ and a conditional measure $P_{z|e,\theta}$ to Z for every e in E and every θ in Θ. If the decision maker does proceed in this way, then the measures required for analysis in extensive form—the marginal measure $P_{z|e}$ on Z and the conditional measure $P''_{\theta|z}$ on Θ—must be computed from the measures directly assigned by the decision maker. In this chapter and the next we shall deal with two important technical problems involved in this procedure: (1) the selection of a *prior measure* P'_θ which will both express the decision maker's best judgment about Θ and at the same time be mathematically tractable, and (2) the use of *sufficient statistics* as an aid in the computation of $P''_{\theta|z}$. These two problems will actually be discussed in inverse order for reasons which will be amply clear in due course.

2.1.1. Simplifying Assumptions

Since it is not our purpose in this monograph to seek generality for its own sake, our subsequent discussion will make use of some assumptions which greatly simplify the analysis without any loss of practical applicability.

1. State Space Θ. We assume that the state θ can be described by an r-tuple of real numbers $(\theta_1, \cdots, \theta_r)$ and that the state space Θ can be represented as either a discrete set or an interval set in Euclidean r-space.†

2. Prior Measure on Θ. We assume that the prior measure P'_θ on the state space Θ possesses either a mass function or a density function (or possibly both). In other words, we assume that if Θ_o is any measurable subset of Θ, then either there exists a function D' on Θ (the prime denoting *prior*) such that

$$P'_\theta\{\Theta_o\} = \sum\nolimits_{\Theta_o} D'(\theta) \qquad \text{or} \qquad P'_\theta\{\Theta_o\} = \int_{\Theta_o} D'(\theta)\, d\theta$$

or else there exist two functions D'_1 and D'_2 such that

$$P'_\theta\{\Theta_o\} = c_1 \sum\nolimits_{\Theta_o} D'_1(\theta) + c_2 \int_{\Theta_o} D'_2(\theta)\, d\theta \; .$$

† The reader must be careful to distinguish between the r-tuple $(\theta_1, \cdots, \theta_r)$ here defined, in which θ_i is the ith component of one particular vector state θ, and the set $\{\theta_1, \cdots, \theta_r\}$ of Section 1.5, in which θ_i was the ith possible state in the state space Θ. In the remainder of this monograph, subscripts on θ or z will always distinguish components rather than members of a set.

Since the three cases behave virtually identically in virtually all respects, we shall usually discuss only the case where P'_θ has a density function, leaving it to the reader to match our results for this case with the corresponding results for the two others.

3. Sample Space Z. We assume that if the sample space $Z = \{z\}$ is temporarily restricted for the purpose of some argument to the potential outcomes of a particular experiment e, then the outcomes $\{z\}$ can be represented as a subset of a Euclidean space.

4. Measures on Z. If $P_{z|e,\theta}$ is the conditional measure on Z for given (e, θ) and P'_θ is the prior measure on Θ, the marginal measure on Z for the given e is

$$P_{z|e} \equiv P_z\{\cdot|e\} = E'_\theta \, P_z\{\cdot|e, \tilde{\theta}\}$$

where E'_θ is taken with respect to P'_θ. We assume that either (a) $P_{z|e,\theta}$ is discrete for *every* θ in Θ *and* $P_{z|e}$ is discrete, so that both measures possess mass functions, or else (b) $P_{z|e,\theta}$ possesses a density function for *every* θ in Θ *and* $P_{z|e}$ possesses a density function. [This assumption would *not* be satisfied if, for example, P'_θ were continuous while $P_{z|e,\theta}$ conditionally put a mass of 1 on $z = \theta$.]

If the conditional measure has a mass function, we shall denote by $\ell(z|\theta)$ the probability given θ that e results in z; because our discussions of such probabilities will in general be restricted to some particular experiment e, we suppress the condition e in our notation. If the conditional measure has a density function, we shall denote by $\ell(z|\theta)$ the value of this density function at z for given θ (and e). In either case, we shall use the word *likelihood* to denote the value $\ell(z|\theta)$ taken on by the mass or density function for given z and θ (and e); and we assume that, for any fixed z in Z, $\ell(z|\cdot)$ as a function on Θ is continuous except for at most a finite number of discontinuities (which may depend on z).

We define the *marginal likelihood* of the experimental outcome z given a particular prior density D' by

$$\ell^*(z|D') \equiv \int_\Theta \ell(z|\theta) \, D'(\theta) \, d\theta \; , \tag{2-1}$$

and we shall say that z lies in the *spectrum* of D' if $\ell^*(z|D') > 0$.

2.1.2. Bayes' Theorem; Kernels

If the prior distribution of the random variable $\tilde{\theta}$ has a density function D' and if the experimental outcome z is in the spectrum of D', then it follows from Bayes' theorem that the posterior distribution of $\tilde{\theta}$ has a density function D'' whose value at θ for the given z is

$$D''(\theta|z) = D'(\theta) \, \ell(z|\theta) \, N(z) \tag{2-2a}$$

where $N(z)$ is simply the normalizing constant defined by the condition

$$\int_\Theta D''(\theta|z) \, d\theta = N(z) \int_\Theta D'(\theta) \, \ell(z|\theta) \, d\theta = 1 \; . \tag{2-2b}$$

▶ To prove (2-2) when $\ell(\cdot|\theta)$ is discrete, let Θ_o be any measurable subset of Θ. Then by Bayes' theorem

(1) $$P''_\theta\{\Theta_o|z\} = \frac{P_{\theta,z}\{\Theta_o, z|e\}}{P_z\{z|e\}} .$$

Defining $N(z)$ by

$$[N(z)]^{-1} = P_z\{z|e\}$$

and recalling that

$$P_{\theta,z}\{\Theta_o, z|e\} = \int_{\Theta_o} \ell(z|\theta)\, D'(\theta)\, d\theta$$

we may substitute in (1) to obtain

(2) $$P''_\theta\{\Theta_o|z\} = \int_{\Theta_o} D'(\theta)\, \ell(z|\theta)\, N(z)\, d\theta ,$$

and the integrand is by definition the posterior density of $\tilde{\theta}$.

To prove (2-2) when $\ell(\cdot|\theta)$ is continuous, we must show that, for any interval $(z, z + dz)$, which we label $Z_o(dz)$ to show the dependence on dz,

(3) $$\lim_{dz\to 0} P''_\theta\{\Theta_o|Z_o(dz)\} = \frac{\int_{\Theta_o} D'(\theta)\, \ell(z|\theta)\, d\theta}{\int_{\Theta} D'(\theta)\, \ell(z|\theta)\, d\theta} .$$

By Bayes' theorem

(4) $$P''_\theta\{\Theta_o|Z_o\} = \frac{P_{\theta,z}\{\Theta_o, Z_o|e\}}{P_z\{Z_o|e\}} .$$

Because both P'_θ and $P_{z|e,\theta}$ have densities,

$$P_{\theta,z}\{\Theta_o, Z_o|e\} = \int_{\Theta_o}\int_{Z_o} D'(\theta)\, \ell(\zeta|\theta)\, d\zeta\, d\theta ;$$

and by the mean-value theorem this may be written

(5) $$P_{\theta,z}\{\Theta_o, Z_o|e\} = \int_{\Theta_o} D'(\theta)\, \ell(z'_\theta|\theta)\, |dz|\, d\theta$$

where z'_θ is some z in $Z_o(dz)$ and $|dz|$ is the volume of $Z_o(dz)$. Similarly

(6) $$P_z(Z_o|e) = \int_\Theta\int_{Z_o} D'(\theta)\, \ell(\zeta|\theta)\, d\zeta\, d\theta = \int_\Theta D'(\theta)\, \ell(z''_\theta|\theta)\, |dz|\, d\theta$$

where $z''_\theta \in Z_o(dz)$. Substituting (5) and (6) in (4), taking the limit as $dz \to 0$, and noticing that as $dz \to 0$ both z'_θ and z''_θ go to z, we obtain (3), which was to be proved. ◀

If the density function of $\tilde{\theta}$ is D, where D denotes *either* a prior or a posterior density, and if K is another function on Θ such that

$$D(\theta) = \frac{K(\theta)}{\int_\Theta K(\theta)\, d\theta} , \tag{2-3}$$

i.e., if the ratio $K(\theta)/D(\theta)$ is a constant as regards θ, we shall write

$$D(\theta) \propto K(\theta) \tag{2-4}$$

and say that K is a (not "the") *kernel* of the density of θ.

If the likelihood of z given θ is $\ell(z|\theta)$, and if ρ and κ are functions on Z such that for all z and θ

$$\ell(z|\theta) = \kappa(z|\theta)\, \rho(z) , \tag{2-5}$$

i.e., if the ratio $\kappa(z|\theta)/\ell(z|\theta)$ is a constant as regards θ, we shall say that $\kappa(z|\theta)$ is a (not "the") *kernel* of the likelihood of z given θ and that $\rho(z)$ is a *residue* of this likelihood.

Letting K′ denote a kernel of the *prior* density of $\tilde{\theta}$, it follows from the definitions of K and ℓ and of the symbol $\propto$ that the Bayes formula (2-2) can be written

$$\begin{aligned} D''(\theta|z) &= D'(\theta)\, \ell(z|\theta)\, N(z) \\ &= K'(\theta) \left[\int_\Theta K'(\theta)\, d\theta\right]^{-1} \kappa(z|\theta)\, \rho(z)\, N(z) \\ &\propto K'(\theta)\, \kappa(z|\theta)\ . \end{aligned} \tag{2-6}$$

The value of the constant of proportionality for the given z,

$$\rho(z)\, N(z) \left[\int_\Theta K'(\theta)\, d\theta\right]^{-1} ,$$

can always be determined by the condition

$$\int_\Theta D''(\theta|z)\, d\theta = 1\ .$$

Example. Let the state θ be the intensity λ of a Poisson process and let the sample outcome z be the fact that r successes were observed in time t, so that the likelihood of $z = (r, t)$ for given $\theta = \lambda$ is given by the Poisson mass function:

$$\ell(z|\theta) = \frac{e^{-\lambda t}(\lambda t)^r}{r!}\ ; \tag{2-7}$$

and let the prior density of $\tilde{\lambda}$ be gamma with parameters r' and t', the primes denoting prior,

$$D'(\lambda) = \frac{e^{-\lambda t'}(\lambda t')^{r'-1}}{(r'-1)!}\, t'\ . \tag{2-8}$$

Then we may take as kernels

$$\begin{aligned} \kappa(z|\theta) &= e^{-\lambda t}\, \lambda^r\ , \\ K'(\theta) &= e^{-\lambda t'}\, \lambda^{r'-1}\ ; \end{aligned}$$

and by (2-6) the posterior density of $\tilde{\theta}$ given the observed $z = (r, t)$ is

$$D''(\lambda|z) \propto e^{-\lambda(t'+t)}\, \lambda^{r'+r-1} \equiv e^{-\lambda t''}\, \lambda^{r''-1} \tag{2-9}$$

where r'' and t'' are implicitly defined, the double primes denoting "posterior". The normalizing constant for D″ is the reciprocal of

$$\int e^{-\lambda t''}\, \lambda^{r''-1}\, d\lambda = (r''-1)!/t''^{r''}\ ;$$

it can be found more easily by merely observing that the kernel (2-9) of the posterior density is of the same form as the kernel of the prior density (2-8), from which it follows that the normalizing constant for (2-9) must be of the same form as the normalizing constant in (2-8).

2.2. Sufficiency

There will in general be more than one way of describing the outcome of an experiment, and therefore the definition of the set $Z = \{z\}$ of all possible outcomes depends on the decision maker's judgment concerning the features of the outcome which may be relevant to his decision problem. Once he has formalized his decision problem, it is always possible to define a very extensive set $\{z\}$ which is certainly rich enough to include all relevant information, but the question will then arise whether there exists a set of descriptions which is simpler and therefore easier to manipulate and yet contains all the information which is actually relevant to the decision problem. Thus given a particular model of some production process generating good pieces (g) and defectives (d) it may be obvious that a sample of five pieces from this process can be adequately described by a 5-tuple such as $ggdgd$ which records the "values" of the five pieces in the order in which they were produced, but at the same time it may seem possible that all the *relevant* information can be conveyed by a 2-tuple such as (3, 5) recording that 3 of the 5 pieces produced were good. Letting y denote a possible abridged description of a particular outcome, the question is whether y is a *sufficient* description of this outcome.

2.2.1. Bayesian Definition of Sufficiency

We shall assume that any abridged description y consists of an r-tuple of real numbers and can therefore be represented as a point in a Euclidean space Y, and we shall denote by $\tilde{y}$ the mapping (or random variable) which sends Z into Y. The event that the random variable $\tilde{y}$ assumes the value y will be abbreviated $\tilde{y} = y$; this event is thus equivalent to the event comprising the set of all z such that $\tilde{y}(z) = y$, i.e., the set

$$\{z: \tilde{y}(z) = y\} \equiv \tilde{y}^{-1}(y) \ . \tag{2-10}$$

The conditional measure $P_{z|e,\theta}$ on Z determines the conditional measure on Y given (e, θ), so that given any prior density or mass function D' on Θ and any particular value y of $\tilde{y}$ the posterior density or mass function can be obtained by the use of Bayes' theorem. The posterior density or mass function so calculated will be denoted $D''(\cdot|\tilde{y} = y)$.

Clearly the coarser information y about a particular experimental outcome will lead to exactly the same conclusions as the more refined information z if $D''(\cdot|\tilde{y} = y)$ is identical to $D''(\cdot|\tilde{z} = z)$ for all z in $\tilde{y}^{-1}(y)$. We are then entitled to say that y is a "sufficient" description of the experimental outcome; and if any y in the range Y of $\tilde{y}$ is sufficient, we are entitled to say that the mapping $\tilde{y}$ itself is sufficient. Formally, we give the following

Definition: The mapping $\tilde{y}$ from Z into Y is *sufficient* if for *any* prior density or mass function D' and *any* z in the spectrum of D'

$$D''(\cdot|\tilde{y} = y) = D''(\cdot|\tilde{z} = z) \qquad \text{where} \qquad y = \tilde{y}(z) \ . \tag{2-11}$$

Where no confusion can result, we shall also use the expression *sufficient statistic* to denote either (1) a random variable $\tilde{y}$ which satisfies this definition or (2) a particular value y of such a random variable, i.e., a sufficient description of a particular experimental outcome.

2.2.2. Identification of Sufficient Statistics

The definition of sufficiency which we have just given does not in itself enable us to *find* sufficient sets $\{y\}$ of abridged descriptions of the experimental outcome (or sufficient mappings $\tilde{y}$ of Z into Y); this is accomplished by examination of the kernel of the likelihood of the "complete" descriptions $\{z\}$. We saw in (2-6) that if $\kappa(z|\theta)$ is a kernel of the likelihood $\ell(z|\theta)$, then the posterior distribution of $\tilde{\theta}$ given z can be determined by consideration of the kernel $\kappa(z|\theta)$ just as well as by consideration of the complete likelihood $\ell(z|\theta)$. In other words, we learn absolutely nothing useful by distinguishing among zs whose kernels are equal; and it follows that a mapping $\tilde{y}$ from Z into Y is sufficient if, given any y in Y, all z in $\tilde{y}^{-1}(y)$ have the same kernel. Such mappings can be found by use of the following

Theorem. Let $\tilde{y}$ map Z into Y. If the likelihood function ℓ on $Z \times \Theta$ can be factored as the product of a kernel function k on $Y \times \Theta$ and a residue function ρ on Z,

$$\ell(z|\theta) = \mathrm{k}[\tilde{y}(z)|\theta]\, \rho(z) \, , \tag{2-12}$$

then $\tilde{y}$ is sufficient in the sense of (2-11).

▶ For any measurable sets Θ_o and Z_o the posterior probability of Θ_o given Z_o is

$$\mathrm{P}_\theta''\{\Theta_o|Z_o\} = \mathrm{E}^*\, \mathrm{P}_\theta''\{\Theta_o|\tilde{z}\} \, ,$$

where E* denotes expectation with respect to the measure on Z conditional on Z_o but unconditional as regards $\tilde{\theta}$. Now take $Z_o = \tilde{y}^{-1}(y)$, so that the left-hand side of this equation becomes $\mathrm{P}_\theta''\{\Theta_o|\tilde{y} = y\}$ while by (2-6) and (2-12) we have for the quantity following the operator E* on the right-hand side

$$\mathrm{P}_\theta''\{\Theta_o|z\} \propto \int_{\Theta_o} \mathrm{D}'(\theta)\, \mathrm{k}[\tilde{y}(z)|\theta]\, d\theta \, .$$

Since this quantity is obviously the same for all $z \in \tilde{y}^{-1}(y)$, i.e., for all z such that $\tilde{y}(z) = y$, we may suppress the E* and write

$$\mathrm{P}_\theta''\{\Theta_o|\tilde{y} = y\} = \mathrm{P}_\theta''\{\Theta_o|z\} \, , \qquad z \in \tilde{y}^{-1}(y) \, ,$$

showing that $\tilde{y}$ satisfies the definition (2-11) of sufficiency. ◀

Example. Let the "complete" description of an experimental outcome be $z = (x_1, \cdots, x_n)$ where the xs are the observed values of n independent random variables with identical densities

$$(2\pi)^{-\frac{1}{2}}\, e^{-\frac{1}{2}(x-\mu)^2} \, ;$$

then the likelihood of z given $\theta = \mu$ is

$$\ell(z|\theta) = \Pi_{i=1}^n \left[(2\pi)^{-\frac{1}{2}}\, e^{-\frac{1}{2}(x_i-\mu)^2}\right] = (2\pi)^{-\frac{1}{2}n}\, e^{-\frac{1}{2}\Sigma(x_i-\mu)^2} \, .$$

If we define the (vector) function $\tilde{y} = (\tilde{y}_1, \tilde{y}_2) = (\tilde{m}, \tilde{n})$ by

$$n = \text{number of random variables observed} \, ,$$

$$m = \frac{1}{n} \Sigma\, x_i \, ,$$

the likelihood can be written

$$\ell(z|\theta) = e^{-\frac{1}{2}n(m-\mu)^2} \cdot (2\pi)^{-\frac{1}{2}n}\, e^{-\frac{1}{2}\Sigma(x_i-m)^2} \, .$$

The first factor is the kernel $k[\tilde{y}(z)|\theta] = k(m, n|\mu)$; the remaining factors constitute the residue $\rho(z)$ because they do not contain $\theta = \mu$.

Observe that we do not need to know the actual likelihood of y given θ in order to find $D''(\cdot|\tilde{y} = y)$; all that we need to know is the kernel $k(y|\theta)$ obtained by factoring the likelihood of z, since by (2-11) and (2-12) we may put the Bayes formula (2-6) in the form

$$D''[\theta|\tilde{y}(z)] = D''(\theta|z) \propto D'(\theta)\, \kappa(z|\theta) = D'(\theta)\, k[\tilde{y}(z)|\theta] \; . \qquad (2\text{-}13)$$

Thus in the example just discussed: if D' is the prior density of $\tilde{\mu}$, then the posterior density is

$$D''(\mu|m, n) \propto D'(\mu)\, k(m, n|\mu) = D'(\mu)\, e^{-\frac{1}{2}n(m-\mu)^2} \; .$$

It is true, of course, that the actual likelihood of y given θ will be of the form

$$k(y|\theta)\, r(y) \; ,$$

but *after e is fixed* the residue r is irrelevant. We shall see later that r is *not* irrelevant when the problem is *to choose the e* which is to be performed.

2.2.3. *Equivalence of the Bayesian and Classical Definitions of Sufficiency*

Classically, sufficiency is usually *defined* in terms of the factorability condition (2-12). We prefer the definition (2-11) because it is more easily extended to cover the important concept of partial or marginal sufficiency which we shall define and discuss in a moment, but before going on to this new subject we remark for the sake of completeness that the two definitions of "ordinary" sufficiency are completely equivalent. That classical sufficiency implies Bayesian sufficiency is shown by the theorem containing (2-12); that Bayesian sufficiency implies classical sufficiency is shown by the following

> *Theorem:* If $\tilde{y}$ is sufficient in the sense of (2-11), then there exist a function k on $Y \times \Theta$ and a function ρ on Z such that the sample likelihood can be factored as in (2-12):
>
> $$\ell(z|\theta) = k[\tilde{y}(z)|\theta]\, \rho(z) \; .$$

▶ We prove this theorem by (1) showing that if $\tilde{y}$ is sufficient, it is possible to find a kernel function κ_* such that $\kappa_*(z|\theta)$ has the same value for all z in any one $\tilde{y}^{-1}(y)$, and then (2) defining $k[\tilde{y}(z)|\theta] = \kappa_*(z|\theta)$.

Consider a prior density D'_* such that $D'_*(\theta) > 0$ for all $\theta \in \Theta$. Given the assumptions in Section 2.1.1 above, we lose no generality if we restrict Z to all z such that

$$\ell^*(z|D'_*) \equiv \int D'_*(\theta)\, \ell(z|\theta)\, d\theta > 0 \; .$$

In accordance with the general definition (2-5) of kernels define the particular kernel and the corresponding residue

$$(1) \qquad \kappa_*(z|\theta) \equiv \frac{\ell(z|\theta)}{\ell^*(z|D'_*)} \; , \qquad \rho(z) \equiv \ell^*(z|D'_*) \; .$$

We now have by Bayes' theorem

$$\text{(2)} \qquad D_*''(\theta|z) = \frac{D_*'(\theta)\, \ell(z|\theta)}{\int D_*'(\theta)\, \ell(z|\theta)\, d\theta} = D_*'(\theta) \frac{\ell(z|\theta)}{\ell^*(z|D_*')} = D_*'(\theta)\, \kappa_*(z|\theta) \;;$$

observe that with this kernel we have equality rather than proportionality between the two sides.

Next consider any z_1 and z_2 such that $\tilde{y}(z_1) = \tilde{y}(z_2)$. Since $\tilde{y}$ is sufficient, we have by the definition (2-11) of sufficiency

$$D_*''(\theta|y) = D_*''(\theta|z_1) = D_*''(\theta|z_2) \;.$$

By (2) above this implies that

$$D_*'(\theta)\, \kappa_*(z_1|\theta) = D_*'(\theta)\, \kappa_*(z_2|\theta) \;;$$

and since $D_*'(\theta) > 0$ for all θ, this in turn implies that

$$\text{(3)} \qquad \kappa_*(z_1|\theta) = \kappa_*(z_2|\theta) \;, \qquad \text{all } \theta.$$

Finally, define k by choosing some one $z \in \tilde{y}^{-1}(y)$ for every $y \in Y$ and setting

$$\text{(4)} \qquad k(y|\theta) \equiv \kappa_*(z|\theta) \;.$$

By (3), equation (4) will be true for *every* $z \in Z$; and we may therefore substitute (4) in (1) to obtain

$$\ell(z|\theta) = k[\tilde{y}(z)|\theta)\, \ell^*(z|D_*') = k[\tilde{y}(z)|\theta]\, \rho(z)$$

as was to be proved. ◀

2.2.4. *Nuisance Parameters and Marginal Sufficiency*

In the previous section we defined the conditions under which a description y would be a sufficient summarization of the experimental evidence concerning the *complete* state parameter θ. In many situations, however, some of the components of θ are *nuisance parameters* in the sense that they are irrelevant to the utility $u(e, z, a, \theta)$ for any (e, z, a, θ) and enter the problem only as parameters of the conditional probability measure $P_{z|e,\theta}$. If in such a situation we partition θ into (θ_1, θ_2) where θ_2 represents the nuisance parameters, it is clear that *after* the experiment has been conducted and both e and z are fixed the choice of a *terminal* act a will depend, not on the complete posterior distribution of $\tilde{\theta} = (\tilde{\theta}_1, \tilde{\theta}_2)$, but only on the marginal posterior distribution of $\tilde{\theta}_1$; and we are therefore interested in condensed descriptions $\{y\}$ of the possible experimental outcomes that summarize all the experimental evidence which is *relevant to* θ_1. Formally, we define "partial" or "marginal" sufficiency as follows:

Definition. Let θ be expressible as a doublet (θ_1, θ_2), so that $\Theta = \Theta_1 \times \Theta_2$, and let C be a class of prior distributions on Θ. Then $\tilde{y}$ is *marginally sufficient for* $\tilde{\theta}_1$ *relative to* C if for *any* prior distribution in C and *any* z in the spectrum of that distribution the marginal posterior distribution of $\tilde{\theta}_1$ is the same given y as given z.

In the applications the most important class C is that in which $\tilde{\theta}_1$ and $\tilde{\theta}_2$ are *independent*, so that the prior density can be written

$$D'(\theta_1, \theta_2) = D_1'(\theta_1)\, D_2'(\theta_2) \;. \tag{2-14}$$

Relative to this class of prior distributions, $\tilde{y}$ is marginally sufficient for $\tilde{\theta}_1$ if the likelihood of z given θ can be factored

$$\ell(z|\theta_1, \theta_2) = \text{k}[\tilde{y}(z)|\theta_1]\, \rho(z|\theta_2) \ . \tag{2-15}$$

For letting $\tilde{y}(z) = y$, we have by (2-2)

$$\text{D}''(\theta_1, \theta_2|z) = \text{D}_1'(\theta_1)\, \text{D}_2'(\theta_2)\, \text{k}(y|\theta_1)\, \rho(z|\theta_2)\, N(z) \ , \tag{2-16}$$

and integrating out θ_2 we get

$$\text{D}''(\theta_1|z) \propto \text{D}_1'(\theta_1)\, \text{k}(y|\theta_1) \ .$$

It is perhaps worth emphasizing that even though the *likelihood* can be factored as in (2-15), $\tilde{y}$ will *not* in general be marginally sufficient for $\tilde{\theta}_1$ unless the *prior* density can also be factored as in (2-14). The factor $\rho(z|\theta_2)$ in (2-16) will alter the prior distribution of $\tilde{\theta}_2$; and if $\tilde{\theta}_1$ is dependent on $\tilde{\theta}_2$, this effect will be transmitted to the distribution of $\tilde{\theta}_1$.

2.3. Noninformative Stopping

2.3.1. Data-Generating Processes and Stopping Processes

In many situations the complete description of an experiment can usefully be decomposed into two parts. The experiment consists of observing random variables $\tilde{x}_1, \cdots, \tilde{x}_i, \cdots$ successively generated by some *data-generating* process, and the description of this process constitutes the first part of the description of the experiment. The number of random variables observed depends on some criterion which may or may not be of a probabilistic nature; we shall say that the end of the experiment is caused by a *stopping process* which generates this criterion, and the description of this stopping process constitutes the second part of the description of the experiment. The stopping process will be deterministic if, for example, it is definitely known before the experiment is begun that neither more nor less than the first n random variables will be observed; it will be probabilistic if, for example, the experiment is to be continued until the experimenter's budget of time or money is exhausted and the cost of any observation is itself a random variable. We shall see, however, that in many situations there is more than one way of defining the "random variables" generated by the *data-generating* process and that a *stopping* process which is "probabilistic" on one definition of these random variables may be "deterministic" on another, equally valid definition.

2.3.2. Likelihood of a Sample

Consider a *data-generating process* which generates a discrete-time sequence of random variables $\tilde{x}_1, \cdots, \tilde{x}_i, \cdots$ not necessarily independent and not necessarily discrete-valued. With a slight loss of generality we shall assume that a probability measure is assigned to the $\tilde{x}$s via conditional likelihood (mass or density) functions. Letting θ_1 denote the (scalar or vector) state parameter which characterizes this process, the conditional likelihood of x_i given the previous observations $x_1, \cdots x_{i-1}$ and the parameter θ_1 can then be written

$$f(x_i|x_1, \cdots, x_{i-1}; \theta_1) \ , \tag{2-17}$$

and the likelihood that the *first* n elements of any sequence of n or more elements generated by this process will be $z = (x_1, \cdots, x_n)$ is

$$h(z|\theta_1) = f(x_1|\theta_1)\, f(x_2|x_1; \theta_1) \cdots f(x_n|x_1, \cdots, x_{n-1}; \theta_1) \; . \tag{2-18}$$

Next consider a *stopping process* such that, given a particular value of the parameter θ_1 which characterizes the data-generating process and a particular value of another (scalar or vector) state parameter θ_2, the probability that a first observation x_1 will be made is

$$\phi(1|\theta_1, \theta_2) \; , \tag{2-19a}$$

while given these same two parameters and a particular set $(x_1, \cdots, x_k)$ of observations already made, the probability that at least one more observation will be made before the experiment terminates is

$$\phi(k + 1|x_1, \cdots, x_k; \theta_1, \theta_2) \; . \tag{2-19b}$$

The likelihood that an experiment involving these two processes will result in the sequence of *exactly* n elements $z = (x_1, \cdots, x_n)$ is obviously

$$\begin{aligned} \ell(z|\theta_1, \theta_2) = {} & \phi(1|\theta_1, \theta_2)\, f(x_1|\theta_1) \cdot \phi(2|x_1; \theta_1, \theta_2)\, f(x_2|x_1; \theta_1) \\ & \cdot \phi(3|x_1, x_2; \theta_1, \theta_2) \cdots f(x_n|x_1, \cdots, x_{n-1}; \theta_1) \\ & \cdot [1 - \phi(n + 1|x_1, \cdots, x_n; \theta_1, \theta_2)] \; , \end{aligned}$$

so that if we define $h(z|\theta_1)$ as in (2-18) and

$$\begin{aligned} s(n|z; \theta_1, \theta_2) = {} & \phi(1|\theta_1, \theta_2) \cdots \phi(n|x_1, \cdots, x_{n-1}; \theta_1, \theta_2) \\ & \cdot [1 - \phi(n + 1|x_1, \cdots, x_n; \theta_1, \theta_2)] \; , \end{aligned} \tag{2-20}$$

we may write

$$\ell(z|\theta_1, \theta_2) = h(z|\theta_1)\, s(n|z; \theta_1, \theta_2) \; . \tag{2-21}$$

The likelihood of the sample depends *both* on the data-generating process through $h(z|\theta_1)$ *and* on the stopping process through $s(n|z; \theta_1, \theta_2)$.

2.3.3. Noninformative Stopping Processes

In most situations where an experiment can be decomposed in the way just described, θ_2 is a nuisance parameter in the sense of Section 2.2.4: the utility of $(e, z, a, \theta_1, \theta_2)$ depends on the state parameter θ_1 which characterizes the data-generating process but does *not* depend on the state parameter θ_2 which partially characterizes the stopping process. If this is true, then after the experiment has been conducted and both e and z are fixed the rational choice of an act a will depend only on the marginal posterior distribution of $\tilde{\theta}_1$ and not on the complete posterior distribution of $(\tilde{\theta}_1, \tilde{\theta}_2)$.

We therefore now inquire into the conditions under which the factor $s(n|z; \theta_1, \theta_2)$ in the sample likelihood (2-21) can be disregarded in determining the marginal posterior distribution of $\tilde{\theta}_1$—i.e., into the conditions under which the nature of the stopping process is totally irrelevant to $\tilde{\theta}_1$. In the light of the discussion of marginal sufficiency in Section 2.2.4, one set of sufficient conditions is virtually self-evident. If

1. $s(n|z; \theta_1, \theta_2)$ does not actually depend on θ_1,

2. $\tilde{\theta}_1$ and $\tilde{\theta}_2$ are independent *a priori*,

then the factor $s(n|z; \theta_1, \theta_2)$ in the likelihood (2-21) will have no effect on the distribution of $\tilde{\theta}_1$. When this is true, the factor $h(z|\theta_1)$, which depends only on the data-generating process, can be taken as a *kernel of the likelihood for* $\tilde{\theta}_1$, and the stopping process will be said to be *noninformative for* $\tilde{\theta}_1$.

▶ To prove that the two conditions just cited permit us to determine the marginal posterior distribution of $\tilde{\theta}_1$ by use of $h(z|\theta_1)$ alone, we observe that they permit us to write

$$\ell(z|\theta_1, \theta_2) = h(z|\theta_1)\, s(n|z; \theta_1, \theta_2) = h(z|\theta_1)\, s(n|z; \theta_2) \;,$$

$$\mathrm{D}'(\theta_1, \theta_2) = \mathrm{D}_1'(\theta_1)\, \mathrm{D}_2'(\theta_2) \;.$$

By (2-2) we then have for the joint posterior density of $(\tilde{\theta}_1, \tilde{\theta}_2)$

$$\mathrm{D}''(\theta_1, \theta_2|z) \propto \mathrm{D}_1'(\theta_1)\, \mathrm{D}_2'(\theta_2)\, h(z|\theta_1)\, s(n|z; \theta_2) \;,$$

and integrating out θ_2 we obtain

$$\mathrm{D}''(\theta_1|z) \propto \mathrm{D}_1'(\theta_1)\, h(z|\theta_1)$$

as was to be proved. ◀

Quite obviously, if $h(z|\theta_1)$ is a kernel for $\tilde{\theta}_1$ of $\ell(z|\theta_1, \theta_2)$ and if $h(z|\theta_1)$ can be factored

$$h(z|\theta_1) = \kappa(z|\theta_1)\, \rho(z) \;,$$

then $\kappa(z|\theta_1)$ is also *a* kernel for $\tilde{\theta}_1$ of $\ell(z|\theta_1, \theta_2)$. If furthermore there exists a function $\tilde{y}$ on Z such that, for any z in Z,

$$h(z|\theta_1) = \mathrm{k}(\tilde{y}(z)|\theta_1)\, \rho(z) \;,$$

then $\tilde{y}$ can be said to be *marginally sufficient for* $\tilde{\theta}_1$.

Stopping in Continuous Time. So far we have considered the problem of stopping only in the case where the data generating process is a *discrete* stochastic process generating $\{\tilde{x}_i, i = 1, 2, \cdots\}$. Since for all practical purposes it is intuitively clear that the essence of our results for this case can be extended to the case where the stochastic process is defined in terms of a continuous (time) parameter, generating $\{\tilde{z}_t, t \geq 0\}$ rather than $\{\tilde{x}_i, i \geq 1\}$, we shall not give a formal statement or proof of the extension.

Example 1. Let the data-generating process be a Bernoulli process generating independent random variables $\tilde{x}_1, \cdots, \tilde{x}_i, \cdots$ with identical mass functions

$$f(x|\theta_1) = \theta_1^x(1 - \theta_1)^{1-x} \;, \qquad x = 0, 1 \;;$$

and consider the following stopping processes (ways of determining the number of observations n):

1. The experimenter by an act of free will decides before the experiment begins that he will observe exactly n trials. The stopping probabilities (2-19) are thus

$$\phi(k+1|x_1, \cdots, x_k; \theta_1, \theta_2) = \begin{cases} 1 & \text{if} \quad k < n \ , \\ 0 & \text{if} \quad k = n \ , \end{cases}$$

and by (2-20)

$$s(n|z; \theta_1, \theta_2) = 1 \ .$$

2. The experimenter decides to observe the process until r 1's have occurred (or has \$$r$ to spend and each observation costs \$1 which is refunded if $\tilde{x} = 0$). The stopping probabilities are

$$\phi(k+1|x_1, \cdots, x_k; \theta_1, \theta_2) = \begin{cases} 1 & \text{if} \quad \Sigma_{i=1}^k x_i < r \ , \\ 0 & \text{if} \quad \Sigma_{i=1}^k x_i = r \ , \end{cases}$$

and

$$s(n|z; \theta_1, \theta_2) = 1 \ .$$

Because $s(n|z; \theta_1, \theta_2)$ is simply a constant, these two stopping processes are *necessarily* noninformative and

$$h(z|\theta_1) = \Pi_{i=1}^n [\theta_1^{x_i}(1-\theta_1)^{1-x_i}] = \theta_1^{\Sigma x_i}(1-\theta_1)^{n-\Sigma x_i} \tag{2-22}$$

is a kernel for $\tilde{\theta}_1$ of the outcome $z = (x_1, \cdots, x_n)$.

In this example the first of the two stopping processes is "deterministic" while the second is "probabilistic". Notice, however, that instead of regarding the Bernoulli *data-generating* process as consisting of a sequence of *trials* each of which may be either a success ($\tilde{x}_i = 1$) or a failure ($\tilde{x}_i = 0$), we can regard it as consisting of a sequence of *intervals* each of which is measured by the number of trials $n_i = 1, 2, \cdots, \infty$ required to obtain one success. If we do so regard it, then it is the second stopping process described above which is "deterministic", since the number r of intervals which will be observed is fixed in advance, while the first stopping process is "probabilistic".

Example 2. Consider the same data-generating process as before but assume that the stopping process is the following. The person operating the data-generating process knows the true value of $\tilde{\theta}_1$ and says that if $\tilde{\theta}_1 \leq \frac{1}{2}$ he will stop the process after the 10th x has been generated but that if $\tilde{\theta}_1 > \frac{1}{2}$ he will generate 20 xs; the experimenter observes the process until it stops. The likelihood of $z = (x_1, \cdots, x_n)$ is then

$$\ell(z|\theta_1, \theta_2) = h(z|\theta_1) \, s(n|z; \theta_1)$$

where $h(z|\theta_1)$ is defined by (2-22) and

$$s(n|z; \theta_1) \equiv \begin{cases} 1 \text{ if } n = 10, \theta_1 \leq \frac{1}{2} \ , \\ 1 \text{ if } n = 20, \theta_1 > \frac{1}{2} \ , \\ 0 \text{ otherwise} \ . \end{cases}$$

The stopping process is necessarily *informative.*

Example 3. Consider the same data-generating process as before but assume that a coin is tossed before each trial is observed and that the experiment is stopped as soon as tails occurs. Letting θ_2 denote the probability of tails we have for the stopping probabilities

$$\phi(k+1|x_1, \cdots, x_k; \theta_1, \theta_2) = 1 - \theta_2 \ ;$$

the likelihood of $(x_1, \cdots, x_n)$ is

$$\ell(z|\theta_1, \theta_2) = h(z|\theta_1)\, \theta_2(1 - \theta_2)^n \ .$$

The factor $h(z|\theta_1)$ can be taken as a kernel for $\tilde{\theta}_1$ *provided that* $\tilde{\theta}_1$ *and* $\tilde{\theta}_2$ are *independent a priori;* on this condition, the stopping process is noninformative for $\tilde{\theta}_1$.

Example 4. A production process generates items which may have either or both of two possible defects d_1 and d_2. For $i = 1, 2, \cdots$ and $j = 1, 2$, let

$$x_{ij} = \begin{cases} 1 & \text{if the } i\text{th item does possess defect } d_j \ . \\ 0 & \text{if the } i\text{th item does not possess defect } d_j \end{cases}$$

The process thus generates vector-valued random variables $(\tilde{x}_{11}, \tilde{x}_{12}), \cdots (\tilde{x}_{i1}, \tilde{x}_{i2}), \cdots$; we assume that it behaves as a double Bernoulli process with parameter $\theta = (\theta_1, \theta_2)$ and that given θ the components $(\tilde{x}_{i1}, \tilde{x}_{i2})$ are conditionally independent for all i. An experiment is conducted by observing the process until the r_2th defect of type d_2 is observed; this turns out to occur on the nth trial and the n trials turn out to contain r_1 defects of type d_1. The likelihood of this outcome is

$$\ell(z|\theta_1, \theta_2) = h(z|\theta_1)\, s(n|z; \theta_1, \theta_2)$$

where

$$h(z|\theta_1) = \theta_1^{r_1}(1 - \theta_1)^{n-r_1}, \; z \equiv (x_{11}, \cdots, x_{n1})$$

$$s(n|z; \theta_1, \theta_2) = \frac{(n-1)!}{(r_2 - 1)!(n - r_2)!}\, \theta_2^{r_2}(1 - \theta_2)^{n-r_2} \ .$$

Suppose now that a decision problem turns only on $\tilde{\theta}_1$ in the sense that an act a must be chosen and $u(e, z, a, \theta_1, \theta_2)$ does not actually depend on θ_2 for any a in A. Then $h(z|\theta_1)$ is a kernel for $\tilde{\theta}_1$ of $\ell(z|\theta_1, \theta_2)$ and (r_1, n) is marginally sufficient for $\tilde{\theta}_1$ *provided that* $\tilde{\theta}_1$ *and* $\tilde{\theta}_2$ *are independent a priori;* with this proviso, the stopping process is noninformative for $\tilde{\theta}_1$. If on the contrary $\tilde{\theta}_1$ is *not* independent of $\tilde{\theta}_2$, then $s(n|z; \theta_1, \theta_2)$ must also be taken into account in determining the marginal posterior distribution of $\tilde{\theta}_1$: the stopping process is *indirectly informative* for $\tilde{\theta}_1$ even though it does not *depend* on θ_1.

2.3.4. Contrast Between the Bayesian and Classical Treatments of Stopping

The essence of our discussion of noninformative stopping can be summed up informally as follows. In Bayesian analysis, the decision maker asks what action is reasonable to take in the light of the available evidence about the state θ of the world. If this evidence happens to include knowledge of the outcome z_o of some experiment, this knowledge is incorporated with the decision maker's other information by looking at the likelihood of z_o given each of the possible states θ which may prevail; the process does *not* involve looking at the implications of any zs which were *not* observed. Since moreover the description z_o of the experimental outcome includes all information required to compute the likelihood of that outcome given every possible θ, there is no need to know anything further about the experiment e which produced the outcome—such information would bear only on the likelihoods of the zs which were *not* observed.

Thus suppose that a sample from a Bernoulli process consists of 10 trials the third and last of which were successes while the remainder were failures. Unless the decision maker sees some *specific logical connection* between the value of the process parameter $\tilde{p}$ and the fact that the experiment terminated after 10 trials and 2 successes had been observed, his posterior distribution of $\tilde{p}$ will be the same *whatever* the reason for termination may have been, and that part of the complete description of e which states this reason is therefore *totally irrelevant.*

In classical analysis, on the contrary, it is meaningless to ask what action is reasonable given only the *particular* information *actually* available concerning the state of the world. The choice of action to be taken after a particular z_o has been observed can be evaluated only in relation to requirements placed on a complete decision rule which stipulates what action should be taken for every z which *might* have been observed; and for this reason classical analysis usually depends critically on a complete description of the experiment which *did* produce z_o because this same e *might* have produced some other z.

Thus suppose that the decision maker wants to make a minimum-variance unbiassed estimate of the process parameter p on the basis of the sample described just above. If he knows that the person who actually conducted the experiment decided in advance to observe 10 trials, he must estimate

$$\hat{p}_1 = \frac{r}{n} = \frac{2}{10} \ ,$$

whereas if he knows that the experimenter decided in advance to observe 2 successes, he must estimate

$$\hat{p}_2 = \frac{r-1}{n-1} = \frac{1}{9} \ ;$$

and this is true even if the decision maker is absolutely convinced (a) that the true value of the process parameter p cannot possibly have had any effect on the decision made by the experimenter, and (b) that the decision made by the experimenter cannot possibly have had any effect on the true value of the process parameter p.

Again, if the decision maker wishes to test the null hypothesis that $p \geq \frac{1}{2}$ against the alternative hypothesis that $p < \frac{1}{2}$, then if he knows that $n = 10$ was predetermined by the experimenter, he must compute the binomial probability

$$\mathrm{P}\{\tilde{r} \leq 2 | p = \tfrac{1}{2}, n = 10\} = .0547$$

and conclude that the experimental evidence against the null hypothesis is *not* significant at level .05; whereas if he knows that it was $r = 2$ on which the experimenter decided in advance, he must calculate

$$\mathrm{P}\{\tilde{n} \geq 10 | p = \tfrac{1}{2}, r = 2\} = \mathrm{P}\{\tilde{r} < 2 | p = \tfrac{1}{2}, n = 9\} = .0195$$

and conclude that the evidence *is* significant at level .05.

If the stopping process is not well defined—and this is more common than not in situations where the experimental budget is fixed but the cost in time or money of each observation is uncertain—then it is impossible in classical theory to make unbiassed estimates. Hypotheses are usually tested under such circumstances by means of conditional tests, but it should be observed that when the stopping process is not well defined the choice of the condition is completely arbitrary and therefore *the result of the test is completely arbitrary.* If a sample from a Bernoulli process consists of 10 observations only the third and last of which are good, we *may* think of the process as generating random variables $\tilde{x}_1, \cdots, \tilde{x}_i, \cdots$ where $x = 0, 1$ and apply a test conditional on the fact that 10 observations were taken, thus obtaining the level of significance .0547. We may however equally well think of the process as generating random variables $\tilde{n}_i$ where $n = 1, 2, 3, \cdots$ is the number of trials required to obtain a success; in this case the sample consists of *two* observations $n_1 = 3$, $n_2 = 7$, and a test conditional on the fact that 2 observations were taken leads to the level of significance .0195.

2.3.5. *Summary*

To summarize the discussion in Section 2.3, suppose that a Bernoulli process has yielded a sample consisting of r successes and $(n - r)$ failures. (1) As far as *the implications of this particular sample* are concerned, it will *usually* be quite immaterial in the further analysis of the problem whether r was predetermined and $\tilde{n}$ left to chance, or n was predetermined and $\tilde{r}$ left to chance, or neither r nor n was predetermined and the experiment was terminated in some quite other way. Even though the actual likelihood of this particular sample may well depend on the stopping process, the likelihoods for all noninformative stopping processes have a common kernel and therefore all lead to the same posterior distribution. (2) On the other hand, we shall also be concerned with the problems of *experimental design* as they look *before* any sample has actually been taken, and then we *shall* want to ask what can be expected to happen *if* we predetermine r rather than n, and so forth. The reader must keep in mind throughout the remainder of this monograph that the extent to which the complete description of e enters a Bayesian analysis depends on the exact nature of the problem being analyzed.

CHAPTER 3

Conjugate Prior Distributions

3.1. Introduction; Assumptions and Definitions

Unless the state space Θ contains a *very* limited number of possible states $\{\theta\}$, it will usually be simply impossible for the decision maker to assign a prior probability to each θ individually and then verify the consistency of these assignments and make adjustments where necessary. In most applied problems the number of possible states will be extremely large if not infinite—think of the number of possible levels of demand for a given product or of the number of possible yields with a given fertilizer—and the decision maker will be forced to assign the required probabilities by specifying a limited number of summary measures of the distribution of $\tilde{\theta}$ and then filling in the details by some kind of "reasonable" short-cut procedure. Thus the decision maker may feel that his "best estimate" of demand is $\hat{\theta} = 2500$ and may feel sure enough of this estimate to be willing to bet even money that $\tilde{\theta}$ is between 2000 and 3000, but at the same time he may feel that further refinement is not worth the trouble and may be *willing to act consistently* with the implications of any "reasonably" smooth and symmetric distribution of $\tilde{\theta}$ which has a mode at $\theta = 2500$ and which assigns probability $\frac{1}{2}$ to the interval $2000 < \theta < 3000$.

An obvious way of finding a specific distribution to meet specifications of this kind is to start by selecting some *family* of distributions defined by a mathematical formula containing a certain number of adjustable parameters and then to select the specific member of this family which meets the decision maker's quantitative specifications by giving the proper numerical values to these parameters. Thus in the example just cited, a Normal distribution with mean 2500 and standard deviation 746 may fully satisfy the decision maker's requirements; or in a situation where $\tilde{\theta}$ represents a process fraction defective and is therefore certain to have a value in the interval [0, 1] the decision maker's requirements may be fully satisfied by some member of the beta family, all members of which have the convenient property of restricting the domain of $\tilde{\theta}$ to the interval in question.

3.1.1. Desiderata for a Family of Prior Distributions

The fact that the decision maker cannot specify every detail of his prior distribution by direct assessment means that there will usually be considerable latitude in the choice of the *family* of distributions to be used in the way just described even though the selection of a particular member *within* the chosen family will usually be wholly determined by the decision maker's expressed beliefs or betting odds. Our objective in the present chapter is to aid the useful exploitation of this latitude by showing how it is possible in certain commonly occurring situations to find a family F of distributions which at least comes very close to satisfying the following desiderata.

1. F should be analytically *tractable* in three respects: (a) it should be reasonably easy to determine the posterior distribution resulting from a given prior distribution and a given sample; (b) it should be possible to express in convenient form the expectations of some simple utility functions with respect to any member of F; (c) F should be closed in the sense that if the prior is a member of F, the posterior will also be a member of F.
2. F should be *rich*, so that there will exist a member of F capable of expressing the decision maker's prior information and beliefs.
3. F should be parametrizable in a manner which can be readily *interpreted*, so that it will be easy to verify that the chosen member of the family is really in close agreement with the decision maker's prior judgments about Θ and not a mere artifact agreeing with one or two quantitative summarizations of these judgments.

To help guide the reader through the discussion that follows we shall outline very briefly the general procedure which we shall use in generating families of prior distributions that satisfy these desiderata in the special but very important case where the sample observations are independent (conditional, of course, on knowing the state parameters) and admit of a sufficient statistic of fixed dimensionality. Denoting this sufficient statistic by $\tilde{y}$ and its range by Y, we shall show that it is possible to construct a family F of prior distributions for $\tilde{\theta}$ where each member of F is *indexed* by an element of Y. In addition, we shall show that if we choose for $\tilde{\theta}$ a particular member of F indexed by y', say, and if a sample then yields a sufficient statistic y, the posterior distribution will also belong to F and will be indexed by some element of Y, say y''. The binary operation of combining y' and y to compute $y'' = y' * y$ will be examined in great detail for several data-generating processes, and the family F indexed by Y will be shown in many cases to be tractable, rich, and interpretable.

3.1.2. *Sufficient Statistics of Fixed Dimensionality*

We consider only data-generating processes which generate independent, identically distributed random variables $\tilde{x}_1, \cdots, \tilde{x}_i, \cdots$ such that, for any n and any $(x_1, \cdots, x_n)$, there exists a sufficient statistic

$$\tilde{y}_n(x_1, \cdots, x_n) = y = (y_1, \cdots, y_j, \cdots, y_s)$$

where y_j is a (possibly restricted) real number and the dimensionality s of y does not depend on n. More formally, we assume throughout the remainder of this chapter that

1. If the likelihood of a sample $(x_1, \cdots, x_n)$ is $\ell_n(x_1, \cdots, x_n|\theta)$, then the joint likelihood of any two samples $(x_1, \cdots, x_p)$ and $(x_{p+1}, \cdots, x_n)$ is

$$\ell_p(x_1, \cdots, x_p|\theta) \cdot \ell_{n-p}(x_{p+1}, \cdots, x_n|\theta) \; . \tag{3-1}$$

2. Given any n and any sample $(x_1, \cdots, x_n)$ there exists a function k and an s-tuple of real numbers $\tilde{y}_n(x_1, \cdots, x_n) = y = (y_1, \cdots, y_s)$, where s does not depend on n, such that

$$\ell_n(x_1, \cdots, x_n|\theta) \propto \mathrm{k}(y|\theta) \ . \tag{3-2}$$

The families of prior distributions which we shall discuss in this chapter exploit the second of these assumptions directly and also exploit the following easily proved theorem which rests on the two assumptions together.

Theorem. Let $y^{(1)} = \tilde{y}_p(x_1, \cdots, x_p)$ and let $y^{(2)} = \tilde{y}_{n-p}(x_{p+1}, \cdots, x_n)$. Then it is possible to find a binary operation $*$ such that

$$y^{(1)} * y^{(2)} = y^* \equiv (y_1^*, \cdots, y_s^*) \tag{3-3a}$$

has the properties

$$\ell_n(x_1, \cdots, x_n|\theta) \propto \mathrm{k}(y^*|\theta) \ , \tag{3-3b}$$

$$\mathrm{k}(y^*|\theta) \propto \mathrm{k}(y^{(1)}|\theta) \cdot \mathrm{k}(y^{(2)}|\theta) \ . \tag{3-3c}$$

The point lies of course in the fact that y^* can be computed from $y^{(1)}$ and $y^{(2)}$ alone, *without* knowledge of $(x_1, \cdots, x_n)$.

Proof. Given $y^{(1)}$ and $y^{(2)}$ we can always find *some* p-tuple $(x_1', \cdots, x_p')$ and *some* $(n-p)$-tuple $(x_{p+1}', \cdots, x_n')$ such that $\tilde{y}_p(x_1', \cdots, x_p') = y^{(1)}$ and $\tilde{y}_{n-p}(x_{p+1}', \cdots, x_n') = y^{(2)}$. Having found these fictitious experimental outcomes we may define the operation $*$ by

$$y^{(1)} * y^{(2)} \equiv \tilde{y}_n(x_1', \cdots, x_p', x_{p+1}', \cdots, x_n') \tag{3-4}$$

because, by assumptions (3-1) and (3-2),

(a)
$$\ell_n(x_1, \cdots, x_n|\theta) = \ell_p(x_1, \cdots, x_p|\theta) \cdot \ell_{n-p}(x_{p+1}, \cdots, x_n|\theta) \propto \mathrm{k}(y^{(1)}|\theta) \cdot \mathrm{k}(y^{(2)}|\theta) \ ,$$

(b)
$$\ell_n(x_1', \cdots, x_n'|\theta) = \ell_p(x_1', \cdots, x_p'|\theta) \cdot \ell_{n-p}(x_{p+1}', \cdots, x_n'|\theta) \propto \mathrm{k}(y^{(1)}|\theta) \cdot \mathrm{k}(y^{(2)}|\theta) \ ,$$

(c)
$$\ell_n(x_1', \cdots, x_n'|\theta) \propto \mathrm{k}(y^*|\theta) \ .$$

The theorem follows immediately from these results.

The *usefulness* of the theorem will depend largely on the fact that *when y is of fixed dimensionality* we can usually find a very simple $*$ and show that it is equivalent to $*$ as defined by (3-4). It can be shown that if (and only if) the common density of the individual observations $\tilde{x}$ is of the exponential form† and meets certain regularity conditions, then it is possible to find sufficient statistics of fixed dimensionality for which the $*$ operation consists of simple component-by-component addition, i.e., for which the ith component of $y^{(1)} * y^{(2)}$ is simply

$$y_i^* = y_i^{(1)} + y_i^{(2)} \ , \qquad i = 1, \cdots, s \ .$$

We shall make no use of this fact, however, partly because we do not wish to restrict our discussion to the exponential class but principally because even within the

† Cf. B. O. Koopman, "On Distributions Admitting a Sufficient Statistic," *Trans. Amer. Math. Society* 39 (1936) 399–409.

exponential class we shall often find it more convenient to work with nonadditive sufficient statistics.

Example 1. Consider a Bernoulli process with parameter $\theta = p$ generating independent random variables $\tilde{x}_1, \cdots, \tilde{x}_i, \cdots$ with identical mass functions $p^x(1-p)^{1-x}$ where $x = 0, 1$. If we define

$$n = \text{number of random variables observed} ,$$

$$r = \Sigma x_i ,$$

then the likelihood of a sample $(x_1, \cdots, x_n)$ is obviously

$$\ell_n(x_1, \cdots, x_n|\theta) = p^{\Sigma x_i}(1-p)^{n-\Sigma x_i} = p^r(1-p)^{n-r} :$$

the statistic

$$y = (y_1, y_2) = (r, n)$$

is sufficient and is of dimensionality $s = 2$ independent of n. Furthermore, given that two samples described by (r_1, n_1) and (r_2, n_2) have been observed, we can verify that

$$y^* = y^{(1)} * y^{(2)} = (r_1, n_1) * (r_2, n_2) = (r_1 + r_2, n_1 + n_2) \tag{3-5}$$

has the properties (3-3b) and (3-3c) either (1) by arguing that if we "pool" *any* two samples described by (r_1, n_1) and (r_2, n_2) the pooled sample will contain $n_1 + n_2$ values $r_1 + r_2$ of which are equal to 1, or (2) by multiplying

$$p^{r_1}(1-p)^{n_1-r_1} \cdot p^{r_2}(1-p)^{n_2-r_2} = p^{r_1+r_2}(1-p)^{n_1+n_2-r_1-r_2} .$$

In this example, the operation $*$ consists of simple component-by-component addition.

Example 2. Consider an Independent Normal process with parameter $\theta = (\mu, h)$ generating independent random variables $\tilde{x}_1, \cdots, \tilde{x}_i, \cdots$ with identical densities

$$(2\pi)^{-\frac{1}{2}} e^{-\frac{1}{2}h(x-\mu)^2} h^{\frac{1}{2}} , \qquad -\infty < x < \infty ;$$

and define

$$n = \text{number of random variables observed} ,$$

$$m = \frac{1}{n} \Sigma x_i ,$$

$$v = \begin{cases} \dfrac{1}{n-1} \Sigma (x_i - m)^2 & \text{if } n > 1 , \\ 0 & \text{if } n = 1 . \end{cases}$$

It is shown in Section 11.5.1 that the likelihood of a sample $(x_1, \cdots, x_n)$ from this process is

$$\ell_n(x_1, \cdots, x_n|\theta) = (2\pi)^{-\frac{1}{2}n} e^{-\frac{1}{2}h\Sigma(x_i-\mu)^2} h^{\frac{1}{2}n}$$
$$\propto e^{-\frac{1}{2}h[n(m-\mu)^2 + (n-1)v]} h^{\frac{1}{2}n} :$$

the statistic

$$y = (y_1, y_2, y_3) = (m, v, n)$$

is sufficient and of dimensionality $s = 3$ independent of n. The reader can verify that

$$y^{(1)} * y^{(2)} \equiv (m_1, v_1, n_1) * (m_2, v_2, n_2) = (m^*, v^*, n^*) \tag{3-6a}$$

where

$$n^* = n_1 + n_2 \ , \qquad m^* = \frac{1}{n^*}(n_1 m_1 + n_2 m_2) \ , \tag{3-6b}$$

$$v^* = \frac{1}{n^* - 1}\{[(n_1 - 1)v_1 + n_1 m_1^2] + [(n_2 - 1)v_2 + n_2 m_2^2] - n^* m^{*2}\} \ .$$

In this example the * operation is simple addition for only one component; but if we had chosen the sufficient statistic

$$y = (y_1, y_2, y_3) \equiv (\Sigma\, x_i, \Sigma\, x_i^2, n)$$

the * operation would have been simple addition for all three components.

Example 3. As an example of a process which is not of the exponential class, consider a process generating independent random variables $\tilde{x}_1, \cdots, \tilde{x}_i, \cdots$ each of which has a uniform distribution over the interval $[0, \theta]$ where $\theta > 0$, and define

$$n = \text{number of random variables observed} \ ,$$

$$M = \max\{x_i\} \ .$$

It is easy to see that the likelihood of a sample $(x_1, \cdots, x_n)$ can be written

$$\ell_n(x_1, \cdots, x_n|\theta) = \theta^{-n}\,\delta(\theta - M) \qquad \text{where} \qquad \delta(a) = \begin{cases} 0 \text{ if } a < 0 \ , \\ 1 \text{ if } a \geq 0 \ ; \end{cases}$$

the statistic

$$y = (y_1, y_2) = (M, n)$$

is sufficient and of dimensionality $s = 2$ independent of n. It is also easy to verify that

$$y^{(1)} * y^{(2)} = (M_1, n_1) * (M_2, n_2) = (\max\{M_1, M_2\}, n_1 + n_2) \ :$$

the * operation on the component M is not addition and it is impossible to find a sufficient statistic such that * *does* consist of component-by-component addition.

3.2. Conjugate Prior Distributions

3.2.1. Use of the Sample Kernel as a Prior Kernel

Instead of considering the kernel function k defined by (2-12) as a function $k(\cdot|\theta)$ with parameter θ on the reduced sample space Y, we can consider it as a function $k(y|\cdot)$ with parameter y on the state space Θ. We shall now examine the conditions under which a *density* function $f(\cdot|y)$ with parameter y on the state space Θ can be derived by defining

$$f(\theta|y) = N(y)\, k(y|\theta) \tag{3-7}$$

where N is a function of y still to be determined. Such a density $f(\cdot|y)$ on Θ will be called a *natural conjugate with parameter* y of the kernel function k.

All that is required for a function $f(\cdot|y)$ defined on Θ to be a proper density function is (1) that it be everywhere nonnegative and (2) that its integral over Θ be 1. Since $k(\cdot|\theta)$ is a kernel function on Y for all θ in Θ, $k(y|\theta)$ is necessarily nonnegative for all (y, θ) in $Y \times \Theta$; and $f(\theta|y)$ as defined by (3-7) therefore satisfies the first requirement automatically if y is restricted to Y and $N(y)$ is positive. It follows that *if* the integral of $\mathrm{k}(y|\cdot)$ over Θ exists, we can define $N(y)$ by

$$[N(y)]^{-1} = \int_{\Theta} \mathrm{k}(y|\theta)\, d\theta \tag{3-8}$$

and $f(\cdot|y)$ as defined by (3-7) will be a proper density function on Θ.

Example 1. It is shown in Section 11.3.1 that the likelihood of observations $(x_1, \cdots, x_n)$ on an Independent Normal process with known precision $h > 0$ and unknown mean $\tilde{\theta} = \tilde{\mu}$ can be expressed in terms of the sufficient statistic

$$y = (m, n) = \left(\frac{1}{n}\Sigma\, x_i,\, n\right)$$

by the kernel

$$k(y|\theta) = e^{-\frac{1}{2}hn(m-\mu)^2} .$$

Since the integral of $\mathrm{k}(y|\cdot)$ over $\Theta = (-\infty, +\infty)$ for fixed y in Y is

$$\int_{-\infty}^{\infty} e^{-\frac{1}{2}hn(m-\mu)^2}\, d\mu = (2\pi)^{\frac{1}{2}}\, (hn)^{-\frac{1}{2}} , \qquad \begin{array}{l} -\infty < m < \infty , \\ n = 1, 2, \cdots , \end{array}$$

we may take the function $f_N(\cdot|m, hn)$ of μ with parameter (m, n) defined by

$$f_N(\mu|m, hn) = (2\pi)^{-\frac{1}{2}}\, (hn)^{\frac{1}{2}}\, e^{-\frac{1}{2}hn(m-\mu)^2} \qquad \begin{array}{l} -\infty < m < \infty , \\ n = 1, 2, \cdots , \end{array}$$

as a density function for $\tilde{\mu}$.

Example 2. To see that we may not simply *assume* that the integral of $\mathrm{k}(y|\cdot)$ *over the state space* Θ will converge for any parameter y in the range Y of the statistic $\tilde{y}$, consider the rectangular process discussed in Example 3 of Section 3.1.2, where it was shown that the likelihood of observations $(x_1, \cdots, x_n)$ can be expressed in terms of the sufficient statistic

$$y = (M, n) = (\max\, \{x_i\},\, n)$$

by the kernel

$$\mathrm{k}(y|\theta) = \theta^{-n}\, \delta(\theta - M) , \qquad \theta > 0 .$$

Although $n = 1$ is in the range of the statistic $\tilde{y}$, the integral of $\mathrm{k}(y|\cdot)$ over $\Theta = (0, \infty)$ for fixed $y = (M, n)$,

$$\int_0^M 0\, d\theta + \int_M^{\infty} \theta^{-n}\, d\theta ,$$

converges only if n is *greater* than 1.

3.2.2. The Posterior Distribution When the Prior Distribution Is Natural-Conjugate

We have seen in Section 2.2.2 that if the *prior* density of θ is $\mathrm{D}'(\cdot)$, if $y = \tilde{y}(z)$ is a sufficient statistic, and if $\mathrm{k}(y|\theta)$ is a kernel of the likelihood of z given θ, then the posterior density of $\tilde{\theta}$ given y is $\mathrm{D}''(\cdot|y)$ as defined by

$$\mathrm{D}''(\theta|y) \propto \mathrm{D}'(\theta)\, \mathrm{k}(y|\theta) .$$

We have also seen, in Section 3.1.2, that if y is of fixed dimensionality we can always find an operation $*$ such that

$$k(y^{(1)}|\theta) \cdot k(y^{(2)}|\theta) \propto k(y^{(1)} * y^{(2)}|\theta) .$$

It follows immediately that if D′ is a natural conjugate of k with parameter y' in the range Y of the statistic $\tilde{y}$, i.e., if

$$D'(\theta) \propto k(y'|\theta) , \qquad y' \in Y ,$$

and if a sample yields a sufficient statistic y, then

$$D''(\theta|y) \propto k(y'|\theta)\, k(y|\theta) \propto k(y' * y|\theta) ; \tag{3-9}$$

the kernel of the prior density combines with the sample kernel in exactly the same way that two sample kernels combine.

Example. We return to the first example in Section 3.2.1, an Independent Normal process with known precision h and unknown mean $\tilde{\theta} = \tilde{\mu}$, and recall that in terms of the sufficient statistic

$$y = (m, n) = \left(\frac{1}{n}\Sigma\, x_i, n\right)$$

the likelihood of $(x_1, \cdots, x_n)$ is proportional to

$$k(y|\theta) = e^{-\frac{1}{2}hn(m-\mu)^2} .$$

The reader can readily verify that if we have two statistics $y' = (m', n')$ and $y = (m, n)$, then by the definition (3-4) of the $*$ operation

$$y'' \equiv y' * y = \left(\frac{n'm' + nm}{n' + n}, n' + n\right) \equiv (m'', n'') . \tag{3-10}$$

By (3-9) we may therefore infer that if the *prior* density of $\tilde{\theta} = \tilde{\mu}$ is

$$D'(\theta) = f_N(\mu|m', hn') \propto e^{-\frac{1}{2}hn'(m'-\mu)^2}$$

and the parameter (m', n') is in the range of the statistic $(\tilde{m}, \tilde{n})$, then the density *posterior* to observing a sample with sufficient statistic (m, n) will be

$$D''(\theta|y) = f_N(\mu|m'', hn'') \propto e^{-\frac{1}{2}hn''(m''-\mu)^2} . \tag{3-11}$$

This result obtained via the $*$ operation can of course be verified by using Bayes' theorem (2-13) to write

$$D''(\theta|y) = D''(\mu|m, n) \propto e^{-\frac{1}{2}hn'(m'-\mu)^2} e^{-\frac{1}{2}hn(m-\mu)^2} = e^{-\frac{1}{2}hn'(m'-\mu)^2 - \frac{1}{2}hn(m-\mu)^2}$$

and completing the square in the exponent.

3.2.3. Extension of the Domain of the Parameter

The result (3-9) indicates immediately that families of natural-conjugate priors with parameter y in the range Y of the statistic $\tilde{y}$ are very likely to have the properties of *tractability* which we listed as desirable at the beginning of this chapter. The family will be *closed:* a Normal prior yields a Normal posterior, and so forth. The posterior for given prior and sample will be *easy to find* if the operation $*$ is simple; we have already seen that it *is* simple for samples from Bernoulli, Normal, and rectangular processes, and we know that it *can* be simple for any process of the exponential class. Finally, the fact that the likelihood functions from which these priors are derived have been well studied and tabulated means that it will often be possible to evaluate expectations of utility functions in terms of tabled functions.

On the other hand, the *richness* of families of natural-conjugate densities can obviously be very greatly increased by extending the domain of the parameter y to include *all* values for which (1) $\mathrm{k}(y|\theta)$ is nonnegative, all θ, and (2) the integral of k over Θ converges. We are clearly not *obliged* to restrict the parameter y to the range Y of the statistic $\tilde{y}$ if these two conditions are met by all ys in some larger domain; and we shall now show by examples that an extended family of densities obtained in this way is often just as tractable as the original family.

Example 1. Consider again the Independent Normal process with known precision h and unknown mean $\tilde{\theta} = \tilde{\mu}$ which was considered in the first example in Section 3.2.1 and again in the example in Section 3.2.2. Since the integral

$$\int_\Theta \mathrm{k}(y|\theta)\, d\theta = \int_{-\infty}^{\infty} e^{-\frac{1}{2}hn(m-\mu)^2}\, d\mu$$

converges for all real $n > 0$ and not just for the values $n = 1, 2, \cdots$ in the range of the *statistic* $\tilde{n}$, we may take the entire half line $n > 0$ as the domain of the *parameter* n. The natural-conjugate family in which n had the domain $\{1, 2, \cdots\}$ was *convenient* because, as we saw in (3-11), a prior density

$$f_N(\mu|m', hn') \propto e^{-\frac{1}{2}hn'(m'-\mu)^2}$$

yielded a posterior density of the same form with parameter

$$(m'', n'') = (m', n') * (m, n)$$

given by the simple $*$ operation defined by (3-10). It is easy to verify by application of Bayes' theorem that the same operation holds when the domain of n is extended to $n > 0$, even though proof by the interpretation (3-4) is no longer possible.

Example 2. Consider again the Bernoulli process with parameter $\theta = p$ discussed in the first example of Section 3.1.2, where the likelihood of the sample outcome $y = (r, n)$ was shown to have the kernel

$$\mathrm{k}(y|\theta) = p^r(1-p)^{n-r} .$$

It is well known that the integral of $\mathrm{k}(y|\cdot)$ over $\Theta = [0, 1]$ converges for all real $r > -1$ and $n > r - 1$ to the complete beta function,

$$\int_0^1 p^r(1-p)^{n-r}\, dp = B(r+1, n-r+1) ,$$

and we can therefore obtain a richer family of densities by allowing the parameter $y = (r, n)$ to have any value in the domain thus defined instead of restricting it to the range of the statistic $(\tilde{r}, \tilde{n})$, which includes only the integers such that $r \geq 0$ and $n \geq \max\{1, r\}$.

Purely for notational convenience we shall actually make this extension, not by allowing the parameter (r, n) to take on negative values, but by writing the density in the form

$$f_\beta(p|r, n) = \frac{1}{B(r, n-r)} p^{r-1}(1-p)^{n-r-1} , \qquad n > r > 0 . \qquad (3\text{-}12)$$

Placing primes on the parameters r and n of this density to indicate a prior density and letting (r, n) denote the actual sample outcome it is easy to verify by application of Bayes' theorem that the posterior density will be of the same form as the prior with parameter

$$(r'', n'') = (r', n') * (r, n) = (r' + r, n' + n) \; ;$$

the extended family remains closed and the $*$ operation is the same as it was when applied in (3-5) to two *statistics* (r_1, n_1) and (r_2, n_2).

3.2.4. *Extension by Introduction of a New Parameter*

The results of the last section can be summarized by saying that a natural-conjugate family can often be enriched without loss of tractability by extending the domain of the parameter y beyond the range of the statistic $\tilde{y}$; the essential reason why the enrichment is "free" is that existing tables serve as well or virtually as well for the extended as for the original family. We shall now examine another way of enriching a conjugate family without loss of tractability. As before, we do so by examining the natural conjugate density to see whether additional flexibility can be introduced without any change in the essential mathematical properties of the function; but this time what we look for is an opportunity to introduce an additional parameter rather than an opportunity to extend the domain of an existing parameter. We shall first try to make the general idea clear by an artificial example, and we shall then go on to exploit the idea by applying it to the Independent Normal process with both mean and precision unknown.

As an artificial example, suppose that instead of considering the Bernoulli process in its general form we had considered a Bernoulli process constrained to stop as soon as but not before the tenth trial is produced. For such a process the single statistic r = number of successes is sufficient, the likelihood of a sample is proportional (actually equal) to

$$p^r(1 - p)^{10-r} \; , \qquad r \geq 0 \; ,$$

and the natural conjugate density would have this same expression as its kernel. Consideration of this kernel as a function of p would however immediately reveal that, in the notation of (3-12),

$$p^r(1 - p)^{10-r} \propto f_\beta(p|r + 1, 12) \; ;$$

and since we know that the *general* (two-parameter) beta function is just as tractable as this special case, we would realize at once that the family could be enriched at no cost by (a) introducing another parameter, n, and (b) redefining the parameter r, so that the conjugate family becomes

$$f_\beta(p|r, n) \propto p^{r-1}(1 - p)^{n-r-1} \; , \qquad n > r > 0 \; .$$

We now turn to the really important application of this idea, to the Independent Normal process with both mean $\tilde{\mu}$ and precision $\tilde{h}$ unknown. As we remarked in the second example of Section 3.1.2, it is shown in Section 11.5.1 that the statistics

$$n = \text{number of random variables observed} \;,$$

$$m = \frac{1}{n}\Sigma\, x_i \;,$$

$$v = \begin{cases} \dfrac{1}{n-1}\Sigma(x_i - m)^2 & \text{if } n > 1 \;, \\ 0 & \text{if } n = 1 \;, \end{cases}$$

are sufficient for a sample from such a process and the sample likelihood is proportional to

$$e^{-\frac{1}{2}hn(m-\mu)^2}\, h^{\frac{1}{2}} \cdot e^{-\frac{1}{2}h(n-1)v}\, h^{\frac{1}{2}(n-1)} \;.$$

Considering this expression as defining a function of μ and h we observe that when $n > 1$ and $v > 0$ it is the product of (1) a Normal conditional density of $\tilde{\mu}$ given h and (2) a gamma marginal density of $\tilde{h}$. Both these distributions are well known and tractable, and further investigation reveals (as shown in Section 11.5.5) that the marginal density of $\tilde{\mu}$ is the tractable Student density.

Now this natural-conjugate joint density has only three free parameters, m, v, and n, and this gives us very little flexibility in assigning a prior distribution to *two* unknown quantities, $\tilde{\mu}$ and $\tilde{h}$. We immediately observe, however, that if we replace the expression $n - 1$ in the two places in which it occurs by a new, fourth parameter ν, the mathematical forms of the two factors in the joint density will be unchanged; and from this it follows that the form of the marginal density of $\tilde{\mu}$ will also be unchanged. In other words, no tractability would be lost if we took

$$e^{-\frac{1}{2}hn(\mu-m)^2}\, h^{\frac{1}{2}} \cdot e^{-\frac{1}{2}h\nu v}\, h^{\frac{1}{2}\nu} \;, \qquad \nu, v > 0 \;,$$

as the kernel of the conjugate density.

We now observe, however, that the complete integral of this kernel with respect to (μ, h) exists not only for $\nu > 0$, $v > 0$ but also for $0 > \nu > -1$, $v < 0$, so that the family can be still further enriched by extending the domain of the parameters ν and v beyond the range of the statistics ν and v. For convenience of notation, however, we accomplish this same end by *redefining* the parameters ν and v and writing the "Normal-gamma" conjugate density in the form

$$f_{N\gamma}(\mu, h|m, v, n, \nu) \propto e^{-\frac{1}{2}hn(m-\mu)^2}\, h^{\frac{1}{2}} \cdot e^{-\frac{1}{2}h\nu v}\, h^{\frac{1}{2}\nu-1} \;, \qquad \begin{aligned} -\infty < \mu < \infty \;, \\ h > 0 \;, \\ -\infty < m < \infty \;, \\ v, n, \nu > 0 \;. \end{aligned}$$

It is shown in Section 11.5.3 that the * operation for obtaining the posterior parameters from the prior parameters and the sample statistics remains simple: instead of the original operation (3-6) we have

$$n'' = n' + n \;, \qquad m'' = \frac{1}{n''}(n'm' + nm) \;, \qquad \nu'' = \nu' + \nu + 1 \;,$$

$$v'' = \frac{1}{\nu''}\left[(\nu'v' + n'm'^2) + (\nu v + nm^2) - n''m''^2\right] \;.$$

3.2.5. *Conspectus of Natural-Conjugate Densities*

To summarize the results obtained thus far we now give a conspectus of the natural-conjugate families for all the data-generating processes examined in this monograph. Full discussion of all but one of these processes will be found in Part III of this monograph.

Bernoulli Process

Defined as generator of *either* $\tilde{x}$s with mass function

$$p^x(1-p)^{1-x} , \qquad x = 0, 1 , \quad 0 < p < 1 ,$$

or $\tilde{n}$s with mass function

$$(1-p)^{n-1}\,p , \qquad n = 1, 2, \cdots , \quad 0 < p < 1 .$$

Sufficient Statistics:

$$\left.\begin{aligned} r &= \Sigma x_i , \\ n &= \text{number of } \tilde{x}\text{s observed} ; \end{aligned}\right\} \quad \text{or} \quad \left\{\begin{aligned} r &= \text{number of } \tilde{n}\text{s observed} , \\ n &= \Sigma n_i . \end{aligned}\right.$$

Likelihood of sample: proportional to

$$p^r(1-p)^{n-r} .$$

Conjugate prior density (beta):

$$f_\beta(p|r', n') \propto p^{r'-1}(1-p)^{n'-r'-1} , \qquad 0 < p < 1 , \quad n' > r' > 0 .$$

Posterior density:

$$f_\beta(p|r'', n'') \qquad \text{where} \begin{cases} r'' = r' + r , \\ n'' = n' + n . \end{cases}$$

Graphs of typical members of this family will be found in Section 7.3.2.

Poisson Process

Defined as generator of $\tilde{x}$s with density

$$e^{-\lambda x}\,\lambda , \qquad x \geq 0 , \quad \lambda > 0 .$$

Sufficient statistics:

$$r = \text{number of } \tilde{x}\text{s observed} , \qquad t = \Sigma\, x_i .$$

Likelihood of sample: proportional to

$$e^{-\lambda t}\,\lambda^r .$$

Conjugate prior density (gamma-1):

$$f_{\gamma 1}(\lambda|r', t') \propto e^{-\lambda t'}\,\lambda^{r'-1} , \qquad \lambda \geq 0 , \quad t', r' > 0 .$$

Posterior density:

$$f_{\gamma 1}(\lambda | r'', t'') \qquad \text{where} \begin{cases} r'' = r' + r \ , \\ t'' = t' + t \ . \end{cases}$$

Graphs of typical members of this family will be found in Section 7.6.2.

Rectangular Process

Defined as generator of $\tilde{x}$s with density

$$\frac{1}{\theta} \ , \qquad \begin{array}{c} 0 \le x \le \theta \ , \\ \theta > 0 \ . \end{array}$$

Sufficient statistics:

$$n = \text{number of } \tilde{x}\text{s observed} \ , \qquad M = \max \{x_i\} \ .$$

Likelihood of sample: proportional to

$$\theta^{-n} \, \delta(\theta - M) \qquad \text{where} \qquad \delta(a) = \begin{cases} 0 \text{ if } a < 0 \ , \\ 1 \text{ if } a \ge 0 \ . \end{cases}$$

Conjugate prior density (hyperbolic; not discussed in Part III):

$$f_H(\theta | M', n') \propto \theta^{-n'} \, \delta(\theta - M') \ , \qquad \begin{array}{c} \theta > 0 \ , \\ M' \ge 0 \ , \qquad n' > 1 \ . \end{array}$$

Posterior density:

$$f_H(\theta | M'', n'') \qquad \text{where} \begin{cases} M'' = \max \{M', M\} \ , \\ n'' = n' + n \ . \end{cases}$$

Independent Normal Process

Defined as generator of $\tilde{x}$s with density

$$(2\pi)^{-\frac{1}{2}} \, e^{-\frac{1}{2}h(x-\mu)^2} \, h^{\frac{1}{2}} \ , \qquad \begin{array}{c} -\infty < x < \infty \ , \\ -\infty < \mu < \infty \ , \qquad h > 0 \ . \end{array}$$

μ Known, h Unknown

Sufficient statistics:

$$\nu = \text{number of } \tilde{x}\text{s observed} \ , \qquad w = \frac{1}{\nu} \Sigma (x_i - \mu)^2 \ .$$

Likelihood of sample: proportional to

$$e^{-\frac{1}{2}h\nu w} \, h^{\frac{1}{2}\nu} \ .$$

Conjugate prior density (gamma-2):

$$f_{\gamma 2}(h | v', \nu') \propto e^{-\frac{1}{2}h\nu' v'} \, h^{\frac{1}{2}\nu' - 1} \ , \qquad \begin{array}{c} h > 0 \ , \\ v', \nu' > 0 \ . \end{array}$$

Posterior density:

$$f_{\gamma 2}(h | v'', \nu'') \qquad \text{where} \begin{cases} \nu'' = \nu' + \nu \ , \\ v'' = \dfrac{1}{\nu''} (\nu' v' + \nu w) \ . \end{cases}$$

Graphs of typical members of the implied inverted-gamma-2 density of $\tilde{\sigma} \equiv \tilde{h}^{-\frac{1}{2}}$ will be found in Section 7.7.2.

h Known, μ Unknown

Sufficient statistics:

$$n = \text{number of } \tilde{x}\text{s observed} \ , \qquad m = \frac{1}{n}\Sigma\, x_i \ .$$

Likelihood of sample: proportional to

$$e^{-\frac{1}{2}hn(m-\mu)^2} \ .$$

Conjugate prior density (Normal):

$$f_N(\mu|m', hn') \propto e^{-\frac{1}{2}hn'(\mu-m')^2} \ , \qquad \begin{array}{ll} -\infty < \mu < \infty \ , & h > 0 \ , \\ -\infty < m' < \infty \ , & n' > 0 \ . \end{array}$$

Posterior density:

$$f_N(\mu|m'', hn'') \qquad \text{where} \begin{cases} n'' = n' + n \ , \\ m'' = \dfrac{1}{n''}(n'm' + nm) \ . \end{cases}$$

Both μ and h Unknown

Sufficient statistics:

$$n\text{: number of } \tilde{x}\text{s observed} \ ,$$

$$m = \frac{1}{n}\Sigma\, x_i \ ,$$

$$\nu = n - 1 \ (\text{redundant})\dagger \ ,$$

$$v = \begin{cases} \dfrac{1}{\nu}\Sigma(x_i - m)^2 & \text{if } \nu > 0 \ , \\ 0 & \text{if } \nu = 0 \ . \end{cases}$$

Likelihood of sample: proportional to

$$e^{-\frac{1}{2}hn(m-\mu)^2}\, h^{\frac{1}{2}} \cdot e^{-\frac{1}{2}h\nu v}\, h^{\frac{1}{2}\nu} \ .$$

Conjugate prior density (Normal-gamma):

$$f_{N\gamma}(\mu, h|m', v', n', \nu') \propto e^{-\frac{1}{2}hn'(\mu-m')^2}\, h^{\frac{1}{2}}\, e^{-\frac{1}{2}h\nu'v'}\, h^{\frac{1}{2}\nu'-1} \ , \qquad \begin{array}{c} -\infty < \mu < \infty \ , \\ h > 0 \ , \\ -\infty < m' < \infty \ , \\ v', n', \nu' > 0 \ . \end{array}$$

Posterior density:

$$f_{N\gamma}(\mu, h|m'', v'', n'', \nu'')$$

where

$$n'' = n' + n \ , \qquad m'' = \frac{1}{n''}(n'm' + nm) \ , \qquad \nu'' = \nu' + \nu + 1 \ ,$$

$$v'' = \frac{1}{\nu''}[(\nu'v' + n'm'^2) + (\nu v + nm^2) - n''m''^2] \ .$$

Marginal densities (prior or posterior):

† By *redundant* statistics we mean statistics whose values are wholly determined by the values of the nonredundant sufficient statistics and which are introduced purely for mathematical convenience.

$$f_S(\mu|m,n/v,\nu) \propto [\nu + (\mu - m)^2\, n/v]^{-\frac{1}{2}(\nu+1)} \qquad \text{(Student)},$$
$$f_{\gamma 2}(h|v,\,\nu) \propto e^{-\frac{1}{2}h\nu v}\, h^{\frac{1}{2}\nu-1} \qquad \text{(gamma-2)}\ .$$

Graphs of typical members of the implied inverted-gamma-2 marginal density of $\tilde{\sigma} \equiv \tilde{h}^{-\frac{1}{2}}$ will be found in Section 7.7.2.

Independent Multinormal Process

Defined as generator of $r \times 1$ vector $\tilde{\mathbf{x}}$s with density

$$(2\pi)^{-\frac{1}{2}r}\, e^{-\frac{1}{2}h(\mathbf{x}-\boldsymbol{\mu})^t\, \boldsymbol{\eta}\, (\mathbf{x}-\boldsymbol{\mu})}\, h^{\frac{1}{2}r}\ , \qquad \begin{array}{l} -\infty < \mathbf{x} < \infty\ , \\ -\infty < \boldsymbol{\mu} < \infty\ , \\ h > 0\ , \qquad |\boldsymbol{\eta}| = 1\ , \end{array}$$

where $\boldsymbol{\mu}$ is an $r \times 1$ vector and $\boldsymbol{\eta}$ is an $r \times r$ positive-definite symmetric matrix.

$\boldsymbol{\eta}$ and h Known, $\boldsymbol{\mu}$ Unknown

Sufficient statistics:

n: number of $\tilde{\mathbf{x}}$s observed ,

$$\mathbf{m} = \frac{1}{n}\,\Sigma\, \mathbf{x}^{(j)}\ .$$

$$\mathbf{n} \equiv n\, \boldsymbol{\eta} \text{ (redundant)}\ .$$

Likelihood of sample: proportional to

$$e^{-\frac{1}{2}h(\mathbf{m}-\boldsymbol{\mu})^t\, \mathbf{n}\, (\mathbf{m}-\boldsymbol{\mu})}\ .$$

Conjugate prior density (Normal):

$$f_N^{(r)}(\boldsymbol{\mu}|\mathbf{m}', h\mathbf{n}') \propto e^{-\frac{1}{2}h(\boldsymbol{\mu}-\mathbf{m}')^t\, \mathbf{n}'\, (\boldsymbol{\mu}-\mathbf{m}')}\ , \qquad \begin{array}{r} -\infty < \boldsymbol{\mu} < \infty\ , \\ h > 0\ , \\ -\infty < \mathbf{m}' < \infty\ , \end{array}$$

where $\mathbf{n}'$ is an $r \times r$ positive-definite symmetric matrix.

Posterior density:

$$f_N^{(r)}(\boldsymbol{\mu}|\mathbf{m}'', h\mathbf{n}'') \qquad \text{where} \qquad \begin{cases} \mathbf{n}'' = \mathbf{n}' + \mathbf{n}\ , \\ \mathbf{m}'' = \mathbf{n}''^{-1}(\mathbf{n}'\mathbf{m}' + \mathbf{n}\,\mathbf{m})\ . \end{cases}$$

$\boldsymbol{\eta}$ Known, $\boldsymbol{\mu}$ and h Unknown

Sufficient statistics:

n: number of $\tilde{\mathbf{x}}$s observed, $\quad \mathbf{n} = n\, \boldsymbol{\eta}$ (redundant), $\quad \nu = n - r$ (redundant) ,

$$\mathbf{m} = \frac{1}{n}\Sigma\, \mathbf{x}^{(j)}\ ,$$

$$v = \begin{cases} \dfrac{1}{\nu}\Sigma(\mathbf{x}^{(j)} - \mathbf{m})^t(\mathbf{x}^{(j)} - \mathbf{m}) & \text{if } \nu > 0\ , \\ 0 & \text{if } \nu = 0\ . \end{cases}$$

Likelihood of sample: proportional to

$$e^{-\frac{1}{2}h(\mathbf{m}-\boldsymbol{\mu})^t\, \mathbf{n}\, (\mathbf{m}-\boldsymbol{\mu})}\, h^{\frac{1}{2}r} \cdot e^{-\frac{1}{2}h\nu v}\, h^{\frac{1}{2}\nu}\ .$$

Conjugate prior density (Normal-gamma):

$$f_{N\gamma}^{(r)}(\boldsymbol{\mu}, h|\boldsymbol{m}', v', \mathbf{n}', \nu') \propto e^{-\frac{1}{2}h(\boldsymbol{\mu}-\boldsymbol{m}')^t\mathbf{n}'(\boldsymbol{\mu}-\boldsymbol{m}')}\, h^{\frac{1}{2}r} \cdot e^{-\frac{1}{2}h\nu'v'}\, h^{\frac{1}{2}\nu'-1}$$

where

$$-\infty < \boldsymbol{\mu} < \infty\ , \qquad h > 0\ , \qquad -\infty < \boldsymbol{m}' < \infty\ , \qquad v', \nu' > 0\ ,$$

and $\mathbf{n}'$ is an $r \times r$ positive-definite symmetric matrix.

Posterior density:

$$f_{N\gamma}^{(r)}(\boldsymbol{\mu}|\boldsymbol{m}'', v'', \mathbf{n}'', \nu'')$$

where

$$\mathbf{n}'' = \mathbf{n}' + \mathbf{n}\ , \qquad \boldsymbol{m}'' = \mathbf{n}''^{-1}(\mathbf{n}'\boldsymbol{m}' + \mathbf{n}\,\boldsymbol{m})\ , \qquad \nu'' = \nu' + \nu + r\ ,$$

$$v'' = \frac{1}{\nu''}\,[(\nu'v' + \boldsymbol{m}'^t\,\mathbf{n}'\boldsymbol{m}') + (\nu v + \boldsymbol{m}^t\,\mathbf{n}\,\boldsymbol{m}) - \boldsymbol{m}''^t\,\mathbf{n}''\boldsymbol{m}'']\ .$$

Marginal densities (prior or posterior):

$$f_S^{(r)}(\boldsymbol{\mu}|\boldsymbol{m}, \mathbf{n}/v, \nu) \propto [\nu + (\boldsymbol{\mu} - \boldsymbol{m})^t(\mathbf{n}/v)(\boldsymbol{\mu} - \boldsymbol{m})]^{-\frac{1}{2}(\nu+r)} \qquad \text{(Student)}\ ,$$

$$f_{\gamma 2}(h|v, \nu) \propto e^{-\frac{1}{2}h\nu v}\, h^{\frac{1}{2}\nu-1} \qquad \text{(gamma-2)}\ .$$

Normal Regression Process

Defined as generator of independent random variables $\tilde{y}_1, \cdots, \tilde{y}_i, \cdots$ such that for any i the density of $\tilde{y}_i$ is

$$(2\pi)^{-\frac{1}{2}} \exp[-\tfrac{1}{2}h(y_i - \Sigma_{j=1}^r x_{ij}\,\beta_j)^2]\, h^{\frac{1}{2}}\ , \qquad \begin{array}{c} -\infty < y_i < \infty\ , \\ -\infty < \beta_j < \infty\ , \\ h > 0\ , \end{array} \qquad \text{all } j\ ,$$

where the xs are known numbers.

Letting n denote the number of $\tilde{y}$s observed, we simplify notation by defining the vectors $\boldsymbol{\beta}$ and $\boldsymbol{y}$ and the matrix $\mathbf{X}$ by

$$\boldsymbol{\beta} = [\beta_1 \quad \cdots \quad \beta_r]^t\ , \qquad \boldsymbol{y} = [y_1 \quad \cdots \quad y_n]^t\ ,$$

$$\mathbf{X} = \begin{bmatrix} x_{11} & \cdots & x_{1j} & \cdots & x_{1r} \\ \cdots & \cdots & \cdots & \cdots & \cdots \\ x_{i1} & \cdots & x_{ij} & \cdots & x_{ir} \\ \cdots & \cdots & \cdots & \cdots & \cdots \\ x_{n1} & \cdots & x_{nj} & \cdots & x_{nr} \end{bmatrix}.$$

h Known, $\boldsymbol{\beta}$ Unknown

Sufficient statistics:

$$\mathbf{n} = \mathbf{X}^t\,\mathbf{X}$$

and any $\boldsymbol{b}$ satisfying

$$\mathbf{X}^t\,\mathbf{X}\,\boldsymbol{b} = \mathbf{X}^t\,\boldsymbol{y}\ .$$

Likelihood of sample: proportional to

$$e^{-\frac{1}{2}h(\boldsymbol{b}-\boldsymbol{\beta})^t\,\mathbf{n}\,(\boldsymbol{b}-\boldsymbol{\beta})}\ .$$

Conjugate prior density (Normal):

$$f_N^{(r)}(\boldsymbol{\beta}|\mathbf{b}', h\mathbf{n}') \propto e^{-\frac{1}{2}h(\boldsymbol{\beta}-\mathbf{b}')^t\mathbf{n}'(\boldsymbol{\beta}-\mathbf{b}')} , \qquad \begin{array}{r} -\infty < \boldsymbol{\beta} < \infty , \\ -\infty < \mathbf{b}' < \infty , \\ h > 0 , \end{array}$$

where $\mathbf{n}'$ is an $r \times r$ positive-definite symmetric matrix.

Posterior density:

$$f_N^{(r)}(\boldsymbol{\beta}|\mathbf{b}'', h\mathbf{n}'') \qquad \text{where} \begin{cases} \mathbf{n}'' = \mathbf{n}' + \mathbf{n} , \\ \mathbf{b}'' = \mathbf{n}''^{-1}(\mathbf{n}'\mathbf{b}' + \mathbf{n}\,\mathbf{b}) . \end{cases}$$

Both $\boldsymbol{\beta}$ *and* h *Unknown*

Sufficient statistics:

$$\mathbf{n} = \mathbf{X}^t\mathbf{X} , \qquad p = \text{rank } \mathbf{n} \text{ (redundant)} , \qquad \nu = n - p \text{ (redundant)} ,$$

any $\mathbf{b}$ satisfying

$$\mathbf{X}^t\mathbf{X}\,\mathbf{b} = \mathbf{X}^t\mathbf{y} ,$$

and

$$v = \frac{1}{\nu}(\mathbf{y} - \mathbf{X}\,\mathbf{b})^t(\mathbf{y} - \mathbf{X}\,\mathbf{b}) .$$

Likelihood of sample: proportional to

$$e^{-\frac{1}{2}h(\mathbf{b}-\boldsymbol{\beta})^t\mathbf{n}(\mathbf{b}-\boldsymbol{\beta})}\,h^{\frac{1}{2}p} \cdot e^{-\frac{1}{2}h\nu v}\,h^{\frac{1}{2}\nu} .$$

Conjugate prior density (Normal-gamma):

$$f_{N\gamma}(\boldsymbol{\beta}, h|\mathbf{b}', v', \mathbf{n}', \nu') \propto e^{-\frac{1}{2}h(\boldsymbol{\beta}-\mathbf{b}')^t\mathbf{n}'(\boldsymbol{\beta}-\mathbf{b}')}\,h^{\frac{1}{2}r} \cdot e^{-\frac{1}{2}h\nu'v'}\,h^{\frac{1}{2}\nu'-1}$$

where

$$-\infty < \boldsymbol{\beta} < \infty , \qquad h > 0 ,$$
$$-\infty < \mathbf{b}' < \infty , \qquad \nu', v' > 0 ,$$

and $\mathbf{n}'$ is an $r \times r$ positive-definite symmetric matrix.

Posterior density:

$$f_{N\gamma}(\boldsymbol{\beta}, h|\mathbf{b}'', v'', \mathbf{n}'', \nu'')$$

where

$$\mathbf{n}'' = \mathbf{n}' + \mathbf{n} , \qquad \mathbf{b}'' = \mathbf{n}''^{-1}(\mathbf{n}'\mathbf{b}' + \mathbf{n}\,\mathbf{b}) , \qquad \nu'' = \nu' + \nu + p ,$$

$$v'' = \frac{1}{\nu''}[(\nu'v' + \mathbf{b}'^t\mathbf{n}'\mathbf{b}') + (\nu v + \mathbf{b}^t\mathbf{n}\,\mathbf{b}) - \mathbf{b}''^t\mathbf{n}''\mathbf{b}''] .$$

Marginal densities (prior or posterior):

$$f_S^{(r)}(\boldsymbol{\beta}|\mathbf{b}, \mathbf{n}/v, \nu) \propto [\nu + (\boldsymbol{\beta} - \mathbf{b})^t(\mathbf{n}/v)(\boldsymbol{\beta} - \mathbf{b})]^{-\frac{1}{2}(\nu+r)} \qquad \text{(Student)} ,$$

$$f_{\gamma 2}(h|v, \nu) \propto e^{-\frac{1}{2}h\nu v}\,h^{\frac{1}{2}\nu-1} \qquad \text{(gamma-2)} .$$

3.3. Choice and Interpretation of a Prior Distribution

3.3.1. Distributions Fitted to Historical Relative Frequencies

In some applications there will exist a substantial amount of "objective" evidence on which to base a prior distribution of the state $\tilde{\theta}$; specifically, extensive experience may have made available a "solid" frequency distribution of values taken on by $\tilde{\theta}$ in the past, it may be clear that $\tilde{\theta}$ is itself generated by a fairly stable random process, and therefore most reasonable men might agree that the probability distribution to be assigned to $\tilde{\theta}$ on the present occasion should closely match the frequency distribution of actual historical values. In such cases it is easy to select a close-fitting member of the family of natural conjugates and the only important question remaining for subjective decision is whether or not this distribution fits the "true" distribution closely enough to warrant its use as a convenient approximation. It is possible, of course, to raise questions about the best method of fitting the conjugate distribution, and it is virtually impossible to give conclusive theoretical answers to these questions; but what experience we have had with concrete examples convinces us that in the great majority of applications the method of fitting will have absolutely no *material* effect on the results of the final analysis of the decision problem at hand. Examples of the remarkable insensitivity of results to even substantial changes in the prior distribution are given in Section 5.6.4 below.

3.3.2. Distributions Fitted to Subjective Betting Odds

The problem of assessing a prior distribution is much more difficult and interesting in those situations where no solid "objective" basis for the assessment exists. In such situations the prior distribution represents simply and directly the betting odds with which the responsible person wishes his final decision to be consistent; and, as we have already pointed out, the psychological difficulty of assigning such odds is so great that it will rarely if ever be possible for the decision maker to specify his "true" distribution in more detail than by a few summary measures such as the mean, the mean deviation, or a few fractiles. Provided that these specifications are met by some member of the family of natural conjugates, the decision maker will usually be as willing to act consistently with the implications of this distribution as with the implications of any other distribution meeting the same specifications; but it is of course a responsibility of the statistical consultant who is about to base an analysis on this distribution to point out to the decision maker any extreme implications of the fitted distribution, and it is also his responsibility to suggest to the decision maker those summary measures which are easiest to assess subjectively for a given family of distributions.

To illustrate the choice of a prior distribution in this manner, suppose first that a decision problem turns on the true mean effect $\tilde{\theta} = \tilde{\mu}$ of a proposed change in package design on sales per customer or on the true effect $\tilde{\mu}$ of a proposed change in teaching method on the mean score obtained by students on a particular test; and suppose in either case that sales to individual customers or scores obtained by individual students can be treated as Normally distributed with known precision h (known variance $1/h$), so that the natural-conjugate distribution of $\tilde{\mu}$ is Normal. It is then natural to ask the decision maker first (a) what is his "best guess" $\hat{\mu}$ at the true value of $\tilde{\mu}$ and (b) whether his subjective distribution of $\tilde{\mu}$ is reasonably symmetric about this value. If the answer to (b) is "yes", then a natural way to obtain the dispersion of the distribution is to ask the decision maker (c) to specify a symmetric interval about $\hat{\mu}$ such that he would be willing to bet at the same odds for or against the proposition that the true μ lies within this interval. The statistician can then determine the Normal distribution which meets these specifications and can verify its suitability by calculating the probabilities $P\{\tilde{\mu} > \mu\}$ for a few values of μ or by calculating the values μ such that $P\{\tilde{\mu} > \mu\} = \alpha$ for a few values of α and asking the decision maker whether he accepts these implied odds.

As a second example, suppose that someone wishes to assign a distribution to the random variable $\tilde{\theta} = \tilde{p}$ describing the Democratic vote in the next election as a fraction of the total vote. It may be very difficult for this person to assign a subjective *expected value* to $\tilde{p}$, but he may be quite able to express his judgment of the political situation by assigning *probability* $\frac{3}{4}$ (say) to the proposition that the fraction $\tilde{p}$ will be greater than 50% and probability $\frac{1}{2}$ to the proposition that it will be greater than 53%. If this prior opinion is later to be combined with evidence obtained from a straw ballot, then (abstracting for our present purposes from practical difficulties of sampling and response bias) the sample observations will be almost exactly binomially distributed and the conjugate family is the beta family with density

$$f_\beta(p|r', n') \propto p^{r'-1}(1-p)^{n'-r'-1} .$$

With a little trial-and-error calculation the statistician can determine that the member of this family with parameter ($r' = 62.9$, $n' = 118.5$) agrees quite closely with the specified odds; and he can then verify the suitability of the distribution by computing a few additional implied probabilities in one or the other of the ways suggested in our previous example.

Finally, suppose that a decision problem turns on the precision $\tilde{\theta} = \tilde{h}$ of an Independent Normal process generating $\tilde{x}$s with common density

$$f_N(x|\mu, h) \propto e^{-\frac{1}{2}h(x-\mu)^2}\, h^{\frac{1}{2}}$$

Few people are used to thinking in terms of the precision $\tilde{h}$ and it may therefore be very difficult to elicit any kind of *direct* opinion about $\tilde{h}$; but it may be quite possible to obtain betting odds on the process standard deviation $\tilde{\sigma} = \tilde{h}^{-\frac{1}{2}}$ and these odds can then be used to find the appropriate member of the conjugate gamma-2 family with density

$$f_{\gamma 2}(h|v', \nu') \propto e^{-\frac{1}{2}h\nu' v'}\, h^{\frac{1}{2}\nu'-1} .$$

If the decision maker assigns probability $\frac{1}{2}$ (say) to the proposition that $\tilde{\sigma} > 5$ and probability $\frac{1}{4}$ to the proposition that $\tilde{\sigma} > 10$, then as shown in Section 11.1.4 it is very easy with the aid of Figure 7.5 to determine that the member of this family with parameter ($v' = 12$, $\nu' = 1$) agrees closely with the stated odds.

3.3.3. Comparison of the Weights of Prior and Sample Evidence

Even when the decision maker is well satisfied *a priori* with a distribution fitted to his subjective expectations or betting odds, he may become thoroughly dissatisfied with the implications of this distribution when he sees them reflected in the posterior distribution calculated after a sample has actually been taken.

Suppose, for example, that after the bettor discussed in the previous section has assigned the distribution $f_\beta(p|62.9, 118.5)$ to the fraction voting Democratic in the next election, a simple random sample of 100 voters yields reliable information that only 45 of them will vote Democratic. The posterior distribution of $\tilde{p}$ will have density $f_\beta(p|62.9 + 45, 118.5 + 100) = f_\beta(p|107.9, 218.5)$, and this implies that the odds in favor of the Democrats are now 6 to 4, i.e., that $P\{\tilde{p} > .5\} = .6$. It is quite conceivable, however, that a person who was originally willing to bet 3 to 1 on the Democrats will still be willing to bet at nearly if not quite these same odds rather than at 6 to 4, arguing that in his opinion so small a sample carries very little weight in comparison with the analysis of the general political situation on which the original 3-to-1 odds were based. It is also quite conceivable that he will react in the opposite way, arguing that abstract political reasoning does not convince him nearly so well as solid evidence on the actual intentions of 100 voters, and accordingly he may now much prefer giving 3-to-1 odds *against* the Democrats to giving 6-to-4 odds in their favor.

In either case, what has happened is simply that the person in question has made two different, mutually contradictory evaluations of the "weight" of the prior evidence; and in either case it is his next responsibility to reconcile these two evaluations in whatever way seems best to him. If he finally decides that his prior odds accurately represent his reasoned evaluation of the general political situation, then he *should* still be willing to give odds of 6 to 4 on the Democrats. If on the contrary he finally decides that he is now willing actually to give odds of 3 to 1 against the Democrats, then he has implicitly reevaluated the weight which should be assigned to the prior evidence relative to the sample.

The possibility that posterior subjective feelings may not agree with calculated posterior probabilities suggests that without waiting for an actual sample to give rise to difficulty it will usually be well to check subjective prior betting odds against *hypothetical* sample outcomes before beginning the actual analysis of the decision problem; and this in turn suggests that in some situations it may actually be better to reverse the procedure, making the initial fit of the prior distribution agree with attitudes posterior to some hypothetical sample or samples and then checking by looking at the implied betting odds. Thus if our bettor on elections had said (a) that without any sample evidence he would bet 3 to 1 on the Democrats, but (b) that the presence of only 45 Democratic votes in a sample of 100 would lead him to reverse these odds, then from the implied right-tail cumulative probabilities

$$G_\beta(.5|r', n') = \tfrac{3}{4} , \qquad G_\beta(.5|r' + 45, n' + 100) = \tfrac{1}{4} ,$$

the statistician could calculate $r' \doteq 9.4$, $n' \doteq 16$ to be the appropriate prior parameters, implying a prior standard deviation of .12 where the distribution $f_\beta(p|62.9, 118.5)$ implied a standard deviation less than .05.

It is sometimes argued that a procedure which selects the prior distribution to agree with the desired posterior distribution violates the basic logic of Bayes' theorem, but in our opinion this argument rests on nothing but an excessively literal interpretation of the words "prior" and "posterior". As we pointed out in

Section 1.1.2, the decision maker's real task is to assign a measure to the entire possibility space $\Theta \times Z$ *in whatever way enables him to make the most effective use of his experience and knowledge;* and in Section 1.4.3 we examined a situation where it was reasonable to start by actually assigning a complete set of "posterior" probabilities to Θ and then to compute the "prior" *from* the "posterior" probabilities. We go even further than merely rejecting the argument that prior probabilities *must* be assigned without consideration of posterior probabilities; we maintain that it is *irresponsible* to hold to an original prior distribution if after a real sample has been taken the resulting posterior distribution disagrees with the decision maker's best judgment. Since it is usually difficult to intuit directly concerning the relative weights to be given to prior and sample evidence, it may be that the decision maker will *usually* do well to place his primary reliance on prior expectations and betting odds; but even when this is true, calculation of the distribution posterior to a hypothetical sample will provide a useful check.

3.3.4. "Quantity of Information" and "Vague" Opinions

In some situations the decision maker will feel that the information available prior to obtaining actual sample evidence is virtually non-existent. There has been a great deal of discussion concerning the representation of this state of mind by an appropriate prior distribution, and we shall now briefly review some of the principal arguments which have been advanced. Because the logic is slippery we shall proceed by first discussing an example which makes as good a case as can be made for the existence of distributions which express the absence of any definite prior convictions and then discussing further examples which bring out the difficulties involved.

Normal Prior Distributions. As we have already seen repeatedly, a sample from an Independent Normal process with unknown mean $\tilde{\mu}$ but known precision h can be summarized by the sufficient statistic

$$y = (m, n) = \left(\frac{1}{n} \Sigma x_i, n\right) , \tag{3-13a}$$

the natural-conjugate prior density is

$$f_N(\mu|m', hn') \propto e^{-\frac{1}{2}hn'(m'-\mu)^2} , \qquad \begin{array}{c} -\infty < m' < \infty , \\ n' > 0 , \end{array} \tag{3-13b}$$

and the parameters of the posterior density are

$$(m'', n'') = (m', n') * (m, n) = \left(\frac{n'm' + nm}{n' + n}, n' + n\right) . \tag{3-13c}$$

The fact that the component n of the sample statistic (m, n) corresponds to the number of observations on the process makes it natural to think of this component as a measure of the "size" of the sample or of the "weight" of the sample evidence or of the "quantity of information" in the sample, and the measure thus defined has the obviously desirable property of being additive from sample to sample. Once we have made this observation, it is very tempting to interpret the parameters n' and n'' of the prior and posterior distributions of $\tilde{\mu}$ as measuring the *quantity of information about* μ contained in those distributions; the posterior information n'' will then be the sum of the prior and sample informations.

If the component n of the sample statistic (m, n) is to be interpreted as measuring the *quantity* of information in the sample, then it seems natural to interpret the component m as summarizing the *import* of this information; or since the expected value of $\tilde{m}$ given μ is equal to μ, we naturally tend to think of m as an "estimate" of μ based on n units of information. It then becomes natural to think of the prior parameter m' as an estimate of μ based on n' units of *prior* information, especially since the prior expected value of $\tilde{\mu}$ is m', as can easily be shown. The whole approach seems to hang together logically when we observe that by (3-13c) the *posterior* parameter or "estimate" m'' can be interpreted as a weighted average of the prior and sample estimates, the weights being the respective "quantities of information" n' and n.

Our inclination to interpret the prior and posterior distributions in this way becomes even stronger when we examine the behavior of the prior distribution of $\tilde{\mu}$ as the parameter n' approaches the lower bound of its domain $(0, \infty)$. It is well known that the variance of the distribution (3-13b) is $1/hn'$, so that as n' approaches 0 the variance of $\tilde{\mu}$ increases without bound; and it is also easy to show that as n' approaches 0 the distribution of $\tilde{\mu}$ approaches *uniformity* in the sense that the ratio of the probabilities of any two intervals of equal length approaches unity.† An interpretation of n' as a measure of the quantity of information underlying or summarized by the decision maker's prior "estimate" m' thus accords very well with two instinctive feelings: (1) that large variance is equivalent to great uncertainty and that great uncertainty is equivalent to little information, and (2) that when one knows very little about the value of some quantity, it is not unreasonable to assign roughly the same probability to every possible value of this quantity (strictly speaking, equal probabilities to equal intervals).

Finally, we observe that as n' approaches 0 the parameters m'' and n'' of the *posterior* distribution as given by (3-13c) approach the values of the sample statistics m and n respectively; and it is very tempting to interpret this limiting posterior distribution as being a distribution of $\tilde{\mu}$ which is *wholly determined by the sample evidence* and completely independent of the decision maker's vague prior opinions. Notice that the *posterior* distribution with parameter $n'' = n' + n$ exists even for $n' = 0$ despite the fact that no limiting prior distribution exists.

Beta Prior Distributions. Unfortunately the interpretation of the prior distribution in terms of an "estimate" summarizing a certain "quantity of information" becomes much less clear-cut as soon as we look closely at examples other than the Normal; all the important difficulties can be illustrated by considering as simple a distribution as the beta. We have already seen that the beta distribution is the natural conjugate for a Bernoulli process with unknown parameter $\tilde{p}$; we remind the reader that for this process a sufficient statistic is

$$y = (r, n) = (\Sigma x_i, n) \ , \tag{3-14a}$$

† Alternatively, let M_1 be the interval (μ_1, μ_2) and let M_2 be the interval $(\mu_1 + k, \mu_2 + k)$. Then as $n' \to 0$ the conditional probabilities $P\{M_1|M_1 \cup M_2\}$ and $P\{M_2|M_1 \cup M_2\}$ approach equality for all μ_1, μ_2, and k.

that the beta prior density can be written

$$f_\beta(p|r', n') \propto p^{r'-1}(1-p)^{n'-r'-1} , \qquad n' > r' > 0 , \tag{3-14b}$$

and that the parameters of the posterior density are

$$(r'', n'') = (r', n') * (r, n) = (r' + r, n' + n) . \tag{3-14c}$$

At first glance it may seem as natural in this case as in the Normal to regard the number of trials n as measuring the "size" of or "quantity of information" in a sample. It has already been pointed out, however (in Section 2.3.4), that instead of thinking of a Bernoulli process in terms of random variables $\tilde{x}_i = 0, 1$ which indicate whether the ith trial was a success or a failure, we can just as well think of it in terms of random variables $\tilde{n}_i = 1, 2, \cdots$ which indicate the number of trials required to secure the ith success; and we have seen that if we do look at the process in this way, then it is the number of successes r which is the "natural" measure of the sample size rather than the number of trials n. Clearly, then, we can *not* think of *either* component of the statistic (r, n) as *the* measure of the information in a sample from a Bernoulli process, and it follows at once that it would make no sense to think of either component of the parameter (r', n') as *the* measure of the information underlying a prior distribution.

It next occurs to us to try to remedy this situation by arguing that the choice of statistic (r, n) to summarize the sample and of the parameter (r', n') to describe the prior distribution is arbitrary even though convenient for practical computations and that interpretations will be easier if some other choice is made. Not only is neither component of (r, n) an apparently clear *measure of information* like n in the Normal case; neither component is a natural *estimate* of the process parameter like m in the Normal case. It would seem, however, that if we summarize the sample by the number of trials n and the *sample mean* m, defined exactly as in the Normal case by

$$m \equiv \frac{1}{n} \Sigma\, x_i = \frac{r}{n} ,$$

then we may recover all the simplicity of interpretation that we seemed to have in the Normal case. The expected value of $\tilde{m}$ given p is p, so that m as here defined is just as natural an "estimate" of p as m is of μ in the Normal case; and since for given m an increase in n implies an increase in r, the sample size or quantity of information would seem to be unambiguously measured by the component n of the statistic (m, n).

If we now wish to interpret the prior and posterior distributions in an analogous manner, we will substitute $m'n'$ for r' in the formula (3-14b) for the prior density, thus obtaining

$$\text{D}'(p) \propto p^{n'm'-1}(1-p)^{n'(1-m')-1} , \qquad \begin{aligned} 0 < m' &< 1 , \\ n' &> 0 , \end{aligned}$$

and we will think of m' as a prior "estimate" of p summarizing n' units of prior information. The implications are analogous to the Normal case and seem reasonable in several respects. (1) Since the expected value of $\tilde{p}$ is m', m' is a "natural" estimate of p. (2) Since the posterior parameters are given by

$$n'' = n' + n \ , \qquad m'' = \frac{n'm' + nm}{n' + n} \ ,$$

we can interpret the posterior "estimate" m'' as a weighted average of the prior and sample "estimates", the weights being the "quantities of information" summarized by these estimates. (3) As the prior information n' approaches 0, the posterior parameter (m'', n'') approaches the sample statistic (m, n), so that we are again tempted to think of the limiting posterior distribution as one which is wholly determined by the sample information and independent of any individual's vague prior opinions.

Closer examination reveals serious difficulties with these apparently simple interpretations, however. As n' approaches 0 with the "estimate" m' held fixed, we find that the beta distribution (unlike the Normal) *does* approach a proper limiting distribution: namely (as shown in Section 9.1.3) a two-point distribution with a mass of m' on $p = 1$ and a mass of $(1 - m')$ on $p = 0$. Now this limiting distribution cannot in *any* sense be considered "vague". On the contrary it is completely *prejudicial* in the sense that no amount of sample information can alter it to place any probability whatever on the entire open interval $(0, 1)$. A single sample success will annihilate the mass at $p = 0$, and a single failure will annihilate the mass at $p = 1$; but a sample containing both successes and failures will give the meaningless result $0/0$ as the posterior density at all p in $(0, 1)$ and also at the extreme values 0 and 1 themselves.

Even if we stop short of the actual limiting two-point distribution and consider the implications of a beta distribution with a very small but nonzero n', we find that we cannot say that such a distribution represents "very little information" or expresses opinions which are in any sense "vague". As n' approaches 0, the distribution assigns a probability approaching 0 to any interval $[\epsilon_1, 1 - \epsilon_2]$ where ϵ_1 and ϵ_2 are arbitrarily small positive constants; and it requires a very great deal of information in the ordinary sense of the word to persuade a reasonable man to act in accordance with such a distribution even if the probability assigned to the interval is *not* strictly 0. Long experience with a particular production process or with very similar processes may persuade such a man to bet at long odds that the fraction defective on the next run will be very close to 0 or very close to 1, but he is *not* likely to be willing to place such bets if he is completely unfamiliar with the process.

The Arbitrariness of Parametrization. Actually, the difference between Normal and beta conjugate families which we have just emphasized is apparent rather than real, since what we have said applies to only one among an infinity of possible parametrizations of the families in question. A Bernoulli process can be characterized by $\pi \equiv \log [p/(1 - p)]$ just as well as by p itself. If it is so characterized, then the conjugate prior density becomes

$$\mathrm{D}'(\pi) \propto \frac{e^{\pi r'}}{(1 + e^{\pi})^{n'}} \ , \qquad \begin{array}{c} -\infty < \pi < \infty \ , \\ n' > r' > 0 \ ; \end{array}$$

and as n' and $r' < n'$ approach 0 in any manner whatever, this density behaves exactly like the Normal density discussed above: it becomes more and more uniform over the entire real line but does not approach any proper density as a limit. And vice versa: as the Normal distribution of $\tilde{\mu}$ approaches uniformity over the real line, the implied distribution of $\tilde{\eta} \equiv 1/\tilde{\mu}$ approaches a mass of 1 at the point $\tilde{\eta} = 0$.

Thus in general: if the distribution of one particular set of parameters $\tilde{\theta}$ has very great variance, the distributions of other, equally legitimate sets of parameters will have very little variance; so that if we choose one particular set of parameters and assign to it a very broad prior distribution, it must be because we have *reason* to assign such a distribution to these particular parameters and not to others. *The notion that a broad distribution is an expression of vague opinions is simply untenable.* There is no justification for saying that a man who is willing to bet at odds of a million to one that $1/\tilde{\mu} = 0 \pm .0001$ holds an opinion which is either more or less vague or diffuse than that of a man who is willing to bet at the same odds that $\tilde{\mu} = 0 \pm .0001$. There is no justification for saying that a man is *either* "definite" or "vague" if he assigns the distribution $f_\beta(p|1,100)$ with standard deviation .01 to the mean number $\tilde{p}$ of successes per Bernoulli trial and thereby assigns an infinite standard deviation to the mean number $1/\tilde{p}$ of trials per success.

Notice, however, that although we cannot distinguish meaningfully between "vague" and "definite" prior distributions, we *can* distinguish meaningfully between prior distributions which can be *substantially modified by a small number of sample observations from* a particular *data-generating process* and those which cannot. Thus consider an Independent Normal process with known precision $h = 1$ and unknown mean $\tilde{\mu}$, so that if a Normal prior distribution with parameter (m', n') is assigned to $\tilde{\mu}$ the posterior distribution will be Normal with parameter

$$m'' = \frac{n'm' + nm}{n' + n} , \qquad n'' = n' + n .$$

Given that the prior distribution of $\tilde{\mu}$ is to be of the Normal form, willingness to bet at odds of a million to one that $\tilde{\mu} = 0 \pm .0001$ implies $m' = 0$, $n' \doteq 2.4 \times 10^9$, so that almost no conceivable amount of sample information could make the posterior parameters differ noticeably from the prior, whereas willingness to bet at odds of a million to one that $1/\tilde{\mu} = 0 \pm .0001$ implies $m' = 0$, $n' \doteq 1.6 \times 10^{-20}$, so that a single sample observation will almost wholly determine the parameters of the posterior distribution.

3.3.5. Sensitivity Analysis

Our last example suggests that *in some decision problems* prior opinions and prior betting odds may have little or no effect on the decision actually chosen. Thus for example, we often find situations in which, *after a sample has once been taken*, the posterior expected utility of every a in A is practically the same for every prior distribution that the decision maker *is willing to consider*. The question is sometimes delicate, however, and to avoid giving the impression that it is simpler than it really is, we shall illustrate with an example involving the beta rather than the excessively well-behaved Normal distribution.

Suppose that utility depends on the parameter $\tilde{p}$ of a Bernoulli process; consider three extremely different prior distributions: (1) the rectangular distribution $f_\beta(p|1, 2)$, (2) the extremely U-shaped distribution $f_\beta(p|.01, .02)$, and (3) another extremely U-shaped distribution which places a mass of .49 on $p = 0$, a mass of .49 on $p = 1$, and distributes the remaining probability .02 uniformly over the interval (0, 1); suppose that, after one of these distributions has been chosen, a fairly large number n of observations are made on the process; and let r denote the number of successes in this sample.

1. Assume first that $0 < r < n$. Then the posterior distributions corresponding to the first and third priors will be identical, having density $f_\beta(p|r + 1, n + 2)$ and mean $(r + 1)/(n + 2)$, while the posterior corresponding to the second prior will have density $f_\beta(p|r + .01, n + .02)$ and mean $(r + .01)/(n + .02)$. In this case the first and third priors are absolutely equivalent for *any* action problem; and in the very common type of problem where the utility of any a in A depends only on the *mean* of the distribution of $\tilde{p}$ (cf. Chapter 5A) the second prior will usually be *practically* equivalent to the other two if both r and $(n - r)$ are at all large compared to 1.

2. Assume next that $r = 0$. Then the first two priors yield posterior densities $f_\beta(p|1, n + 2)$ and $f_\beta(p|.01, n + .02)$ respectively while the third yields a posterior distribution which places a mass $(n + 1)/(n + 1.04)$ on the point $p = 0$ and distributes the remaining probability over (0, 1) with density $[.04/(n + 1.04)]f_\beta(p|1, n + 2)$. In many action problems each of these three posteriors would have implications *totally different* from each of the others; in particular, the means of the three posterior distributions are respectively $1/(n + 2)$, $.01/(n + .02)$, and $.04/(n^2 + 3.04n + 2.08)$, the first of which may be many times as large as the second.

Thus even when the sample n is large the criticality of the choice of a prior distribution depends crucially on the value of the sample r. When $0 < r < n$ in our example, a decision maker who feels sure that his prior information should lead to some distribution "between" the rectangular and an extreme U simply does not need to worry further: he can select the terminal act which is optimal under the posterior distribution corresponding to *any* prior in the range he is considering and be sure that for all practical purposes he has maximized his expected utility with respect to any prior he *might* have chosen. He can in fact simplify his problem still further by observing that all possible posteriors are *practically* equivalent to the posterior density $f_\beta(p|r, n)$ corresponding to the improper prior $f_\beta(p|0, 0)$ and spare himself the trouble of interpolating for fractional r and n in tables of the beta distribution. When on the contrary $r = 0$, the decision maker may be *forced* to make a definite choice of the prior distribution with which he wishes his act to be consistent, no matter how vague or confused his initial psychological state may be. In short, sensible men will always observe the principle of *de minimis*, but what are *minima* depends on circumstances.

One particular point is so often overlooked that it requires explicit mention. In a great many applications of great practical importance, the question is *not* one of deciding what to do *after* a sample has already been taken; it is one of deciding *whether or not any sample should be taken at all and if so how large the sample should be.* We cannot emphasize too strongly that, in situations where it is possible to take at most one sample or where each successive sample involves a substantial "stage cost" independent of the sample size, answers to questions of this kind must *always* depend *critically* on the prior distribution. In such situations it is therefore a vital and unavoidable responsibility of the decision maker to adopt a prior distribution which represents his best judgment as accurately as possible, however much he may wish that this judgment could be based on more evidence than is actually available.

3.3.6. Scientific Reporting

Finally, let us look briefly at the implications of the discussion above for the vexed question of the proper way in which to report a "scientific" experiment, by which we presume is meant an experiment carried out with no *immediate* action problem in mind. An obvious "solution" to this problem is simply to report the data or—provided that there is no substantial doubt about the nature of the data-generating process—to report the sufficient statistics; but it has long been urged that the reporter should add *some* indication of the "reliability" of the data or the sufficient statistics, and if the proper measure of reliability is to be given we must ask what purpose the measure will serve.

It seems to us that the answer to this question is not hard to find. It is true, of course, that the *ultimate* use to which a measure of reliability will be put depends upon the user. If the reported values of certain physical constants are uncertain, a physicist may want a measure of uncertainty because he wants to decide whether or not to invest effort in improvement of the basic data before putting the data to any use whatever, whereas an engineer examining the same values may want a measure of uncertainty in order to allow a suitable margin of safety in the equipment he is designing. In either case, however, the user is not interested in uncertainty out of idle curiosity or because he wants to verbalize about it; he is interested because *he has a decision problem to solve, to solve it he* must *take account of his uncertainty about some quantity* θ, *and therefore he must, implicitly if not explicitly, assign a probability distribution to* $\tilde{\theta}$. We are, of course, using the word "decision" in its broadest interpretation, but we believe that it *is* very clearly a *decision* problem to choose between alternatives such as (1) consider some scientific theory as having been substantiated to the extent that the expenditure of further effort on research would be unwarranted, and (2) reserve judgment about the validity of the theory, thereby implying that further research *should* be done before the theory is accepted or rejected.

Because the distribution which the user of any report will ultimately assign to $\tilde{\theta}$ *must* depend in part on evidence external to the data being reported, the reporter cannot in general supply the user with the distribution he requires; but what he *can* do is to report in such a way that the user has ready access to those aspects of the data which will enter into his own assignment of a distribution to $\tilde{\theta}$. What is completely intolerable is to report in such a way that the sufficient statistics cannot be recovered by the user; and while it is probably unnecessary to point out that it is inexcusable to report a level of significance or a confidence interval without also reporting sample size, it does seem necessary to point out that even if sample size *is* reported the user cannot in general recover the information he needs from a statement that the data was or was not significant at some predetermined level or from an asymmetric or randomized confidence interval. And even if a report *is* full enough to make it possible to recover the sufficient statistics, the user is not well served if he is forced to calculate these numbers which might have been given to him directly.

Although the only way in which the user of a scientific report can obtain a really adequate analysis of his decision problem, whatever it may be, is to take the sufficient statistics and then make the calculations required by his particular problem, it is true that in many situations the reporter can foresee (a) that many users will want to know whether or not an unknown scalar $\tilde{\theta}$ is above or below some "standard" value θ_o and (b) that many users would if pressed assign to $\tilde{\theta}$ a prior distribution with very substantial variance. In such cases the reporter may be able to spare *some* users some arithmetic by adding to his report of the sufficient statistics a statement of the posterior probability $P\{\tilde{\theta} < \theta_o\}$ based on a rectangular prior distribution, warning the user at the same time that this probability is inappropriate unless the user's own prior distribution of this particular parameter is at least roughly rectangular. If no clear-cut "standard" θ_o exists, the entire cumulative posterior distribution based on a rectangular prior can be reported either graphically or by listing quartiles, deciles, or other fractiles.†

In the very common case where the sufficient statistics are the sample size and the value of an unbiassed, approximately Normally distributed "estimator", a statement of the value and the sampling variance of the estimator can of course be directly interpreted as the mean and variance of the posterior distribution of the estimated parameter given a rectangular prior distribution. Unsophisticated users of statistical reports usually *do* interpret probability statements about statistics as if they were probability statements about parameters because their intuition tells them that probability statements about parameters are directly relevant to their problem while probability statements about statistics are not. What we are suggesting is that the sophisticated reporter should make clear to such users the conditions under which their relevant interpretations are valid, rather than insist on valid interpretations which are not directly relevant to the users' problems.

Finally, let us remark that if the reporter believes that he has substantial information outside the evidence of the particular experiment which he is reporting, no reasonable canon of scientific reporting should force him to suppress this information or any conclusions he believes can be based on it. If he feels that the evidence obtained from this one experiment does not completely overwhelm his own personal prior distribution, and if he believes that his prior judgments are well enough founded to be of interest to the reader of the report, then he can—and we believe he should—add to his report a posterior distribution based on his own explicitly stated and defended prior distribution.

† It is perhaps worth remarking that the posterior probability of "extreme" propositions of the type $\tilde{\theta} = \theta_0$ exactly are necessarily so sensitive to the prior distribution that it is completely impossible to make a useful general-purpose statement of posterior probability.

3.4. Analysis in Extensive Form When the Prior Distribution and Sample Likelihood are Conjugate

The basic principle of the extensive form of analysis as set forth in Section 1.2.1 is to evaluate every e in E by computing

$$u^*(e) = \mathrm{E}_{z|e} \max_a \mathrm{E}''_{\theta|z}\, u(e, \tilde{z}, a, \tilde{\theta}) \tag{3-15}$$

and then to select an e which maximizes u^*. The application of the analysis was illustrated in Sections 1.2.2 and 1.4.3 by simple examples in which these computations could be carried out numerically, but in the majority of practical applications numerical analysis would be extremely difficult if not impossible because Θ and Z are extremely large if not infinite. We shall now show, however, that analytical computation is often very easy when:

1. The experimental outcome z admits of a sufficient statistic y of fixed dimensionality;
2. The prior measure P'_θ on Θ is conjugate to the conditional measures $P_{y|e,\theta}$ on the reduced sample space Y.

In the present section we shall indicate in a brief and general way how the catalogue of distributions and integrals which constitutes Part III of this monograph can be used to evaluate u^* when y is generated by one of the data-generating processes listed in Section 3.2.5 above and the prior distribution is conjugate to the likelihood of y. We shall then go on in Part II to develop still simpler methods of analysis for cases where the utility function $u(\cdot, \cdot, \cdot, \cdot)$ meets certain conditions which are often met in practice.

3.4.1. Definitions of Terminal and Preposterior Analysis

Given the existence of a sufficient statistic y, the definition (3-15) of $u^*(e)$ can be rewritten in the form

$$u^*(e) = \mathrm{E}_{y|e} \max_a \mathrm{E}''_{\theta|y}\, u(e, \tilde{y}, a, \tilde{\theta}) \;. \tag{3-16}$$

Our discussion of the evaluation of (3-16) for any given e will be divided into two parts:

1. Evaluation for a particular y of
$$u^*(e, y) \equiv \max_a \mathrm{E}''_{\theta|y}\, u(e, y, a, \tilde{\theta}) \;. \tag{3-17}$$
2. Repetition of this procedure for all y in Y followed by computation of
$$u^*(e) \equiv \mathrm{E}_{y|e}\, u^*(e, \tilde{y}) \;. \tag{3-18}$$

Evaluation of (3-17) will be called *terminal analysis* because it deals with the evaluation of and choice among terminal acts *after* the experiment has (actually or hypothetically) already been conducted and the outcome y observed; *it includes as a special case the evaluation of and choice among terminal acts when the experiment and outcome are the dummies e_0 and y_0 which consist of no experiment and no observation at all.* Evaluation of (3-18) will be called *preposterior analysis* because it involves looking at the decision problem as it appears *before* the experiment has been conducted and taking the *prior* expected value of all possible *posterior* expected utilities $u^*(e, y)$. In terms of the game trees discussed in Section 1.2, terminal analysis is concerned with the two outer levels of branches (the last two moves in the game) while preposterior analysis is concerned with the two inner levels (the first two moves in the game) and can take place only after the two outer levels have been averaged out and folded back.

Each step in the analysis will be illustrated by application to the following example. A choice must be made between two agricultural processes, one of which will yield a known profit K_1 while the profit of the other depends on an unknown quantity μ in a linear manner, $K_2 + k_2\mu$. It is possible to take observations $\tilde{x}$ on μ which are Normally distributed with known precision; the density of any $\tilde{x}$ is

$$f_N(x|\mu, h) = (2\pi)^{-\frac{1}{2}} e^{-\frac{1}{2}h(x-\mu)^2} h^{\frac{1}{2}} . \tag{3-19}$$

If n observations $(x_1, \cdots, x_n)$ are made and the process with profit K_1 is chosen (act a_1), the decision maker's utility will be

$$u[e, (x_1, \cdots, x_n), a_1, \mu] = -\exp(k_s n - K_1) \ ; \tag{3-20a}$$

if the same experiment is conducted but the process with unknown profit is chosen (act a_2), his utility will be

$$u[e, (x_1, \cdots, x_n), a_2, \mu] = -\exp(k_s n - K_2 - k_2\mu) \ . \tag{3-20b}$$

(One interpretation of these utilities would be that the cost of sampling is proportional to n and that utility to the decision maker of v' units of money can be approximated over a suitable range by a function of the form

$$u'(v') = c_1(1 - e^{c_2 - c_3 v'}) \ , \tag{3-21a}$$

which can be reduced without loss of generality to the form

$$u(v) = -e^{-v} \tag{3-21b}$$

by linear transformations of the scales on which money and utility are measured. We use this function here simply as a mathematical illustration; its applicability to a real problem involves acceptance by the decision maker of the very restrictive implication that addition of a constant amount to every prize in a lottery increases the value of the lottery by that same amount.)

3.4.2. Terminal Analysis

1. *Computation of the Posterior Distribution of $\tilde{\theta}$.* The first step in terminal analysis as defined by (3-17) is to use the prior measure P'_θ and the conditional measure $P_{y|e,\theta}$ to determine the posterior measure $P''_{\theta|y}$ for the particular y which has (actually or hypothetically) been observed. We have already seen in Section 3.2.5 that the conjugate prior distribution for any data-generating process considered in this monograph possesses a density function, that the posterior distribution possesses a density of the same mathematical form as the prior, and that it is possible to compute the parameter of the posterior distribution from the prior parameter and the sample statistic by a simple algebraic * operation; no analysis is required.

The conspectus in Section 3.2.5 gives the * operation as well as definitions of the sufficient statistics and conjugate densities for all the data-generating processes there listed. In our example, if the decision maker assigns to $\tilde{\mu}$ a Normal prior distribution with parameter $y' = (m', n')$,

$$D'(\mu) = f_N(\mu|m', hn') \propto e^{-\frac{1}{2}hn'(\mu - m')^2} ,$$

and if he then observes a statistic $y = (m, n)$—i.e., if the mean of n experimental observations turns out to be m—his posterior distribution will be Normal with density

$$D''(\mu|m) = f_N(\mu|m'', hn'') \propto e^{-\frac{1}{2}hn''(\mu - m'')^2} \tag{3-22a}$$

where

$$(m'', n'') = (m', n') * (m, n) = \left(\frac{m'n' + mn}{n' + n}, n' + n\right) . \tag{3-22b}$$

2. *Computation of the Posterior Utility of an Act.* The next step in terminal analysis as defined by (3-17) is the use of the posterior distribution to compute the expected utility of each a in A. Here the feasibility of the analysis will obviously depend on the choice of mathematical function used to approximate the decision maker's "true" utility, but with a little ingenuity it will often be possible to find a tractable function which gives a good approximation *over the range of consequences which have any substantial probability of occurrence in the particular decision problem at hand.*

In our example, we make use of the fact (easily verified by completion of squares) that for any z_1 and $z_2 > z_1$

$$\int_{z_1}^{z_2} e^{a+bz} f_N(z|m, H)\, dz = \exp\left(a + bm + \frac{b^2}{2H}\right) \int_{z_1}^{z_2} f_N\left(z|m + \frac{b}{H}, H\right) dz \tag{3-23}$$

to evaluate the expectations of the utilities given by (3-20):

$$u^*[e, (m, n), a_1] = \int_{-\infty}^{\infty} -\exp(k_s n - K_1)\, f_N(\mu|m'', hn'')\, d\mu$$

$$= -\exp(k_s n - K_1) , \tag{3-24a}$$

$$u^*[e, (m, n), a_2] = \int_{-\infty}^{\infty} -\exp(k_s n - K_2 - k_2\mu)\, f_N(\mu|m'', hn'')\, d\mu$$

$$= -\exp\left(k_s n - K_2 - k_2 m'' + \frac{1}{2hn''} k_2^2\right) . \tag{3-24b}$$

Because of the great variety of possible utility functions we have made no attempt to give a complete catalogue of integrals of this sort in Part III, but we do call the reader's attention to the fact that the expectations of *linear* and *quadratic* utilities can be evaluated by use of the moment formulas given in Chapters 7 and 8 for every distribution occurring in the analysis of the data-generating processes listed in Section 3.2.5 above.

3. *Choice of an Optimal Act.* The last step in terminal analysis as defined by (3-17) is to select an a which maximizes $u^*(e, y, \cdot)$, and this will ordinarily be relatively easy if explicit formulas have been found for all $u^*(e, y, a)$.

In our example we can compare the utilities (3-24) of a_1 and a_2 by writing

$$\frac{u^*[e, (m, n), a_1]}{u^*[e, (m, n), a_2]} = \exp\left(-K_1 + K_2 + k_2 m'' - \frac{1}{2hn''} k_2^2\right)$$

and observing that this ratio will be equal to or less than 1 if

$$m'' \le \frac{1}{k_2}\left(K_1 - K_2 + \frac{1}{2hn''} k_2^2\right) . \tag{3-25}$$

Since the utilities of both acts are *negative* for all m'', this implies that a_1 is an optimal act if m'' satisfies (3-25); if m'' is greater than or equal to the right-hand side of (3-25), a_2 is an optimal act.

3.4.3. *Preposterior Analysis*

1. *Repetition of Terminal Analysis for all y.* The first step in preposterior analysis as defined by (3-18) is the repetition of terminal analysis as described in the previous section for every y in Y, and it is here that the use of conjugate prior distributions makes its greatest contribution to the feasibility of the analysis. A *single* terminal analysis can be carried out in almost any problem, but if the mathematical form of $P''_{\theta|y}$ depends on y, the task of repeating this analysis for every y in Y will be prohibitive unless Y is extremely small. If, however, the prior distribution is conjugate to the likelihood of y, so that the mathematical form of $P''_{\theta|y}$ is the same for all y and the *only* effect of a particular y is on the parameter $y'' = y' * y$, then formulas which give $u^*(e, y)$ for any y (and y'') give it for all y (and y'') which may result from the e in question or from any other e whatever.

Thus in our example, formulas (3-24) for $u^*(e, y, a_1)$ and $u^*(e, y, a_2)$ and condition (3-25) for determination of the optimal act hold unchanged for every possible $y'' = (m'', n'')$ and thus for every possible $y = (m, n)$.

2. *Computation of the Marginal Distribution of $\tilde{y}$.* The next step in preposterior analysis is the determination of the measure $P_{y|e}$ with respect to which the expectation called for by (3-18) is to be carried out. We again call attention to the fact that it is only at this point in analysis in extensive form that it becomes necessary to know anything about the *distribution* of $\tilde{y}$, since terminal analysis requires knowledge of only the *kernel* of the likelihood of y and (as we pointed out in Section 2.2.2) this kernel can be found without actually finding the distribution of $\tilde{y}$ or even knowing anything whatever about the e which produces y.

The marginal distribution of $\tilde{y}$ given e which is called for by (3-18) is usually most easily found by first determining the conditional distribution of $\tilde{y}$ given e *and* θ from the nature of the experiment e, the nature of the data-generating process with which we are dealing, and the definition of $\tilde{y}$ in terms of the random variables generated by the process. Then letting $D_c(\cdot|e, \theta)$ denote the mass or density function of this *conditional* distribution of $\tilde{y}$ and letting D' denote the prior density of $\tilde{\theta}$, we can obtain the *marginal* mass or density function D_m of $\tilde{y}$ for the given e,

$$D_m(y|e) = \int_\Theta D_c(y|e, \theta)\, D'(\theta)\, d\theta .$$

In Part III we have derived both the conditional and marginal distributions of $\tilde{y}$ for one or more kinds of experiments on each of the data-generating processes listed in Section 3.2.5 above; thus for the Bernoulli process the functions are given both for the case where the number of trials n is predetermined, the number of successes $\tilde{r}$ being left to chance, and for the case where r is predetermined and $\tilde{n}$ is left to chance. These results are catalogued in Table 3.1 (page 75). The first and second columns name the data-generating process and the conjugate prior distribution as defined in Section 3.2.5. The third column shows the nature of the experiment by listing the components of the sufficient statistic with a tilde over those which are random variables, the others being predetermined by the experimental design. The next two columns give respectively the conditional and marginal distributions of the random components of $\tilde{y}$ with references to the sections in Part III where these distributions are discussed; the first reference in each case is to the section where the distribution is derived while the second is to the section where its properties are described and tables of the distribution are cited. The last column of the table will be explained in a moment.

If in our example the experiment consists of taking a predetermined number n of observations on the Normal process defined by (3-19), then as shown in Section 11.4.1 the *conditional* distribution, given $\theta = \mu$, of the random component $\tilde{m}$ of the sufficient statistic $\tilde{y} = (\tilde{m}, n)$ has the density

$$D_c(m|n, \mu) = f_N(m|\mu, hn) \propto e^{-\frac{1}{2}hn(m-\mu)^2} .$$

The *unconditional* density of $(\tilde{m}, n)$ is then, as shown in Section 11.4.2,

$$D_m(m|n) = \int_{-\infty}^{\infty} f_N(m|\mu, hn)\, f_N(\mu|m', hn')\, d\mu = f_N(m|m', hn_u) \qquad \text{(3-26a)}$$

where

$$\frac{1}{n_u} = \frac{1}{n'} + \frac{1}{n} . \qquad \text{(3-26b)}$$

3. *Computation of the Expected Utility of an Experiment.* The third step in preposterior analysis as defined by (3-18) is the use of the marginal distribution of $\tilde{y}$ to compute

$$u^*(e) = E_{y|e}\, u^*(e, \tilde{y}) \qquad \text{where} \qquad u^*(e, y) = \max_a u^*(e, y, a) .$$

When explicit formulas for all $u^*(e, y, a)$ have been found in terminal analysis, it will sometimes be possible to find a single explicit formula for $u^*(e, y)$; in other cases it will be necessary to express $u^*(e, y)$ by a whole set of formulas for $u^*(e, y, a)$, each of which gives $u^*(e, y)$ for those y for which the a in question is optimal. Thus in our example $u^*(e, y)$ is given by (3-24a) for all $y = (m, n)$ such that condition (3-25) for the optimality of a_1 is met; it is given by (3-24b) for all other y.

4. *The Distribution of $\tilde{y}''$.* Because $u^*(e, y)$ is obtained by expectation with respect to the posterior distribution of $\tilde{\theta}$,

$$u^*(e, y) = \max_a E''_{\theta|y}\, u(e, y, a, \tilde{\theta}) ,$$

a *formula* ϕ for $u^*(e, y)$ obtained by carrying out this expectation analytically will in general have as an argument the posterior parameter $y'' = y' * y$ as well as the statistic y itself,

$$u^*(e, y) = \phi[y, y''(y)] ;$$

and in many cases (of which our example is one) the formula will involve *only* y''.

Table 3.1

Distributions of Statistics and Posterior Parameters

Process	Prior Distribution	Experiment	Distribution of Statistic: Conditional	Distribution of Statistic: Marginal	Distribution of Posterior Parameter
Bernoulli	beta	$\tilde{r}\|n$	binomial (9.2.2; 7.1)	beta-binomial (9.2.3; 7.11)	—
		$\tilde{n}\|r$	Pascal (9.3.2; 7.2)	beta-Pascal (9.3.3; 7.11)	—
Poisson	gamma-1	$\tilde{r}\|t$	Poisson (10.3.2; 7.5)	negative-binomial (10.3.3; 7.10)	—
		$\tilde{t}\|r$	gamma-1 (10.2.2; 7.6.2)	inverted-beta-2 (10.2.3; 7.4.2)	—
Normal	h known; $\tilde{\mu}$ Normal	$\tilde{m}\|n$	Normal (11.4.1; 7.8.2)	Normal (11.4.2; 7.8.2)	Normal (11.4.3; 7.8.2)
	μ known; $\tilde{h}$ gamma-2	$\tilde{w}\|\nu$	gamma-2 (11.2.1; 7.6.4)	inverted-beta-2 (11.2.2; 7.4.2)	inverted-beta-1 (11.2.3; 7.4.1)
	Normal-gamma	$\tilde{m}, \tilde{v}\|n, \nu$	*See* 11.6.1	11.6.2, 11.6.3	11.7.1, 11.7.2
Multinormal	h known; $\tilde{\boldsymbol{\mu}}$ Normal	$\tilde{\mathbf{m}}\|\mathbf{n}$	Normal (12.2.1; 8.2)	Normal (12.2.2; 8.2)	Normal (12.3.1; 8.2)
	Normal-gamma	$\tilde{\mathbf{m}}, \tilde{v}\|\mathbf{n}, \nu$	*See* 12.5.1	12.5.2, 12.5.3	12.6.1, 12.6.2
Regression	h known, $\tilde{\boldsymbol{\beta}}$ Normal	$\tilde{\mathbf{b}}\|\mathbf{n}$	Normal (13.3.1; 8.2)	Normal (13.3.2; 8.2)	Normal (13.4.1; 8.2)
	Normal-gamma	$\tilde{\mathbf{b}}, \tilde{v}\|\mathbf{n}, \nu$	*See* 13.6.2	13.6.3	13.7.1, 13.7.2

In order to compute

$$u^*(e) = \mathrm{E}_{y|e}\, u^*(e, \tilde{y})$$

we can of course substitute $y' * y$ for y'' in the formula $\phi[y, y''(y)]$ as originally derived and then sum or integrate over y,

$$u^*(e) = \sum \phi(y, y' * y)\, \mathrm{D}_m(y|e) \qquad \text{or} \qquad \int \phi(y, y' * y)\, \mathrm{D}_m(y|e)\, dy \;;$$

but when the original ϕ does not even contain y explicitly, it will be easier to proceed by first deriving the distribution of the random variable $\tilde{y}'' = y' * \tilde{y}$ and then summing or integrating over y'',

$$u^*(e) = \sum \phi(y'')\, \mathrm{D}(y''|e) \qquad \text{or} \qquad \int \phi(y'')\, \mathrm{D}(y''|e)\, dy'' \;.$$

Thus in our example we can derive the distribution of $\tilde{y}'' = (\tilde{m}'', n'')$ by first solving (3-22b) to obtain

$$m = \frac{(n' + n)\, m'' - n'm'}{n}$$

and substituting this result in the formula for the unconditional density (3-26) of $\tilde{y} = (\tilde{m}, n)$ to obtain

$$\mathrm{D}(m''|n) = f_N(m''|m', hn^*) \;, \qquad \frac{1}{n^*} = \frac{1}{n'} - \frac{1}{n''} \;.$$

Then letting m''_c denote the "critical value" of m'' which corresponds to exact equality in the optimality condition (3-25) we have by (3-24) for the experiment e_n which consists of taking n observations

$$u^*(e_n) = \int_{-\infty}^{m''_c} -\exp\,(k_s n - K_1)\, f_N(m''|m', hn^*)\, dm''$$

$$+ \int_{m''_c}^{\infty} -\exp\left(k_s n - K_2 - k_2 m'' + \frac{1}{2hn''}\, k_2^2\right) f_N(m''|m', hn^*)\, dm'' \;.$$

By use of (3-23) the integrals can easily be evaluated explicitly in terms of the tabulated cumulative unit-Normal function.

In Part III we have derived the distributions of $\tilde{y}''$ for most of the experiments listed in Table 3.1 (page 75); the last column of the table names the distribution and gives references, first to the derivation and then to the discussion of properties. The experiments for which no distribution of $\tilde{y}''$ is given are those where the * operation is simple addition, i.e. where $y'' = y' + y$, and it is therefore a trivial matter to reexpress a formula $\phi(y, y'')$ for $u^*(e, y)$ in terms of y alone.

PART TWO

EXTENSIVE-FORM ANALYSIS WHEN SAMPLING AND TERMINAL UTILITIES ARE ADDITIVE

CHAPTER 4

Additive Utility, Opportunity Loss, and the Value of Information: Introduction to Part II

4.1. Basic Assumptions

In the first chapter of this monograph we described the extensive form of analysis and explained how it can be applied numerically in problems where the spaces E, Z, A, and Θ are all discrete and reasonably small. In Chapter 3 we showed how even when these spaces are very large or infinite the extensive form can often be applied analytically provided that two conditions are met:

1. The experimental outcome z can be described by a sufficient statistic y of fixed dimensionality;
2. The prior measure P'_θ on Θ is conjugate to the conditional measures $P_{z|e,\theta}$ on the sample space Z.

In Part II of this monograph we devote our attention to the development of special versions of the extensive form which greatly simplify the analysis of certain problems in which both the above conditions are met and in addition

3. The utility function $u(\cdot, \cdot, \cdot, \cdot)$ on $E \times Z \times A \times \Theta$ can be expressed as the sum of a function $u_s(\cdot, \cdot)$ on $E \times Z$ and a function $u_t(\cdot, \cdot)$ on $A \times \Theta$:

$$u(e, z, a, \theta) = u_s(e, z) + u_t(a, \theta) \ , \qquad \text{all } e, z, a, \theta \ ; \qquad (4\text{-}1)$$

the reader can observe that our analysis will hold with only trivial modifications if the function $u_s(\cdot, \cdot)$ on $E \times Z$ is generalized to a function on $E \times Z \times \Theta$. In the present chapter, which is merely an introduction to Chapters 5 and 6, we examine certain implications of this new assumption concerning the utility function without reference to assumptions (1) and (2) concerning the probability measures on $\Theta \times Z$.

4.2. Applicability of Additive Utilities

The assumption that utility can be decomposed according to (4-1) will be valid at least as a very good approximation in a wide variety of practical decision problems; we shall first explain its rationale in problems in which all consequences are purely monetary, but we shall then show that the assumption will hold in a great many problems where the consequences are only partially measurable in money or not measurable in money at all.

In the great majority of decision problems in which the consequence of every possible (e, z, a, θ) can be described completely enough for practical purposes by the net amount of money which the decision maker will receive or pay out, this net amount will be expressible as the algebraic *sum* of (1) the cash flow due to performing an experiment e which results in the outcome z, and (2) the cash flow due to taking action a when the prevailing state is θ. If in addition the *utility of money* to the decision maker is *linear* over the whole range of cash flows which are possible in the decision problem at hand, we can set the utility of any cash flow numerically equal to the cash flow itself. The terminal utility $u_t(a, \theta)$ in (4-1) will then be either the profit or the negative of the cost of the terminal act while the sampling utility $u_s(e, z)$ will be the negative of the cost of sampling.

As we have already said, measurability in money is by no means a necessary condition for utility to be decomposable according to (4-1). The decomposition will be possible whenever (1) the consequence of (e, z) and the consequence of (a, θ) are measurable in *some* common unit or *numéraire* such that the "total" consequence of (e, z, a, θ) is the sum of the two "partial" consequences and (2) the utility of this *numéraire* to the decision maker is linear over the whole range of consequences involved in the problem at hand. Thus the director of a clinic who must ultimately decide whether or not to adopt a new drug in place of an old one may well feel that the consequence of either terminal act is measured almost entirely by the number of patients cured as a result; and if so, then he will probably also feel that at least the most important part of the "cost" of further experimentation with the new drug is measured by its effect on the number cured among the patients involved in the experiment. If "number cured" does thus constitute a common *numéraire*, it may well be that the *utility* of this *numéraire* to the decision maker is at least close to linear over a fairly wide range.

Even when the possible consequences of an act are complex and cannot be completely *described* in terms of any single *numéraire*, monetary or other, the most effective way of assigning utilities may nevertheless be to start by *scaling* the actual consequences in terms of a single *numéraire;* utilities can then be assigned to the consequences indirectly, via a utility function assigned directly to the *numéraire*. Thus in an industrial setting, an act may under certain circumstances result not only in a certain cash payment but also in a serious administrative annoyance which has no "objective" monetary equivalent; but if the decision maker feels that he would be willing to pay about D dollars cash to avoid this annoyance, he can subtract D from the purely monetary consequence of the act under each state θ under which the annoyance will result and then assign utilities to the "adjusted" monetary consequences via his utility function for money. Evaluation of the utilities by this indirect procedure will probably be much easier than evaluation by direct consideration of a large number of hypothetical gambles all involving mixed monetary and nonmonetary consequences.

Similarly in our drug example, the new drug may differ from the old not only as regards the *numéraire* "patients cured" but also as regards certain unpleasant or dangerous side effects. The most effective way of handling this complication, however, may well be to start by scaling these side effects in terms of the cure rate,

Letting θ_o denote the known fraction cured by the old drug and $\tilde{\theta}$ the unknown fraction cured by the new drug, the director can ask himself what choice he would make if he *did* know the true value of $\tilde{\theta}$. If he decides that because of its side effects he would not adopt the new drug unless $\theta - \theta_o \geq c$, say, and if N is the expected number of patients affected by the decision, the consequence of accepting the new drug is measured by $(\theta - c)\,N$, the consequence of continuing to use the old drug is measured by $\theta_o N$ as before, and the "adjustment" of the consequence of adopting the new drug should leave the director's utility as linear as it was before.

In many cases it will be possible to find a common *numéraire* even though the natural descriptions of the consequences of (e, z) have nothing at all in common with the natural descriptions of the consequences of (a, θ). Thus a scientist who wishes to estimate some physical constant θ may feel that whatever the error $(a - \theta)$ in his estimate a may be, he would be willing to make $10k$ more observations if he could be sure that by so doing he would reduce $(a - \theta)^2$ by k units. If so, the *consequence* of any (e, z, a, θ) can be measured by the sum of the actual number of observations plus $10(a - \theta)^2$; and if given any $k > 0$ the scientist would also be willing to make $10k$ more observations in order to reduce the *expected* value of $(\tilde{a} - \tilde{\theta})^2$ by k units, then the *utility* of (e, z, a, θ) can be expressed as $u_s(e, z) + u_t(a, \theta)$ where $u_s(e, z)$ is the negative of the number of actual observations and $u_t(a, \theta)$ is the negative of $10(a - \theta)^2$.

4.3. Computation of Expected Utility

When the utility of any (e, z, a, θ) can be expressed as the sum of a sampling utility $u_s(e, z)$ and a terminal utility $u_t(a, \theta)$, the expected utility of an experiment e can be expressed as the sum of the expected utility of sampling and the expected terminal utility. For by (1-3) and (4-1)

$$\begin{aligned} u^*(e) &\equiv \mathrm{E}_{z|e} \max_a \mathrm{E}''_{\theta|z}[u_s(e, \tilde{z}) + u_t(a, \tilde{\theta})] \\ &= \mathrm{E}_{z|e}[u_s(e, \tilde{z}) + \max_a \mathrm{E}''_{\theta|z}\, u_t(a, \tilde{\theta})] \ , \end{aligned} \tag{4-2}$$

and we may therefore write

$$u^*(e) = u_s^*(e) + u_t^*(e) \tag{4-3a}$$

where

$$u_s^*(e) \equiv E_{z|e}\, u_s(e, \tilde{z}) \ , \tag{4-3b}$$

$$u_t^*(e) \equiv E_{z|e} \max_a \mathrm{E}''_{\theta|z}\, u_t(a, \tilde{\theta}) \ . \tag{4-3c}$$

This decomposition very materially simplifies the computation of $u^*(e)$.

Computation of the expected sampling utility $u_s^*(e)$ is usually very easy, since $u_s(e, \tilde{z})$ is usually either a known number independent of $\tilde{z}$ or a very simple function of $\tilde{z}$ whose expected value is easy to compute once the (marginal) distribution of $\tilde{z}$ has been determined; the required distributions are given in Part III and indexed in Table 3.1 (page 75) for the data-generating processes and conjugate prior distributions listed in Section 3.2.5.

Computation of the expected terminal utility $u_t^*(e)$ will of course require the successive steps of terminal and preposterior analysis defined in Section 3.4.1, but even so the work is much easier when we have a function of only a and $\tilde{\theta}$ to deal with rather than a function which depends on e and $\tilde{z}$ as well. First, the posterior expected terminal utility for *given* (e, z),

$$\max_a \text{E}''_{\theta|z}\, u_t(a, \tilde{\theta}) \;,$$

will depend on (e, z) only through the measure $\text{P}''_{\theta|z}$ and not through the utility function itself; and if we also assume that the mathematical form of $\text{P}''_{\theta|z}$ is fixed because P'_θ is conjugate to the likelihood of a sufficient statistic $\tilde{y}(z)$, the expected terminal utility will depend on (e, z) only through the parameter $y'' = y' * y$. Since the expected utility of terminal action *without* experimentation will be merely the special case $y'' = y'$, we can omit all reference to (e, z) in *terminal analysis* as defined by (3-17) and discuss simply the evaluation of $\max_a \text{E}_\theta\, u_t(a, \tilde{\theta})$ without specifying whether the expectation is with respect to a "prior" or "posterior" measure. Second, we shall see in Chapters 5 and 6 that when we have a function of only a and $\tilde{\theta}$ to deal with we can often devise special methods which greatly reduce the burden of *preposterior* analysis as defined by (3-18).

Before taking up these special methods, however, we shall digress briefly in the remainder of this chapter to define and discuss the concepts of *opportunity loss* and *value of information*, both of which are useful and instructive in any problem where the utilities of terminal action and of sampling are separable and additive. Specifically, we shall show that when terminal and sampling utilities are additive (1) it is often much easier for the decision maker to *assess* the opportunity loss of every a in A given every θ in Θ than it is to assess the corresponding utilities, and (2) computation of expected opportunity loss often enables the statistician to find *upper bounds on optimal sample size* which greatly reduce the labor required to find the optimum. We shall then go on to isolate and examine the value of a particular piece of information concerning Θ, a concept whose usefulness as an analytical tool will become fully apparent in Chapter 5.

Because these concepts of opportunity loss and value of information involve nothing more abstruse than decomposition of conditional and expected utility into sums of economically meaningful parts, we suggest that in a first reading of Sections 4.4 and 4.5 attention be paid primarily to the economic sense of these decompositions rather than to their formal derivations; a summary of definitions and results will be found in Section 4.5.3.

4.4. Opportunity Loss

Dropping all special assumptions for the moment, we shall first give a definition of opportunity loss which applies in *any* decision problem and show how any decision problem can be analyzed in terms of opportunity loss rather than utility. We shall then show how this new concept is *useful* when terminal and sampling utilities are additive.

4.4.1. *Definition of Opportunity Loss*

Let e_0 denote the "null" experiment corresponding to immediate choice of a terminal act, let z_0 be the dummy outcome of e_0, and define a_θ by

$$u(e_0, z_0, a_\theta, \theta) \geq u(e_0, z_0, a, \theta) \ , \qquad \text{all } a \epsilon A \ . \tag{4-4}$$

In other words: *if* θ is known and *if* a terminal act is to be chosen immediately, then the decision maker's utility will be maximized by the choice of a_θ. If now instead of choosing a_θ without experimentation the decision maker performs e, observes z, chooses a, and thereby enjoys $u(e, z, a, \theta)$, we shall say that he has "lost the opportunity" of enjoying $u(e_0, z_0, a_\theta, \theta)$ and has thereby suffered an *opportunity loss* amounting to

$$l(e, z, a, \theta) \equiv u(e_0, z_0, a_\theta, \theta) - u(e, z, a, \theta) \ . \tag{4-5}$$

In most decision problems $u(e_0, z_0, a_\theta, \theta)$ is at least as great as any possible $u(e, z, a, \theta)$ and our use of the word "loss" to denote the quantity defined by (4-5) reflects this fact, but the use which we shall make of opportunity loss does not depend on this inequality and situations do occur in which it is possible and even natural to define the spaces E and A and the spaces Z and Θ in such a way that (4-5) is negative for some (e, z, a, θ).

4.4.2. *Extensive-Form Analysis Using Opportunity Loss Instead of Utility*

Writing (4-5) in the form

$$u(e, z, a, \theta) = u(e_0, z_0, a_\theta, \theta) - l(e, z, a, \theta) \tag{4-6}$$

and observing that $u(e_0, z_0, a_\theta, \theta)$ is a function of θ alone and not of the decision variables e and a, we see that *maximization of expected utility is equivalent to minimization of expected opportunity loss.* Instead of labelling the end branches of the decision tree with their utilities $u(e, z, a, \theta)$ and then making choices of a and e which maximize expected utility as we work back down the tree (cf. Section 1.2.1), we can label the end branches with their opportunity losses $l(e, z, a, \theta)$ and work back making choices of a and e which minimize expected opportunity loss. Thus after e has been performed and z observed, the expected opportunity loss of any a in A is

$$l^*(e, z, a) = \text{E}''_{\theta|z}\, l(e, z, a, \tilde\theta) \ ;$$

since the decision maker is free to choose an a which minimizes this quantity, "the" opportunity loss of being at the "position" (e, z) is

$$l^*(e, z) = \min_a l^*(e, z, a) = \min_a \text{E}''_{\theta|z}\, l(e, z, a, \tilde\theta) \ ;$$

before z is known the expected value of this opportunity loss is

$$l^*(e) = \text{E}_{z|e}\, l^*(e, \tilde z) = E_{z|e} \min_a \text{E}''_{\theta|z}\, l(e, \tilde z, a, \tilde\theta) \ ; \tag{4-7}$$

and since the decision maker is free to choose an e which minimizes this quantity, his expected opportunity loss is

$$l^* = \min_e l^*(e) = \min_e \text{E}_{z|e} \min_a \text{E}''_{\theta|z}\, l(e, \tilde z, a, \tilde\theta) \ . \tag{4-8}$$

To see the relationship between *expected* opportunity loss and *expected* utility, we first define the quantity

$$U \equiv \mathrm{E}'_\theta\, u(e_0, z_0, \tilde{a}_\theta, \tilde{\theta}) \tag{4-9}$$

and observe that this is equivalent to

$$U = \mathrm{E}_{z|e}\, \mathrm{E}''_{\theta|z}\, u(e_0, z_0, \tilde{a}_\theta, \tilde{\theta})$$

because $\mathrm{E}_{z|e}\, \mathrm{E}''_{\theta|z} \equiv \mathrm{E}_{\theta,z|e}$ and $\mathrm{E}_{\theta,z|e}$ is equivalent to E'_θ when the function being expected does not depend on $\tilde{z}$. Then by (1-3), (4-6), and (4-7),

$$\begin{aligned} u^*(e) &\equiv E_{z|e} \max_a \mathrm{E}''_{\theta|z}\, u(e, \tilde{z}, a, \tilde{\theta}) = \mathrm{E}_{z|e} \max_a \mathrm{E}''_{\theta|z}[u(e_0, z_0, \tilde{a}_\theta, \tilde{\theta}) - l(e, \tilde{z}, a, \tilde{\theta})] \\ &= \mathrm{E}_{z|e}\, \mathrm{E}''_{\theta|z}\, u(e_0, z_0, \tilde{a}_\theta, \tilde{\theta}) - \mathrm{E}_{z|e} \min_a \mathrm{E}''_{\theta|z}\, l(e, \tilde{z}, a, \tilde{\theta}) \\ &= U - l^*(e)\ . \end{aligned} \tag{4-10}$$

The expected utility of any experiment is simply a constant less its expected opportunity loss.

4.4.3. Opportunity Loss When Terminal and Sampling Utilities Are Additive

We next show that when $u(e, z, a, \theta)$ is the sum of a part $u_t(a, \theta)$ due to terminal action and a part $u_s(e, z)$ due to sampling, opportunity loss can be similarly decomposed; but to facilitate the interpretation of our result we first define the *cost of sampling* to be the negative of the utility of sampling,

$$c_s(e, z) \equiv -u_s(e, z)\ , \tag{4-11}$$

so that (4-1) can be written in the form

$$u(e, z, a, \theta) = u_t(a, \theta) - c_s(e, z)\ ; \tag{4-12}$$

and we specify that utility shall be measured on a scale such that

$$c_s(e_0, z_0) = 0\ , \qquad \text{so that} \qquad u(e_0, z_0, a, \theta) = u_t(a, \theta)\ . \tag{4-13}$$

Now substituting (4-12) and (4-13) in the definition (4-5) of opportunity loss we obtain

$$l(e, z, a, \theta) = u_t(a_\theta, \theta) - u_t(a, \theta) + c_s(e, z)\ ;$$

and if we also define the *terminal opportunity loss*†

$$l_t(a, \theta) \equiv u_t(a_\theta, \theta) - u_t(a, \theta)\ , \tag{4-14}$$

we can write

$$l(e, z, a, \theta) = l_t(a, \theta) + c_s(e, z)\ . \tag{4-15}$$

In other words, we can regard the opportunity loss of (e, z, a, θ) as the sum of two parts: (1) the cost of experimenting instead of making an immediate terminal choice. and (2) the opportunity loss due to making a "wrong" terminal choice after the experiment has been conducted. Observe that substitution of (4-13) in (4-4) yields

$$u_t(a_\theta, \theta) \geq u_t(a, \theta)\ , \qquad \text{all } a \in A\ , \tag{4-16}$$

so that *terminal* opportunity loss as defined by (4-14) is *necessarily* nonnegative.

† The quantity $l_t(a, \theta)$ is called by some writers the decision maker's "regret" at having chosen a rather than a_θ. The word "loss" is used by some writers to denote the negative of the terminal utility $u_t(a, \theta)$, by others to denote $l_t(a, \theta)$, i.e. what we call terminal *opportunity* loss.

Whenever the opportunity loss of (e, z, a, θ) can be expressed as the sum of terminal opportunity loss and sampling cost, the *expected* opportunity loss of an experiment e can be expressed as the sum of the expected values of these quantities. For if we define the expected terminal opportunity loss† of e by

$$l_t^*(e) \equiv \mathrm{E}_{z|e} \min_a \mathrm{E}''_{\theta|z}\, l_t(a, \tilde{\theta}) \tag{4-17}$$

and the expected cost of sampling by

$$c_s^*(e) \equiv \mathrm{E}_{z|e}\, c_s(e, \tilde{z}) \ , \tag{4-18}$$

then substituting (4-15) in (4-7) we have

$$\begin{aligned} l^*(e) &\equiv \mathrm{E}_{z|e} \min_a \mathrm{E}''_{\theta|z}\, l(e, \tilde{z}, a, \tilde{\theta}) = \mathrm{E}_{z|e} \min_a \mathrm{E}''_{\theta|z}[l_t(a, \tilde{\theta}) + c_s(e, \tilde{z})] \\ &= \mathrm{E}_{z|e} \min_a \mathrm{E}''_{\theta|z}\, l_t(a, \tilde{\theta}) + \mathrm{E}_{z|e}\, c_s(e, \tilde{z}) \\ &\equiv l_t^*(e) + c_s^*(e) \ . \end{aligned} \tag{4-19}$$

We have seen in (4-3) that the expected *utility* of e can be similarly decomposed,

$$u^*(e) = u_t^*(e) + u_s^*(e) = u_t^*(e) - c_s^*(e) \ , \tag{4-20}$$

and as we might suspect there is a simple direct relation between $u_t^*(e)$ and $l_t^*(e)$. Substituting the right-hand sides of (4-19) and (4-20) in (4-10) we have

$$u_t^*(e) = U - l_t^*(e) \tag{4-21}$$

where by (4-9) and (4-13)

$$U = \mathrm{E}'_\theta\, u_t(\tilde{a}_\theta, \tilde{\theta}) \ . \tag{4-22}$$

An economic interpretation will be attached to U in Section 4.5.1 below.

4.4.4. *Direct Assessment of Terminal Opportunity Losses*

The *calculations* required to minimize the expected value of

$$l(e, \tilde{z}, a, \tilde{\theta}) = l_t(a, \tilde{\theta}) + c_s(e, \tilde{z}) \tag{4-23}$$

are in general neither more nor less difficult than the calculations required to maximize the expected value of

$$u(e, \tilde{z}, a, \tilde{\theta}) = u_t(a, \tilde{\theta}) - c_s(e, \tilde{z}) \ , \tag{4-24}$$

but in many situations analysis in terms of opportunity loss will have a very real advantage because the decision maker will find it much easier to *assess* $l_t(a, \theta)$ than to assess $u_t(a, \theta)$ for all (a, θ) in $A \times \Theta$. Thus in a problem of inventory control where the utility of money is linear, it may be very difficult to evaluate the net profit and hence the utility of stocking a units when there is a demand for θ units because the net profit depends *inter alia* on the costs of purchasing, receiving, processing accounts payable, and so forth, and it is usually very hard to trace the share of these costs which is really attributable to any particular order; but it may well be clear that these costs are virtually independent of the quantity ordered, and if so it will be relatively easy to compute for any θ the *difference* $l_t(a, \theta)$ between the net profit with an arbitrary stock a and the profit with the optimal stock $a_\theta = \theta$.

† Called the Bayes or average "risk" by some writers.

Observe that assessment of $l(e, z, a, \theta)$ will *not* be easier than assessment of $u(e, z, a, \theta)$ in any situation where these quantities cannot be decomposed as in (4-23) and (4-24).

4.4.5. Upper Bounds on Optimal Sample Size

Provided that utility and therefore opportunity loss *can* be decomposed as in (4-23) and (4-24), *analysis* in terms of expected opportunity loss will sometimes greatly facilitate the selection of an optimal e whether or not it is easier to *assess* the conditional opportunity losses $l_t(a, \theta)$ than to assess the corresponding utilities $u_t(a, \theta)$. The problems in which this may be true are those in which the experiments in E can be ordered in a sequence $\{e_n\}$ such that $u_t^*(e_n)$ and $c_s^*(e_n)$ are both increasing functions of n; and such problems are very common in the applications.

Our objective in such a problem will of course be to find the value n° of n which maximizes

$$u^*(e_n) = u_t^*(e_n) - c_s^*(e_n)$$

or equivalently minimizes

$$l^*(e_n) = l_t^*(e_n) + c_s^*(e_n) \ .$$

In a very few problems of this kind n° will be defined by an equation which can be solved either explicitly or by a simple iterative procedure—examples will be given later in this monograph—but more often n° can be found only by actually computing $u^*(e_n)$ or $l^*(e_n)$ for a number of values of n and thus tracing the behavior of $u^*(e_n)$ or $l^*(e_n)$ as a function of n. In such a situation, suppose that

$$l^*(e_{n'}) = l_t^*(e_{n'}) + c_s^*(e_{n'})$$

has been computed for some value n' of n. We know by (4-10) that the increase in utility which will result from using any n *larger* than n' is

$$\begin{aligned} u^*(e_n) - u^*(e_{n'}) &= l^*(e_{n'}) - l^*(e_n) \\ &= [l_t^*(e_{n'}) - l_t^*(e_n)] - [c_s^*(e_n) - c_s^*(e_{n'})] \ ; \end{aligned}$$

and since $l_t(a, \theta)$ is necessarily nonnegative by (4-16) and therefore $l_t^*(e_n)$ is necessarily nonnegative for all n, we may write

$$u^*(e_n) - u^*(e_{n'}) \leq l_t^*(e_{n'}) - [c_s^*(e_n) - c_s^*(e_{n'})] \ .$$

The left-hand side of this expression will certainly be negative if the right-hand side is negative, and therefore n cannot possibly be optimal unless it satisfies

$$c_s^*(e_n) - c_s^*(e_{n'}) \leq l_t^*(e_{n'}) \ . \tag{4-25}$$

In other words: *given any arbitrarily chosen n' with its associated sampling cost and terminal opportunity loss, it will never pay to* increase *the expenditure on sampling by more than the terminal opportunity loss of $e_{n'}$.* The *largest* n which satisfies (4-25) is an *upper bound* on the values of n for which we need to evaluate $u^*(e_n)$ or $l^*(e_n)$.

In practice, the best procedure will usually be to start the analysis of a decision problem by determining the upper bound corresponding to the terminal opportunity loss of the "null" experiment e_0. Since $c_s^*(e_0) = 0$ we obtain the constraint

$$c_s^*(e_n) \leq l_t^*(e_0) \tag{4-26}$$

on possibly optimal n without having to compute the term $c_s^*(e_{n'})$ which appears in (4-25); and since the distribution of $\tilde{\theta}$ "posterior" to e_0 is simply the "prior" distribution of $\tilde{\theta}$, the definition (4-17) of $l_t^*(e)$ gives us

$$l_t^*(e_0) = \min_a \mathrm{E}'_\theta \, l_t(a, \tilde{\theta}) \ ,$$

which is in general much easier to compute than

$$l_t^*(e_{n'}) = \mathrm{E}_{z|e_{n'}} \min_a \mathrm{E}''_{\theta|z} \, l_t(a, \tilde{\theta}) \ , \qquad n' > 0 \ .$$

Once an initial upper bound has been found by use of (4-26), the fact that even a very small experiment $e_{n'}$ often has a terminal opportunity loss $l_t^*(e_{n'})$ which is very much smaller than $l_t^*(e_0)$ implies that we can often obtain a much smaller upper bound by next evaluating (4-25) for some n' which is much closer to 0 than to the upper bound given by (4-26). Thus if we seek to locate n° by tracing the function $l_t^*(e_n)$, each computation not only serves to establish a point on the curve but may also bring a new reduction in the domain over which it has to be traced; and the computations are no more difficult than those required to trace the function $u_t^*(e_n)$ without any hope of discovering reduced upper bounds on n°. We remind the reader that if the decision maker originally assigns terminal utilities rather than terminal opportunity losses to the various possible (a, θ), we are *not* obliged to determine a_θ for every θ and then to compute $l_t(a, \theta) = u_t(a_\theta, \theta) - u_t(a, \theta)$ for every possible (a, θ) in order to compute $l_t^*(e)$. For by (4-21),

$$l_t^*(e) = U - u_t^*(e) \ ;$$

and by (4-22) $U \equiv \mathrm{E}'_\theta \, u_t(\tilde{a}_\theta, \tilde{\theta})$ is a constant which need be computed only once in the analysis of any decision problem.

4.5. The Value of Information

In the extensive form of analysis as we have used it hitherto, an experiment e has always been evaluated by computing the "absolute" utility $u^*(e, z)$ for every possible z and then taking a weighted average of these absolute utilities. When sampling and terminal utilities are additive an alternative procedure is available and we shall see in Chapter 5 that under certain circumstances this alternative procedure is very advantageous. For each z we can compute, not the absolute utility $u^*(e, z)$, but the *increase* in utility which would result if the decision maker learned that $\tilde{z} = z$ and therefore altered his prior choice of an act a; and we can then take a weighted average of these utility increases. The increase in utility which results or would result from learning that $\tilde{z} = z$ will be called the *value of the information z.*

4.5.1. The Value of Perfect Information

Given the additivity assumption, we may regard equation (4-16),

$$u_t(a_\theta, \theta) \geq u_t(a, \theta) \ , \qquad \text{all } a \in A \ ,$$

as the *definition* of a_θ: a_θ is an act, which, if chosen, would maximize the decision maker's *realized* terminal utility. Let us similarly define a' to be an act which maximizes the decision maker's *expected* utility under his prior distribution of $\tilde{\theta}$, satisfying

$$E'_\theta\, u_t(a', \tilde{\theta}) \geq E'_\theta\, u_t(a, \tilde{\theta}) \,, \qquad \text{all } a \in A \,. \tag{4-27}$$

Now let us imagine an ideal experiment e_∞ with a known cost $c_s^*(e_\infty)$ which is capable of yielding *exact* or *perfect information* concerning the true state θ, and let us suppose that the decision maker wishes to choose between e_∞ and the "null" experiment e_0 (i.e., immediate choice of a terminal act.) If he purchases the perfect information on θ, he will choose a_θ and his net utility will be

$$u(e_\infty, z_\infty, a_\theta, \theta) = u_t(a_\theta, \theta) - c_s^*(e_\infty) \,,$$

whereas if he acts without this information he will choose a' and his net utility will be

$$u(e_0, z_0, a', \theta) = u_t(a', \theta) - 0 \,.$$

The former of these two quantities will be greater than the latter if $c_s^*(e_\infty)$ is less than

$$v_t(e_\infty, \theta) \equiv l_t(a', \theta) = u_t(a_\theta, \theta) - u_t(a', \theta) \,, \tag{4-28}$$

and therefore we may regard $v_t(e_\infty, \theta) = l_t(a', \theta)$ as the *conditional value of perfect information* given a particular θ. We shall henceforth refer to this quantity as the CVPI of θ and regard it as defined by (4-28).

The idea of CVPI can be visualized by reference to Figure 4.1, where it is assumed that $A = \{a_1, a_2, a_3\}$. Assuming that a_2 is optimal under the decision maker's prior distribution, $a' = a_2$, perfect information that $\theta = \theta_*$ would lead the decision maker to choose a_1 rather than a' and thereby increase his utility by the amount $l_t(a_2, \theta_*)$ shown on the graph. Perfect information that $\theta = \theta^*$ would leave the decision maker's original choice of a_2 unchanged, and the value of the information would therefore be $0 = l_t(a_2, \theta^*)$.

The CVPI as defined by (4-28) can be evaluated only conditionally or after the fact; but before the fact we can compute its *expected* value, which we shall call the *expected value of perfect information* or EVPI: remembering that e_0 will lead to a' we have

$$v_t^*(e_\infty) \equiv E'_\theta\, v_t(e_\infty, \tilde{\theta}) = E'_\theta\, l_t(a', \tilde{\theta}) \equiv l_t^*(e_0) \equiv E'_\theta[u_t(\tilde{a}_\theta, \tilde{\theta}) - u_t(a', \tilde{\theta})] \,. \tag{4-29}$$

Observe that since the terminal utilities of action without information and of action with perfect information are respectively

$$u_t^*(e_0) \equiv \max_a E'_\theta\, u_t(a, \tilde{\theta}) = E'_\theta\, u_t(a', \tilde{\theta}) \tag{4-30}$$

and

$$u_t^*(e_\infty) \equiv E'_\theta \max_a u_t(a, \tilde{\theta}) = E'_\theta\, u_t(\tilde{a}_\theta, \tilde{\theta}) \,, \tag{4-31}$$

we may write

$$v_t^*(e_\infty) \equiv l_t^*(e_0) = u_t^*(e_\infty) - u_t^*(e_0) \,. \tag{4-32}$$

The quantity $u_t^*(e_\infty)$, which was given the designation U in (4-22), may be called the *prior expectation of the utility of terminal action posterior to receipt of perfect information.*

Graphically, the prior expected utility of any terminal act a_i, $i = 1, 2, 3$, can be interpreted as the weighted-average height of the corresponding utility curve in Figure 4.1, the weights being given by the prior measure P'_θ; the quantity $u_t^*(e_\infty)$ can be interpreted as the weighted-average height of the upper envelope of these lines; the prior expected opportunity loss of any a_i is given by either the difference

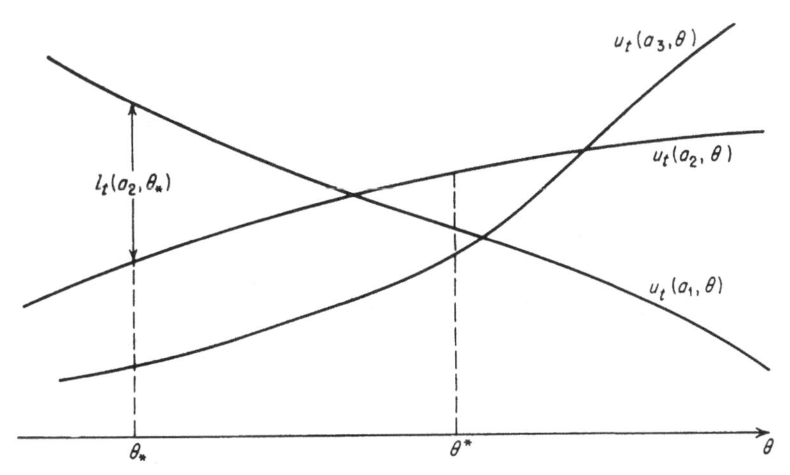

Figure 4.1

CVPI When $a' = a_2$

between the average height of the envelope and the average height of the ith curve or by the weighted average of the local differences in height; and a' is an act which minimizes this difference of averages or average difference.

4.5.2. The Value of Sample Information and the Net Gain of Sampling

We now go on to show that we can regard the value of the information to be obtained from a *real* experiment e in a way which is closely analogous to the way in which we have just looked at the value of *perfect* information obtained from the ideal experiment e_∞.

In (4-27) we defined a' to be an act which is optimal under the decision maker's *prior* distribution,

$$E'_\theta\, u_t(a', \tilde\theta) = \max_a E'_\theta\, u_t(a, \tilde\theta) \;, \qquad \text{all } a \in A \;;$$

we now define a_z to be an act which is optimal under the *posterior* distribution determined by the outcome z of some real experiment e,

$$E''_{\theta|z}\, u_t(a_z, \tilde\theta) = \max_a E''_{\theta|z}\, u_t(a, \tilde\theta) \;, \qquad \text{all } a \in A \;. \tag{4-33}$$

Now if instead of choosing a' without experimentation the decision maker performs e, observes z, and then chooses a_z, he increases his *terminal* utility as evaluated after the fact by

$$v_t(e, z) \equiv E''_{\theta|z}\, u_t(a_z, \tilde\theta) - E''_{\theta|z}\, u_t(a', \tilde\theta) \;; \tag{4-34}$$

we shall call this the *conditional value of the sample information* z or CVSI of z.

The conditional value of sample information can be given a graphical representation nearly identical to the representation of the value of perfect information in Figure 4.1. In Figure 4.2 we show for each act in $A = \{a_1, a_2, a_3\}$ the *posterior expected* terminal utility $E''_{\theta|z}\, u_t(a, \tilde\theta)$ as a function of the experimental outcome z. If the optimal act under the *prior* distribution would be $a' = a_2$ but if instead of

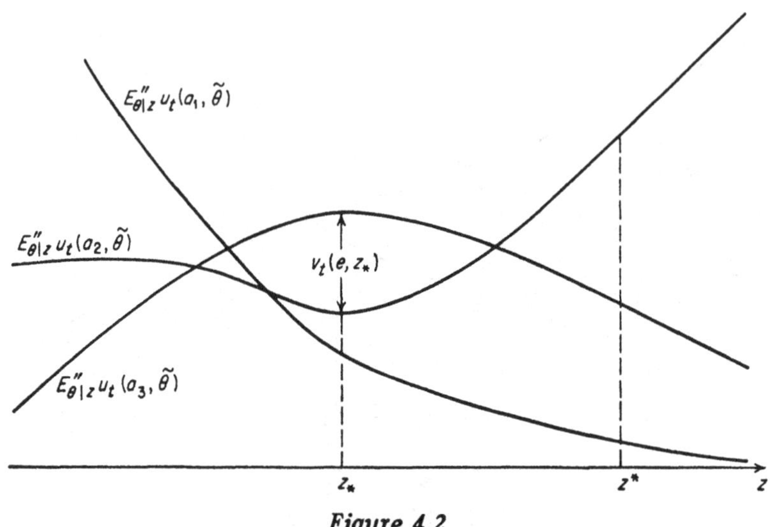

Figure 4.2

CVSI When $a' = a_2$

choosing a' immediately the decision maker performs e, observes z_*, and chooses a_3, he has increased his terminal utility by the amount $v_t(e, z_*)$ shown on the graph. If the experimental outcome were z^*, his original choice of a_2 would be left unchanged and the value of the information would have been 0.

The CVSI as defined by (4-34) can be evaluated only conditionally on z or after z is known; but before z is known the decision maker can compute the *expected value of sample information* or EVSI of any given e,

$$v_t^*(e) \equiv E_{z|e}\, v_t(e, \tilde z) \ . \tag{4-35}$$

The economic significance of this quantity is due to the fact that *the expected terminal utility of any experiment is the expected utility of immediate terminal action augmented by the expected value of the sample information:*

$$u_t^*(e) = u_t^*(e_0) + v_t^*(e) \ . \tag{4-36}$$

Graphically, $u_t^*(e)$ can be interpreted as the weighted-average height of the upper envelope of the curves of posterior expected utility in Figure 4.2, the weights being given by the marginal measure $P_{z|e}$; the expected terminal utility $u_t^*(e_0)$ of terminal action *without* experimentation is the weighted-average height of the curve corresponding to a'; and the expected value of sample information is the difference between these two averages or the weighted average of the local differences between the heights of the two curves. Formally, we have by (4-3c), (4-33), (4-34), (4-29), (4-30), and (4-35)

$$u_t^*(e) \equiv \mathrm{E}_{z|e} \max_a \mathrm{E}''_{\theta|z}\, u_t(a, \tilde\theta) = \mathrm{E}_{z|e}\, \mathrm{E}''_{\theta|z}\, u_t(\tilde a_z, \tilde\theta)$$
$$= \mathrm{E}_{z|e}[\mathrm{E}''_{\theta|z}\, u_t(a', \tilde\theta) + v_t(e, \tilde z)] = u_t^*(e_0) + v_t^*(e) \ ,$$

as asserted by (4-36).

Similarly it is apparent and can easily be proved that *the expected terminal opportunity loss of any experiment e is the expected opportunity loss of immediate terminal action diminished by the expected value of the sample information:*

$$l_t^*(e) = l_t^*(e_0) - v_t^*(e) = v_t^*(e_\infty) - v_t^*(e) \ . \tag{4-37}$$

The *expected net gain of sampling* or ENGS of a particular experiment e is now naturally defined to be the expected value of the sample information less the expected cost of obtaining it,

$$v^*(e) \equiv v_t^*(e) - c_s^*(e) \ . \tag{4-38}$$

It follows by (4-36) that the *net* utility of a decision to perform any experiment is

$$u^*(e) = u_t^*(e) - c_s^*(e) = u_t^*(e_0) + v^*(e) = u^*(e_0) + v^*(e) \tag{4-39}$$

and thus that the *general decision problem may be viewed as maximization of net gain* rather than maximization of utility or minimization of opportunity loss.

4.5.3. Summary of Relations Among Utilities, Opportunity Losses, and Value of Information

With No Special Assumptions

Def. $u(e, z, a, \theta)$: utility of performing e, observing z, and choosing a, when θ is true.

Def. $u^*(e)$: expected utility of experiment e;

$$u^*(e) \equiv \mathrm{E}_{z|e} \max_a \mathrm{E}''_{\theta|z}\, u(e, \tilde z, a, \tilde\theta) \ .$$

Def. e_0: null experiment, i.e. a terminal act is chosen without experimentation.

Def. z_0: dummy outcome of e_0.

Def. a_θ: an optimal act given θ;

$$u(e_0, z_0, a_\theta, \theta) \geq u(e_0, z_0, a, \theta) \ , \qquad \text{all } a \in A \ .$$

Def. $l(e, z, a, \theta)$: opportunity loss of (e, z, a, θ);

$$l(e, z, a, \theta) \equiv u(e_0, z_0, a_\theta, \theta) - u(e, z, a, \theta) \ .$$

Def. $l^*(e)$: expected opportunity loss of e;

$$l^*(e) \equiv \mathrm{E}_{z|e} \min_a \mathrm{E}''_{\theta|z}\, l(e, \tilde z, a, \tilde\theta) \ .$$

Def. U: $U \equiv \mathrm{E}'_\theta\, u(e_0, z_0, \tilde a_\theta, \tilde\theta)$.

Result: $u^*(e) = U - l^*(e)$.

When Terminal and Sampling Utilities are Additive

$$u(e, z, a, \theta) = u_t(a, \theta) - c_s(e, z) \ ; \qquad c_s(e_0, z_0) = 0 \ .$$

Def. $l_t(a, \theta)$: terminal opportunity loss of (a, θ);

$$l_t(a, \theta) \equiv u_t(a_\theta, \theta) - u_t(a, \theta) \ .$$

Result: $l(e, z, a, \theta) = l_t(a, \theta) + c_s(e, z)$.

Def. $l_t^*(e)$: prior expectation of terminal opportunity loss posterior to e;

$$l_t^*(e) \equiv E_{z|e} \min_a E''_{\theta|z}\, l_t(a, \tilde\theta) \ .$$

Def. $c_s^*(e)$: expected cost of experiment e;

$$c_s^*(e) \equiv E_{z|e}\, c_s(e, \tilde z) \ .$$

Result: $l^*(e) = l_t^*(e) + c_s^*(e)$.

Def. e_∞: ideal experiment which would result in exact knowledge of the true state θ.

Def. $u_t^*(e_\infty)$: $u_t^*(e_\infty) \equiv E'_\theta \max_a u_t(a, \tilde\theta) = E'_\theta\, u_t(\tilde a_\theta, \tilde\theta) = U$.

Result: $u_t^*(e) = u_t^*(e_\infty) - l_t^*(e)$.

Def. e_{n°: optimal experiment.

Result: $c_s^*(e_{n^\circ}) - c_s^*(e_n) \leq l_t^*(e_n)$, all n .

Def. a': optimal act under prior distribution;

$$E'_\theta\, u_t(a', \tilde\theta) \geq E'_\theta\, u_t(a, \tilde\theta) \ , \qquad \text{all } a \in A \ .$$

Def. $v_t(e_\infty, \theta)$: conditional value of perfect information (CVPI) given θ:

$$v_t(e_\infty, \theta) \equiv u_t(a_\theta, \theta) - u_t(a', \theta) \ .$$

Result: $v_t(e_\infty, \theta) = l_t(a', \theta)$.

Def. $v_t^*(e_\infty)$: expected value of perfect information (EVPI);

$$v_t^*(e_\infty) \equiv E'_\theta\, l_t(a', \tilde\theta) \equiv l_t^*(e_0) \ .$$

Def. $v_t(e, z)$: conditional value of sample information (CVSI) given z;

$$v_t(e, z) \equiv \max_a E''_{\theta|z}\, u_t(a, \tilde\theta) - E''_{\theta|z}\, u_t(a', \tilde\theta) \ .$$

Def. $v_t^*(e)$: expected value of sample information (EVSI) given e;

$$v_t^*(e) \equiv E_{z|e}\, v_t(e, \tilde z) \ .$$

Result:

$$u_t^*(e) = u_t^*(e_0) + v_t^*(e) \ ,$$

$$l_t^*(e) = l_t^*(e_0) - v_t^*(e) \ .$$

Def. $v^*(e)$: expected net gain of sampling (ENGS) of e;

$$v^*(e) = v_t^*(e) - c_s^*(e) \ .$$

Result:

$$u^*(e) = u_t^*(e_0) + v^*(e) \ ,$$

$$l^*(e) = l_t^*(e_0) - v^*(e) \ .$$

CHAPTER 5A

Linear Terminal Analysis

5.1. Introduction

In many situations where the utility of (e, z, a, θ) can be expressed as the sum of a utility due to (e, z) plus a utility due to (a, θ), it will also be true that $u_t(a, \cdot)$ considered as a function of θ will be *linear in θ for every a in A*. Thus if a choice must be made between acceptance (a_1) and rejection (a_2) of a lot of purchased parts, the utility of a_1 may be of the form $K_1 - k_1\theta$ where θ is the unknown fraction defective while the utility of a_2 may be some fixed amount K_2 independent of θ. Or if a choice must be made between continuing to use an old drug with known cure rate θ_o and adopting a new drug with unknown cure rate θ and undesirable side effects, the utilities of the two acts may, as we saw in Section 4.2, be respectively $N\theta_o$ and $N(\theta - c)$.

In other situations u_t may be a linear function, not of the parameter θ in terms of which it is customary to describe the state, but of some function of θ. Thus suppose that a choice must be made between an old production process which generates a known fraction defective θ_o and a new production process which generates an unknown fraction θ, and suppose further that defectives are scrapped at a total loss. If the cost of manufacturing one piece is k, if N good pieces must be manufactured, and if the utility of money is linear, the utilities of the old and the new processes are respectively $\left(-\frac{kN}{1-\theta_o}\right)$ and $\left(-\frac{kN}{1-\theta}\right)$ and the latter is not linear in θ; but if we characterize the two processes, not by the ratio θ of defectives to total pieces, but by the ratio $\rho = \frac{1}{1-\theta}$ of total pieces to good pieces, then the utilities of both acts are obviously linear in ρ.

5.1.1. *The Transformed State Description ω*

In this chapter we shall derive special methods of analysis for problems of the kind illustrated by these examples, i.e., problems where the terminal utility *of every a in A* can be expressed as a linear function of either the customary state description θ or some simple function W of θ. Formally, we shall consider problems in which

There exists a mapping W from the state space Θ into a new space Ω, carrying θ into $\omega = W(\theta)$, such that

$$u_t(a, \theta) = K_a + k_a\,\omega\ , \qquad \text{all } a \in A\ , \qquad \text{all } \theta \in \Theta\ . \tag{5-1}$$

In general, k_a and ω are respectively a $1 \times q$ matrix and a $q \times 1$ vector, although some special analytical results will be given in this chapter for the case where k_a and ω are scalar. K_a is of course always scalar.

Nuisance Parameters. We have seen in Section 2.2.4 that when the state θ is vector valued, $\theta = (\theta_1, \cdots, \theta_r)$, some of the components of θ may be *nuisance parameters* in the sense that they are irrelevant to the utility of any (e, z, a, θ) and enter the problem only as parameters of the conditional probability measures $P_{z|e,\theta}$. Given the assumption of additive utilities, a nuisance parameter is a component of θ which is irrelevant to $u_t(a, \theta)$ and enters the problem only as a parameter of $P_{z|e,\theta}$.

The presence of nuisance parameters does not really create any special problem as regards the linearity of terminal utility, since if $u_t(a, \theta)$ is linear in a component θ_1 and independent of θ_2 it is obviously linear in θ. We can however simplify notation and comprehension by always defining the transformed state description ω as free of nuisance components. Thus if utility is linear in the unknown mean $\theta_1 = \mu$ of an Independent Normal process but independent of its unknown precision $\theta_2 = h = 1/\tilde{\sigma}^2$, we shall define $\omega = \mu$ just as we would define $\omega = \mu^2$ if utility were linear in μ^2 but independent of h.

5.1.2. Terminal Analysis

When Θ is mapped onto Ω, any measure P_θ on Θ—prior *or* posterior—induces a corresponding measure P_ω on Ω. Using the same symbol u_t to denote the terminal utility function on $A \times \Omega$ that we used to denote the corresponding function on $A \times \Theta$, we have for the *expected* terminal utility of any act

$$E_\omega\, u_t(a, \tilde{\omega}) = u_t(a, \bar{\omega}) = K_a + k_a\bar{\omega} \tag{5-2}$$

where

$$\bar{\omega} \equiv E_\omega(\tilde{\omega}) \; . \tag{5-3}$$

When terminal utility is linear in ω, the mean of the distribution of $\tilde{\omega}$ is sufficient for evaluation of and choice among terminal acts. In Figure 5.1 we give a geometric interpretation of this result for the case where Ω is the real line and $A = \{a_1, a_2, a_3\}$.

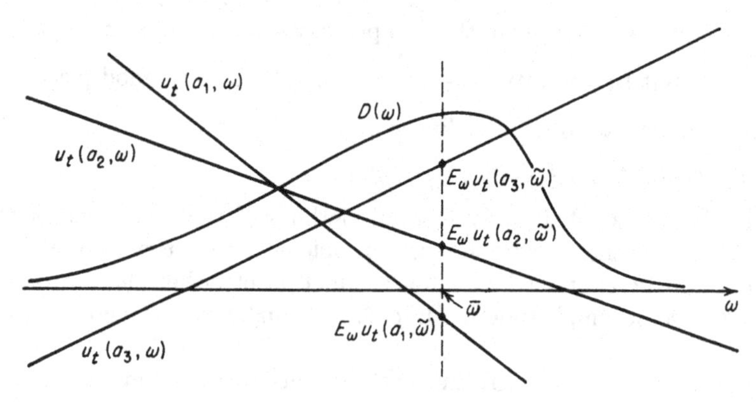

Figure 5.1

Expectation of Linear Terminal Utilities

Terminal analysis under the assumptions of this chapter is thus completely trivial once the appropriate distribution of $\tilde{\omega}$—prior or posterior—has been obtained, and in many practical applications the required distribution is very easy to obtain. Very often ω will be the "natural" process parameter θ itself or the nonnuisance component of θ; and even when the distribution of $\tilde{\omega}$ has to be derived from the distribution of $\tilde{\theta}$, the derivation is often very easy. Two examples of such derived distributions have been worked out by way of example in Part III of this monograph:

1. If a beta distribution is assigned to the parameter $\tilde{\theta} = \tilde{p}$ which describes the mean number of successes per trial generated by a Bernoulli process, then $\tilde{\omega} = 1/\tilde{p}$ or $\tilde{\omega} = 1/(1 - \tilde{p})$ has a very tractable inverted-beta-1 distribution (Sections 9.1.4 and 7.4.1).
2. If a gamma distribution is assigned to the parameter $\tilde{\theta} = \tilde{\lambda}$ which measures the mean number of events per unit of time generated by a Poisson process, then $\tilde{\omega} = 1/\tilde{\lambda}$ has a very tractable inverted-gamma-1 distribution (Sections 10.1.4 and 7.7.1).

When ω is not identical to θ there is of course no real advantage in finding the distribution of $\tilde{\omega}$ if all that we want to do is choose among terminal acts—it would usually be just as easy to expect the formula for the terminal utility in terms of θ over the distribution of $\tilde{\theta}$ itself. The advantage of the linearization lies in the simplification it makes in the computation of the EVPI and in preposterior analysis, as we shall now see.

5.2. Expected Value of Perfect Information When ω is Scalar

When a decision problem is to be analyzed in terms of the random variable $\tilde{\omega}$, let a_ω denote an act which is optimal for given ω, satisfying

$$u_t(a_\omega, \omega) \geq u_t(a, \omega) \ , \qquad \text{all } a \in A \ ; \tag{5-4}$$

define the terminal opportunity loss of any a to be

$$l_t(a, \omega) \equiv u_t(a_\omega, \omega) - u_t(a, \omega) \ ; \tag{5-5}$$

define a' to be an act which is optimal under the *prior* distribution of $\tilde{\omega}$, satisfying

$$\text{E}'_\omega\, u_t(a', \tilde{\omega}) \geq \text{E}'_\omega\, u_t(a, \tilde{\omega}) \ , \qquad \text{all } a \in A \ ; \tag{5-6}$$

define the conditional value of perfect information or CVPI to be

$$v_t(e_\infty, \omega) \equiv l_t(a', \omega) = u_t(a_\omega, \omega) - u_t(a', \omega) \ ; \tag{5-7}$$

and define the expected value of perfect information or EVPI to be

$$v_t^*(e_\infty) \equiv l_t^*(e_0) = \text{E}'_\omega\, l_t(a', \tilde{\omega}) \ . \tag{5-8}$$

For a discussion of the *interpretation* of these quantities, see Sections 4.4 and 4.5 above; we now turn our attention to their *evaluation* in the case where the terminal utility of every a in A is linear in a *scalar* ω.

5.2.1. *Two-Action Problems*

Assume that $A = \{a_1, a_2\}$ and that Ω is the real line, write

$$u_t(a_i, \omega) = K_i + k_i\, \omega \ , \qquad i = 1, 2 \ , \tag{5-9}$$

and assume that the acts have been labelled so that $k_1 < k_2$. (We exclude the trivial case $k_1 = k_2$ in which at least one of the two acts is optimal for *all* ω.) The two utility functions are graphed in Figure 5.2, and it is apparent that in any

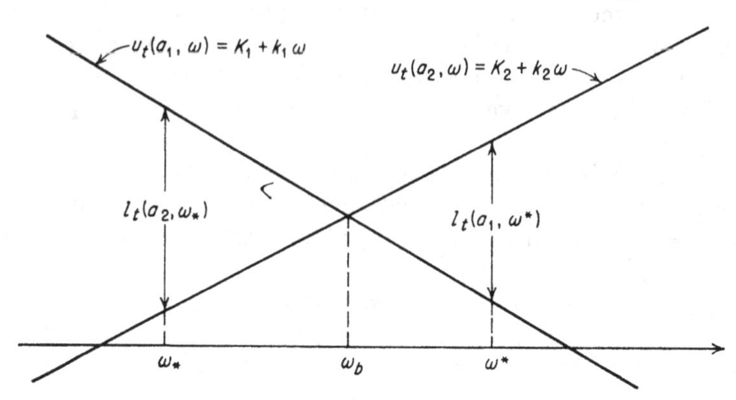

Figure 5.2

Terminal Utilities and Opportunity Losses: A = $\{a_1, a_2\}$

problem of this sort there will exist a *breakeven value* ω_b satisfying

$$K_1 + k_1\,\omega_b = K_2 + k_2\,\omega_b \tag{5-10}$$

and that the optimal acts *for given* ω are

$$a_\omega = \begin{cases} a_1 & \text{if} \quad \omega \leq \omega_b \ , \\ a_2 & \text{if} \quad \omega \geq \omega_b \ . \end{cases}$$

Recalling that the linearity of (5-9) implies that

$$\text{E}_\omega\, u_t(a_i, \tilde\omega) = u_t(a_i, \bar\omega) = K_i + k_i\,\bar\omega \ ,$$

we see that the optimal act under the *prior* distribution of $\tilde\omega$ is

$$a' = \begin{cases} a_1 & \text{if} \quad \bar\omega' \leq \omega_b \ , \\ a_2 & \text{if} \quad \bar\omega' \geq \omega_b \ . \end{cases}$$

The conditional value of perfect information is therefore

$$\text{CVPI} = \begin{cases} l_t(a_1, \omega) = |k_2 - k_1| \max\{\omega - \omega_b, 0\} & \text{if} \quad \bar\omega' \leq \omega_b \ , \\ l_t(a_2, \omega) = |k_2 - k_1| \max\{\omega_b - \omega, 0\} & \text{if} \quad \bar\omega' \geq \omega_b \ ; \end{cases}$$

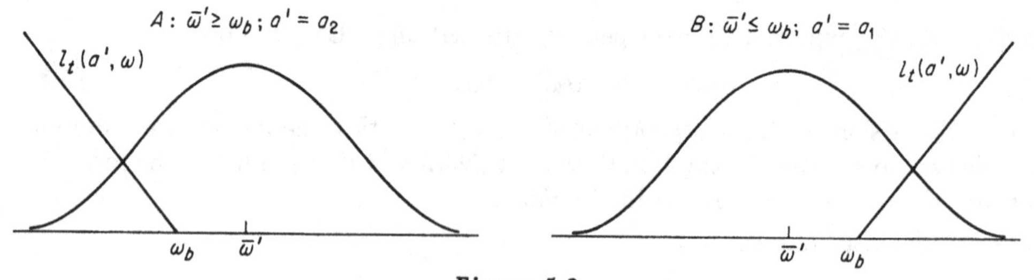

Figure 5.3

Conditional Value of Perfect Information: A = $\{a_1, a_2\}$

the CVPI function is graphed in Figure 5.3. It is immediately apparent that the *expected* value of perfect information or EVPI is

$$v_t^*(e_\infty) \equiv l_t^*(e_0) = \begin{cases} \mathrm{E}'_\omega\, l_t(a_1, \tilde{\omega}) = |k_2 - k_1| \displaystyle\int_{\omega_b}^{\infty} (\omega - \omega_b)\, \mathrm{D}'(\omega)\, d\omega & \text{if} \quad \bar{\omega}' \leq \omega_b\,, \\ \mathrm{E}'_\omega\, l_t(a_2, \tilde{\omega}) = |k_2 - k_1| \displaystyle\int_{-\infty}^{\omega_b} (\omega_b - \omega)\, \mathrm{D}'(\omega)\, d\omega & \text{if} \quad \bar{\omega}' \geq \omega_b\,. \end{cases} \tag{5-11}$$

If now we define the left- and right-hand *linear-loss integrals* under the distribution of any random variable $\tilde{x}$, discrete or continuous, by

$$\begin{aligned} L_x^{(l)}(a) &\equiv \int_{-\infty}^{a} (a - x)\, d\mathrm{P}_x\,, \\ L_x^{(r)}(a) &\equiv \int_{a}^{\infty} (x - a)\, d\mathrm{P}_x\,, \end{aligned} \tag{5-12}$$

we can write for the EVPI in our present problem

$$v_t^*(e_\infty) \equiv l_t^*(e_0) = \begin{cases} |k_2 - k_1|\, L_\omega^{(r)}(\omega_b) & \text{if} \quad \bar{\omega}' \leq \omega_b\,, \\ |k_2 - k_1|\, L_\omega^{(l)}(\omega_b) & \text{if} \quad \bar{\omega}' \geq \omega_b\,. \end{cases} \tag{5-13}$$

Notice that formulas (5-13) give the expected *terminal opportunity losses* of acts a_1 and a_2 *whether or not* the acts in question are optimal under uncertainty.

5.2.2. Finite-Action Problems

Assume as before that Ω is the real line and that $u_t(a_i, \omega) = K_i + k_i\omega$ but assume now that $A = \{a_1, a_2, \cdots, a_s\}$ where s may be any finite number, and assume that the acts have been labelled so that $k_1 < k_2 < \cdots < k_s$. We may also assume without loss of generality that every a in A is optimal for *some* ω, since the fact that $\mathrm{E}_\omega u_t(a, \tilde{\omega}) = u_t(a, \bar{\omega})$ implies that an act like $a^\square$ in Figure 5.4

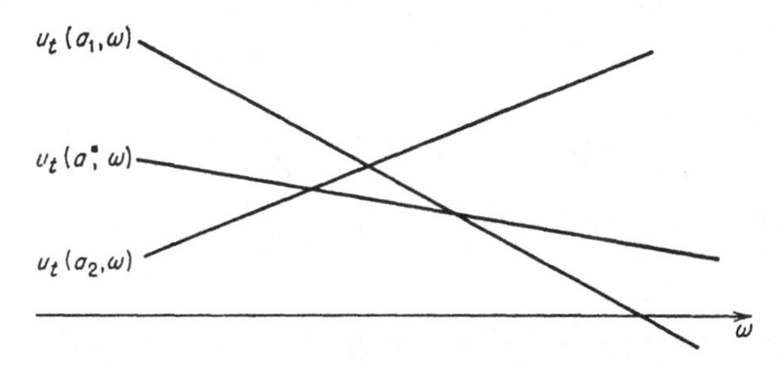

Figure 5.4

An Inadmissible Act

which is not optimal for *any* ω is not optimal for any P_ω, prior or posterior, and would therefore never be chosen in any case.† It follows that the utilities of the acts in A will appear as in Figure 5.5 and that there will exist $s - 1$ "breakeven values" $\omega_1, \omega_2, \cdots, \omega_{s-1}$ such that a_1 is optimal for ω in the interval $(-\infty, \omega_1]$,

† Notice also that if three or more lines meet in a point, all but two of the corresponding acts are inadmissible.

a_2 is optimal for ω in $[\omega_1, \omega_2]$, $\cdots$, a_s is optimal for ω in $[\omega_{s-1}, \infty)$. If we define $\omega_0 = -\infty$ and $\omega_s = \infty$, we may say more briefly that a_p is *optimal for* ω if and only if $\omega_{p-1} \leq \omega \leq \omega_p$; and the linearity of u_t then implies that a_q is *optimal under uncertainty* (prior or posterior) if and only if $\omega_{q-1} \leq \bar{\omega} \leq \omega_q$.

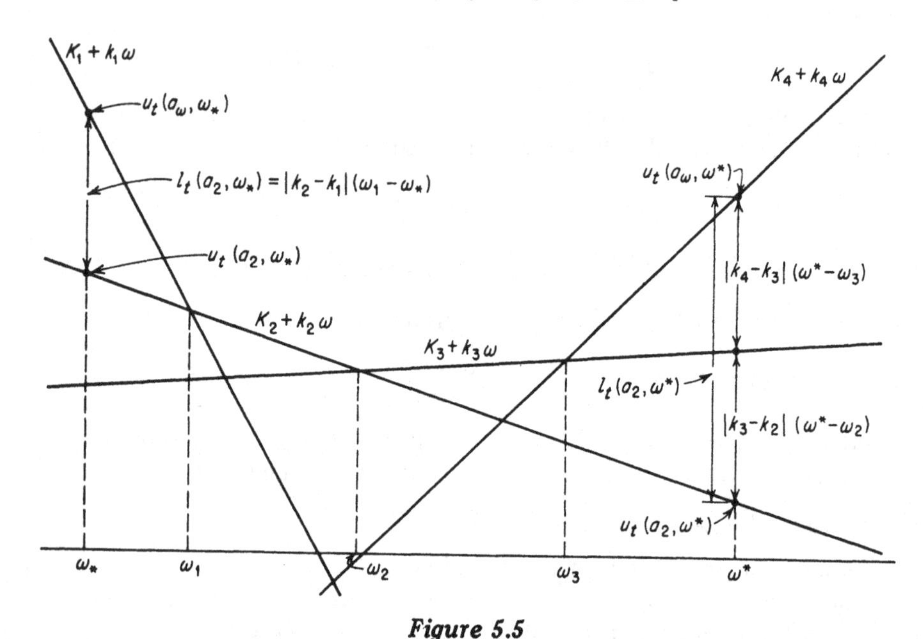

Figure 5.5

Terminal Utilities and Opportunity Losses: A = {a₁, a₂, a₃, a₄}

On these assumptions, the *prior* expected terminal opportunity loss of *any* act a_p—the EVPI if a_p is *optimal* under the prior distribution of $\tilde{\omega}$—can easily be shown to be

$$E'_\omega\, l_t(a_p, \tilde{\omega}) = \sum_{i=1}^{p-1} |k_{i+1} - k_i| L_\omega^{(l)}(\omega_i) + \sum_{i=p}^{s-1} |k_{i+1} - k_i| L_\omega^{(r)}(\omega_i) \ , \quad \text{(5-14)}$$

$L^{(l)}$ and $L^{(r)}$ being defined by (5-12).

▶ Since the basic idea of the proof is extremely simple but the proof is hard to present formally because of notational complexities, we shall merely show that (5-14) is valid for a_2 in the problem illustrated by Figure 5.5. It is clear from the geometry of Figure 5.5 that in this case the *conditional* terminal opportunity loss of a_2 (the CVPI if a_2 is optimal under the prior distribution of $\tilde{\omega}$) is

$$l_t(a_2, \omega) = \begin{cases} |k_2 - k_1|(\omega_1 - \omega) & \text{if} \quad -\infty < \omega \leq \omega_1 \ , \\ 0 & \text{if} \quad \omega_1 \leq \omega \leq \omega_2 \ , \\ |k_3 - k_2|(\omega - \omega_2) & \text{if} \quad \omega_2 \leq \omega \leq \omega_3 \ , \\ |k_3 - k_2|(\omega - \omega_2) + |k_4 - k_3|(\omega - \omega_3) & \text{if} \quad \omega_3 \leq \omega < \infty \ . \end{cases}$$

Expecting these conditional values we obtain the expected terminal opportunity loss of a_2 (the EVPI if a_2 is optimal under the prior distribution of $\tilde{\omega}$):

$$\mathrm{E}'_\omega l_t(a_2, \tilde\omega) = \int_{-\infty}^{\omega_1} |k_2 - k_1|(\omega_1 - \omega)\, \mathrm{D}'(\omega)\, d\omega$$
$$+ \int_{\omega_2}^{\infty} |k_3 - k_2|(\omega - \omega_2)\, \mathrm{D}'(\omega)\, d\omega + \int_{\omega_3}^{\infty} |k_4 - k_3|(\omega - \omega_3)\, \mathrm{D}'(\omega)\, d\omega \;;$$

and comparing the definitions (5-12) of the linear loss integrals we see that this result can be written

$$\mathrm{E}'_\omega\, l_t(a_2, \tilde\omega) = |k_2 - k_1|\, L_\omega^{(l)}(\omega_1) + |k_3 - k_2|\, L_\omega^{(r)}(\omega_2) + |k_4 - k_3|\, L_\omega^{(r)}(\omega_3)$$

in agreement with (5-14) for $s = 4$, $p = 2$. ◀

5.2.3. Evaluation of Linear-Loss Integrals

The linear-loss integrals which occur in formulas (5-13) and (5-14) for the EVPI can often be evaluated in terms of well tabulated functions when $\tilde\omega$ is a simple function of $\tilde\theta$ and $\tilde\theta$ has one of the conjugate distributions listed in Section 3.2.5. In Table 5.1 we index the cases which have been worked out in Part III of this monograph.

Table 5.1

Linear-Loss Integrals

Process	"Natural" Parameter θ	Distribution of $\tilde\theta$	ω	Loss Integrals in Terms of	Reference Section
Bernoulli	p	beta	$\{p,\ \rho = 1/p\}$	cumulative beta or binomial function	{9.1.3, 9.1.4}
Poisson	λ	gamma-1	$\{\lambda,\ \mu = 1/\lambda\}$	cumulative gamma or Poisson function	{10.1.3, 10.1.4}
Normal	μ, h	{h known, $\tilde\mu$ Normal}	μ	tabulated function L_{N*}	11.3.2
		{Normal-gamma}	μ	Student density and cumulative functions	11.5.5
Multinormal	$\boldsymbol{\mu}, h$	—	$\Sigma c_i \mu_i$	same as univariate	8.2.3, 8.3.2
Regression	$\boldsymbol{\beta}, h$	—	$\Sigma c_i \beta_i$	same as Multinormal	8.2.3, 8.3.2

5.2.4. Examples

There follow four examples intended to illustrate the calculation of EVPI in concrete situations. Because problems of quality control present the essential features of decision problems with a minimum of extraneous complexities, three of the four examples are from that area; one example from outside this area is given to remind the reader that identically the same features will often be present in totally different functional applications.

Example 1. Pieces are produced in lots of size N by a machine which behaves as a Bernoulli process, the parameter p being the fraction defective. After a lot is produced, it can either be accepted (a_1), meaning that all pieces are sent directly to the assembly department, or screened 100% (a_2), meaning that all defectives are removed and replaced by good pieces from an inspected stock. It costs S to screen the lot plus an amount c_s for each defective replaced; it costs $c_a(> c_s)$ to remove and replace a defective which reaches assembly. The utility of money is linear, and therefore we may adopt a scale such that the terminal utilities are

$$u_t(a_1, p) = -c_a N p \ , \qquad u_t(a_2, p) = -c_s N p - S \ .$$

The breakeven value defined by (5-10), which we shall call π for typographical convenience, is

$$\pi \equiv p_b = \frac{S}{(c_a - c_s)N} \ ;$$

a_1 is optimal if $p \leq \pi$, a_2 if $p \geq \pi$; and we define

$$k \equiv |k_2 - k_1| = |c_a - c_s|N \ .$$

If a beta prior distribution of type (9-9) with parameter (r', n') and therefore with mean $\bar{p}' = r'/n'$ is assigned to $\tilde{p}$, the expected value of perfect information or expected opportunity loss of immediate terminal action is, by (5-13),

$$v_t^*(e_\infty) \equiv l_t^*(e_0) = \begin{cases} kL_p^{(l)}(\pi) & \text{if} \quad \bar{p}' \geq \pi \ , \\ kL_p^{(r)}(\pi) & \text{if} \quad \bar{p}' \leq \pi \ . \end{cases}$$

The linear-loss integrals in these formulas are given by (9-12) and (7-23) as

$$L_p^{(l)}(\pi) = \pi \, I_\pi(r', n' - r') - \bar{p}' I_\pi(r' + 1, n' - r') \ ,$$
$$L_p^{(r)}(\pi) = \bar{p}' I_{1-\pi}(n' - r', r' + 1) - \pi \, I_{1-\pi}(n' - r', r') \ ,$$

where I is the "incomplete beta function" tabulated by Pearson (cf. Section 7.3.1).

Example 1a. Same as Example 1 except that a tumbling operation costing T per lot can be expected to eliminate one fifth of the defectives. There are thus *four* possible acts:

a_1: accept as is,
$a^\square$: tumble and then accept,
a_2: screen as is,
a_3: tumble and then screen;

the reason for the peculiar numbering will appear in a moment. Assuming that

$$N = 1000 \ , \qquad S = 20 \ , \qquad T = 10 \ , \qquad c_a = 1.0 \ , \qquad c_s = .4 \ ,$$

we have for the terminal utilities

$$\begin{aligned} u_t(a_1, p) &= -c_a N p = -1000\, p \ , \\ u_t(a^\square, p) &= -T - c_a N(.8p) = -10 - 800\, p \ , \\ u_t(a_2, p) &= -S - c_s N p = -20 - 400\, p \ , \\ u_t(a_3, p) &= -T - S - c_s N(.8p) = -30 - 320\, p \ . \end{aligned}$$

These utilities are graphed in Figure 5.6, where it is apparent that $a^\square$ can be immediately rejected because there is *no* p for which it is optimal; cf. Section 5.2.2 and Figure 5.4. The breakeven values for the three "admissible" acts are

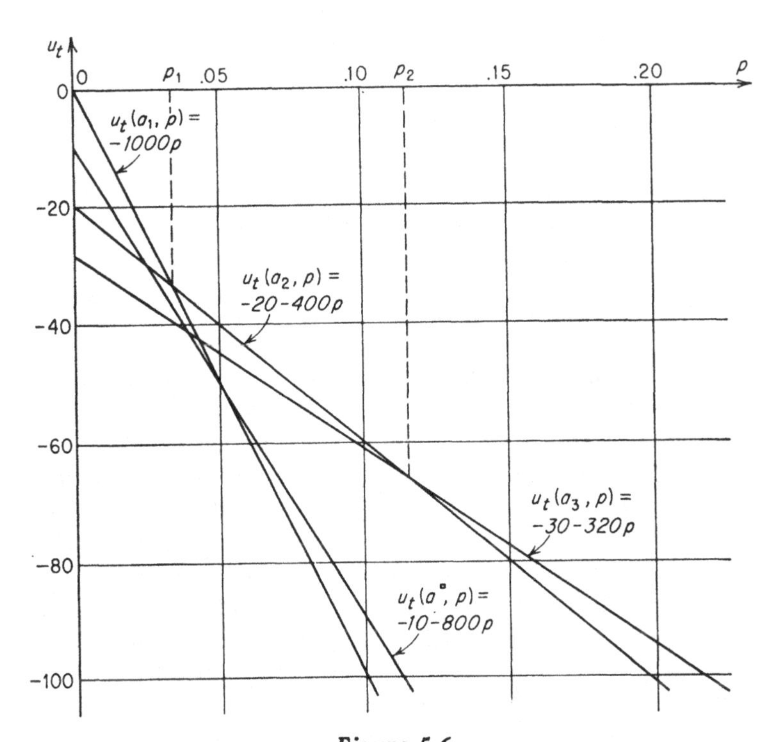

Figure 5.6

Terminal Utilities in Example 1a

$$p_1 = \frac{S}{(c_a - c_s)N} = \frac{20}{600} = .033 \; , \qquad p_2 = \frac{T}{.2c_sN} = \frac{10}{80} = .125 \; ;$$

and

$$|k_2 - k_1| = |-c_sN + c_aN| = 600 \; , \qquad |k_3 - k_2| = |-.8c_sN + c_sN| = 80 \; .$$

By (5-14) we have for the EVPI

$$v_t^*(e_\infty) \equiv l_t^*(e_0) = \begin{cases} 600\, L_p^{(r)}(.033) + 80\, L_p^{(r)}(.125) & \text{if} \quad 0 \leq \bar{p}' \leq .033 \; , \\ 600\, L_p^{(l)}(.033) + 80\, L_p^{(r)}(.125) & \text{if} \quad .033 \leq \bar{p}' \leq .125 \; , \\ 600\, L_p^{(l)}(.033) + 80\, L_p^{(l)}(.125) & \text{if} \quad .125 \leq \bar{p}' \leq 1 \; ; \end{cases}$$

the linear-loss integrals $L_p^{(l)}$ and $L_p^{(r)}$ are given in terms of Pearson's beta function by the formulas at the end of Example 1.

Example 2. A drug which has been in use for a long time is known to cure a fraction π of patients treated; very limited experience with a new drug leads the director of the only clinic investigating the drug to assign to its cure rate $\tilde{p}$ a beta distribution of type (9-9) with parameter (r', n'); its mean $\bar{p}' = r'/n'$ is less than π and the director is therefore thinking of stopping further use of the new drug immediately, but if he does he will be unable to obtain any further information concerning its true effectiveness. Perfect information concerning the new drug *could* reveal that $\tilde{p} > \pi$ and lead the director to adopt it in place of the old drug. Letting

a_1 denote a terminal decision in favor of the old drug and a_2 a terminal decision in favor of the new drug, assume that for the reasons discussed in Section 4.2 the director assigns utilities

$$u_t(a_1, p) = N\pi \ , \qquad u_t(a_2, p) = Np \ ,$$

where N is the expected number of patients who would be affected by an immediate terminal decision. Because $\bar{p}' < \pi$, the optimal act under the prior distribution of $\tilde{p}$ is a_1 and the EVPI is given by the first formula in (5-13):

$$v_t^*(e_\infty) \equiv l_t^*(e_0) = N\, L_p^{(r)}(\pi) = N[\bar{p}' I_{1-\pi}(n' - r', r' + 1) - \pi\, I_{1-\pi}(n' - r', r')] \ .$$

In other words: if in fact $p > \pi$, discarding the new drug at this point can be expected to reduce the number of cures by $N(p - \pi)$; this loss is of course conditional on the unknown p, but in the director's best judgment the *unconditional* expected value of the potential loss is $l_t^*(e_0)$ as given by the formula just above.

Example 3. The standard method of producing a certain part behaves as a Bernoulli process with known fraction defective p_o. Some experiments with a proposed new method lead to the assignment of a beta prior distribution of type (9-9) with parameter (r', n') to its unknown fraction defective $\tilde{p}$. Manufacture of one piece, good or defective, costs c_1 with the old process, c_2 with the new process; N *good* pieces must be manufactured by one process or the other; and the utility of money is linear. If we define $\rho = 1/(1 - p)$, the terminal utilities of the acts "retain old process" (a_1) and "adopt new process" (a_2) can be written

$$u_t(a_1, \rho) = -c_1N/(1 - p_o) = -c_1N\rho_o \ ,$$
$$u_t(a_2, \rho) = -c_2N/(1 - p) = -c_2N\rho \ ;$$

the breakeven value is

$$\rho_b = c_1\rho_o/c_2 \ ;$$

a_1 is optimal if $\rho \geq \rho_b$, a_2 if $\rho \leq \rho_b$; and we define

$$k \equiv |-c_2N - 0| = c_2N \ .$$

The EVPI or expected opportunity loss of immediate terminal action is

$$v_t^*(e_\infty) \equiv l_t^*(e_0) = \begin{cases} k\, L_\rho^{(l)}(\rho_b) & \text{if} \quad \bar{\rho}' \geq \rho_b \ , \\ k\, L_\rho^{(r)}(\rho_b) & \text{if} \quad \bar{\rho}' \leq \rho_b \ , \end{cases}$$

where as shown in Section 9.1.4

$$\bar{\rho}' = \frac{n' - 1}{r' - 1} \ .$$

If we define

$$\pi = 1/\rho_b \ ,$$

then by (9-16), (7-18), and (7-19) the linear-loss integrals can be evaluated in terms of Pearson's incomplete beta function I:

$$L_\rho^{(l)}(\rho_b) = \rho_b\, I_{1-\pi}(n' - r', r') - \bar{\rho}' I_{1-\pi}(n' - r', r' - 1) \ ,$$
$$L_\rho^{(r)}(\rho_b) = \bar{\rho}' I_\pi(r' - 1, n' - r') - \rho_b\, I_{1-\pi}(r', n' - r') \ .$$

Example 4. One batch of raw material yields a variable amount x of a certain chemical and the product sells for an amount sx. The standard method of production has a mean yield μ_o per batch. It is believed that this yield can be increased by preliminary beneficiation of the raw material; the required equipment costs B and operation costs b per batch processed. The exact degree to which beneficiation will increase the yield is uncertain, however; the responsible manager assigns to the mean yield $\tilde{\mu}$ of the beneficiated batches a Normal prior distribution with mean $\bar{\mu}'$ and variance $\breve{\mu}'$. He wishes to choose between a_1, immediate purchase of the equipment, and a_2, postponement of a terminal decision for one year, after which he is sure that accurate information on the effect of beneficiation will be available. The number of batches to be processed during this year is N and the utility of money is linear, so that the terminal utilities can be written

$$u_t(a_1, \mu) = Ns\mu - Nb - B \ , \qquad\qquad u_t(a_2, \mu) = Ns\mu_o \ .$$

The breakeven value is

$$\mu_b = \mu_o + \frac{Nb + B}{Ns} \ ;$$

a_1 is optimal if $\mu \geq \mu_b$, a_2 if $\mu \leq \mu_b$; and we define

$$k = |k_2 - k_1| = |0 - Ns| = Ns \ .$$

The risk involved in an immediate choice of a terminal act is measured by the EVPI, which is given by (5-13) as

$$v_t^*(e_\infty) \equiv l_t^*(e_0) = \begin{cases} k\, L_\mu^{(l)}(\mu_b) & \text{if} \quad \bar{\mu}' \geq \mu_b \ , \\ k\, L_\mu^{(r)}(\mu_b) & \text{if} \quad \bar{\mu}' \leq \mu_b \ . \end{cases}$$

In order to evaluate the linear-loss integrals by use of (11-25), we first write the decision maker's prior distribution of $\tilde{\mu}$ in the notation of (11-21):

$$\mathrm{D}'(\mu) = f_N(\mu|m', hn') \propto e^{-\frac{1}{2}hn'(\mu - m')^2}$$

where in the present context h is simply an arbitrary positive constant and

$$m' \equiv \bar{\mu}' \ , \qquad n' \equiv \frac{1}{h\breve{\mu}'} \ .$$

We then have by (11-25)

$$v_t^*(e_\infty) \equiv l_t^*(e_0) = k\, L_{N^*}(u)/\sqrt{hn'} \ , \qquad u = |\mu_b - m'|\sqrt{hn'} \ ;$$

the function L_{N^*} is tabulated in Table II at the end of this monograph.

5.3. Preposterior Analysis

Our assumption that sampling and terminal utilities are additive implies, as we saw in (4-3), that the *expected* utility of an experiment can also be broken down into a sum of the expected utility of sampling and the expected terminal utility,

$$u^*(e) = u_s^*(e) + u_t^*(e)$$

where (in terms now of ω rather than θ)

$$u_t^*(e) = E_{z|e} \max_a E''_{\omega|z}\, u_t(a, \tilde{\omega}) \ . \tag{5-15}$$

We now turn our attention to evaluation of $u_t^*(e)$ when $u_t(a, \cdot)$ is linear in ω.

5.3.1. *The Posterior Mean as a Random Variable*

The assumption that $u_t(a, \cdot)$ is linear in (either a scalar or vector-valued) ω implies that the posterior expected terminal utility of any act depends only on the mean of the posterior distribution,

$$E''_{\omega|z}\, u_t(a, \tilde{\omega}) = u_t(a, \bar{\omega}''_z) \qquad \text{where} \qquad \bar{\omega}''_z \equiv E''_{\omega|z}(\tilde{\omega}) \ , \tag{5-16}$$

so that "the" expected terminal utility after observing z is $\max_a u_t(a, \bar{\omega}''_z)$ and is thus a function of $\bar{\omega}''_z$ alone. We shall call this function ψ, defining

$$\psi(\bar{\omega}''_z) \equiv \max_a u_t(a, \bar{\omega}''_z) = \max_a (K_a + k_a\, \bar{\omega}''_z) \ . \tag{5-17}$$

To perform the preposterior analysis of any e as defined by (5-15) we must evaluate

$$u_t^*(e) = E_{z|e}\, \psi(\tilde{\bar{\omega}}''_z) \ . \tag{5-18}$$

Now formula (5-17) for ψ is in terms of $\bar{\omega}''_z$ rather than z and would have to be reexpressed in terms of z before we could evaluate (5-18) by direct use of the measure $P_{z|e}$. It is often much simpler to turn the problem around and proceed much as we did in paragraph 4 of Section 3.4.3 when we had to find the expectation of $\phi[y''(\tilde{y})]$ with respect to $P_{y|e}$: we can treat the posterior mean *explicitly* as a random variable $\tilde{\bar{\omega}}''_z$, obtain its distribution $P_{\bar{\omega}''|e}$ from the distribution $P_{z|e}$ of $\tilde{z}$, and then compute

$$u_t^*(e) = E_{z|e}\, \psi(\tilde{\bar{\omega}}''_z) = E_{\bar{\omega}''|e}\, \psi(\tilde{\bar{\omega}}'') \ , \tag{5-19}$$

where we drop the subscript from $\tilde{\bar{\omega}}''_z$ in the last member because $\tilde{z}$ plays no role once $P_{\bar{\omega}''|e}$ has been determined.

The distribution $P_{\bar{\omega}''|e}$ will be called the *prior distribution of the posterior mean;* it is also called the *preposterous* distribution. The theory of this distribution will be discussed in Section 5.4, after we have completed our discussion of the basic economics of problems with linear terminal utility in Section 5.3.2.

5.3.2. *The Expected Value of Sample Information*

1. *General Theory.* In Section 4.5 we defined a' to be an act which is optimal under the *prior* distribution of $\tilde{\theta}$, satisfying

$$E'_\theta\, u_t(a', \tilde{\theta}) = \max_a E'_\theta\, u_t(a, \tilde{\theta}) \ ,$$

and we defined a_z to be an act which is optimal under the *posterior* distribution corresponding to a particular experimental outcome z, satisfying

$$E''_{\theta|z}\, u_t(a_z, \tilde{\theta}) = \max_a E''_{\theta|z}\, u_t(a, \tilde{\theta}) \ .$$

We then defined the CVSI or value of the sample information z by

$$v_t(e, z) = E''_{\theta|z}\, u_t(a_z, \tilde{\theta}) - E''_{\theta|z}\, u_t(a', \tilde{\theta}) \ ,$$

defined the EVSI or expected value of sample information to be obtained from e by

$$v_t^*(e) = E_{z|e}\, v_t(e, \tilde{z}) \ ,$$

and showed that the terminal utility of e is the utility of immediate terminal action augmented by the EVSI of e:

$$u_t^*(e) = u_t^*(e_0) + v_t^*(e) \ .$$

This way of looking at $u_t^*(e)$ is particularly instructive when the utility function $u_t(a, \cdot)$ on Ω is linear in $\omega = W(\theta)$. Remembering that this linearity implies that $\text{E}_\omega\, u_t(a, \tilde{\omega}) = u_t(a, \bar{\omega})$, we see that a' can now be defined by

$$u_t(a', \bar{\omega}') = \max_a u_t(a, \bar{\omega}') \ ; \tag{5-20}$$

and remembering that z uniquely determines $\bar{\omega}''$ we can see that a_z can be denoted by $a_{\bar{\omega}''}$ and defined by

$$u_t(a_{\bar{\omega}''}, \bar{\omega}'') = \max_a u_t(a, \bar{\omega}'') \ . \tag{5-21}$$

The CVSI or value of the "sample information" $\bar{\omega}''$ is accordingly

$$v_t(e, \bar{\omega}'') = u_t(a_{\bar{\omega}''}, \bar{\omega}'') - u_t(a', \bar{\omega}'') = \max_a u_t(a, \bar{\omega}'') - u_t(a', \bar{\omega}'') \ , \tag{5-22}$$

where we use the same symbol v_t for the CVSI function on $A \times \Omega$ that we use for the CVSI function on $A \times Z$ because the meaning will always be clear from the context. Comparing the definition (5-7) of the CVPI function, we see that

$$v_t(e, \bar{\omega}'') = l_t(a', \bar{\omega}'') \ : \tag{5-23}$$

when terminal utility is linear in ω, the CVSI function is formally identical to the CVPI function. It follows immediately that the *expected* value of sample information is given by

$$v_t^*(e) = \text{E}_{\bar{\omega}''|e}\, v_t(e, \tilde{\bar{\omega}}'') = \text{E}_{\bar{\omega}''|e}\, l_t(a', \tilde{\bar{\omega}}'') \ , \tag{5-24}$$

where the substitution of $\text{E}_{\bar{\omega}''|e}$ for $\text{E}_{z|e}$ is explained in Section 5.3.1.

2. *Scalar ω, Finite A.* The formal analogy between EVSI as given by (5-24) and EVPI as given by (5-8) implies at once that EVSI for the case where $\tilde{\omega}$ is scalar and A is finite is given by the same formula (5-14) which gives EVPI for this case,

$$v_t^*(e) = \textstyle\sum_{i=1}^{p-1} |k_{i+1} - k_i| L_{\bar{\omega}''}^{(l)}(\omega_i) + \sum_{i=p}^{s-1} |k_{i+1} - k_i| L_{\bar{\omega}''}^{(r)}(\omega_i) \ , \qquad a_p = a' \ , \tag{5-25}$$

the only difference being that in this case the linear-loss integrals are to be evaluated with respect to the distribution of $\tilde{\bar{\omega}}''$ rather than the distribution of $\tilde{\omega}$. When the number s of acts in A is 2, this formula reduces to (5-13):

$$v_t^*(e) = \begin{cases} |k_2 - k_1| L_{\bar{\omega}''}^{(r)}(\omega_b) & \text{if} \quad \bar{\omega}' \leq \omega_b \ , \\ |k_2 - k_1| L_{\bar{\omega}''}^{(l)}(\omega_b) & \text{if} \quad \bar{\omega}' \geq \omega_b \ . \end{cases} \tag{5-26}$$

For a geometrical interpretation of these results, return to the problem with $s = 4$ illustrated in Figure 5.5. The figure graphs the *conditional* terminal utilities $u_t(a, \cdot)$ as functions of ω, but because $\text{E}_{\omega|z}''\, u_t(a, \tilde{\omega}) = u_t(a, \bar{\omega}'')$, we may relabel the horizontal axis $\bar{\omega}''$ instead of ω and interpret the "curves" as graphs of $\text{E}_{\omega|z}''\, u_t(a, \tilde{\omega})$; observe that this does *not* change the "breakeven values" ω_1, ω_2, and ω_3, which depend only on the constants $K_1, \cdots, K_4, k_1, \cdots, k_4$. We have already seen that if *no* experiment is performed, the decision maker will choose the terminal act for which the *prior* expected utility $\text{E}_\omega'\, u_t(a, \tilde{\omega}) = u_t(a, \bar{\omega}')$ is greatest; suppose that $\omega_1 \leq \bar{\omega}' \leq \omega_2$ and that this act is therefore a_2. If an experiment *is* performed, then after observing z the decision maker will compute $\bar{\omega}''$ and choose the terminal act for which the *posterior* expected utility $\text{E}_{\omega|z}''\, u_t(a, \tilde{\omega}) = u_t(a, \bar{\omega}'')$ is greatest;

suppose that $\bar{\omega}'' = \omega_*$ and that as indicated by the figure the decision maker therefore chooses a_1. Then as evaluated after the fact, the decision maker's utility is $u_t(a_1, \omega_*)$ rather than $u_t(a_2, \omega_*)$ and we may say that the value of the sample information has been

$$u_t(a_1, \omega_*) - u_t(a_2, \omega_*) = l_t(a_2, \omega_*) \ ;$$

just as $l_t(a_2, \omega_*)$ is the value of "perfect" information that $\tilde{\omega} = \omega_*$ if before receiving this information the decision maker would have chosen a_2. Generalizing, we see that the CVSI for *any* $\bar{\omega}''$ is $l_t(a', \bar{\omega}'')$, and it follows at once that the EVSI for any e is $\mathrm{E}_{\bar{\omega}''|e}\, l_t(a', \tilde{\bar{\omega}}'')$.

5.4. The Prior Distribution of the Posterior Mean $\tilde{\bar{\omega}}''$ for Given *e*

We now turn our attention to the actual distribution of the random variable $\tilde{\bar{\omega}}''$, but before beginning our formal treatment we shall try to guide the reader by some heuristic observations. (1) If no sample is taken at all, the "posterior" distribution of $\tilde{\omega}$ will be identical to the "prior" distribution and therefore the posterior mean $\bar{\omega}''$ will be equal to the prior mean $\bar{\omega}'$; the "distribution" of $\tilde{\bar{\omega}}''$ consists of a mass of 1 at the point $\bar{\omega}'$. (2) Under appropriate conditions an infinitely large sample would yield exact knowledge of the true ω; and therefore before such a sample is taken, the distribution of $\tilde{\bar{\omega}}''$ is identical to the prior distribution of $\tilde{\omega}$ itself. (3) It seems reasonable to infer that as the "sample size" increases from 0 to ∞ the distribution of $\tilde{\bar{\omega}}''$ will in most cases *spread out* from a single mass point at $\bar{\omega}'$ toward the prior distribution of $\tilde{\omega}$ as a limit. In other words, *the effect of sample size on the distribution of* $\tilde{\bar{\omega}}''$ *should be (loosely speaking) the* opposite *of its effect on the sampling distribution of the sample mean for given* ω.

5.4.1. Mean and Variance of $\tilde{\bar{\omega}}''$

We now turn to formal treatment of the theory of the distribution of $\tilde{\bar{\omega}}''$ and give some general results which apply in all cases where *the mean and variance of the prior distribution of* $\tilde{\omega}$ *exist.* We begin with two theorems which will prove very useful in the applications:

Theorem 1. For any e, the expected value of $\tilde{\bar{\omega}}''$ is the mean of the prior distribution of $\tilde{\omega}$:

$$\mathrm{E}(\tilde{\bar{\omega}}'') = \mathrm{E}'(\tilde{\omega}) \ . \tag{5-27}$$

Theorem 2. For any e, the variance of $\tilde{\bar{\omega}}''$ is the variance of the prior distribution of $\tilde{\omega}$ *less* the expected variance of the posterior distribution of $\tilde{\omega}$ given z:

$$\mathrm{V}(\tilde{\bar{\omega}}'') = \mathrm{V}'(\tilde{\omega}) - E_{z|e}\, \mathrm{V}''_{\omega|z}(\tilde{\omega}) \ . \tag{5-28}$$

In general, $\tilde{\omega}$ is an $r \times 1$ vector, the expected values of (5-27) are $r \times 1$ vectors, and the "variances" of (5-28) are $r \times r$ matrices of (actual or expected) variances and covariances; we use the word "variance" rather than "covariance" for the entire matrix to avoid having to use special terminology for the case $r = 1$.

▶ The proof of (5-27) is trivial. Recalling that for any given prior distribution $\bar{\omega}''$ is uniquely determined by z and that $E_{z|e} E''_{\omega|z} = E_{\omega,z|e}$,

$$E(\tilde{\bar{\omega}}''_z) = E_{z|e} E''_{\omega|z}(\tilde{\omega}) = E_{\omega,z|e}(\tilde{\omega}) = E'_\omega(\tilde{\omega}) \equiv \bar{\omega}' \ .$$

To prove (5-28) we first prove the following lemma: Let $\tilde{x}$ and $\tilde{y}$ be jointly distributed and define

$$\tilde{\bar{x}}_y \equiv E_{x|y}(\tilde{x}) \ , \qquad \bar{\bar{x}} \equiv E(\tilde{x}) = E_y E_{x|y}(\tilde{x}) \ ;$$

then

$$V(\tilde{x}) = E_y V_{x|y}(\tilde{x}) + V_y E_{x|y}(\tilde{x}) \ .$$

For

$$\begin{aligned} V(\tilde{x}) &= E(\tilde{x} - \bar{\bar{x}})^2 = E_y E_{x|y}(\tilde{x} - \tilde{\bar{x}}_y + \tilde{\bar{x}}_y - \bar{\bar{x}})^2 \\ &= E_y E_{x|y}(\tilde{x} - \tilde{\bar{x}}_y)^2 + E_y(\tilde{\bar{x}}_y - \bar{\bar{x}})^2 + 2E_y(\tilde{\bar{x}}_y - \bar{\bar{x}}) E_{x|y}(\tilde{x} - \tilde{\bar{x}}_y) \\ &= E_y V_{x|y}(\tilde{x}) + V_y E_{x|y}(\tilde{x}) \end{aligned}$$

as asserted. The result can be interpreted as saying that the variance of $\tilde{x}$ is the sum of (1) the mean conditional variance of $\tilde{x}$ about the regression surface of $\tilde{x}$ on $\tilde{y}$, plus (2) the variance of the regression surface about its mean $E(\tilde{x})$.

The theorem (5-28) is now proved by rewriting the lemma in the form

$$V_y E_{x|y}(\tilde{x}) = V(\tilde{x}) - E_y V_{x|y}(\tilde{x})$$

and substituting ω for x and z for y to obtain

$$V_{z|e} E''_{\omega|z}(\tilde{\omega}) \equiv V_{z|e}(\tilde{\bar{\omega}}''_z) \equiv V(\tilde{\bar{\omega}}'') = V'(\tilde{\omega}) - E_{z|e} V''_{\omega|z}(\tilde{\omega}) \ . \quad . \quad ◀$$

5.4.2. Limiting Behavior of the Distribution

Now consider a sequence of experiments $\{e_n\} = e_0, e_1, e_2, \cdots$ where n is in some sense a measure of the "sample size" or of the "amount of information" to be obtained from e_n. The marginal distribution $P_{z|e}$ of the experimental outcome $\tilde{z}$ depends of course on n, and since $\tilde{\bar{\omega}}''$ is a function of $\tilde{z}$ its distribution $P_{\bar{\omega}''}\{\cdot|e_n\}$ also depends on n. We now wish to investigate the behavior of the distribution of $\tilde{\bar{\omega}}''$ as n increases, and we shall therefore display the dependence by writing $\tilde{\bar{\omega}}''_n$ for the random variable. We shall consider only sequences $\{e_n\}$ for which the corresponding sequence $\{\tilde{\bar{\omega}}''_n\}$, considered as a sequence of "estimators" in the classical sense, is *consistent in squared error* in the sense of the following

Definition. The sequence $\{\tilde{\bar{\omega}}''_n\}$ is *consistent in squared error* over the subset $\Omega_o \subset \Omega$ if for all ω in Ω_o

(a) $$\lim_{n\to\infty} E(\tilde{\bar{\omega}}''_n|\omega) = \omega \ , \tag{5-29a}$$

(b) $$\lim_{n\to\infty} V(\tilde{\bar{\omega}}''_n|\omega) = 0 \ . \tag{5-29b}$$

▶ When $\tilde{\omega}$ and therefore $\tilde{\bar{\omega}}''_n$ is an $r \times 1$ vector, the second part of the definition requires all elements of the $r \times r$ variance matrix $V(\tilde{\bar{\omega}}''_n)$ to vanish in the limit, but it suffices to verify this for the diagonal elements (the marginal variances of the scalar components of $\tilde{\bar{\omega}}''_n$) because by the Schwarz inequality

$$|E(\tilde{x} - \bar{x})(\tilde{y} - \bar{y})| \le [V(\tilde{x})\cdot V(\tilde{y})]^{\frac{1}{2}} .$$

for any two scalar random variables $\tilde{x}$ and $\tilde{y}$. ◀

We have already seen in (5-27) that the *mean* of the distribution of $\tilde{\bar{\omega}}''_n$ is $\bar{\omega}'$ regardless of the value of n; the effect of n on the *variance* of $\tilde{\bar{\omega}}''_n$ is given by the following

Theorem 3. Let $\{\tilde{\bar{\omega}}''_n\}$ be consistent in squared error over the spectrum of the prior distribution of $\tilde{\omega}$. Then as $n \to \infty$ the variance of $\tilde{\bar{\omega}}''_n$ approaches the prior variance of $\tilde{\omega}$:

$$\lim_{n\to\infty} V(\tilde{\bar{\omega}}''_n) = V'(\tilde{\omega}) . \tag{5-30}$$

▶ By the lemma in the proof of (5-28),

$$V(\tilde{\bar{\omega}}''_n) = E'_\omega V(\tilde{\bar{\omega}}''_n|\tilde{\omega}) + V'_\omega E(\tilde{\bar{\omega}}''_n|\tilde{\omega}) .$$

Taking the limit as $n \to \infty$ and then interchanging the order of the operations lim and E in the first term on the right and of the operations lim and V in the second term we have

$$\lim_{n\to\infty} V(\tilde{\bar{\omega}}''_n) = E'_\omega \lim_{n\to\infty} V(\tilde{\bar{\omega}}''_n|\tilde{\omega}) + V'_\omega \lim_{n\to\infty} E(\tilde{\bar{\omega}}''_n|\tilde{\omega}) .$$

The theorem then follows by (5-29). ◀

A much more basic result is contained in

Theorem 4. Let $\{\tilde{\bar{\omega}}''_n\}$ be consistent in squared error over the spectrum of the prior distribution of $\tilde{\omega}$. Then the sequence converges in distribution to $\tilde{\omega}$; i.e.,

$$\lim_{n\to\infty} P\{\tilde{\bar{\omega}}''_n \le c\} = P'\{\tilde{\omega} \le c\} \tag{5-31}$$

for any c which is a continuity point of the prior cumulative distribution function of $\tilde{\omega}$.

▶ Given any $r \times 1$ vector ω in the spectrum of P'_ω, the sequence of $r \times 1$ vectors $\{\tilde{\bar{\omega}}''_n\}$ converges to ω in mean square by hypothesis, and it follows by Chebysheff's inequality that $\{\tilde{\bar{\omega}}''_n\}$ *converges to* ω *in probability*, i.e., that for any $r \times 1$ vector $\epsilon > 0$ however small

$$\lim_{n\to\infty} P\{\omega - \epsilon < \tilde{\bar{\omega}}''_n < \omega + \epsilon|\omega\} = 1 .$$

From this it follows that, for any $r \times 1$ vector c whatever,

$$\lim_{n\to\infty} P\{\tilde{\bar{\omega}}''_n \le c|\omega\} = \begin{cases} 1 & \text{if} \quad \omega_i < c_i , \quad \text{all } i, \\ 0 & \text{if} \quad \omega_i > c_i , \quad \text{any } i. \end{cases}$$

Now writing

$$\lim_{n\to\infty} P\{\tilde{\bar{\omega}}''_n \le c\} = \lim_{n\to\infty} E'_\omega P\{\tilde{\bar{\omega}}''_n \le c|\tilde{\omega}\} = E'_\omega \lim_{n\to\infty} P\{\tilde{\bar{\omega}}''_n \le c|\tilde{\omega}\}$$

and using our previous result for the limit on the right we have

$$P'\{\tilde\omega < c\} \le \lim_{n\to\infty} P\{\tilde{\bar\omega}_n'' \le c\} \le P'\{\tilde\omega \le c\} \ ,$$

and (5-31) follows because the left- and right-hand members of this expression are equal when c is a continuity point of P'_ω.

We could have taken an alternate route to Theorems 3 and 4 by first proving that $\tilde{\bar\omega}_n''$ converges *unconditionally* to $\tilde\omega$ in mean square and then using theorems in probability theory relating different modes of convergence. ◀

5.4.3. *Limiting Behavior of Integrals When ω is Scalar*

We now specialize to the case where Ω is the real line and give two results concerning the limiting behavior of the integral of a linear function of $\tilde{\bar\omega}_n''$ with respect to the distribution of $\tilde{\bar\omega}_n''$. We first define the *incomplete first moment* or *partial expectation over* $(-\infty, c]$ of any *scalar* random variable $\tilde u$ with measure P_u to be

$$E_{-\infty}^c(\tilde u) = \int_{-\infty}^c u\, dP_u \ . \tag{5-32}$$

We then have

Theorem 5. If $\tilde\omega$ is scalar, then for any real c

$$\lim_{n\to\infty} E_{-\infty}^c(\tilde{\bar\omega}_n'') = E_{-\infty}^c(\tilde\omega) \ . \tag{5-33}$$

Corollary. If $\tilde\omega$ is scalar, then for any real c

$$\lim_{n\to\infty} L_{\bar\omega_n''}^{(l)}(c) = L_\omega^{(l)}(c) \ ; \qquad \lim_{n\to\infty} L_{\bar\omega_n''}^{(r)}(c) = L_\omega^{(r)}(c) \tag{5-34}$$

where the L functions are defined by (5-12).

▶ The proof of Theorem 5 follows from the Helly-Bray theorem which can be found in M. Loève, *Probability Theory*, Van Nostrand, 1955, pages 180–183. Then if we write the definitions (5-12) of the L functions in the form

$$L_u^{(l)}(c) = \int_{-\infty}^c (c - u)\, dP_u = c\, P_u\{\tilde u \le c\} - E_{-\infty}^c(\tilde u) \ ,$$

$$L_u^{(r)}(c) = \int_c^\infty (u - c)\, dP_u = E(\tilde u) - E_{-\infty}^c(\tilde u) - c\, P_u\{\tilde u \ge c\} \ ,$$

the corollary follows at once from (5-31), (5-33), and (5-27). ◀

5.4.4. *Exact Distributions of* $\tilde{\bar\omega}''$

The distribution of $\tilde{\bar\omega}''$ has been worked out in Part III of this monograph for the processes and ωs listed in Table 5.1 (page 99) and for all the experiments listed in Table 3.1 (page 75); since $\tilde{\bar\omega}''$ is a function of the sample outcome $\tilde z$ and the distribution $P_{z|e}$ depends on e, the distribution of $\tilde{\bar\omega}''$ will obviously be of one type if, say, the number of trials is predetermined in sampling from a Bernoulli process, of another if the number of successes is predetermined. The results are summarized in Table 5.2, where references are to formulas rather than to sections. The column

Table 5.2

Distribution of $\tilde{\bar{\omega}}''$

Process	Prior Distribution	Experiment	$\tilde{\omega}$	Distribution of $\tilde{\bar{\omega}}''$: P	V	E, L	Approximation Parameters
Bernoulli	beta	$\tilde{r}\|n$	$\tilde{p}$	(9-20) hypergeometric	$\frac{n}{n+n'}$	(9-21b) (9-22)	(9-23)
			$1/\tilde{p}$	(9-25) hypergeometric	(9-29)	(9-27b) (9-28)	(9-30)
		$\tilde{n}\|r$	$\tilde{p}$	(9-34) hypergeometric	(9-38)	(9-36b) (9-37)	(9-39)
			$1/\tilde{p}$	(9-41) hypergeometric	$\frac{r}{r+r'-1}$	(9-43b) (9-44)	(9-45)
Poisson	gamma	$\tilde{r}\|t$	$\tilde{\lambda}$	(10-36) binomial	$\frac{t}{t+t'}$	(10-37b) (10-38)	(10-39)
			$1/\tilde{\lambda}$	(10-42) binomial	(10-45)	(10-43b) (10-44)	(10-46)
		$\tilde{t}\|r$	$\tilde{\lambda}$	(10-21) beta	$\frac{r}{r+r'+1}$	(10-22b) (10-23)	(10-24)
			$1/\tilde{\lambda}$	(10-27) beta	$\frac{r}{r+r'-1}$	(10-28c) (10-29)	(10-30)
Normal	h known, $\tilde{\mu}$ Normal	$\tilde{m}\|n$	$\tilde{\mu}$	(11-32) Normal	$\frac{n}{n+n'}$	(11-24b) (11-25)	—
	Normal-gamma	$\tilde{m}, \tilde{v}\|n, \nu$	$\tilde{\mu}$	(11-67) Student	$\frac{n}{n+n'}$	(11-49c) (11-50)	—
Multinormal	—	—	$\Sigma c_i\tilde{\mu}_i$	*same as univariate*			
Regression	—	—	$\Sigma c_i\tilde{\beta}_i$	*same as univariate*			

headed P cites the formula for the distribution of $\tilde{\bar{\omega}}''$ and names the most familiar function in terms of which cumulative probabilities can be evaluated. Except for the Bernoulli process, the function cited has been well tabulated; computation and approximation in the Bernoulli case is discussed in Section 7.11.2. The column headed V gives the ratio $V(\tilde{\bar{\omega}}'')/V'(\tilde{\omega})$ of the variance of $\tilde{\bar{\omega}}''$ to the prior variance of $\tilde{\omega}$. The column headed E, L cites (1) the formula for the *partial* expectation of $\tilde{\bar{\omega}}''$ as defined by (5-32) and (2) formulas for the linear-loss integrals $L^{(l)}_{\bar{\omega}''}$ and $L^{(r)}_{\bar{\omega}''}$ as defined by (5-12). In the case of the Bernoulli and Poisson processes, these formulas involve cumulative probabilities of the type listed in the column headed P; in the other cases, the formulas involve one Normal or Student density and one Normal or Student cumulative probability.

5.4.5. *Approximations to the Distribution of* $\tilde{\bar{\omega}}''$

If we let η denote the predetermined component of the sufficient statistic in Table 5.2, the sequence $\{\tilde{\bar{\omega}}''_\eta\}$ is consistent in squared error in every case in the table and therefore (a) the distribution of $\tilde{\bar{\omega}}''_\eta$ approaches the prior distribution of $\tilde{\omega}$ as η increases, and (b) the first two moments, the partial expectation, and the loss integrals under the distribution of $\tilde{\bar{\omega}}''_\eta$ approach the corresponding quantities under the prior distribution of $\tilde{\omega}$. This implies that when η is "large" we will obtain accurate results if we *approximate the exact distribution of* $\tilde{\bar{\omega}}''$ *by an appropriately chosen distribution belonging to the same conjugate family as the prior distribution of* $\tilde{\omega}$. Notice that we can *not* expect good results from a Normal approximation unless the prior distribution of $\tilde{\omega}$ is Normal. The last column of Table 5.2 cites formulas for the parameters of a distribution of the conjugate family which will have the same first two moments as the exact distribution of $\tilde{\bar{\omega}}''$.

5.4.6. *Examples*

To illustrate the use of the information indexed in Table 5.2 we shall compute the value of the information which could be obtained by sampling before reaching a terminal decision in some of the situations for which we computed the expected value of perfect information (or expected opportunity loss of an immediate terminal decision) in Section 5.2.4. In each case we shall assume a prior distribution of the same type that we assumed in Section 5.2.4.

Example 1. In the situation of Example 1 of Section 5.2.4, assume that $\bar{p}'$ is *below* the breakeven value p_b, so that the optimal immediate terminal act would be acceptance. The expected value of the information which could be obtained by sampling the lot before making a terminal decision is then given by (5-26) as

$$v_t^*(e) = k\, L^{(r)}_{\bar{p}''}(p_b) \;.$$

The value of the loss integral L in this expression depends on the distribution of $\tilde{\bar{p}}''$ and this depends in turn on the experimental design; we shall consider two possible designs.

1a. Binomial Sampling. If a predetermined number n of pieces is to be drawn from the lot, the number of defectives $\tilde{r}$ being left to chance, then $\tilde{\bar{p}}''$ has the distribution (9-20) and by (9-22b) and (7-82)

$$L^{(r)}_{\bar{p}''}(p_b) = \bar{p}'\, G_h(r_c|n, n', r_c + r') - p_b\, G_h(r_c|n, n' - 1, r_c + r' - 1) \quad \text{(5-35a)}$$

where r_c is defined by

$$\frac{r' + r_c}{n' + n} = p_b \tag{5-35b}$$

and $G_h(s|S, F, \nu)$ is the hypergeometric probability of obtaining s or more successes in ν drawings without replacement from a finite population containing S successes and F failures.

Exact evaluation of the first of the two cumulative probabilities in this formula involves computing and summing no more than $r' + 1$ individual hypergeometric terms, the second no more than r' terms; but even this usually moderate labor is unnecessary if $\bar{p}'$ and p_b are small, since condition (7-86) will then be met and we can use the approximation

$$L_{\tilde{p}''}^{(r)}(p_b) = \bar{p}'\, G_b\left(r_c \middle| \frac{n}{n + n'}, r_c + r'\right) - p_b\, G_b\left(r_c \middle| \frac{n}{n + n' - 1}, r_c + r' - 1\right) \tag{5-36}$$

where $G_b(r|\pi, \nu)$ is the binomial probability of r or more successes in ν Bernoulli trials when the probability of a success on any trial is π.

1b. Pascal Sampling. If the sample is to be taken by drawing pieces from the lot until a predetermined number r of defectives has been found, the number $\tilde{n}$ of drawings being left to chance, then $\bar{\tilde{p}}''$ has the distribution (9-34) rather than (9-20), and by (9-37b) and (7-82)

$$L_{\tilde{p}''}^{(r)}(p_b) = \bar{p}'\, G_h(r|n_c, n', r + r') - p_b\, G_h(r|n_c, n' - 1, r + r' - 1)$$

where n_c is defined by

$$\frac{r' + r}{n' + n_c} = p_b\ .$$

Exact evaluation of the first cumulative hypergeometric probability involves computing and summing no more than $r' + 1$ terms, the second no more than r'; if $\bar{p}'$ and p_b are both small, we may use the binomial approximation

$$L_{\tilde{p}''}^{(r)}(p_b) = \bar{p}'\, G_b\left(r \middle| \frac{n_c}{n_c + n'}, r + r'\right) - p_b\, G_b\left(r \middle| \frac{n_c}{n_c + n' - 1}, r + r' - 1\right)\ .$$

Example 2. In the situation of Example 2 of Section 5.2.4, suppose that the director of the clinic proposes to try the new drug on n more patients before making up his mind definitely whether or not to return to exclusive use of the old drug. Recalling that we have already assumed that $\bar{p}'$ is *less* than the known cure rate $\pi = p_b$ of the old drug, we have by the same reasoning as in Example 1 above

$$v_t^*(e) = N\, L_{\tilde{p}''}^{(r)}(p_b)$$

where N is the expected number of patients affected by the terminal decision and the value of the loss integral is given in terms of hypergeometric cumulative probabilities by (5-35) above.

Suppose now that

$$\pi \equiv p_b = .500\ , \qquad r' = 23\ , \qquad n' = 51\ , \qquad \bar{p}' = r'/n' = .451\ .$$

(These values of r' and n' describe the distribution which the director would *now* assign to $\tilde{p}$ if, starting from an original rectangular distribution of $\tilde{p}$, he had *already* tried the new drug on 49 patients and found that 22 were cured.) Suppose further that the proposed *new* test is to involve

$$n = 50$$

patients. Under these assumptions, condition (7-86) for the use of the binomial approximation (5-36) to the hypergeometric probabilities in (5-35) is definitely *not* met; but on the other hand the fact that the sample size n is large means that we *are* probably justified in using a direct beta approximation to the distribution of $\tilde{\bar{p}}''$ itself. By (9-23) a beta distribution with parameters

$$n^* = \frac{n + n'}{n}(n' + 1) - 1 = \frac{50 + 51}{50} 52 - 1 = 104 \ ,$$

$$r^* = \frac{n^*}{n'} r' = \frac{104}{51} 23 = 47 \ ,$$

will have the same mean and variance as the exact distribution of $\tilde{\bar{p}}''$, and we may then use (9-12b) and (7-23) to evaluate

$$\begin{aligned} L^{(r)}_{\bar{p}''}(\pi) &= \pi\, I_{1-\pi}(n^* - r^*, r^* + 1) - \bar{p}'\, I_{1-\pi}(n^* - r^*, r^*) \\ &= .5\, I_{.5}(57, 48) - .455\, I_{.5}(57, 47) \ . \end{aligned}$$

The required values of I are beyond the range of Pearson's tables, but they can be expressed in terms of binomial probabilities by use of (7-20) and doing so we obtain

$$\begin{aligned} L^{(r)}_{\bar{p}''}(\pi) &= .5\, G_b(57|.5, 104) - .455\, G_b(57|.5, 103) \\ &= (.5 \times .1888) - (.455 \times .1622) = .0206 \ . \end{aligned}$$

The expected value of the sample information is thus the equivalent of $.0206N$ patients cured. As we pointed out in Section 4.2, the "cost" of this information to a person with the utilities assumed in this example lies in the fact that, given his *present* information about the new drug, he can expect only $50\,\bar{p}' = 50 \times .455 = 22.7$ of the experimentally treated patients to be cured whereas he could expect $50\pi = 50 \times .5 = 25$ of them to be cured if they were treated with the old drug. The cost of the proposed experiment is thus 2.3 patients cured, and the *net gain* to be expected from it is therefore $(.0206N - 2.3)$ patients cured.

Example 3. In the situation of Example 4 of Section 5.2.4, assume that the decision maker can have raw material beneficiated by an outside contractor. The contractor charges more than it would cost the decision maker to do the work himself if he bought the necessary equipment, but the decision maker is thinking of having a small number n of lots treated by the contractor so that he can measure their yields x before deciding whether or not to invest in the equipment. Assume that the lot-to-lot variance of the yields $\tilde{x}$ of *individual* untreated batches is

$$V(\tilde{x}|\mu) = 1/h$$

where h is a known number, and assume that it is virtually certain that the treatment will have no appreciable effect on this variance even though it does affect the mean $\mu = E(\tilde{x})$. Letting $\bar{\mu}'$ and $\breve{\mu}'$ denote the mean and variance of the decision maker's prior distribution of $\tilde{\mu}$ and defining the parameters m' and n' of this distribution by

$$m' = \bar{\mu}' , \qquad n' = \frac{1}{h\breve{\mu}'} ,$$

we have by (5-26), (11-32), and (11-25)

$$v_t^*(e) = k(hn^*)^{-\frac{1}{2}} L_{N^*}(u)$$

where

$$\frac{1}{n^*} = \frac{1}{n'} - \frac{1}{n'+n} , \qquad u = |\mu_b - m'|\sqrt{hn^*} ;$$

the function L_{N^*} is tabulated in Table II at the end of this monograph.

5.5. Optimal Sample Size in Two-Action Problems When the Sample Observations are Normal and Their Variance is Known

It is usually fairly easy to compute the expected *cost* $c_s^*(e_n)$ of performing an experiment of size n, and when we also have a convenient formula for the expected *value* $v_t^*(e_n)$ of the information to be obtained from the experiment, we can always find the value $n°$ of n which maximizes the expected *net gain*

$$v^*(e_n) = v_t^*(e_n) - c_s^*(e_n)$$

by actually tracing the behavior of $v^*(e_n)$ as a function of n; cf. Section 4.5.2. In one very important case, however, we can write an equation for the optimal sample size which can be solved either by a reasonably simple iterative procedure or still more easily by reference to published charts. This case, which we shall now examine in detail, is that of two-action problems where terminal utility is linear in the mean of an Independent Normal process whose precision h (or variance $1/h$) is known and where the cost of sampling is linear in the sample size.

5.5.1. *Definitions and Notation*

Let μ denote the mean of an Independent Normal process with known precision, i.e., a process generating independent random variables $\tilde{x}_1, \cdots, \tilde{x}_i, \cdots$ with identical densities

$$D(x|\mu, h) = f_N(x|\mu, h) \propto e^{-\frac{1}{2}h(x-\mu)^2} \tag{5-37a}$$

where h is known; and let the prior density of $\tilde{\mu}$ be

$$D'(\mu) = f_N(\mu|m', hn') \propto e^{-\frac{1}{2}hn'(\mu-m')^2} . \tag{5-37b}$$

Let $A = \{a_1, a_2\}$, let

$$u_t(a_i, \mu) = K_i + k_i\mu , \qquad i = 1, 2 , \tag{5-37c}$$

and define the breakeven value μ_b and the terminal loss constant k_t by

$$\mu_b \equiv \frac{K_1 - K_2}{k_2 - k_1} , \qquad k_t \equiv |k_2 - k_1| . \tag{5-37d}$$

Assume temporarily that the known or expected cost of sampling is *proportional* to the number of observations n,

$$c_s^*(e_n) = k_s n ; \tag{5-37e}$$

we shall generalize later to include a fixed element in sampling cost, i.e., to the case where $c_s^*(e_n) = K_s + k_s n$. Then by (4-38) and (5-26) in conjunction with (11-32) and (11-25) the expected net gain of an experiment e_n which consists of observing $(x_1, \cdots, x_n)$ is

$$v^*(e_n) = v_t^*(e_n) - c_s^*(e_n) = k_t(hn^*)^{-\frac{1}{2}} L_{N^*}(D^*) - k_s n \tag{5-38a}$$

where

$$\frac{1}{n^*} \equiv \frac{1}{n'} - \frac{1}{n' + n} , \qquad D^* \equiv |\mu_b - m'|(hn^*)^{\frac{1}{2}} . \tag{5-38b}$$

The really essential features of the problem are best brought out by putting the net gain $v^*(e_n)$ into a dimensionless form and by looking at certain ratios involving the sample size n and the prior parameter n' rather than at n itself; and the interpretation of these ratios will be easier if we first define the sampling and prior standard deviations

$$\sigma_\epsilon \equiv h^{-\frac{1}{2}} , \qquad \sigma'_\mu \equiv (hn')^{-\frac{1}{2}} . \tag{5-39}$$

We then define the ratios

$$\rho \equiv \frac{n}{n'} = \frac{\sigma'^2_\mu}{\sigma^2_\epsilon/n} , \tag{5-40a}$$

$$D' \equiv |\mu_b - m'|(hn')^{\frac{1}{2}} = \frac{|\mu_b - m'|}{\sigma'_\mu} , \tag{5-40b}$$

$$\lambda \equiv k_t(n'^3h)^{-\frac{1}{2}} k_s^{-1} = \frac{k_t \sigma_\epsilon}{k_s} \left[\frac{\sigma'_\mu}{\sigma_\epsilon}\right]^3 ; \tag{5-40c}$$

to facilitate typography we define the supplementary symbol

$$\theta \equiv \left[\frac{\rho}{\rho + 1}\right]^{\frac{1}{2}} = \left[\frac{\sigma'^2_\mu}{\sigma'^2_\mu + \frac{1}{n}\sigma^2_\epsilon}\right]^{\frac{1}{2}} ; \tag{5-40d}$$

and we then consider the *dimensionless net gain* of e_n

$$g(\rho; D', \lambda) \equiv \frac{v^*(e_n)}{k_s n'} = \lambda\, \theta\, L_{N^*}(D'/\theta) - \rho . \tag{5-41}$$

The first term on the right is the dimensionless EVSI; the second is the dimensionless cost of sampling or sample size.

5.5.2. *Behavior of Net Gain as a Function of Sample Size*

Concerning the function $g(\rho; D', \lambda)$ defined by (5-41) we shall prove the following results, treating ρ *as if* it were a continuous variable.

If $D' = 0$, then for any given λ the net gain behaves with increasing ρ in the way shown in Figure 5.7. Because the prior distribution is totally indecisive, the value of additional information at first rises much more rapidly than the cost of obtaining it. The EVSI cannot exceed the EVPI, however, and as it approaches this bound it increases more and more slowly, while the cost of sampling increases steadily in proportion to the sample size. The net gain accordingly reaches a

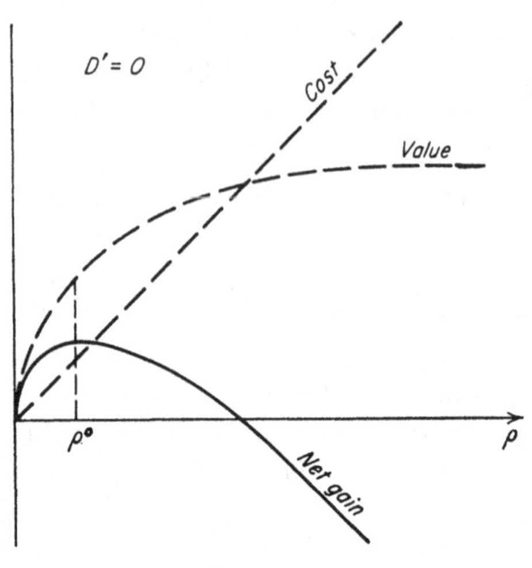

Figure 5.7

maximum and then declines more and more rapidly; the optimal "sample size" ρ° is the value of ρ corresponding to this maximum. The true optimal value of n may of course be 0 if ρ° corresponds to $n < 1$.

If $D' > 0$, the prior distribution definitely favors one of the two terminal acts, so that it is very improbable that a very small sample could reverse this decision, and the EVSI accordingly at first increases very slowly with n. The rate of increase of the EVSI becomes greater as the sample becomes large enough to have an appreciable chance of reversing the decision, but then slows down again as the EVSI approaches the EVPI. Meanwhile the cost of sampling increases steadily; and the net result is that the net gain may behave in any of the ways shown in Figures 5.8 and 5.9, depending on the value of (D', λ). For any given D', there is some value λ_c of λ such that if $\lambda > \lambda_c$ the net gain has a local maximum

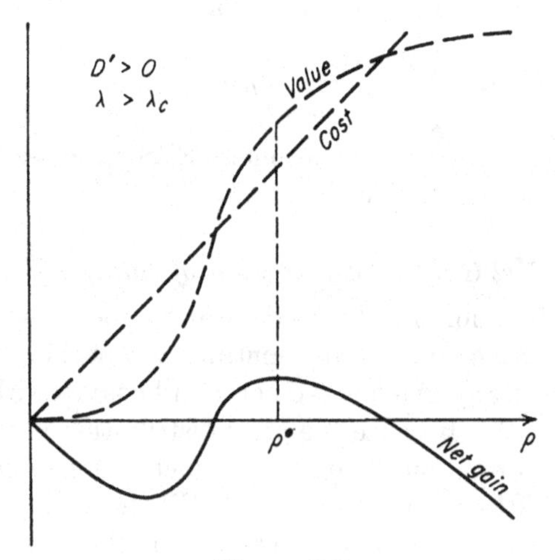

Figure 5.8

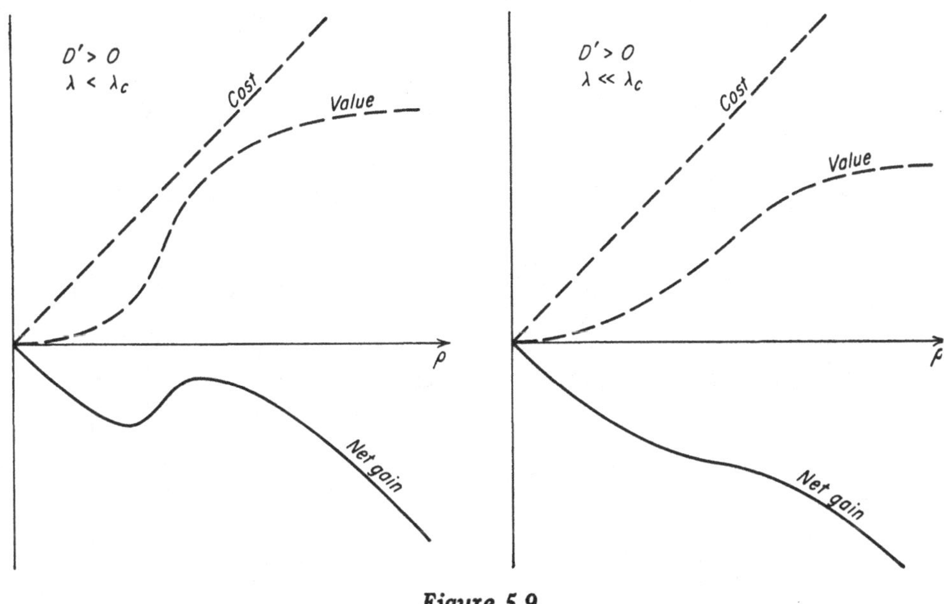

Figure 5.9

which is a true maximum as in Figure 5.8 and the optimal sample size ρ° is the value of ρ which produces this maximum. If $\lambda < \lambda_c$, then as shown in Figure 5.9 the net gain has either a negative local maximum or no local maximum at all and the optimal sample size is 0. The critical value λ_c increases with D' as shown by the graph Figure 5.10 of $Z_c \equiv \lambda_c^{\frac{1}{3}}$.

5.5.3. *Optimal Sample Size*

When $D' = 0$, the optimal "sample size" ρ° is the unique root of

$$\tfrac{1}{2}\lambda\rho^{-\frac{1}{2}}(\rho + 1)^{-\frac{3}{2}}(2\pi)^{-\frac{1}{2}} = 1 \; ; \tag{5-42a}$$

when $D' > 0$ and the optimal sample size is not 0, ρ° is the larger of the two roots of

$$\tfrac{1}{2}\lambda\rho^{-\frac{1}{2}}(\rho + 1)^{-\frac{3}{2}} f_{N^*}(D'/\theta) = 1 \tag{5-42b}$$

where $\theta = \sqrt{\rho/(\rho + 1)}$. The quantity

$$\eta^\circ \equiv \rho^\circ/Z^2 = \rho^\circ\lambda^{-\frac{2}{3}} = \frac{n^\circ}{(k_t\sigma_\epsilon/k_s)^{\frac{2}{3}}} = n^\circ \frac{h^{\frac{1}{3}}}{(k_t/k_s)^{\frac{2}{3}}} \tag{5-43a}$$

is graphed as a function of

$$D_\infty \equiv D' \; , \quad D_\infty = 0(.1).2(.2)3.4, \qquad \text{and} \qquad Z \equiv \lambda^{\frac{1}{3}} \; , \quad .7 \le Z \le 80 \; , \tag{5-43b}$$

in Chart I at the end of this monograph. A nomogram showing $N \equiv \rho^\circ$ as a function of λ and $|X| \equiv D'$ for $70 \le \lambda \le 4000$ and $0 \le |X| \le 2.6$ is given on page 37 of the *Journal of the Royal Statistical Society*, Series B, Volume 18, Number 1 (1956).

For some purposes all that we require is the net gain of an optimal sample for given (D', λ), the actual size of this sample being irrelevant. Chart II at the end of

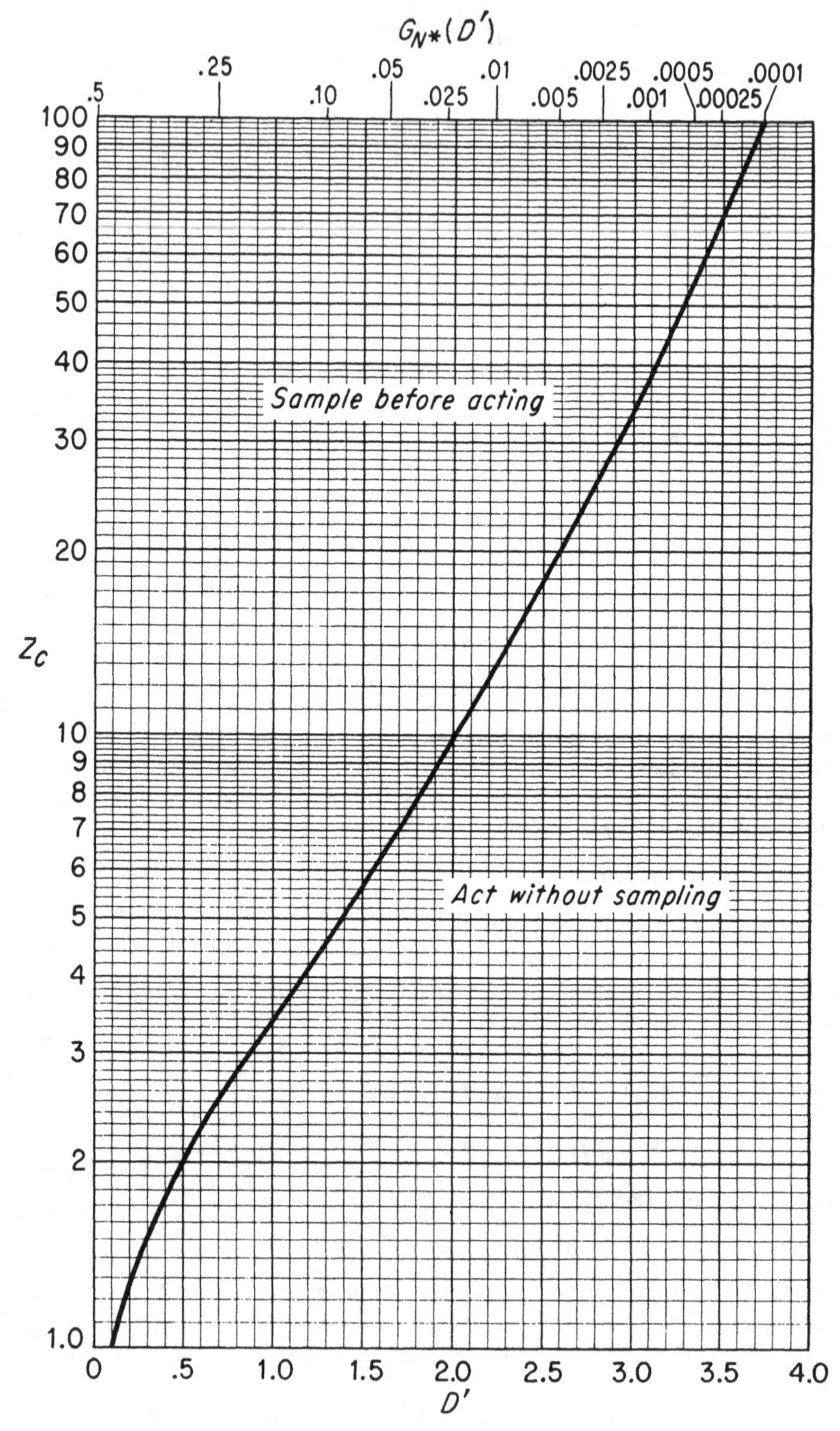

Figure 5.10

Critical Value of $Z \equiv \lambda^{\frac{1}{3}}$ When $K_s = 0$

this monograph, the data for which were computed at the expense of the International Business Machines Corporation under the direction of Arthur Schleifer, Jr., shows the quantity

$$\gamma^\circ \equiv \frac{v^*(e_{n^\circ})}{(k_t^2 \sigma_\epsilon^2 k_s)^{\frac{1}{3}}} = g(\rho^\circ; D', \lambda)\, n' \frac{h^{\frac{1}{3}}}{(k_t/k_s)^{\frac{2}{3}}} \tag{5-43c}$$

as a function of the parameters defined by (5-43b).

▶ The results presented in Sections 5.5.2 and 5.5.3 were first published by Grundy, Healy, and Rees (*JRSS* 18 [1956] pp. 32–48), who derived the equivalent of formula (5-41) for the net gain by using the fiducial distribution resulting from an initial sample as a prior distribution and then applying the normal form of analysis to the problem of determining the optimal size of a second and final sample. The size n_1 of the first sample in their analysis plays the role of the prior parameter n' in ours; their n_2 is our n. For completeness we shall reprove the results here by a line of argument which differs from theirs in a few details.

We start from the dimensionless net gain as given by (5-41),

$$g(\rho; D', \lambda) = \lambda\, \theta\, L_{N^*}(D'/\theta) - \rho \tag{1}$$

where

$$\theta = \left[\frac{\rho}{\rho + 1}\right]^{\frac{1}{2}} \tag{2}$$

and differentiate twice with respect to ρ, thus obtaining

$$g'(\rho; D', \lambda) = \tfrac{1}{2}\lambda\, \rho^{-\frac{1}{2}}(\rho + 1)^{-\frac{3}{2}} f_{N^*}(D'/\theta) - 1 \;, \tag{3}$$

$$g''(\rho; D', \lambda) = \tfrac{1}{4}\lambda[\rho(\rho + 1)]^{-\frac{5}{2}} f_{N^*}(D'/\theta)\{D'^2 + (D'^2 - 1)\rho - 4\rho^2\} \;. \tag{4}$$

For $D' = 0$, we see that $g'(\rho)$ decreases steadily from $+\infty$ at $\rho = 0$ to -1 at $\rho = \infty$, so that $g(\rho; D', \lambda)$ must behave as shown in Figure 5.7. For $D' > 0$, we see that $g'(\rho) = -1$ at $\rho = 0$ and at $\rho = \infty$ while $g''(\rho)$ is positive for all ρ below the unique positive root of the quadratic in curly brackets and negative for all greater ρ; it follows that $g(\rho; D', \lambda)$ may vary with ρ in any of the ways shown in Figures 5.8 and 5.9.

The first-order condition for a local maximum of $g(\rho; D', \lambda)$ is obtained by setting (3) equal to 0,

$$\tfrac{1}{2}\lambda\rho^{-\frac{1}{2}}(\rho + 1)^{-\frac{3}{2}} f_{N^*}(D'/\theta) - 1 = 0 \;. \tag{5}$$

Denoting by ρ° the root of this equation which corresponds to a local *maximum* of $g(\rho; D', \lambda)$ if one exists, we observe that by Figures 5.7, 5.8, and 5.9:

If $D' = 0$, a local maximum always exists and ρ° is the unique root of (5).
If $D' > 0$, a local maximum may or may not exist. If one does exist, the maximizer ρ° is the larger root of (5), but $g(\rho^\circ; D', \lambda)$ may or may not be greater than $g(0; D', \lambda) = 0$.

We can partition all (D', λ) pairs into those for $g(\rho^\circ; D', \lambda) > 0$ and those for which $g(\rho^\circ; D', \lambda) \leq 0$ by first finding the pairs (D'_c, λ_c) for which $g(\rho^\circ; D', \lambda) = 0$ exactly. The computation of λ_c as a function of D'_c is best carried out by treating both λ_c and D'_c as functions of the ratio D'_c/θ°_c where θ°_c is the optimal $\theta = \sqrt{\rho/(\rho + 1)}$ for the given (D'_c, λ_c). Substituting (2) in (3) and setting the result equal to 0 we obtain the *optimality condition*

$$\frac{2\theta}{(1 - \theta^2)^2 f_{N^*}(D'/\theta)} = \lambda \;. \tag{6}$$

Substituting (2) in (1) we obtain the *condition for zero net gain*

$$\frac{\theta}{(1-\theta^2)\,L_{N^*}(D'/\theta)} = \lambda\ . \tag{7}$$

These two equations taken together define the locus of $(\theta_c^\circ, D'_c, \lambda_c)$ as a curve in 3-space. Eliminating λ between them, substituting

$$L_{N^*}(u) \equiv f_{N^*}(u) - u\,G_{N^*}(u)\ ,$$

and defining the function ϕ by

$$\phi(u) \equiv \frac{u\,G_{N^*}(u)}{f_{N^*}(u)}\ , \tag{8}$$

we obtain

$$\phi(D'_c/\theta_c^\circ) = \tfrac{1}{2}(1 + \theta_c^{\circ 2})\ . \tag{9}$$

Computation of λ_c as a function of D'_c can be carried out by inserting values of D'_c/θ_c° into the left side of (9) and for each such value: (a) solving (9) for θ_c° and then computing $D'_c = \theta_c^\circ(D'_c/\theta_c^\circ)$; (b) inserting θ_c° and D'_c/θ_c° into (6) or (7) to determine λ_c.

Since $0 \le \theta_c^\circ \le 1$ by the definition of θ, the right-hand side of (9) is restricted to the interval $[\tfrac{1}{2}, 1]$, and therefore the permissible values of D'_c/θ_c° for this computing procedure are those for which $\tfrac{1}{2} \le \phi(D'_c/\theta_c^\circ) \le 1$. To delimit these permissible values, we first observe that

$$\frac{d}{du}\,\phi(u) = \frac{(u^2+1)\,G_{N^*}(u) - u\,f_{N^*}(u)}{f_{N^*}(u)} > 0$$

for all $u < \infty$, as is proved by the fact that

$$(u^2+1)\,G_{N^*}(u) - u\,f_{N^*}(u)$$

is 0 at $u = \infty$ and that its derivative

$$\frac{d}{du}\,[(u^2+1)\,G_{N^*}(u) - u\,f_{N^*}(u)] = -2L_{N^*}(u) < 0$$

for all $u < \infty$. Then since $\phi(.612) \doteq \tfrac{1}{2}$ and $\phi(\infty) = 1$, we conclude that the permissible values of D'_c/θ_c° are those in the interval $[.612, \infty)$: as D'_c/θ_c° runs from .612 to ∞ the right-hand side of (9) runs from $\tfrac{1}{2}$ to 1 and θ_c° itself runs from 0 to 1. We remark incidentally that the fact that D'_c/θ_c° increases with θ_c° implies directly that D'_c increases with θ_c° and implies via (6) that λ_c increases with θ_c°, from which it follows that λ_c increases with D'_c. The quantity $Z_c = \sqrt[3]{\lambda_c}$ is graphed as a function of D' in Figure 5.10.

Our next task is to lay out a systematic procedure for computation of the optimal θ° for those (D', λ) for which net gain is positive, and as a preliminary step we prove the two following propositions.

(10) For any fixed D' and any $\rho > 0$, net gain increases with λ.

(11) For any fixed D', the optimal ρ° and therefore the optimal θ° increase with λ.

Proposition (10) is obvious on inspection of formula (1). To prove proposition (11), we observe that the optimality condition in the form (5) implicitly defines ρ° as a function of λ with parameter D'. Differentiating this constraining equation totally with respect to λ for fixed D' we obtain, with the aid of (4),

$$\tfrac{1}{2}\rho^{-\frac{1}{2}}(\rho+1)^{-\frac{3}{2}} f_{N^*}(D'/\theta)\left[1 + \tfrac{1}{2}\lambda\,\rho^{-2}(\rho+1)^{-1}\,\{\cdot\}\,\frac{d\rho}{d\lambda}\right] = 0$$

where $\{\cdot\}$ stands for the expression in curly brackets in (4); and since $\{\cdot\}$ must be negative at a local maximum, $d\rho^\circ/d\lambda > 0$ as was to be proved.

Proposition (10) implies that $g(\rho^\circ; D', \lambda) > g(0; D', \lambda)$ for given D' if and only if λ is greater than the critical value λ_c for the D' in question; and proposition (11) then implies that $g(\rho^\circ; D', \lambda) > g(0; D', \lambda)$ for given D' if and only if $\theta^\circ > \theta_c^\circ$ for that D'. Optimal sample size may therefore be computed as a function of λ for given D' by taking a series of values of θ° such that $\theta^\circ > \theta_c^\circ$ and using (6) to find the λ corresponding to each θ°. The value $\theta^\circ = 1$ corresponds, of course, to $\rho^\circ = \infty$.

Finally, the net gain of an optimal sample $g(\rho^\circ; D', \lambda)$ can be computed as a function of D' and λ by inserting the values ρ° and θ° which are optimal for each (D', λ) into formula (1). ◀

5.5.4. *Asymptotic Behavior of Optimal Sample Size*

For any values of the parameters D' and λ an *upper bound* for the optimal sample size ρ° is given by

$$\rho^\circ < \sqrt{\tfrac{1}{2}\lambda\, f_{N^*}(D')} \; . \tag{5-44}$$

As λ increases with D' fixed the ratio between the two sides of this inequality approaches unity, so that for large λ we can approximate the value of ρ° by

$$\rho^\circ \doteq \sqrt{\tfrac{1}{2}\lambda\, f_{N^*}(D')} \; ; \tag{5-45a}$$

for fixed λ the relative error in the approximation increases with D'.

Expressed in natural rather than dimensionless units, formula (5-45a) becomes

$$n^\circ \doteq [\tfrac{1}{2}(k_t\sigma_\epsilon/k_s)(\sigma_\epsilon/\sigma'_\mu)^3 f_{N^*}(D')]^{\frac{1}{2}} \; . \tag{5-45b}$$

If we take σ_ϵ as the unit of measurement for μ, optimal sample size increases with the seriousness of a wrong decision as measured by $k_t\sigma_\epsilon$, decreases with the "decisiveness" of the prior distribution as measured by D', and decreases with the standard deviation of the prior distribution as measured by $\sigma'_\mu/\sigma_\epsilon$. The last relation is of particular interest: while it is suggestive in *some* situations to think of the reciprocal of the prior variance as a measure of "quantity of information," the contrary is true when we are making a decision concerning sample size. In this case a large prior variance—strictly, a large ratio $\sigma'^2_\mu/\sigma^2_\epsilon$—represents a great deal of relevant information, since it amounts to an assertion that μ is almost certainly so far from the breakeven value μ_b in one direction or the other that a very small sample can show with near certainty *on which side* of μ_b the true μ actually lies.† In problems of point estimation, on the other hand, optimal sample size naturally increases with σ'_μ, as we shall see in Chapter 6.

▶ To prove (5-44) and (5-45) we substitute $\sqrt{\rho/(\rho+1)}$ for θ in the optimality condition (5-42b), thus obtaining

$$\rho^{\frac{1}{2}}(\rho+1)^{\frac{3}{2}} = \tfrac{1}{2}\lambda(2\pi)^{-\frac{1}{2}} \exp\left(-\tfrac{1}{2}\frac{\rho+1}{\rho} D'^2\right) = \tfrac{1}{2}\lambda[(2\pi)^{-\frac{1}{2}} e^{-\frac{1}{2}D'^2}]\, e^{-\frac{1}{2}D'^2/\rho}$$

† Observe, however, that when we make these statements about the effect of variation in σ'_μ we are assuming that D' is held fixed. If $\bar{\mu}'$ is held fixed instead of D', then $D' \equiv |\mu_b - \bar{\mu}'|/\sigma'_\mu$ will vary with σ'_μ and the effect of the variation will be different.

and hence

(1) $$\rho^2\left[\left(\frac{\rho+1}{\rho}\right)^{\frac{3}{2}} e^{\frac{1}{2}D'^2/\rho}\right] = \tfrac{1}{2}\lambda\, f_{N^*}(D') \ .$$

Since the larger root ρ° of this equation increases with λ for fixed D', as shown in proposition (11) of the previous proof, and since as ρ increases the factor in square brackets in (1) approaches 1 from above, we see at once that the inequality (5-44) holds for all ρ° and that as ρ° increases the ratio of ρ° to the right-hand side of (5-44) approaches 1. ◀

5.5.5. *Asymptotic Behavior of Opportunity Loss*

We have seen in (4-37) that the prior expectation of the expected *terminal opportunity loss* to be incurred after performing e_n (often called the Bayes risk) is the EVPI diminished by the EVSI of e_n,

$$l_t^*(e_n) = l_t^*(e_0) - v_t^*(e_n) \ .$$

Using (5-13) and (5-26) to obtain the values of these quantities when $u_t(a, \cdot)$ is linear in ω and using (11-32), (11-25), and (5-40) to evaluate the loss integrals under a Normal distribution we have

$$l_t^*(e_n) = k_t(hn')^{-\frac{1}{2}}\,[L_{N^*}(D') - \theta\, L_{N^*}(D'/\theta)] \ , \tag{5-46}$$

and in dimensionless form this becomes

$$\tau(\rho; D', \lambda) \equiv \frac{l_t^*(e_n)}{k_s n'} = \lambda[L_{N^*}(D') - \theta\, L_{N^*}(D'/\theta)] \ . \tag{5-47}$$

An upper bound on the expected terminal opportunity loss of e_n is given by

$$\tau(\rho; D', \lambda) < \tfrac{1}{2}\lambda\, f_{N^*}(D')\,\frac{1}{\rho} \ , \qquad \rho > 1 \ . \tag{5-48a}$$

As ρ increases with D' and λ fixed, the ratio between the two sides of this inequality approaches unity, so that for large ρ

$$\tau(\rho; D', \lambda) \doteq \tfrac{1}{2}\lambda\, f_{N^*}(D')\,\frac{1}{\rho} \ . \tag{5-48b}$$

▶ To prove (5-48a) we first observe that

$$\frac{d}{du} L_{N^*}(u) = -G_{N^*}(u)$$

is an increasing function of u, so that

(1) $$L_{N^*}(u+\epsilon) > L_{N^*}(u) - \epsilon\, G_{N^*}(u) \ .$$

Substituting (1) in (5-47) with $u = D'$ and $\epsilon = \frac{D'}{\theta} - D'$ we obtain

(2) $$\frac{1}{\lambda}\tau(\rho; D', \lambda) = L_{N^*}(D') - \theta\, L_{N^*}(D'/\theta)$$
$$< L_{N^*}(D') - \theta\left[L_{N^*}(D') - D'\frac{1-\theta}{\theta}\, G_{N^*}(D')\right]$$

$$= (1-\theta)[L_{N^*}(D') + D'\,G_{N^*}(D')]$$
$$= (1-\theta)\,f_{N^*}(D') \ .$$

The inequality (5-48a) follows from the fact that for $\rho > 1$

$$(3) \qquad 1 - \theta \equiv 1 - \left(1 + \frac{1}{\rho}\right)^{-\frac{1}{2}} = \tfrac{1}{2}(1/\rho) - \tfrac{3}{8}(1/\rho)^2 + \cdots$$
$$< \tfrac{1}{2}(1/\rho) \ ;$$

and we prove the asymptotic approximation (5-48b) by observing (a) that as $\rho \to \infty$ the ratio of the two sides of (3) approaches unity and (b) that as θ approaches 1 and therefore $\epsilon = D'\left(\frac{1-\theta}{\theta}\right)$ approaches 0 the ratio of the two sides of (1) approaches unity.

It is perhaps worth remarking that reasoning very like that which led to (2) gives a *lower* bound

$$\frac{1}{\lambda}\,\tau(\rho; D', \lambda) > (1-\theta)\,f_{N^*}(D'/\theta) \ . \qquad \blacktriangleleft$$

The approximation (5-48b) is valid for any (D', λ) when ρ is large, whether or not ρ is optimal for the given (D', λ); but because ρ° increases with λ for fixed D', we may approximate the terminal opportunity loss of an *optimal* sample when λ is large by substituting in (5-48b) the approximation (5-45a) to ρ°. We thus obtain

$$\tau(\rho^\circ; D', \lambda) \doteq [\tfrac{1}{2}\lambda\, f_{N^*}(D')]^{\frac{1}{2}} \doteq \rho^\circ \ . \qquad (5\text{-}49a)$$

When λ and therefore ρ° are large, the cost of an optimal sample is equal to its expected terminal opportunity loss. (The same result can be obtained by adding the dimensionless sampling cost ρ to (5-48b), thus obtaining an approximation to the *total* expected opportunity loss when ρ is large, and then minimizing this total with respect to ρ.)

Expressed in natural rather than dimensionless units, formula (5-49a) becomes

$$l_t^* \doteq c_s^* \doteq [\tfrac{1}{2}(k_t\sigma_\epsilon k_s)(\sigma_\epsilon/\sigma'_\mu)\, f_{N^*}(D')]^{\frac{1}{2}} \ . \qquad (5\text{-}49b)$$

Both the terminal opportunity loss and the sampling cost of an optimal sample increase with the seriousness of a wrong decision as measured by $k_t\sigma_\epsilon$ and with the cost k_s of a sample observation; with σ'_μ fixed, they decrease with the "decisiveness" of the prior distribution as measured by D'; and with D' fixed, they decrease with the probability that $\tilde{\mu}$ is far from μ_b as measured by $\sigma'_\mu/\sigma_\epsilon$.

5.5.6. *Fixed Element in Sampling Cost*

All of the above discussion has rested on the assumption (5-37e) that the expected sampling cost is strictly proportional to the sample size, $c_s^*(e_n) = k_s n$. We now generalize to the case where sampling cost also includes a fixed element which is incurred if any sample is taken at all but which is independent of the size of the sample,

$$c_s^*(e_n) = K_s + k_s n \ .$$

The effect of K_s—or rather, of the dimensionless fixed cost $K_s/(k_s n')$—on the curves of net gain in Figures 5.7, 5.8 and 5.9 is obvious: every point except the point at $\rho = 0$ will be depressed by the amount $K_s/(k_s n')$. If a local maximum exists when $K_s = 0$, one will exist when $K_s > 0$ and it will correspond to the same value of ρ in both cases; but a local maximum which gives positive net gain when $K_s = 0$ may give negative net gain when $K_s > 0$. Obviously, therefore: if for given (D', λ) it does *not* pay to sample when $K_s = 0$, then *a fortiori* it does not pay when $K_s > 0$; if it *does* pay to sample when $K_s = 0$, then when $K_s > 0$ the optimal size can be found by finding the size which would be optimal if $K_s = 0$ and computing the net gain of this sample to see whether it is greater or less than the 0 net gain of taking no sample at all.

5.6. Optimal Sample Size in Two-Action Problems When the Sample Observations are Binomial

Convenient exact solutions of the problem of optimal sample size in two-action problems with linear terminal utilities have unfortunately not been found for any data-generating process other than the Independent Normal process with known precision. We shall therefore consider only one other problem of this sort: that of sampling from a Bernoulli process when terminal utility is linear in p, when the number of trials is to be predetermined, and when the cost of sampling is linear in n. This one example will suffice to illustrate both the difficulty of finding an exact solution to nonnormal problems and the possibility of finding an approximate solution by use of the results of our analysis of the Normal problem.

5.6.1. Definitions and Notation

Let p denote the mean of a Bernoulli process, i.e., a process generating independent random variables $\tilde{x}_1, \cdots, \tilde{x}_i, \cdots$ with identical densities

$$D(x|p) = f_b(x|p, 1) = p^x(1-p)^{1-x} \, , \qquad x = 0, 1 \; ; \tag{5-50a}$$

and let the prior density of $\tilde{p}$ be

$$D'(p) = f_\beta(p|r', n') \propto p^{r'-1}(1-p)^{n'-r'-1} \, . \tag{5-50b}$$

Let $A = \{a_1, a_2\}$, let

$$u_t(a_i, p) = K_i + k_i p \, , \qquad i = 1, 2 \; ; \tag{5-50c}$$

define the breakeven value p_b and the terminal loss constant k_t by

$$p_b = \frac{K_1 - K_2}{k_2 - k_1} \, , \qquad k_t = |k_2 - k_1| \; ; \tag{5-50d}$$

and assume that

$$0 < p_b < 1 \, . \tag{5-50e}$$

(If this last condition is not met, one of the two acts is optimal for *all* possible p.)

On these assumptions the EVSI or expected value of the information to be gained from n observations on the process is, by (5-26),

$$v_t^*(e_n) = \begin{cases} k_t\, L_{\bar{p}''}^{(r)}(p_b) & \text{if} \quad \bar{p}' \leq p_b \ , \\ k_t\, L_{\bar{p}''}^{(l)}(p_b) & \text{if} \quad \bar{p}' \geq p_b \ , \end{cases}$$

and the values of the functions $L^{(l)}$ and $L^{(r)}$ are given in terms of the cumulative beta-binomial function by the very similar formulas (9-22a) and (9-22b). All essential features of the problem can be brought out by considering only the case where

$$\bar{p}' = \frac{r'}{n'} \geq p_b \ , \tag{5-50f}$$

implying that the optimal act under the prior distribution is a_2. The *dimensionless* EVSI can then be defined to be

$$h(n) \equiv \frac{1}{k_t}\, v_t^*(e_n) = L_{\bar{p}''}^{(l)}(p_b)$$

$$= p_b\, F_{\beta b}(\rho_n|r', n', n) - \bar{p}'\, F_{\beta b}(\rho_n|r' + 1, n' + 1, n) \tag{5-51a}$$

where $F_{\beta b}$ is the left tail of the beta-binomial function defined by (7-76) and ρ is the *integer-valued* function of n defined by

$$\rho_n = [p_b(n' + n) - r'] \ , \tag{5-51b}$$

$[x]$ denoting the greatest integer in x. For any sample size n, ρ_n is the greatest value of the sample statistic r which will lead to $\bar{p}'' \leq p_b$ and thus to the conclusion that the act a_1 is optimal; recall that a_2 is optimal under the prior distribution by the hypothesis (5-50f). For n small enough, the definition (5-51b) may lead to a negative value of ρ, implying that no possible outcome of the sample can reverse the prior choice; if so, then (5-51a) gives $h(n) = 0 - 0 = 0$ in accordance with common sense.

5.6.2. *Behavior of the EVSI as a Function of n*

In order to understand the behavior of the EVSI as a function of the sample size n, we first observe that as n increases from 0 through integral values the function ρ defined by (5-51b) at first retains some constant value, then increases by 1 and retains this value for a time, then increases by 1 again, and so forth. The effect on the dimensionless EVSI defined by (5-51) of a unit increase in n is given in terms of the beta-Pascal function (7-78) by

$$\Delta h(n) = \left[\frac{r' + \rho_n + 1}{n' + n + 1} - p_b\right] f_{\beta Pa}(n + 1|r', n', \rho_n + 1)$$

$$\text{if } \rho_{n+1} = \rho_n \ , \tag{5-52a}$$

$$\Delta h(n) = \left[\frac{n + 1}{\rho_n + 1} - 1\right]\left[p_b - \frac{r' + \rho_n + 1}{n' + n + 1}\right] f_{\beta Pa}(n + 1|r', n', \rho_n + 1)$$

$$\text{if } \rho_{n+1} = \rho_n + 1 \ , \tag{5-52b}$$

where

$$\Delta h(n) \equiv h(n + 1) - h(n) \ .$$

The first differences given by (5-52a) are always positive and those given by (5-52b) are always nonnegative, corresponding to the fact that one would never refuse to accept free information. As long as ρ remains *constant*, each successive first difference is *smaller* than the previous one,

$$\Delta h(n) < \Delta h(n-1) \qquad \text{if} \qquad \rho_{n+1} = \rho_n = \rho_{n-1} \; ; \tag{5-53a}$$

if $p_b \ll \frac{1}{2}$, each *increase* in ρ is either accompanied or followed by an *increase* in the first difference:

$$\Delta h(n) > \Delta h(n-2) \qquad \text{if} \begin{cases} \rho_{n+1} = \rho_n \; , \\ \rho_{n-2} = \rho_{n-1} = \rho_n - 1 \; . \end{cases} \tag{5-53b}$$

It follows that for $p_b \ll \frac{1}{2}$ the dimensionless EVSI behaves in the way shown by the solid curve in Figure 5.11; the dotted curves will be explained in a moment.

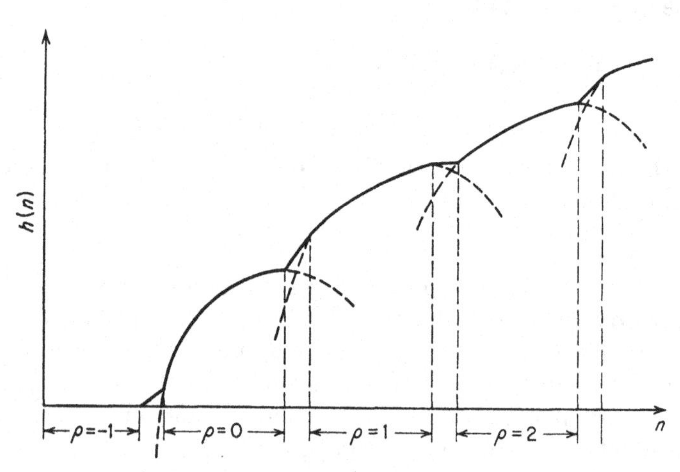

Figure 5.11

Behavior of EVSI with n: Binomial Sampling

▶ To prove (5-52a) we start from formula (5-51) for $h(n)$ and obtain, writing ρ_n for ρ_{n+1},

(1) $$\Delta h(n) = p_b[F_{\beta b}(\rho_n|r', n', n+1) - F_{\beta b}(\rho_n|r', n', n)] - \frac{r'}{n'}[F_{\beta b}(\rho_n|r'+1, n'+1, n+1) - F_{\beta b}(\rho_n|r'+1, n'+1, n)]$$

We then use (7-80) to express $F_{\beta b}$ in terms of $G_{\beta Pa}$ and (7-78) to obtain a recursion relation for $f_{\beta Pa}$:

(2) $$\begin{aligned}\Delta h(n) &= p_b[G_{\beta Pa}(n+2|r', n', \rho_n+1) - G_{\beta Pa}(n+1|r', n', \rho_n+1)] \\ &\quad - \frac{r'}{n'}[G_{\beta Pa}(n+2|r'+1, n'+1, \rho_n+1) - G_{\beta Pa}(n+1|r'+1, n'+1, \rho_n+1)] \\ &= -p_b f_{\beta Pa}(n+1|r', n', \rho_n+1) + \frac{r'}{n'} f_{\beta Pa}(n+1|r'+1, n'+1, \rho_n+1) \\ &= \left[-p_b + \frac{r'+\rho_n+1}{n'+n+1}\right] f_{\beta Pa}(n+1|r', n', \rho_n+1) \; .\end{aligned}$$

This quantity is necessarily positive by the definition (5-51b) of $\rho_{n+1} = \rho_n$.

To prove (5-52b) we start again from (5-51) but this time write $\rho_n + 1$ for ρ_{n+1}, thus obtaining

$$(3)\quad \Delta h(n) = \left[p_b F_{\beta b}(\rho_n + 1|r', n', n+1) - \frac{r'}{n'} F_{\beta b}(\rho_n + 1|r'+1, n'+1, n+1)\right]$$
$$- \left[p_b F_{\beta b}(\rho_n|r', n', n) - \frac{r'}{n'} F_{\beta b}(\rho_n|r'+1, n'+1, n)\right]$$
$$= \Delta_1 + \Delta_2$$

where

$$(4)\quad \Delta_1 = \left[p_b F_{\beta b}(\rho_n|r', n', n+1) - \frac{r'}{n'} F_{\beta b}(\rho_n|r'+1, n'+1, n+1)\right]$$
$$- \left[p_b F_{\beta b}(\rho_n|r', n', n) - \frac{r'}{n'} F_{\beta b}(\rho_n|r'+1, n'+1, n)\right] ,$$

$$(5)\quad \Delta_2 = p_b f_{\beta b}(\rho_n + 1|r', n', n+1) - \frac{r'}{n'} f_{\beta b}(\rho_n + 1|r'+1, n'+1, n+1) .$$

Now (4) is identical to (1) and therefore by (2)

$$(6)\qquad \Delta_1 = -\left[p_b - \frac{r' + \rho_n + 1}{n' + n + 1}\right] f_{\beta Pa}(n+1|r', n', \rho_n + 1) .$$

To evaluate Δ_2 we use (7-76) and (7-78) to obtain the relations

$$f_{\beta b}(\rho_n + 1|r'+1, n'+1, n+1) = \frac{n'(r' + \rho_n + 1)}{r'(n' + n + 1)} f_{\beta b}(\rho_n + 1|r', n', n+1) ,$$

$$f_{\beta b}(\rho_n + 1|r', n', n+1) = \frac{n+1}{\rho_n + 1} f_{\beta Pa}(n+1|r', n', \rho_n + 1) ,$$

and substitute in (5) to obtain

$$(7)\qquad \Delta_2 = \frac{n+1}{\rho_n + 1}\left[p_b - \frac{r' + \rho_n + 1}{n' + n + 1}\right] f_{\beta Pa}(n+1|r', n', \rho_n + 1) .$$

Substituting (6) and (7) in (3) we obtain (5-52b), which is nonnegative by the definition (5-51b) of $\rho_{n+1} = \rho_n + 1$.

To prove (5-53a) we observe that if $\rho_{n+1} = \rho_n = \rho_{n-1}$, then by (2)

$$(8)\quad \Delta h(n) - \Delta h(n-1) = f_{\beta Pa}(n+1|r', n', \rho_n + 1)\left[\frac{r' + \rho_n + 1}{n' + n + 1} - p_b\right]$$
$$- f_{\beta Pa}(n|r', n', \rho_n + 1)\left[\frac{r' + \rho_n + 1}{n' + n} - p_b\right] .$$

By a recursion formula in (7-83)

$$f_{\beta Pa}(n|r', n', \rho_n + 1) = \frac{(n' + n)(n - \rho_n)}{n(n' + n - r' - \rho_n - 1)} f_{\beta Pa}(n+1|r', n', \rho_n + 1) ,$$

and substituting this result in (8) and writing

$$r''_n = r' + \rho_n , \qquad n'' = n' + n ,$$

we obtain

$$(9)\quad \left[\left(\frac{r''_n + 1}{n'' + 1} - p_b\right) - \frac{n''(n - \rho_n)}{n(n'' - r''_n - 1)}\left(\frac{r''_n + 1}{n''} - p_b\right)\right] f_{\beta Pa}(n+1|r', n', \rho_n + 1) .$$

This expression will be negative if the factor in square brackets is negative, and the factor in square brackets is negative because, first,

$$\text{(10)} \qquad \frac{r_n'' + 1}{n''} > \frac{r_n'' + 1}{n'' + 1} > p_b \ ,$$

the first inequality being obvious and the second following from the definition (5-51b) of $\rho_{n+1} = \rho_n$, and second,

$$\text{(11)} \qquad \frac{n''(n - \rho_n)}{n(n'' - r_n'' - 1)} > 1$$

as we shall now prove. Writing the inequality in the form

$$\frac{n - \rho_n}{n} > \frac{n'' - (r_n'' + 1)}{n''}$$

we see that it is equivalent to

$$\frac{\rho_n}{n} < \frac{r_n'' + 1}{n''}$$

and will hold *a fortiori* if

$$\text{(12)} \qquad \frac{\rho_n}{n} \leq \frac{r_n''}{n''} \ .$$

But we know by (5-51b) that

$$\frac{r_n''}{n''} \equiv \frac{r' + \rho_n}{n' + n} \leq p_b \ ,$$

and since $r'/n' \geq p_b$ by the hypothesis (5-50f), (12) follows immediately.

To prove (5-53b) we first rewrite it in the form

$$\frac{\Delta h(n)}{\Delta h(n - 2)} > 1 \ .$$

We next substitute (5-52a) for the numerator and the same expression with $n - 2$ in place of n and $\rho_n - 1$ in place of ρ_n for the denominator, thus putting the assertion (5-53b) in the form

$$\text{(13)} \qquad \frac{\left(\dfrac{r' + \rho_n + 1}{n' + n + 1} - p_b\right) f_{\beta Pa}(n + 1|r', n', \rho_n + 1)}{\left(\dfrac{r' + \rho_n}{n' + n - 1} - p_b\right) f_{\beta Pa}(n - 1|r', n', \rho_n)} > 1 \ .$$

Now using (7-78) to obtain the relation

$$f_{\beta Pa}(n + 1|r', n', \rho_n + 1) = \frac{r_n''(n'' - r_n'' - 1)\, n(n - 1)}{n''(n'' - 1)\, \rho_n(n - \rho_n)} f_{\beta Pa}(n - 1|r', n', \rho_n)$$

and substituting in the numerator of (13) we see that (5-53b) will be true if

$$\text{(14)} \qquad 1 < \frac{r_n'' + 1 - (n'' + 1)\, p_b}{r_n'' - (n'' - 1)\, p_b} \cdot \frac{n'' - 1}{n'' + 1} \cdot \frac{r_n''(n'' - r_n'' - 1)\, n(n - 1)}{n''(n'' - 1)\, \rho_n(n - \rho_n)}$$

$$= \frac{(r_n'' - p_b n'') + (1 - p_b)}{(r_n'' - p_b n'') + p_b} \cdot \frac{r_n''(n'' - r_n'')\, n^2}{n''^2\, \rho_n(n - \rho_n)} \cdot \frac{n''(n'' - r_n'' - 1)(n - 1)}{(n'' + 1)(n'' - r_n'')\, n} \ .$$

The first factor on the right will be greater than 1 for $p_b < \frac{1}{2}$ if its denominator is positive, and the denominator is positive because the fact that $\rho_{n-1} = \rho_n - 1$ implies that

$$\frac{r''}{n''-1} \equiv \frac{r'+\rho_n}{n'+n-1} = \frac{r'+\rho_{n-1}+1}{n'+n-1} > p_b \ .$$

The second factor on the right is never less than 1 when $p_b < \frac{1}{2}$ because by (5-51b) and an argument in the proof of (12) above

$$\frac{\rho_n}{n} \le \frac{r''}{n''} \le p_b < \tfrac{1}{2} \ .$$

The third factor is less than 1, but it will be very close to 1 when n is large and n will be large when $\rho_n > 0$ and p_b is small. Since the first factor will be *much* greater than 1 when p_b is small, the entire expression will clearly be greater than 1 when p_b is small. (How small is small could be answered precisely, but since the condition is tantamount to requiring that the product of the first and third factors on the right-hand side of (14) be greater than 1 we shall omit this cumbersome nicety.) ◀

Some additional insight into the behavior of the EVSI graphed in Figure 5.11 can be obtained by looking at the problem for a moment in terms of normal-form rather than extensive-form analysis. It can easily be shown that if the decision maker takes a sample of size n and then applies a *decision rule* calling for act a_1 if and only if r is less than or equal to an *arbitrary* "critical value" or "acceptance number" c, his expected terminal utility will exceed the expected terminal utility of an immediate choice of a_2 by the amount

$$k_t[p_b\, F_{\beta b}(c|r', n', n) - \bar{p}'\, F_{\beta b}(c|r'+1, n'+1, n)] \ . \tag{5-54}$$

Comparing (5-51), we see that if c is the *optimal* critical value for the given n, i.e., if $c = \rho_n$ as defined by (5-51b), then (5-54) is the EVSI as we have defined the EVSI in Section 4.5. If on the contrary c is *not* optimal for the given n, i.e., if $c \neq \rho_n$, then the value of the decision rule as given by (5-54) is *less* than the EVSI. If we were to plot the value of (5-54) for fixed c as a function of n, it would coincide with the solid curve in Figure 5.11 for those n for which $\rho_n = c$ but for n outside this range the curve would behave as shown by the dotted curve in Figure 5.11; in particular, it would turn *down* as soon as n exceeded the largest value for which c is optimal. If the "acceptance number" c is fixed arbitrarily, too large a sample actually decreases *terminal* utility; the cost of taking the excessive observations is worse than wasted.

▶ To derive (5-54), we first observe that the same reasoning used to derive (5-26) shows that when utility is linear in ω the expected *increase* in utility obtained by using a decision rule of the type discussed in the text instead of simply choosing the terminal act which is optimal under the prior distribution is

$$k_t \sum\nolimits_{\tilde{\omega}'' \le c} (c - \tilde{\omega}'')\, \mathrm{P}\{\tilde{\omega}''\} \qquad \text{if} \qquad \bar{\omega}' \ge \omega_b \ ,$$

and this becomes (5-54) when evaluated by the aid of (9-20) and (9-21b).

To prove that the value of (5-54) actually decreases with n as soon as n exceeds the highest value for which $\rho_n = c$, we need only observe that by (5-52a) the first difference of (5-54) with respect to n is a positive number multiplied by

$$A \equiv \frac{r' + c + 1}{n' + n + 1} - p_b < \frac{r' + c + 1}{n' + n} - p_b \ .$$

If now n is such that

$$\frac{r' + (c + 1)}{n' + n} \leq p_b \ ,$$

i.e., if $\rho_n \geq c + 1$, then obviously $A < 0$, as was to be proved. ◀

5.6.3. *Behavior of the Net Gain of Sampling; Optimal Sample Size*

Let us now assume that the cost of sampling (or the prior expectation of this cost) is proportional to the sample size n:

$$c_s^*(e_n) = k_s n \ ; \tag{5-55}$$

generalization to the case where $c_s^*(e_n) = K_s + k_s n$ is trivial as we saw in Section 5.5.6. We can then define the *dimensionless* net gain to be expected from taking a sample of size n to be

$$g(n) \equiv \frac{1}{k_t} [v_I^*(e_n) - c_s^*(e_n)] = h(n) - \kappa n \tag{5-56a}$$

where

$$\kappa \equiv k_s / k_t \ . \tag{5-56b}$$

A local maximum of g will occur at any value n° of n such that

$$\Delta h(n^\circ - 1) > \kappa > \Delta h(n^\circ) \ , \tag{5-57}$$

where Δh is given by (5-52); but it is apparent from (5-53) or Figure 5.11 that there may be more than one local maximum, and the only way to determine (a) which of these local maxima is the greatest and (b) whether this greatest local maximum is greater than 0 net gain of $n = 0$ is actually to evaluate $g(n^\circ)$ at each n° by use of (5-56) and (5-51).

With a high-speed computer this method of determining optimal sample size is quite feasible. The best method is probably to start from $g(0) = 0$ and then to sum the first differences of g, each difference being obtained from the previous one by recursion. Thus for the case $\bar{p}' \geq p_b$, for which the first differences are given by (5-52), it is not hard to show that $g(n)$ can be computed by starting from the base

$$g(0) = 0 \ , \qquad \phi(0) = \frac{n' - r'}{n'} \ ,$$

$$\rho_0 = -1 \ , \qquad \psi(0) = \frac{r'}{n' + 1} - p_b \ ,$$

and then recursively computing in the following order

$$g(n+1) = g(n) - \kappa + \begin{cases} 0 & \text{if} \quad g(n) = -\kappa n, \ \psi(n) > 0 \ , \\ (\rho_n - n)\, \psi(n)\, \phi(n) & \text{if} \quad g(n) \geq -\kappa n, \ \psi(n) \leq 0 \ , \\ (\rho_n + 1)\, \psi(n)\, \phi(n) & \text{if} \quad g(n) > -\kappa n, \ \psi(n) > 0 \ , \end{cases}$$

$$\phi(n+1) = \begin{cases} \dfrac{(n'+n-r'-\rho_n)(n+1)}{(n-\rho_n+1)(n'+n+1)}\,\phi(n) & \text{if} \quad \psi(n) > 0 \ , \\[2ex] \dfrac{(r'+\rho_n+1)(n+1)}{(\rho_n+2)(n'+n+1)}\,\phi(n) & \text{if} \quad \psi(n) \le 0 \ , \end{cases}$$

$$\rho_{n+1} = \begin{cases} \rho_n & \text{if} \quad \psi(n) > 0 \ , \\ \rho_n + 1 & \text{if} \quad \psi(n) \le 0 \ , \end{cases}$$

$$\psi(n+1) = \frac{r' + \rho_{n+1} + 1}{n' + n + 2} - p_b \ .$$

If on the contrary the computations are to be carried out by hand, evaluation of $g(n)$ for all n up to and beyond the optimal n° will scarcely be feasible unless n° is extremely small. In most cases, it will be much less laborious to start from an approximation to n° obtained by the rough and ready but surprisingly accurate method which we shall describe in the next section. The exact net gain at the approximate optimum can then be computed by using one of the methods discussed in Section 7.11.2 to evaluate the cumulative beta-binomial functions in (5-51), after which recursion formulas can be used to trace the net gain in a limited neighborhood of the approximate optimum.

5.6.4. *A Normal Approximation to Optimal Sample Size*

Because no really convenient way of determining optimal sample size has been found except in the case where the sample observations are Normally distributed with known precision and the prior distribution of the process mean is Normal, it is natural to inquire whether the results obtained for this special case in Section 5.5 above can be used to find an *approximation* to optimal sample size in other cases, since if it can, then the true optimum can be found by computation of the true net gain for a few values of n in the neighborhood of the approximate optimum. We have made no systematic investigation of this question, but experience with a number of numerical examples seems to show empirically that an approximation of this kind will very often be astonishingly close to the true optimum, particularly if we remember that *the comparison should be made, not between the approximate and true optimal* sample sizes, *but between the* expected net gains *of the two sample sizes.*

We shall illustrate one method of approximation and test its accuracy by applying it to the three examples of economically determined optimal sample size discussed by J. Sittig in "The economic choice of a sampling system in acceptance sampling," *Bulletin of the International Statistical Institute* Vol. 33 (International Statistical Conferences 1951, India) Part 5 pp. 51–84. All three involve acceptance sampling from finite lots but are treated by Sittig as if (1) the number of pieces affected by the terminal act were independent of the sample size and (2) sampling were binomial rather than hypergeometric; we shall follow Sittig in both respects, since our purpose is merely to compare approximate with exact *analysis* of a given set of assumptions.†

Sampling Variance. Letting $\tilde{p}$ denote the unknown fraction defective, we see at once that the conditional variance of a single observation $\tilde{x}$ is a random variable $\tilde{p}(1 - \tilde{p})$ in contradiction to the assumption of Section 5.5.1 that the variance $1/h$ of a single observation is a known number. To apply the theory of Section 5.5 we shall set

$$\frac{1}{h} = \mathrm{V}(\tilde{x}|\tilde{p} = \bar{p}') = \bar{p}'(1 - \bar{p}')$$

where $\bar{p}'$ is the mean of the prior distribution of $\tilde{p}$; our justification for this way of handling the difficulty is (a) that it works and (b) that we do not know how to choose among various more "precise" approximations such as $1/h = \mathrm{E}[\tilde{p}(1 - \tilde{p})]$ and $\sqrt{1/h} = \mathrm{E}\sqrt{\tilde{p}(1 - \tilde{p})}$.

Prior Distribution. In all examples to be discussed Sittig assigns to $\tilde{p}$ a beta prior distribution of the form

$$f_\beta(p|1, \nu) \propto (1 - p)^{(\nu - 1) - 1} ,$$

which by (7-22) has mean and variance

$$\mathrm{E}'(\tilde{p}|1, \nu) = \bar{p}' = \frac{1}{\nu} ; \qquad \mathrm{V}'(\tilde{p}|1, \nu) = \frac{\nu - 1}{\nu^2(\nu + 1)} .$$

We shall "approximate" this by a Normal distribution with the same mean and variance,

$$f_N(p|m', hn') \propto e^{-\frac{1}{2}hn'(p - m')^2}$$

where

$$m' = \mathrm{E}'(\tilde{p}) = \bar{p}' = \frac{1}{\nu} , \qquad \frac{1}{h} = \bar{p}'(1 - \bar{p}') = \frac{\nu - 1}{\nu^2} , \qquad n' = \frac{1}{h\,\mathrm{V}'(\tilde{p})} = \nu + 1 .$$

Example 1. An acceptance lot contains 500 brush spindles for floor polishers. Acceptance of a defective ultimately leads to loss of the part to which it is welded; this part costs 3.90 guilders. A rejected lot is screened at a cost of .079 guilders per piece; sampling inspection costs this same amount per piece. The sampling cost is

$$c_s^*(e_n) = k_s n , \qquad k_s = .079 ;$$

the terminal costs (treated as if they were incurred on the whole lot rather than on only the uninspected portion) are

$$\text{cost of acceptance} = K_1 + k_1 p = 3.90 \times 500p = 1950p ,$$

$$\text{cost of rejection} = K_2 + k_2 p = .079 \times 500 = 39.5 ;$$

and from these we can compute

† The first of Sittig's two simplifications is indeed a simplification, but treatment of sampling as binomial rather than hypergeometric amounts to treating the sample as taken from the process which produced the lot and is exact on the assumption that the process average remains constant during the production of any one lot. Cf. Section 1.4.1 above.

$$k_t = |k_2 - k_1| = |0 - 1950| = 1950 \; , \qquad p_b = \frac{39.5}{1950} = .020256 \; .$$

The beta prior distribution of $\tilde{p}$ has mean

$$\bar{p}' = .022 \qquad \text{implying} \qquad \nu = \frac{1}{.022} = 45.45 \; ;$$

the Normal approximation therefore has parameters

$$m' = .022 \; , \qquad n' = 46.45 \; , \qquad \frac{1}{h} = .022 \times .978 = .02152 \; .$$

Both the "true" prior density and its Normal "approximation" are graphed in Figure 5.12.

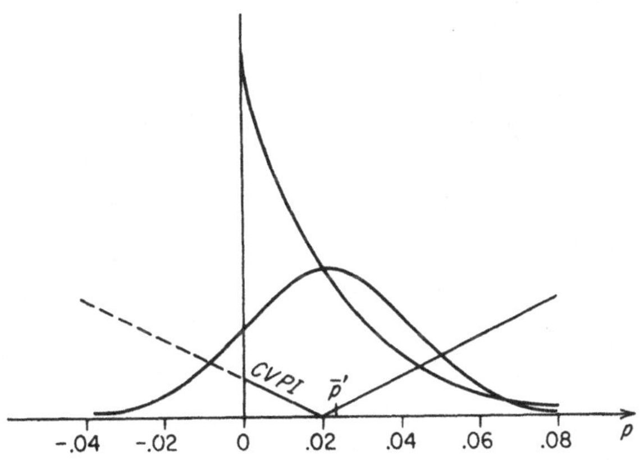

Figure 5.12

Terminal Opportunity Losses and Prior Distribution, Example 1

To find the optimal sample size under the Normal approximation to the real problem we first compute the sampling and prior standard deviations defined by (5-39),

$$\sigma_\epsilon = h^{-\frac{1}{2}} = \sqrt{.02152} = .1467 \; ,$$

$$\sigma_p = (hn')^{-\frac{1}{2}} = \sqrt{.02152/46.45} = .02152 \; ,$$

we then compute the parameters D' and Z defined by (5-40bc) and (5-43b)

$$D' = \frac{|p_b - m'|}{\sigma_p} = \frac{|.020256 - .022000|}{.02152} = .081 \; ,$$

$$Z = \frac{\sigma_p}{\sigma_\epsilon} \sqrt[3]{k_t \sigma_\epsilon / k_s} = \frac{.02152}{.1467} \sqrt[3]{1950 \times .1467/.079} = .1467 \times 15.36 = 2.25 \; ,$$

and find from Chart I at the end of this monograph that

$$\frac{n^\circ}{(k_t\sigma_\epsilon/k_s)^{\frac{2}{3}}} \doteq .169 \; , \qquad \text{so that} \qquad n^\circ \doteq .169(15.36)^2 = 39 \; .$$

In Figure 5.13 we graph as functions of n the exact net gain as given by (5-56) and (5-51) and the net gain as given by the Normal approximation (5-41). The true optimal sample size is $n^\circ = 32$ with a true net gain of 5.80 guilders; the true net gain of the sample $n = 39$ obtained by use of the Normal approximation is 5.66 guilders or 97.6% of the maximum.

Example 2. An acceptance lot contains 500 radio potentiometers. Rejected lots are screened, while acceptance of a defective part ultimately leads to partial

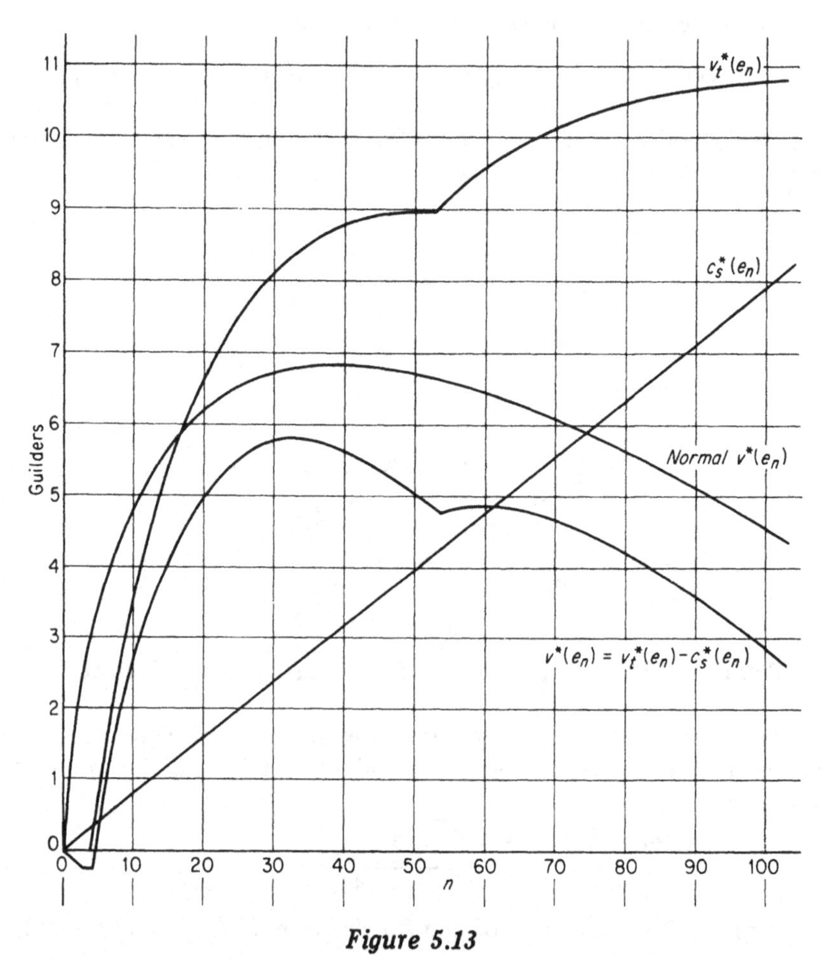

Figure 5.13

Net Gain as a Function of Sample Size, Example 1

disassembly and reassembly of the radio set at a cost 10.8 times as great as the cost of inspecting one part. The prior distribution has density $f_\beta(p|1, 40)$; its mean is $\bar{p}' = 1/40 = .025$ and its standard deviation is $\sigma_p = .0244$. The "true" prior distribution and the Normal approximation are graphed in Figure 5.14.

After we have computed $p_b = .0926$, $k_t = 5400$ in units of the cost of inspecting one part, $\sigma_\epsilon = .156$, $D' = 2.77$, and $Z = 1.48$, we find from Figure 5.10 above that no sample should be taken. Exact calculations show that there in fact exists no sample with a positive net gain; they are easy to carry out because the EVPI or expected value of *perfect* information concerning the true value of $\tilde{p}$ can be shown to be only 2.72 times the cost of a single sample observation and therefore the largest sample which need be considered is $n = 2$. Since the prior mean

$$\bar{p}' = \frac{r'}{n'} = \frac{1}{40} = .025 < .0926 = p_b ,$$

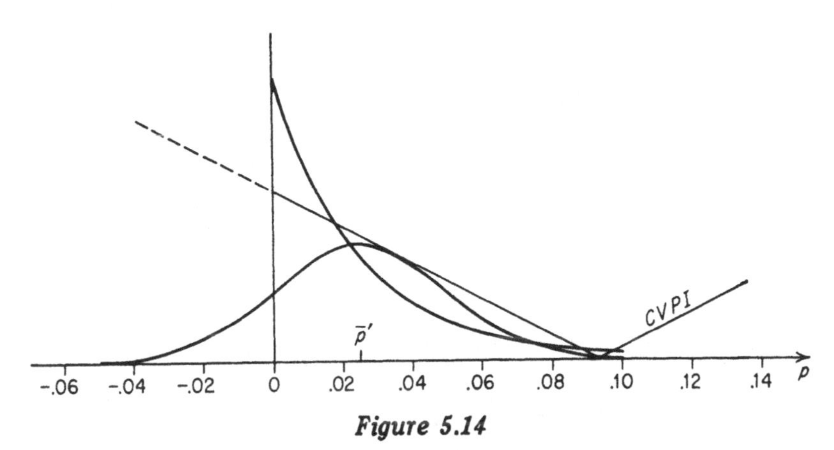

Figure 5.14

Terminal Opportunity Losses and Prior Distribution, Example 2

the optimal *immediate* terminal act is acceptance without screening; if both pieces in a sample of 2 were defective, the posterior mean would be

$$\bar{p}'' = \frac{r''}{n''} = \frac{1+2}{40+2} = .0714 < .0926 = p_b ,$$

so that the optimal act would still be acceptance without screening. Since the largest sample worth consideration cannot lead to a reversal of the initial decision, the EVSI of all samples worth consideration is 0; and therefore *all* samples have a negative net gain.

Example 3. An acceptance lot contains 200 sheets of toughened glass to be used in busses. Testing is destructive; the cost of testing one sheet is essentially the cost of manufacturing the sheet, since the labor involved in the testing is negligible in comparison. The "cost" of passing a defective sheet consists in the resulting danger to drivers and passengers and is evaluated at 10 times the cost of manufacturing one sheet, while rejection of a lot leads to reannealing and re-toughening of the glass at a cost equal to half the total manufacturing cost. The prior distribution has density $f_\beta(p|1, 20)$; its mean is $\bar{p}' = .05$ and its standard deviation is $\sigma_p = .0475$. The "true" prior distribution and its Normal "approximation" are graphed in Figure 5.15.

After we have calculated $p_b = \frac{1}{2}/10 = .05$, $k_t = 10 \times 200 = 2000$ in units of the cost of manufacturing one sheet, $\sigma_\epsilon = .218$, $D' = 0$, and $Z = 1.65$, we find from Chart I at the end of this monograph that

$$n^\circ = .124(2000 \times .218/1)^{\frac{2}{3}} = 7 .$$

In Figure 5.16 we graph as a function of n both the exact net gain of e_n and the net gain as calculated on the basis of the Normal approximation. The true optimal sample size is $n^\circ = 8$ with a net gain of 11.10 times the cost of one sheet of glass; the net gain of the sample $n = 7$ obtained by use of the Normal approximation is 11.00 times the cost of one sheet or 99.1% of the maximum.

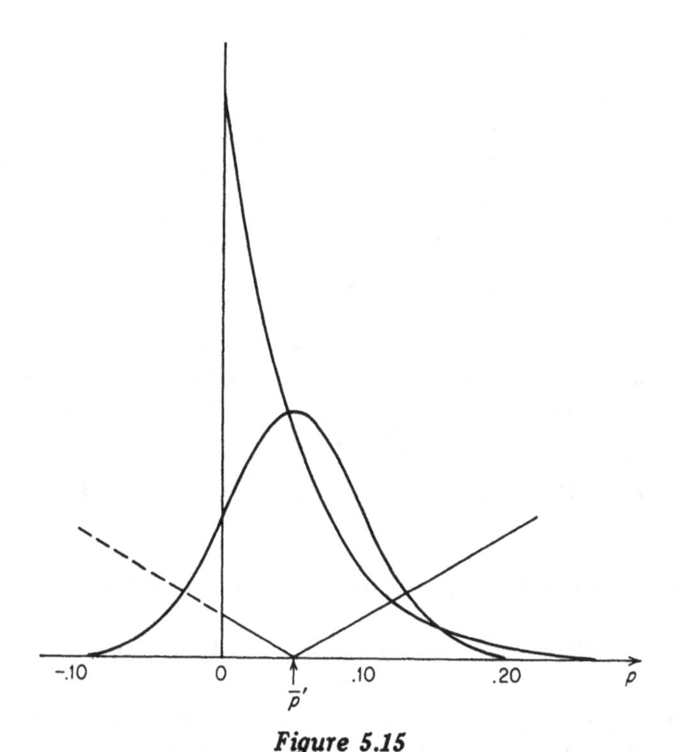

Figure 5.15

Terminal Opportunity Losses and Prior Distribution, Example 3

Why the Approximation Works. In Figures 5.12, 5.14, and 5.15 above we have shown not only the "true" and "approximate" prior distributions for the three examples but also the terminal-opportunity-loss functions $l_t(a, p)$ for *both* possible acts, accept and reject. The CVPI or conditional value of perfect information is $l_t(a', p)$ where a' is the *better* act under the prior distribution; it is given by the left-hand loss line in Figure 5.12, by the right-hand loss line in Figure 5.14, and by either of the two loss lines in Figure 5.15.

Since the CVSI or conditional value of *sample* information is given by the same function as the conditional value of *perfect* information when terminal utility is linear (cf. Section 5.3.2), the EVSI or *expected* value of sample information can be regarded as the integral of the CVPI function weighted by the prior distribution of the posterior mean $\tilde{\bar{p}}''$; and since the distribution of $\tilde{\bar{p}}''$ tends to be of the same general shape as the prior distribution of $\tilde{p}$ (cf. Section 5.4.2), it seems astonishing at first sight that the Normal approximation should lead to anything remotely resembling the true optimal sample size in the example illustrated by Figure 5.12. The mystery can be partially explained, however, by a moment's reflection on the implications of Figure 5.15, where since $\bar{p}' = p_b$ either act is optimal under the prior distribution and therefore *either* of the two l_t functions can be taken as representing the CVPI function. It seems reasonable in this case that the exact

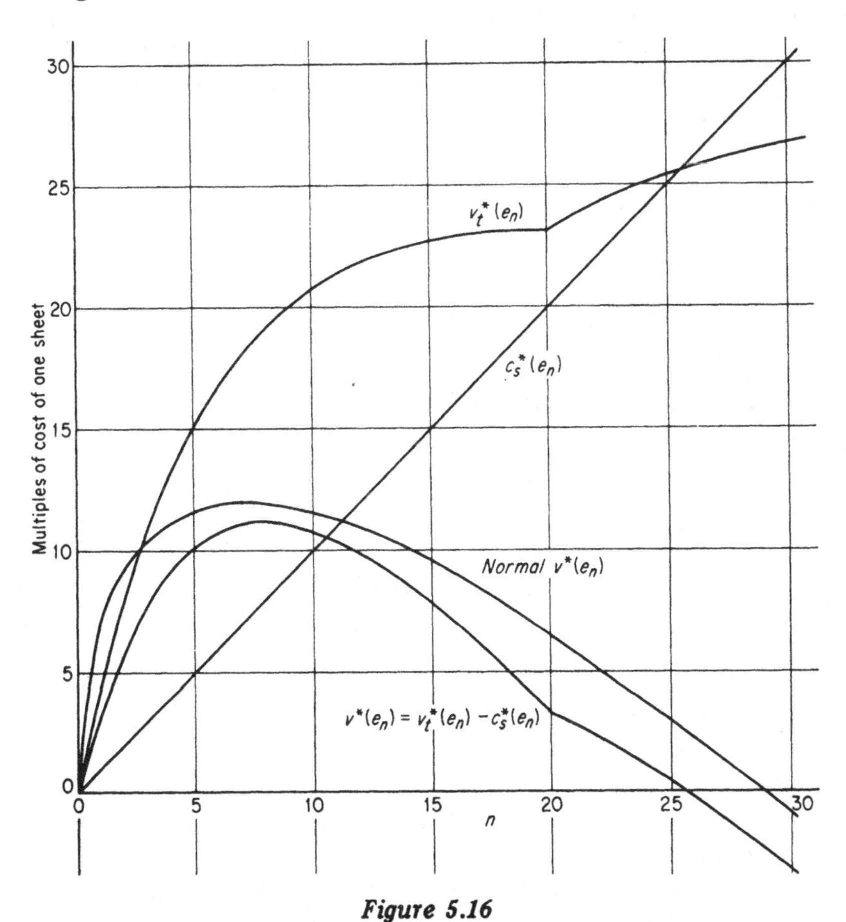

Figure 5.16

Net Gain as a Function of Sample Size, Example 3

and the approximate integral of the *right*-hand l_t function should be of roughly the same magnitude for given n and behave in roughly the same way as n increases; and if the behavior of the approximate integral with n is roughly correct, the value n° of n at which it attains its maximum will be roughly correct.

Returning now to the case of Figure 5.12, it again seems plausible that the integral of the *right*-hand l_t function weighted by the Normal approximation to the distribution of $\tilde{\tilde{p}}''$ should be roughly equal to the integral of this same function weighted by the exact distribution of $\tilde{\tilde{p}}''$. The approximate and exact EVSI's are of course given by the integrals of the *left*-hand l_t functions in this case, and these integrals are not equal to the corresponding right-hand integrals; but the difference between either left-hand integral and the corresponding right-hand integral is a constant independent of sample size as proved below. Consequently the *rate of change with n* of either left-hand integral *is* equal to the rate of change of the corresponding right-hand integral, and then since the *magnitude* of the *right*-hand approximate integral is roughly correct in the neighborhood of the optimal n, it is not too surprising that the *rate of change* of the *left*-hand approximate integral should be roughly correct and therefore that the approximation to optimal n should be roughly correct.

▶ Letting a_1 denote the act which is optimal for low values of p, the left- and right-hand integrals under the prior distribution of the posterior mean are respectively

$$I_l = \mathrm{E}\, l_t(a_2, \tilde{\bar{p}}'') = \mathrm{E}[\max\{u_t(a_1, \tilde{\bar{p}}''), u_t(a_2, \tilde{\bar{p}}'')\} - u_t(a_2, \tilde{\bar{p}}'')]$$

$$I_r = \mathrm{E}\, l_t(a_1, \tilde{\bar{p}}'') = \mathrm{E}[\max\{u_t(a_1, \tilde{\bar{p}}''), u_t(a_2, \tilde{\bar{p}}'')\} - u_t(a_1, \tilde{\bar{p}}'')]$$

and therefore

$$I_l - I_r = \mathrm{E}\, u_t(a_1, \tilde{\bar{p}}'') - \mathrm{E}\, u_t(a_2, \tilde{\bar{p}}'') = u_t(a_1, \bar{p}') - u_t(a_2, \bar{p}') = \text{constant}$$

as was to be proved. ◀

CHAPTER 5B

Selection of the Best of Several Processes

5.7. Introduction; Basic Assumptions

One of the classical problems of statistics is that of choosing the "best" of several "treatments"—e.g., the best of several different fertilizers or the best amount to use of a given fertilizer, the best of several package designs for a commercial product, the best of several educational methods, and so forth. Frequently additional information concerning the "quality" of any one of the treatments can be obtained by sampling, so that the decision maker must decide (1) how large a sample (if any) to take on each of the various treatments, and then (2) which of the treatments to choose after the various sample outcomes are known.†

In the present chapter we shall analyze this problem on the assumptions (1) that the utilities of terminal action and experimentation are additive, and (2) that the terminal utility of adopting any particular treatment is linear in the quality of that particular treatment and independent of the qualities of the rejected treatments. Formally, we assume that

$$u(e, z, a_i, \theta) = u_t(a_i, \theta) - c_s(e, z) \; , \qquad i = 1, 2, \cdots, r \; , \tag{5-58}$$

and that

There exists a mapping W from the state space Θ to a new space Ω, sending θ into $W(\theta) = \omega = (\omega_1, \omega_2, \cdots, \omega_r)$, such that

$$u_t(a_i, \theta) = K_i + k_i\omega_i \; , \qquad i = 1, 2, \cdots r \; . \tag{5-59}$$

The quantity ω_i may represent the mean yield of the ith fertilizer, the mean score on a certain test of students taught by the ith method, and so forth. Specific analytical results will be obtained only for the still further restricted case where sample observations on the ith process are independently Normally distributed with mean ω_i.

5.8. Analysis in Terms of Differential Utility

Analysis of the class of problems to which this chapter is devoted is greatly facilitated by formulating the problems in terms of *utility differences* or *differential utilities*, but because this approach may be useful in other classes of problems as

† For an historical account of the literature, as well as for some very interesting contributions to this problem and its variants, see C. W. Dunnett, "On selecting the largest of k normal population means", *Journal of the Royal Statistical Society* (Series B), 1960.

well we shall first define the concepts and notation without any reference to our linearity assumption (5-59). We do assume, however, that *terminal and sampling utilities are additive* and that $A = \{a_1, \cdots, a_r\}$ is finite.

5.8.1. *Notation: the Random Variables $\tilde{v}$ and $\tilde{\delta}$*

When the state $\tilde{\theta}$ is a random variable, the terminal utility of an act a_i is a random variable which we shall now denote by $\tilde{v}_i$, defining

$$\tilde{v}_i \equiv u_t(a_i, \tilde{\theta}) \; , \qquad i = 1, 2, \cdots, r \; . \tag{5-60a}$$

We also define the vector of the $\tilde{v}$s describing all the as in A,

$$\tilde{v} \equiv (\tilde{v}_1, \cdots, \tilde{v}_r) \; . \tag{5-60b}$$

The measure P'_θ assigned by the decision maker induces a measure P'_v on $\tilde{v}$ with respect to which we can evaluate

$$E'(\tilde{v}) \equiv \bar{v}' = (\bar{v}'_1, \cdots, \bar{v}'_r) \; ,$$

and we shall find it convenient to adopt the convention of numbering the acts in A in such a way that

$$\bar{v}'_r \geq \bar{v}'_i \; , \qquad i = 1, 2, \cdots, r \; ; \tag{5-61}$$

in other words, we assume without loss of generality that *the rth or last act is optimal under the prior distribution.*

Finally, we denote by $\tilde{\delta}_i$ the difference between the terminal utility of a_i and the terminal utility of a_r, defining

$$\tilde{\delta}_i \equiv \tilde{v}_i - \tilde{v}_r \equiv u_t(a_i, \tilde{\theta}) - u_t(a_r, \tilde{\theta}) \; , \qquad i \neq r \; ; \tag{5-62a}$$

and we also define the vector of all the $\tilde{\delta}$s,

$$\tilde{\delta} \equiv (\tilde{\delta}_1, \cdots, \tilde{\delta}_{r-1}) \; . \tag{5-62b}$$

5.8.2. *Analysis in Terms of $\tilde{v}$ and $\tilde{\delta}$*

Terminal Analysis. We have already remarked that the prior measure P'_θ will induce a prior measure P'_v with respect to which we can evaluate

$$E'(\tilde{v}) \equiv \bar{v}' = (\bar{v}'_1, \cdots, \bar{v}'_r)$$

and thus select an act which is optimal under the prior distribution; and obviously an experiment e with outcome z which substitutes for P'_θ the posterior measure $P''_{\theta|z}$ will induce a posterior measure $P''_{v|z}$ which permits us to evaluate

$$E''_{v|z}(\tilde{v}) \equiv \bar{v}'' = (\bar{v}''_1, \cdots, \bar{v}''_r)$$

and thus select an act which is optimal under the posterior distribution. In this notation, terminal analysis reduces to the problem of finding the mean of $\tilde{v}$ as determined by the distribution of $\tilde{\theta}$ and the definition (5-60) of $\tilde{v}$.

Expected Value of Perfect Information. By our convention (5-61) the *prior* expected terminal utility $\bar{v}'_r$ of a_r is at least as great as the prior expected terminal utility of any other a in A, but its *actual* terminal utility may be exceeded by that of one or more other as in A. If the decision maker were to be given perfect information on θ and thus on v, he would choose an act a^* such that

$$u_t(a^*, \theta) = \max \{v_1, \cdots, v_r\}$$

rather than the act a_r which was optimal under the prior distribution. The value of this information is obviously

$$v_t(e_\infty, \theta) \equiv l_t(a_r, \theta) = \max \{v_1, \cdots, v_r\} - v_r$$
$$= \max \{(v_1 - v_r), \cdots, (v_r - v_r)\} ,$$

and by the definition (5-62) of δ this can be written

$$v_t(e_\infty, \theta) \equiv l_t(a_r, \theta) = \max \{\delta_1, \cdots, \delta_{r-1}, 0\} . \tag{5-63}$$

Before the perfect information is received, this quantity is a random variable; but by using the measure P'_δ induced by the measure P'_θ via P'_v the decision maker can compute the *expected value of perfect information*

$$v_t^*(e_\infty) = l_t^*(e_0) = E'_\delta \max \{\tilde{\delta}_1, \cdots, \tilde{\delta}_{r-1}, 0\} . \tag{5-64}$$

The geometry of this computation is shown in Figure 5.17 for the case where $r = 3$ and $\tilde{\delta}$ has a density; the straight lines are "loss contours" along which the CVPI or conditional opportunity loss, $\max \{\delta_1, \delta_2, 0\}$, is constant, while the curved lines are "probability contours" along which the density of $\tilde{\delta} = (\tilde{\delta}_1, \tilde{\delta}_2)$ is constant.

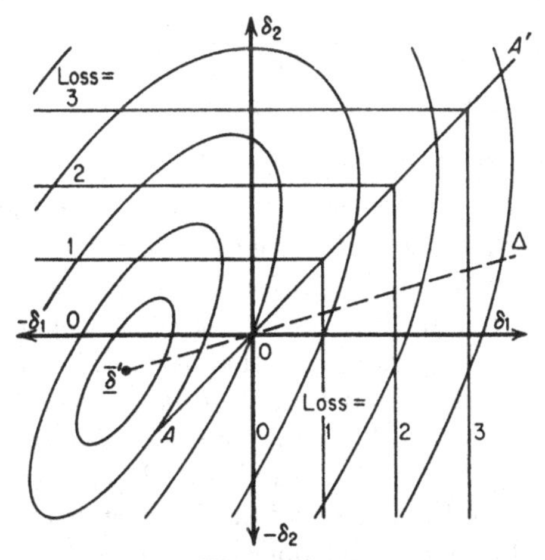

Figure 5.17

Terminal Opportunity Losses and Distribution of δ

Preposterior Analysis; Expected Value of Sample Information. We have already remarked that if the outcome z of some experiment e results in replacement of P'_v by $P''_{v|z}$, the decision maker will select an act such that his expected utility is

$$\max \{\bar{v}_1'', \cdots, \bar{v}_r'')$$

rather than the act a_r which was optimal under P'_v. The resulting increase in utility or value of the information z is obviously

$$v_t(e, z) = \max \{\bar{v}_1'', \cdots, \bar{v}_r''\} - \bar{v}_r'' = \max \{(\bar{v}_1'' - \bar{v}_r''), \cdots, (\bar{v}_r'' - \bar{v}_r'')\} ,$$

and by the definition (5-62) of $\bar{\delta}$ this can be written

$$v_t(e, z) = \max \{\bar{\delta}_1'', \cdots, \bar{\delta}_{r-1}'', 0\} . \tag{5-65}$$

When e is being considered but has not been performed, $\tilde{z}$ is a random variable and therefore for any $i \neq r$

$$\tilde{\delta}_i'' = \tilde{\bar{v}}_i'' - \tilde{\bar{v}}_r'' = \mathrm{E}_{\theta|z}'' \, u_t(a_i, \tilde{\theta}) - \mathrm{E}_{\theta|z}'' \, u_t(a_r, \tilde{\theta})$$

is a random variable with a measure $\mathrm{P}_{\bar{\delta}''|e}$ induced by the marginal measure $\mathrm{P}_{z|e}$. By expecting (5-65) with respect to this measure the decision maker can compute for any e the *expected value of sample information*

$$v_I^*(e) = \mathrm{E}_{\bar{\delta}''|e} \max \{\tilde{\delta}_1'', \cdots, \tilde{\delta}_{r-1}'', 0\} \; . \tag{5-66}$$

Expression (5-66) for the expected value of sample information is formally identical to expression (5-64) for the expected value of perfect information and therefore the geometry of (5-66) is also given by Figure 5.17; the only changes are in the labelling of the axes ($\bar{\delta}''$ instead of δ) and the shape of the probability (but not the loss) contours.

Net Gain of Sampling. Denoting by $c_s^*(e)$ the (prior expected) cost (in utiles) of performing the experiment e, we saw in (4-38) and (4-39) that the decision maker's utility will be maximized if he selects an e which maximizes the net gain of sampling

$$v^*(e) \equiv v_I^*(e) - c_s^*(e) \; . \tag{5-67}$$

By repeating the evaluation of (5-66) for every e in E and also computing $c_s^*(e)$ for each e the decision maker can find an optimal e.

5.8.3. *The Usefulness of Differential Utility*

The theory developed in Section 5.8.2 amounts really to nothing more than a demonstration that maximization of the expected difference between the utility of e and the utility of the optimal immediate terminal act is equivalent to maximization of the utility of e, and the reader may well wonder why we have taken the trouble to reformulate the problem in this way. The answer is given by Figure 5.17: the introduction of the differential utility $\tilde{\delta}$ permits us to reduce the analysis of any decision problem to (1) a series of systematic transformations of probability distributions, from P_θ to P_v to P_δ and from $\mathrm{P}_{z|e}$ to $\mathrm{P}_{\bar{\delta}''|e}$, and (2) expectation with respect to P_δ or $\mathrm{P}_{\bar{\delta}''|e}$ of the fairly "clean" function $\max \{\delta_1, \cdots, \delta_{r-1}, 0\}$. As we have already said, the method is applicable whenever we can obtain the distributions P_δ and $\mathrm{P}_{\bar{\delta}''|e}$, either analytically or in a numerical form which permits a Monte-Carlo evaluation of the expectation. In the remainder of this chapter we shall consider these problems in detail for the most important special case.

5.9. Distribution of $\tilde{\delta}$ and $\tilde{\delta}''$ When the Processes are Independent Normal and $\tilde{v}_i$ is Linear in $\tilde{\mu}_i$

5.9.1. *Basic Assumptions; Notation*

We now return to the basic assumption that

$$v_i \equiv u_t(a_i, \theta) = K_i + k_i \, \omega_i \qquad \text{where} \qquad \omega = W(\theta)$$

and develop general analytical results for the case where $\omega_1, \cdots, \omega_r$ are the means $\mu_1, \cdots, \mu_r$ of r Independent Normal data-generating processes. We assume, in other words, that the ith process ($i = 1, \cdots, r$) generates independent scalar random variables $\tilde{x}_i^{(1)}, \cdots, \tilde{x}_i^{(j)}, \cdots$ with identical densities

$$f_N(x_i|\mu_i, h_i) \propto e^{-\frac{1}{2}h_i(x_i-\mu_i)^2} h_i^{\frac{1}{2}} \tag{5-68}$$

and that the terminal utility of the ith act is

$$v_i = K_i + k_i\mu_i \ . \tag{5-69}$$

As regards the process precisions $h_1, \cdots, h_r$ we shall consider two cases: (1) all h_i are known, and (2) the h_i are unknown but the ratios h_i/h_j are all known. In the present Section 5.9 we shall derive the distributions of the random variable $\tilde{\delta}$ defined by (5-62) and of its posterior mean $\bar{\tilde{\delta}}''$; in the remainder of the chapter we shall discuss the problem of using these distributions to evaluate the EVPI as given by (5-64) and the EVSI as given by (5-66).

We now adopt the vector and matrix notation described in Section 8.0.1 below, using **boldface roman** or vertical Greek letters (either upper or lower case) for matrices, ***boldface italic*** or Porson Greek letters (either upper or lower case) for column vectors, and denoting transposition by a superscript t.

5.9.2. *Conjugate Distributions of $\tilde{\boldsymbol{\mu}}$*

1. *Likelihood of a Sample.* The distribution of $\tilde{\delta}$ is determined by the distribution of $\tilde{\boldsymbol{v}}$ and this in turn by the distribution of $\tilde{\boldsymbol{\mu}}$; and since we wish to use a distribution of $\tilde{\boldsymbol{\mu}}$ which is conjugate to the likelihood of the experimental outcome, our first task is to examine this likelihood. Before doing so we simplify notation by defining the *mean precision* of the r processes

$$h \equiv (h_1 \cdot h_2 \cdots h_r)^{1/r} \ . \tag{5-70}$$

Now consider an experiment with a noninformative stopping process (cf. Section 2.3 above) resulting in n_i observations $(x_i^{(1)}, \cdots, x_i^{(n_i)})$ on the ith process, $i = 1, \cdots, r$. Define for each univariate process

$$\begin{aligned}
n_{ii} &\equiv n_i(h_i/h) \\
m_i &\equiv \begin{cases} \dfrac{1}{n_i}\sum_{j=1}^{n_i} x_i^{(j)} & \text{if} \quad n_i > 0 \ , \\ 0 & \text{if} \quad n_i = 0 \ , \end{cases} \\
\nu_i &\equiv \begin{cases} n_i - 1 & \text{if} \quad n_i > 0 \ , \\ 0 & \text{if} \quad n_i = 0 \ , \end{cases} \\
v_i &\equiv \begin{cases} \dfrac{1}{\nu_i}\sum_{j=1}^{n_i} (x_i^{(j)} - m_i)^2 & \text{if} \quad \nu_i > 0 \ , \\ 0 & \text{if} \quad \nu_i = 0 \ ; \end{cases}
\end{aligned} \tag{5-71}$$

and define the multivariate statistics

$$\mathbf{n} \equiv \begin{bmatrix} n_{11} & & 0 \\ & \ddots & \\ 0 & & n_{rr} \end{bmatrix} ,$$

$$m \equiv [m_1 \quad \cdots \quad m_r]^t \; , \tag{5-72}$$

$$\nu \equiv \Sigma_{i=1}^r \, \nu_i \; ,$$

$$v \equiv \frac{1}{\nu} \Sigma_{i=1}^r \frac{h_i}{h} \nu_i \, v_i \; .$$

$$p \equiv \text{rank}(\mathbf{n}) = \text{number of processes for which } n_i > 0 \; .$$

It is shown in Sections 12.8 and 12.9 of Part III that the kernel of the joint likelihood of the combined samples is

$$(2\pi)^{-\frac{1}{2}(p+\nu)} \cdot e^{-\frac{1}{2}h(m-\mu)^t \mathbf{n} (m-\mu)} \, h^{\frac{1}{2}p} \cdot e^{-\frac{1}{2}h\nu v} \, h^{\frac{1}{2}\nu} \; . \tag{5-73}$$

2. *Conjugate Distribution of $\tilde{\mu}$ When h is Known.* If h is known, the kernel of (5-73)—the only factor which varies with an unknown parameter—is

$$e^{-\frac{1}{2}h(m-\mu)^t \mathbf{n} (m-\mu)} \; ; \tag{5-74}$$

and accordingly the conjugate family for $\tilde{\mu}$ is multivariate Normal as defined by (8-17),

$$f_N^{(r)}(\mu|m, h\mathbf{n}) \propto e^{-\frac{1}{2}h(\mu-m)^t \mathbf{n} (\mu-m)} \; , \tag{5-75}$$

where h is the mean precision of the *processes* as defined by (5-70) and $\mathbf{n}$ is an $r \times r$ positive-definite symmetric matrix. By (12-13) the mean and variance of this distribution are

$$\bar{\mu} \equiv \text{E}(\tilde{\mu}) = m \; , \qquad \check{\mu} \equiv \text{V}(\tilde{\mu}) = (h\mathbf{n})^{-1} \; . \tag{5-76}$$

A *prior* distribution of $\tilde{\mu}$ with parameter $(m', \mathbf{n}')$ would in general be *assessed* by assigning a mean vector $\bar{\mu}'$ and a positive-definite symmetric matrix $\check{\mu}'$ of variances and covariances; m' and $\mathbf{n}'$ would then be *computed* from the relations (5-76). As for the *posterior* distribution, it is shown in Sections 12.8 and 12.9 that if the prior distribution is Normal with parameter $(m', \mathbf{n}')$ and a sample then yields a statistic $(m, \mathbf{n})$, the posterior distribution is again Normal with parameters

$$\mathbf{n}'' = \mathbf{n}' + \mathbf{n} \; , \qquad m'' = \mathbf{n}''^{-1}(\mathbf{n}'m' + \mathbf{n}m) \; . \tag{5-77}$$

3. *Conjugate Distribution of $\tilde{\mu}$ When h is Unknown.* If h is unknown, the kernel of (5-73) is

$$e^{-\frac{1}{2}h(m-\mu)^t \mathbf{n} (m-\mu)} \, h^{\frac{1}{2}p} \cdot e^{-\frac{1}{2}h\nu v} \, h^{\frac{1}{2}\nu} \; ; \tag{5-78}$$

and accordingly the conjugate family for the *joint* distribution of $(\tilde{\mu}, \tilde{h})$ is Normal-gamma as defined by (12-27),

$$f_{N\gamma}^{(r)}(\mu, h|m, v, \mathbf{n}, \nu) \equiv f_N^{(r)}(\mu|m, h\mathbf{n}) \, f_{\gamma 2}(h|v, \nu)$$

$$\propto e^{-\frac{1}{2}h(\mu-m)^t \mathbf{n} (\mu-m)} \, h^{\frac{1}{2}r} \cdot e^{-\frac{1}{2}h\nu v} \, h^{\frac{1}{2}\nu-1} \; , \tag{5-79}$$

where $\mathbf{n}$ is an $r \times r$ positive-definite symmetric matrix and $\nu, v > 0$; because $\mathbf{n}$ is positive-definite and therefore of rank r, the statistic p of (5-78) is replaced by the parameter r in (5-79). It is shown in Sections 12.8 and 12.9 that if the *prior* distribution of $(\tilde{\mu}, \tilde{h})$ is Normal-gamma with parameter $(m', v', \mathbf{n}', \nu')$ and a sample then yields a statistic $(m, v, \mathbf{n}, \nu)$ where $\mathbf{n}$ is of rank p, the posterior distribution of $(\tilde{\mu}, \tilde{h})$ is again Normal-gamma with parameter $(m'', v'', \mathbf{n}'', \nu'')$ where m'' and $\mathbf{n}''$ are given by (5-77) above and

$$\nu'' = \nu' + \nu + p \; ,$$

$$v'' = \frac{1}{\nu''}\left[(\nu' v' + \boldsymbol{m}'^t \mathbf{n}' \boldsymbol{m}') + (\nu v + \boldsymbol{m}^t \mathbf{n}\, \boldsymbol{m}) - \boldsymbol{m}''^t \mathbf{n}'' \boldsymbol{m}''\right] . \tag{5-80}$$

The two factors in (5-79) display the marginal distribution of $\tilde{h}$ and the conditional distribution of $\tilde{\boldsymbol{\mu}}$ given h, but in the applications studied in this chapter what we require is the *marginal distribution of* $\tilde{\boldsymbol{\mu}}$. It is shown in Section 12.4.5 that this distribution is Student as defined by (8-28),

$$f_S^{(r)}(\boldsymbol{\mu}|\boldsymbol{m}, \mathbf{n}/v, \nu) \propto [\nu + (\boldsymbol{\mu} - \boldsymbol{m})^t (\mathbf{n}/v) (\boldsymbol{\mu} - \boldsymbol{m})]^{-\frac{1}{2}(r+\nu)} \; , \tag{5-81}$$

where the parameters will be primed if the density is prior, double primed if it is posterior. By (8-29) the mean and variance of this distribution are

$$\bar{\boldsymbol{\mu}} \equiv \mathrm{E}(\tilde{\boldsymbol{\mu}}) = \boldsymbol{m} \; , \qquad \breve{\boldsymbol{\mu}} \equiv \mathrm{V}(\tilde{\boldsymbol{\mu}}) = v\, \mathbf{n}^{-1} \frac{\nu}{\nu - 2} \; . \tag{5-82}$$

5.9.3. Distribution of $\tilde{\boldsymbol{\delta}}$

In the notation we are now using the basic assumption (5-59) or (5-69) concerning the utility function can be written

$$\tilde{\boldsymbol{v}} \equiv \boldsymbol{K} + \mathbf{k}\, \tilde{\boldsymbol{\mu}} \tag{5-83a}$$

where

$$\tilde{\boldsymbol{v}} \equiv [\tilde{v}_1 \quad \cdots \quad \tilde{v}_r]^t \; , \qquad \boldsymbol{K} \equiv [K_1 \quad \cdots \quad K_r]^t \; ,$$

$$\mathbf{k} \equiv \begin{bmatrix} k_1 & & & 0 \\ & \cdot & & \\ & & \cdot & \\ 0 & & & k_r \end{bmatrix} , \tag{5-83b}$$

and the definition (5-62) of $\tilde{\boldsymbol{\delta}}$ can be written

$$\tilde{\boldsymbol{\delta}} = \mathbf{B}\, \tilde{\boldsymbol{v}} \tag{5-84a}$$

where

$$\mathbf{B} \equiv \begin{bmatrix} 1 & 0 & 0 & \cdots & 0 & -1 \\ 0 & 1 & 0 & \cdots & 0 & -1 \\ 0 & 0 & 1 & \cdots & 0 & -1 \\ \cdots & \cdots & \cdots & \cdots & \cdots & \cdots \\ 0 & 0 & 0 & \cdots & 1 & -1 \end{bmatrix} \quad \text{is} \quad (r - 1) \times r \; . \tag{5-84b}$$

Substituting (5-83a) in (5-84a) we have

$$\tilde{\boldsymbol{\delta}} = \mathbf{B}\, \boldsymbol{K} + \mathbf{B}\, \mathbf{k}\, \tilde{\boldsymbol{\mu}} \; , \tag{5-85}$$

and we can now proceed to obtain the distribution of $\tilde{\boldsymbol{\delta}}$ from the distribution (5-75) or (5-81) of $\tilde{\boldsymbol{\mu}}$.

1. *If h is known* and $\tilde{\boldsymbol{\mu}}$ has the Normal density (5-75) with parameter $(\boldsymbol{m}, \mathbf{n})$, where $\boldsymbol{m}$ and $\mathbf{n}$ may be either primed or double-primed, then by (8-26) the distribution of $\tilde{\boldsymbol{\delta}}$ is Normal with density

$$f_N^{(r-1)}(\boldsymbol{\delta}|\bar{\boldsymbol{\delta}}, h\mathbf{n}_\delta) \tag{5-86}$$

where

$$\bar{\boldsymbol{\delta}} = \mathbf{B}\, \boldsymbol{K} + \mathbf{B}\, \mathbf{k}\, \boldsymbol{m} \; , \qquad \mathbf{n}_\delta^{-1} = (\mathbf{B}\, \mathbf{k})\, \mathbf{n}^{-1} (\mathbf{B}\, \mathbf{k})^t \; . \tag{5-87}$$

2. *If h is unknown* and $\tilde{\mu}$ has the Student density (5-81) with parameter $(m, \mathbf{n}/v, \nu)$, where m, $\mathbf{n}$, v, and ν may be either primed or double-primed, then by (8-32) the distribution of $\tilde{\delta}$ is Student with density

$$f_S^{(r-1)}(\delta|\bar{\delta}, \mathbf{n}_\delta/v, \nu) \tag{5-88}$$

where $\bar{\delta}$ and $\mathbf{n}_\delta$ are given by (5-87).

5.9.4. *Distribution of $\tilde{\bar{\delta}}''$ When All Processes are to be Sampled*

By formula (5-85) for $\tilde{\delta}$ in terms of $\tilde{\mu}$, the posterior mean of $\tilde{\delta}$ treated as a random variable before the sample outcome is known is given by

$$\tilde{\bar{\delta}}'' = \mathbf{B}\,K + \mathbf{B}\,\mathbf{k}\,\tilde{\bar{\mu}}'' \; , \tag{5-89}$$

and we can now proceed to obtain the distribution of $\tilde{\bar{\delta}}''$ from the results obtained in Part III of this monograph for the distribution of $\tilde{\bar{\mu}}''$. We assume that the design of e is such that the value of the statistic $\mathbf{n}$ defined by (5-71) and (5-72) is predetermined; and for the moment we also assume that $\mathbf{n}$ is of full rank, i.e., that at least one observation is to be taken on each of the r processes.

1. *If h is known* and the prior distribution of $\tilde{\mu}$ is Normal with parameter $(m', \mathbf{n}')$, then by Section 12.8 and formulas (12-22) and (8-26) the distribution of $\tilde{\bar{\delta}}''$ is Normal with density

$$f_N^{(r-1)}(\bar{\delta}''|\bar{\delta}', h\mathbf{n}_\delta^*) \tag{5-90}$$

where

$$\begin{aligned} \bar{\delta}' &= \mathbf{B}\,K + \mathbf{B}\,\mathbf{k}\,m' \; , & \mathbf{n}_\delta^{*-1} &= (\mathbf{B}\,\mathbf{k})\,\mathbf{n}^{*-1}(\mathbf{B}\,\mathbf{k})^t \; , \\ \mathbf{n}^* &= \mathbf{n}''\,\mathbf{n}^{-1}\,\mathbf{n}' = \mathbf{n}'\,\mathbf{n}^{-1}\,\mathbf{n}'' \; , & \mathbf{n}^{*-1} &= \mathbf{n}'^{-1} - \mathbf{n}''^{-1} \; . \end{aligned} \tag{5-91}$$

2. *If h is unknown* and the prior distribution of $\tilde{\mu}$ is Student with parameter $(m', \mathbf{n}'/v', \nu')$, then by Section 12.8 and formulas (12-46) and (8-32) the distribution of $\tilde{\bar{\delta}}''$ is Student with density

$$f_S^{(r-1)}(\bar{\delta}''|\bar{\delta}', \mathbf{n}_\delta^*/v', \nu') \tag{5-92}$$

where $\bar{\delta}'$ and $\mathbf{n}_\delta^*$ are given by (5-91).

5.9.5. *Distribution of $\tilde{\bar{\delta}}''$ When Some Processes are not to be Sampled*

If observations are to be taken on only $p < r$ of the r jointly distributed means $\tilde{\mu}_1, \cdots, \tilde{\mu}_r$, then (as shown in Section 12.9) the distribution of $\tilde{\bar{\mu}}''$ is degenerate or singular in the sense that it is confined to a p-dimensional subspace within the r-space on which $\bar{\mu}''$ is defined. For this reason the distribution of

$$\tilde{\bar{\delta}}'' = \mathbf{B}\,K + \mathbf{B}\,\mathbf{k}\,\tilde{\bar{\mu}}''$$

cannot in general be obtained by the method used in Section 5.9.4 just above, but it is still determined by the distribution of $\tilde{\bar{\mu}}''$ and the distribution of $\tilde{\bar{\mu}}''$ can be obtained by making use of the fact that (1) the distribution of those components of $\tilde{\bar{\mu}}''$ corresponding to the processes actually sampled is perfectly well behaved, and (2) the values of these components of $\tilde{\bar{\mu}}''$ determine the values of the remaining components.

Distribution of $\tilde{\bar{\mu}}''$ When the rth Process is not to be Sampled. If the process which is optimal under the prior distribution and which is therefore identified by the subscript r is not to be sampled, we are free to number the other $r - 1$ processes in such a way that it is the *first* p processes which *are* to be sampled. The statistic $\mathbf{n}$ will then be of the form

$$\mathbf{n} = \begin{bmatrix} \mathbf{n}_{11} & 0 \\ 0 & 0 \end{bmatrix} \qquad \text{where} \qquad \begin{matrix} \mathbf{n}_{11} \text{ is } p \times p \ , \\ \mathbf{n} \ \text{ is } r \times r \ , \end{matrix} \tag{5-93}$$

and we partition the random variable $\tilde{\boldsymbol{\mu}}$ and the parameters $\boldsymbol{m}$ and $\mathbf{n}$ (primed or double primed) correspondingly

$$\tilde{\boldsymbol{\mu}} = \begin{bmatrix} \tilde{\boldsymbol{\mu}}_1 \\ \tilde{\boldsymbol{\mu}}_2 \end{bmatrix} , \qquad \boldsymbol{m} = \begin{bmatrix} \boldsymbol{m}_1 \\ \boldsymbol{m}_2 \end{bmatrix} , \qquad \mathbf{n} = \begin{bmatrix} \mathbf{n}_{11} & \mathbf{n}_{12} \\ \mathbf{n}_{21} & \mathbf{n}_{22} \end{bmatrix} . \tag{5-94}$$

If now we define

$$\mathbf{n}_m'^{-1} = (\mathbf{n}'^{-1})_{11} \ , \qquad \mathbf{n}_m^{*-1} = \mathbf{n}_m'^{-1} - (\mathbf{n}_m' + \mathbf{n}_{11})^{-1} \ , \tag{5-95a}$$

then as shown in Section 12.9.6 the distribution of $\tilde{\bar{\boldsymbol{\mu}}}_1''$ is nondegenerate with density

$$D(\bar{\boldsymbol{\mu}}_1'') = \begin{cases} f_N^{(p)}(\bar{\boldsymbol{\mu}}_1''|\boldsymbol{m}_1', h\mathbf{n}_m^*) & \text{if } h \text{ is known} \ , \\ f_S^{(p)}(\bar{\boldsymbol{\mu}}_1''|\boldsymbol{m}_1', \mathbf{n}_m^*/v', \nu') & \text{if } h \text{ is unknown} \ ; \end{cases} \tag{5-95b}$$

and if we also define

$$\mathbf{C}^{(1)} \equiv \begin{bmatrix} \mathbf{I} \\ -\mathbf{n}_{22}'^{-1}\,\mathbf{n}_{21}' \end{bmatrix} \qquad \text{where} \qquad \begin{matrix} \mathbf{I} \ \text{ is } p \times p, \\ \mathbf{C}^{(1)} \text{ is } r \times p \ , \end{matrix} \tag{5-96a}$$

then it follows from (12-73) that the distribution (5-95) of $\tilde{\bar{\boldsymbol{\mu}}}_1''$ determines the distribution of the complete vector $\tilde{\bar{\boldsymbol{\mu}}}''$ through the relation

$$\tilde{\bar{\boldsymbol{\mu}}}'' = \bar{\boldsymbol{\mu}}' + \mathbf{C}^{(1)}(\tilde{\bar{\boldsymbol{\mu}}}_1'' - \bar{\boldsymbol{\mu}}_1') \ . \tag{5-96b}$$

Distribution of $\tilde{\bar{\boldsymbol{\mu}}}''$ When the rth Process is to be Sampled. If the process which is optimal under the prior distribution and which therefore bears the subscript r *is* to be sampled, we are free to number the other $r - 1$ processes in such a way that it is the *last* p processes which are to be sampled. The statistic $\mathbf{n}$ will then be of the form

$$\mathbf{n} = \begin{bmatrix} 0 & 0 \\ 0 & \mathbf{n}_{22} \end{bmatrix} , \tag{5-97}$$

and the reader can readily modify the results of the previous paragraph to show that if we define

$$\mathbf{n}_m'^{-1} = (\mathbf{n}'^{-1})_{22} \ , \qquad \mathbf{n}_m^{*-1} = \mathbf{n}_m'^{-1} - (\mathbf{n}_m' + \mathbf{n}_{22})^{-1} \ , \tag{5-98a}$$

then the distribution of $\tilde{\bar{\boldsymbol{\mu}}}_2''$ is nondegenerate with density

$$D(\bar{\boldsymbol{\mu}}_2'') = \begin{cases} f_N^{(p)}(\bar{\boldsymbol{\mu}}_2''|\boldsymbol{m}_2', h\mathbf{n}_m^*) & \text{if } h \text{ is known} \ , \\ f_S^{(p)}(\bar{\boldsymbol{\mu}}_2''|\boldsymbol{m}_2', \mathbf{n}_m^*/v', \nu') & \text{if } h \text{ is unknown} \ ; \end{cases} \tag{5-98b}$$

and if we define

$$\mathbf{C}^{(2)} \equiv \begin{bmatrix} -\mathbf{n}_{11}'^{-1}\,\mathbf{n}_{12}' \\ \mathbf{I} \end{bmatrix} \qquad \text{where} \qquad \begin{matrix} \mathbf{I} \ \text{ is } p \times p \ , \\ \mathbf{C}^{(2)} \text{ is } r \times p \ , \end{matrix} \tag{5-99a}$$

then the distribution (5-98) of $\tilde{\bar{\boldsymbol{\mu}}}_2''$ determines the distribution of the complete vector $\tilde{\bar{\boldsymbol{\mu}}}''$ through the relation

$$\tilde{\bar{\boldsymbol{\mu}}}'' = \bar{\boldsymbol{\mu}}' + \mathbf{C}^{(2)}(\tilde{\bar{\boldsymbol{\mu}}}_2'' - \bar{\boldsymbol{\mu}}_2') \ . \tag{5-99b}$$

Distribution of $\tilde{\boldsymbol{\delta}}''$. Substituting (5-96b) or (5-99b) in (5-89) we obtain

$$\tilde{\boldsymbol{\delta}}'' = \mathbf{B}(K + \mathbf{k}\,\bar{\boldsymbol{\mu}}') + \mathbf{B}\,\mathbf{k}\,\mathbf{C}^{(i)}(\tilde{\bar{\boldsymbol{\mu}}}_i'' - \bar{\boldsymbol{\mu}}_i')$$
$$= \boldsymbol{\delta}' + \mathbf{B}\,\mathbf{k}\,\mathbf{C}^{(i)}(\tilde{\bar{\boldsymbol{\mu}}}_i'' - \bar{\boldsymbol{\mu}}_i') \qquad (5\text{-}100)$$

where $i = 1$ if it is the first p processes which are to be sampled, $i = 2$ if it is the last p processes which are to be sampled. Concerning the distribution of $\tilde{\boldsymbol{\delta}}''$ as determined by this relation or the equivalent relation (5-89) we have in all cases

$$\mathrm{E}(\tilde{\boldsymbol{\delta}}'') = \boldsymbol{\delta}' \; ;$$

$$\mathrm{V}(\tilde{\boldsymbol{\delta}}'') = \begin{cases} (h\mathbf{n}_\delta^*)^{-1} & \text{if } h \text{ is known}, \\ (\mathbf{n}_\delta^*/v')^{-1}\,\nu'/(\nu' - 2) & \text{if } h \text{ is unknown}, \end{cases} \qquad (5\text{-}101a)$$

where

$$\mathbf{n}_\delta^{*-1} = (\mathbf{B}\,\mathbf{k}\,\mathbf{C}^{(i)})(\mathbf{n}_m'^{-1} - \mathbf{n}_m''^{-1})(\mathbf{B}\,\mathbf{k}\,\mathbf{C}^{(i)})^t \qquad (5\text{-}101b)$$
$$= (\mathbf{B}\,\mathbf{k})(\mathbf{n}'^{-1} - \mathbf{n}''^{-1})(\mathbf{B}\,\mathbf{k})^t \,. \qquad (5\text{-}101c)$$

▶ The formula for the mean follows from (5-27) applied directly to the distribution of $\tilde{\boldsymbol{\delta}}''$. The formula for the variance with $\mathbf{n}_\delta^*$ given by (5-101b) follows from (8-8c) in conjunction with (5-100) and (5-28) applied to the distribution of $\tilde{\bar{\boldsymbol{\mu}}}_i''$. The formula for the variance with $\mathbf{n}_\delta^{*-1}$ given by (5-101c) follows from (8-8c) in conjunction with (5-89) and (5-28) applied to the distribution of $\tilde{\bar{\boldsymbol{\mu}}}''$. ◀

As regards the *details* of the distribution of $\tilde{\boldsymbol{\delta}}''$, however, we must distinguish two cases.

1. *If only one process is not to be sampled*, i.e. if $p = r - 1$, then the $(r - 1) \times (r - 1)$ matrix $\mathbf{B}\,\mathbf{k}\,\mathbf{C}^{(i)}$ in (5-100) is of rank $(r - 1)$, so that by (8-26) or (8-32) the distribution of $\tilde{\boldsymbol{\delta}}''$ is *nondegenerate* with density

$$\mathrm{D}(\boldsymbol{\delta}'') = \begin{cases} f_N^{(r-1)}(\boldsymbol{\delta}''|\boldsymbol{\delta}', h\mathbf{n}_\delta^*) & \text{if } h \text{ is known}, \\ f_S^{(r-1)}(\boldsymbol{\delta}''|\boldsymbol{\delta}', \mathbf{n}_\delta^*/v', \nu') & \text{if } h \text{ is unknown}, \end{cases} \qquad (5\text{-}102)$$

$\mathbf{n}_\delta^*$ being defined by (5-101b) or (5-101c).

2. *If there are two or more processes which are not to be sampled*, i.e. if $p < r - 1$, then the fact that by (5-100) the $(r - 1) \times 1$ vector $\tilde{\boldsymbol{\delta}}''$ is a function of the $p \times 1$ vector $\tilde{\bar{\boldsymbol{\mu}}}_i''$ implies that the distribution of $\tilde{\boldsymbol{\delta}}''$ is *degenerate.* In this case it is more convenient to treat $\tilde{\boldsymbol{\delta}}''$ explicitly as a function of $\tilde{\bar{\boldsymbol{\mu}}}_i''$ than to try to work with an analytical expression for the density of $\tilde{\boldsymbol{\delta}}''$ itself.

5.10. Value of Information and Optimal Sample Size When There are Two Independent Normal Processes

With the theory of Sections 5.8 and 5.9 as background we are now ready to attack the problem of actually evaluating the expected value of perfect information or EVPI and the expected value of sample information or EVSI and the further problem of using our results concerning the EVSI to select the optimal e in E. We remind the reader that by (5-64) and (5-66) the EVPI and EVSI are respectively given by

$$v_t^*(e_\infty) = \text{E}_\delta \max \{\tilde\delta_1, \cdots, \tilde\delta_{r-1}, 0\} = l_t^*(e_0) \,, \tag{5-103a}$$

$$v_t^*(e) = \text{E}_{\delta''|e} \max \{\tilde\delta_1'', \cdots, \tilde\delta_{r-1}'', 0\} \,, \tag{5-103b}$$

and that an optimal e is one which maximizes the net gain of sampling

$$v^*(e) = v_t^*(e) - c_s^*(e) \,. \tag{5-104}$$

In the present section we shall consider these problems for the special case $r = 2$, so that $\tilde{\boldsymbol\delta}$ is actually a scalar $\tilde\delta$ and we have by (5-83), (5-84), and (5-72)

$$\begin{aligned} \mathbf{B\,k} &= [k_1 \quad -k_2] \,, \\ \bar\delta' &= (K_1 + k_1 m_1') - (K_2 + k_2 m_2') \,, \\ \mathbf{n} &= \begin{bmatrix} n_1(h_1/h) & 0 \\ 0 & n_2(h_2/h) \end{bmatrix} . \end{aligned} \tag{5-105}$$

5.10.1. EVPI

1. *If h is known* and the prior distribution of $\tilde{\boldsymbol\mu}$ is Normal with parameter $(\boldsymbol m', \mathbf{n}')$, we have by (5-103a), (5-86), and (11-25b) that

$$\begin{aligned} v_t^*(e_\infty) = l_t^*(e_0) &= \int_{-\infty}^{\infty} \max \{\delta, 0\}\, f_N(\delta|\bar\delta', hn_\delta')\, d\delta \\ &= \int_0^{\infty} \delta\, f_N(\delta|\bar\delta', hn_\delta')\, d\delta = (hn_\delta')^{-\frac{1}{2}}\, L_{N^*}(|\bar\delta'|\sqrt{hn_\delta'}) \end{aligned} \tag{5-106}$$

where by (5-87)

$$n_\delta'^{-1} = (\mathbf{B\,k})\, \mathbf{n}'^{-1}(\mathbf{B\,k})^t \,, \qquad \mathbf{B\,k} = [k_1 \quad -k_2] \,, \tag{5-107}$$

and L_{N^*} is tabulated in Table II at the end of this monograph.

2. *If h is unknown* and the prior distribution of $\tilde{\boldsymbol\mu}$ is Student with parameter $(\boldsymbol m', \mathbf{n}'/v', \nu')$, we have by (5-103a), (5-88), and (11-50b) that

$$v_t^*(e_\infty) = l_t^*(e_0) = (n_\delta'/v')^{-\frac{1}{2}}\, L_{S^*}(|\bar\delta'|\sqrt{n_\delta'/v'}|\nu') \tag{5-108}$$

where n_δ' is defined by (5-107) and L_{S^*} is defined in terms of tabulated univariate Student ordinates and tail areas by

$$L_{S^*}(t|\nu) \equiv \frac{\nu + t^2}{\nu - 1}\, f_{S^*}(t|\nu) - t\, G_{S^*}(t|\nu) \,. \tag{5-109}$$

5.10.2. EVSI

Whether both or only one of the two processes are to be sampled, we have by (5-91) and (5-101c)

$$n_\delta^{*-1} = (\mathbf{B\,k})\{\mathbf{n}'^{-1} - (\mathbf{n}' + \mathbf{n})^{-1}\}(\mathbf{B\,k})^t \,, \qquad \mathbf{B\,k} = [k_1 \quad -k_2] \,; \tag{5-110}$$

in the case where the ith process is to be sampled but the jth is not, this formula reduces by (5-101b) to

$$n_\delta^{*-1} = \left[k_i + k_j \frac{n_{ij}'}{n_{jj}'}\right]^2 \{n_m'^{-1} - (n_m' + n_{ii})^{-1}\} \,, \qquad n_m'^{-1} = (\mathbf{n}'^{-1})_{ii} \,. \tag{5-111}$$

In either case the EVSI is given by

$$v_i^*(e) = \begin{cases} (hn_\delta^*)^{-\frac{1}{2}}\, L_{N^*}(|\bar{\delta}'|\sqrt{hn_\delta^*}) & \text{if } h \text{ is known ,} \\ (n_\delta^*/v')^{-\frac{1}{2}}\, L_{S^*}(|\bar{\delta}'|\sqrt{n_\delta^*/v'}|\nu') & \text{if } h \text{ is unknown .} \end{cases} \tag{5-112}$$

When $\tilde{\mu}$ is Normal (h known), the result follows from (5-103b), (5-90) or (5-102), and (11-25b). When $\tilde{\mu}$ is Student (h unknown), the result follows from (5-103b), (5-92) or (5-102), and (11-50b).

5.10.3. *Optimal Allocation of a Fixed Experimental Budget*

In analyzing the problem of optimal experimentation we shall consider only experiments which have a predetermined "sample size"

$$\mathbf{n} = \begin{bmatrix} n_{11} & 0 \\ 0 & n_{22} \end{bmatrix} = \begin{bmatrix} n_1(h_1/h) & 0 \\ 0 & n_2(h_2/h) \end{bmatrix} , \tag{5-113}$$

n_1 and n_2 being the actual numbers of observations; and we shall assume throughout that the (prior expected) cost of sampling is of the form

$$c_s^*(e) = c_1 n_1 + c_2 n_2 , \tag{5-114}$$

leaving to the reader the easy generalization to the case where fixed costs independent of sample size may be incurred (a) if any observations at all are taken, and/or (b) if any observations are taken on μ_1, and/or (c) if any observations are taken on μ_2. In the present section we show that, whether or not h is known, it is always possible to obtain an explicit solution of the "allocation problem" of optimizing n_1 and n_2 subject to the condition that some predetermined amount A is to be spent on n_1 and n_2 together; formally, we shall optimize n_1 and n_2 subject to the *budgetary constraint*

$$c_1 n_1 + c_2 n_2 = A . \tag{5-115}$$

In the following sections we shall then show how when h is known we can easily determine the unconstrained optima for n_1 and n_2.

Whether h is known or not, we have by (5-112) that the only effect of either n_1 or n_2 on the EVSI is through the quantity n_δ^*; and since both factors in both formulas decrease as n_δ^* increases, our present objective is to minimize n_δ^* subject to the constraint (5-115). This is equivalent to maximizing n_δ^{*-1} and thus by (5-110) to minimizing

$$n_\delta''^{-1} \equiv (\mathbf{B\,k})(\mathbf{n}' + \mathbf{n})^{-1}(\mathbf{B\,k})^t , \qquad \mathbf{B\,k} = [k_1 \quad -k_2] .$$

Substituting herein

$$\mathbf{n}' + \mathbf{n} = \begin{bmatrix} n'_{11} & n'_{12} \\ n'_{21} & n'_{22} \end{bmatrix} + \begin{bmatrix} n_{11} & 0 \\ 0 & n_{22} \end{bmatrix} = \begin{bmatrix} n''_{11} & n'_{12} \\ n'_{21} & n''_{22} \end{bmatrix}$$

and recalling that $\mathbf{n}'$ is symmetric we obtain as the quantity to be minimized

$$n_\delta''^{-1} = \frac{k_1^2 n''_{22} + 2k_1k_2n'_{12} + k_2^2 n''_{11}}{n''_{11}n''_{22} - n'^2_{12}} \tag{5-116}$$

where $n''_{ii} = n'_{ii} + n_{ii}$ and $n_{ii} = n_i h_i/h$ for $i = 1, 2$.

The problem of choosing n_{11} and n_{22} and thus n_1 and n_2 so as to minimize (5-116) subject to the constraints

$$n_{11} \geq 0 \,, \qquad n_{22} \geq 0 \,, \qquad\qquad (5\text{-}117)$$
$$(c_1 h/h_1)\, n_{11} + (c_2 h/h_2)\, n_{22} = A \;,$$

is a special case of a more general convex programming problem analyzed by Ericson,† whose results we shall present without proof.

In a plane with n_{11} and n_{22} axes let R' denote the positive orthant as shown in Figure 5.17*; define

$$c_{ij} \equiv \frac{\sqrt{c_i/h_i}}{k_j} \,, \qquad i, j = 1, 2 \;; \qquad\qquad (5\text{-}118a)$$

and let ℓ^+ and ℓ^- be the two lines defined by

$$(n'_{11} + n_{11})\, c_{11} + n'_{12} c_{12} = \pm[(n'_{22} + n_{22})\, c_{22} + n'_{12} c_{21}] \,, \qquad\qquad (5\text{-}118b)$$

ℓ^+ corresponding to the use of $+$ on the right-hand side and ℓ^- to the use of $-$. Since $c_{ij} > 0$ for all i, j, the line ℓ^+ must intersect R'; the line ℓ^- may or may not intersect R', but it can be proved that the intersection of ℓ^+ and ℓ^- can never lie in R'. We thus have altogether four cases to distinguish: the two shown in Figure 5.17*, and two more obtained by interchanging the labels n_{11} and n_{22} in that figure.

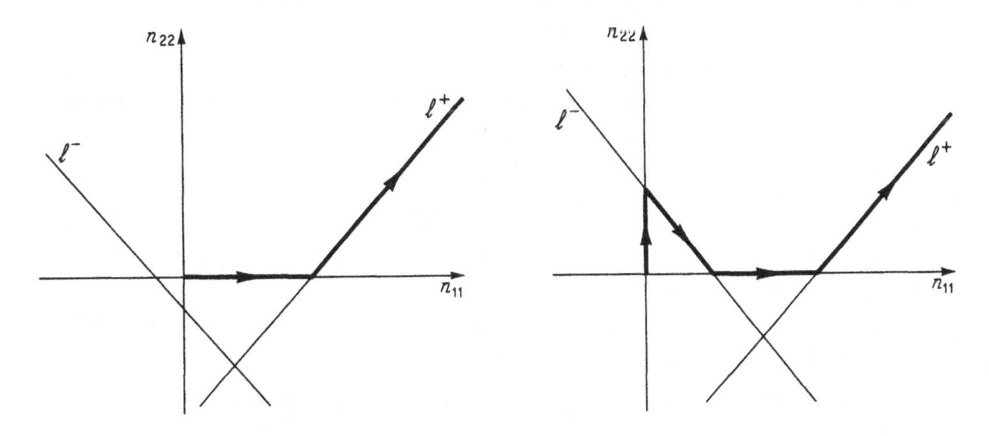

Figure 5.17*

Optimal Sampling Path

In what follows we shall refer explicitly to only the first two of these four cases, leaving it to the reader to make the easy translation to the other two.

The principal result proved by Ericson is that as the sampling budget A of (5-117) increases from 0, the sample-size pairs or "vector sample sizes" (n_{11}, n_{22}) which minimize (5-116) for given A follow a path like one or the other of the two indicated by the heavy lines and arrows in Figure 5.17*. In case (1), where ℓ^- does not intersect R', there exists a number A_1 such that if the budget $A \leq A_1$, then the entire budget should be spent on sampling from process 1, whereas if $A > A_1$, both processes should be sampled with sample sizes given by ℓ^+. In case

† William Ericson, *Optimum Stratified Sampling Using Prior Information,* unpublished doctoral dissertation, Harvard University, 1963.

(2), where ℓ^- does intersect R', there exist numbers $A_1 < A_2 < A_3$ such that if the budget $A \leq A_1$, then only process 2 should be sampled; if $A_1 < A < A_2$, then both processes should be sampled with sample sizes given by ℓ^-; if $A_2 \leq A \leq A_3$, then only process 2 should be sampled; and finally if $A > A_3$, then both processes should be sampled with sample sizes given by ℓ^+.

▶ The proof of these results as given by Ericson (*l.c. supra*) employs the theory of convex nonlinear programming and makes use of the Kuhn-Tucker saddle-point theorem. We merely remark here that if one ignores the nonnegativity constraints on n_{11} and n_{22} and sets up the problem in Lagrangian terms, a quadratic equation results whose two roots correspond to the two critical lines ℓ^+ and ℓ^- defined by (5-118b). ◀

5.10.4. Optimal Sample Size When h is Known and Only One Process is to be Sampled

Having seen that optimal allocation of any given sampling budget will yield a vector sample size (n_{11}, n_{22}) lying somewhere on a path like one of the two shown in Figure 5.17*, we next address ourselves to the problem of determining the optimal sampling budget, or alternatively, the optimal sample point on the optimal sampling path. We shall attack this problem as follows. (1) In the present section, we shall show how to determine a local maximum of the net gain of sampling when the sample is constrained to lie on one of the *axes* in Figure 5.17*—i.e., when only one of the two processes is to be sampled. (2) In the next section, we shall show how to determine a local maximum of the net gain of sampling when the sample is constrained to lie on one of the *lines* ℓ^+ or ℓ^- in Figure 5.17*. (3) In Section 5.10.6 we shall then show how to use these results to determine the sample which yields a global maximum of the net gain.

When only the ith process is to be sampled, we have by (5-112) and (5-114) that the net gain of a sample of size n_i is

$$v^*(e) = v_t^*(e) - c_s^*(e) = (hn_\delta^*)^{-\frac{1}{2}} L_{N^*}(|\bar\delta'|\sqrt{hn_\delta^*}) - c_i n_i \qquad \text{(5-119a)}$$

where by (5-111)

$$n_\delta^{*-1} = A[n_m'^{-1} - (n_m' + n_{ii})^{-1}] , \qquad \text{(5-119b)}$$

$$A \equiv \left[k_i + k_j \frac{n'_{ij}}{n'_{jj}}\right]^2 , \qquad n_m'^{-1} = (\mathbf{n}'^{-1})_{ii} .$$

Defining

$$n_\infty^* = n_m'/A , \qquad \nu_\delta = n_{ii}/A = n_i(h_i/h)/A , \qquad \kappa_s^* = c_i A/(h_i/h) , \qquad \text{(5-120)}$$

we can put (5-119) in the form

$$v^*(e) = (hn_\delta^*)^{-\frac{1}{2}} L_{N^*}(|\bar\delta'|\sqrt{hn_\delta^*}) - \kappa_s^* \nu_\delta \qquad \text{(5-121a)}$$

where

$$n_\delta^{*-1} = n_\infty^{*-1} - (n_\infty^* + \nu_\delta)^{-1} . \qquad \text{(5-121b)}$$

Our choice of the symbol n_∞^* is due to the fact, apparent from (5-121b), that n_δ^* has the value n_∞^* when the sample size n_i and therefore the "adjusted sample size" ν_δ are infinite.

Now regarded as a function of the parameter n_∞^* and the variable ν_δ, the net gain of sampling as given by (5-121) is formally identical to the net gain in the univariate problem described by (5-38),

$$v^*(e) = k_t(hn^*)^{-\frac{1}{2}} L_{N^*}(|\mu_b - m'|\sqrt{hn^*}) - k_s n$$

where

$$n^{*-1} = n'^{-1} - (n' + n)^{-1} \ ,$$

and therefore we can apply the analysis of Section 5.5 to our present problem.

Comparing (5-38) as repeated above with (5-121), we see that the constant k_t of the earlier problem has the value 1 in our present problem (essentially because the random variable $\tilde{\delta}$ is measured in utiles), so that the parameters D' and λ defined by (5-40) become

$$D' = |\bar{\delta}'|\sqrt{hn_\infty^*} \equiv D_\infty \ , \qquad \lambda = (n_\infty^{*3}\, h)^{-\frac{1}{2}}\, \kappa_s^{*-1} \ , \tag{5-122a}$$

where n_∞^* and κ_s^* are defined by (5-120). Once these quantities have been computed, the optimal value ν_δ° of ν_δ can be found by entering Chart I at the end of this monograph with D_∞ and $Z = \lambda^{\frac{1}{3}}$ and reading

$$\rho^\circ/Z^2 = \nu_\delta^\circ(h\kappa_s^{*2})^{\frac{1}{3}} \ , \tag{5-122b}$$

after which the optimal sample size n_{ii}° can be found from (5-120).

If n_{ii}° as thus determined lies on the segment of the n_{ii} axis in Figure 5.17* which belongs to the optimal-allocation path, it is a candidate for the role of global optimizer. If it lies outside the segment of the n_{ii} axis that is on the optimal-allocation path, we know at once that it is *not* the global optimizer since the cost of n_{ii}° can be spent more profitably on some sample that *does* lie on the path.

5.10.5. *Optimal Sample Size When h is Known and Both Processes are to be Sampled According to ℓ^+ or ℓ^-*

We now turn to the problem of locating the vector sample sizes (n_{11}, n_{22}) that correspond to local maxima of the net gain of sampling when (n_{11}, n_{22}) is constrained to lie on one of the two lines ℓ^+ and ℓ^- in Figure 5.17*; and to do so we first define or repeat the definition of

$$n_\delta'^{-1} = (\mathbf{B\,k})\,\mathbf{n}'^{-1}(\mathbf{B\,k})^t \ , \qquad n_\delta''^{-1} = (\mathbf{B\,k})(\mathbf{n}' + \mathbf{n})^{-1}(\mathbf{B\,k})^t \ . \tag{5-123}$$

We next define the "scalar sample size"

$$n_\delta \equiv n_\delta'' - n_\delta' \ , \tag{5-124}$$

calling the reader's attention to the fact that the definition of n_δ is *not* analogous to the definitions of n_δ' and n_δ'', and proceed to show that *if* the vector sample size (n_{11}, n_{22}) lies on one of the lines ℓ^+ or ℓ^- in Figure 5.17*, *then* the cost of a sample e of size n_δ is linear in n_δ—more specifically that

$$c_s^*(e) = K_s^* + k_s^*\, n_\delta \tag{5-125a}$$

where

$$k_s^* = \begin{cases} h(k_1\sqrt{c_1/h_1} + k_2\sqrt{c_2/h_2})^2 & \text{if} \quad (n_{11}, n_{22}) \in \ell^+ \ , \\ h(k_1\sqrt{c_1/h_1} - k_2\sqrt{c_2/h_2})^2 & \text{if} \quad (n_{11}, n_{22}) \in \ell^- \ , \end{cases} \tag{5-125b}$$

and K_s^* is a constant (different for ℓ^+ and ℓ^-) whose value need not be determined for the purpose of our present argument.

▶ To prove (5-125) for ℓ^+, we first write the constraint (5-118b), using the + sign on the right, in the form

$$n_{22}'' = \frac{k_2\sqrt{c_1/h_1}}{k_1\sqrt{c_2/h_2}}\, n_{11}'' + \left[\frac{\sqrt{c_1/h_1}}{\sqrt{c_2/h_2}} - \frac{k_2}{k_1}\right] n_{12}' \,, \tag{1}$$

substitute the right-hand side for n_{22}'' in the formula

$$n_\delta'' = \frac{n_{11}''n_{22}'' - n_{12}'^2}{k_1^2 n_{22}'' + 2k_1k_2n_{12}' + k_2^2 n_{11}''} \tag{2}$$

derived from (5-116), and solve for n_{11}'' in terms of n_δ'', obtaining, after considerable algebra,

$$n_{11}'' = \frac{k_1}{\sqrt{c_1/h_1}}\,(k_1\sqrt{c_1/h_1} + k_2\sqrt{c_2/h_2})\, n_\delta'' + \frac{\sqrt{c_2/h_2}}{\sqrt{c_1/h_1}}\, n_{12}' \,. \tag{3}$$

Each unit of increase in $n_\delta \equiv n_\delta'' - n_\delta'$ thus implies that n_{11}'' increases by the amount

$$\frac{k_1}{\sqrt{c_1/h_1}}\,(k_1\sqrt{c_1/h_1} + k_2\sqrt{c_2/h_2}) \;; \tag{4}$$

and by the constraint (1) this in turn implies that n_{22}'' increases by the amount

$$\frac{k_2}{\sqrt{c_2/h_2}}\,(k_1\sqrt{c_1/h_1} + k_2\sqrt{c_2/h_2}) \,. \tag{5}$$

By (5-114) and (5-113) the costs of unit increases in n_{11}'' and n_{22}'' are respectively c_1h/h_1 and c_2h/h_2, so that *as long as the constraint is satisfied while n_1 and n_2 are increasing* we have

$$\frac{dc_s^*(e)}{dn_\delta} = h(k_1\sqrt{c_1/h_1} + k_2\sqrt{c_2/h_2})^2 \;; \tag{6}$$

and from this result, (5-125) for ℓ^+ follows immediately.

To prove (5-125) for ℓ^-, we follow exactly the same procedure except that we use the − sign in (5-118b). ◀

By (5-112) and (5-125) we now have that the net gain of a sample e of "scalar size" n_δ lying on ℓ^+ or ℓ^- in Figure 5.17* is

$$v^*(e) = (hn_\delta^*)^{-\frac{1}{2}}\, L_{N^*}(|\bar\delta'|\,\sqrt{hn_\delta^*}) - K_s^* - k_s^*\, n_\delta \tag{5-126a}$$

where by (5-110) and (5-123)

$$n_\delta^* = n_\delta'^{-1} - (n_\delta' + n_\delta)^{-1} \,. \tag{5-126b}$$

Regarded as a function of the parameter n_δ' and the variable n_δ, the net gain as given by this expression is formally identical to the net gain in the univariate problem described by (5-38), and therefore we can apply the analysis of Section 5.5 to our present problem just as we applied it in Section 5.10.4 to the problem of sampling from just one of the two processes.

Comparing (5-38) and (5-126), we see that the constant k_t in the former problem again has the value 1 in our present problem, so that the parameters D' and λ defined by (5-40) become

$$D' = |\bar{\delta}'| \sqrt{h n'_\delta} \ , \qquad \lambda = (n'^3_\delta h)^{-\frac{1}{2}} \, k^{*-1}_s \ , \tag{5-127a}$$

where by (5-123) and (5-105)

$$n'^{-1}_\delta = [k_1 \qquad -k_2] \, \mathbf{n}'^{-1} \, [k_1 \qquad -k_2]^t \tag{5-127b}$$

and k^*_s is defined by (5-125b).

Once these quantities have been computed, the optimal value n°_δ of n_δ can be found by entering Chart I at the end of this monograph with $D_\infty = D'$ and $Z = \lambda^{\frac{1}{3}}$ and reading

$$\eta^\circ \equiv \rho^\circ/Z^2 = n^\circ_\delta (h k^{*2}_s)^{\frac{1}{3}} \ . \tag{5-128}$$

After the optimal scalar sample size n°_δ has been computed from (5-128), the corresponding vector size $(n^\circ_{11}, n^\circ_{22})$ can be obtained as follows. *If the sample is constrained to lie on ℓ^+,* then

$$n^\circ_{ii} = \frac{k_i}{\sqrt{c_i/h_i}} S^+(n'_\delta + n^\circ_\delta) + \frac{\sqrt{c_j/h_j}}{\sqrt{c_i/h_i}} n'_{ij} - n'_{ii} \tag{5-129a}$$

where

$$S^+ = k_1 \sqrt{c_1/h_1} + k_2 \sqrt{c_2/h_2} \ ; \tag{5-129b}$$

if the sample is constrained to lie on ℓ^-, then

$$n^\circ_{ii} = (-1)^{i+1} \frac{k_i}{\sqrt{c_i/h_i}} S^-(n'_\delta + n^\circ_\delta) - \frac{\sqrt{c_j/h_j}}{\sqrt{c_i/h_i}} n'_{ij} - n'_{ii} \tag{5-130a}$$

where

$$S^- = k_1 \sqrt{c_1/h_1} - k_2 \sqrt{c_2/h_2} \ . \tag{5-130b}$$

If both n°_{11} and n°_{22} as determined by (5-129) or (5-130) are nonnegative, they correspond to a local optimum on ℓ^+ or ℓ^-; if either one is negative, there is no local optimum on the line in question.

▶ Formulas (5-129) follow from (3) and (1) in the proof of (5-125); formulas (5-130) are similarly derived. ◀

5.10.6. *The General Problem of Optimal Sample Size When h is Known*

We are now ready to describe a general procedure for determination of the optimal vector sample size $(n^\circ_{11}, n^\circ_{22})$ which in turn determines the optimal values for the two actual sample sizes n_1 and n_2 via (5-113).

1. Draw ℓ^+ and ℓ^- and determine the optimal allocation path as in Figure 5.17*.
2. For each segment determine whether a local optimum exists on that segment, using the results of Section 5.10.4 or 5.10.5 as appropriate.

3. If more than one local optimum exists, evaluate the net gain of each via (5-112) and (5-110) and select the largest; if no local optimum exists, the optimal sample size is zero.

5.11. Value of Information When There are Three Independent-Normal Processes

5.11.1. *The Basic Integral in the Nondegenerate Case*

When a choice is to be made among three Independent-Normal processes, $r = 3$, formula (5-64) for the EVPI becomes

$$v_t^*(e_\infty) = l_t^*(e_0) = \int_{-\infty}^{\infty}\int_{-\infty}^{\infty} \max\{\delta_1, \delta_2, 0\}\ \mathrm{D}'(\boldsymbol{\delta})\ d\delta_1\, d\delta_2\ , \qquad (5\text{-}131\text{a})$$

where by (5-86) and (5-88)

$$\mathrm{D}'(\boldsymbol{\delta}) = \begin{cases} f_N^{(2)}(\boldsymbol{\delta}|\bar{\boldsymbol{\delta}}', h\mathbf{n}'_\delta) & \text{if } h \text{ is known ,} \\ f_S^{(2)}(\boldsymbol{\delta}|\bar{\boldsymbol{\delta}}', \mathbf{n}'_\delta/v', \nu') & \text{if } h \text{ is unknown ,} \end{cases} \qquad (5\text{-}131\text{b})$$

$$\mathbf{n}_\delta'^{-1} = (\mathbf{B\,k})\,\mathbf{n}'^{-1}(\mathbf{B\,k})^t\ ; \qquad (5\text{-}131\text{c})$$

and *provided that at least two of the three processes are to be sampled,* we have by formula (5-66) for the EVSI

$$v_t^*(e) = \int_{-\infty}^{\infty}\int_{-\infty}^{\infty} \max\{\bar{\delta}_1'', \bar{\delta}_2'', 0\}\ \mathrm{D}(\bar{\boldsymbol{\delta}}'')\ d\bar{\delta}_1''\, d\bar{\delta}_2'' \qquad (5\text{-}132\text{a})$$

where by (5-90), (5-92), and (5-102)

$$\mathrm{D}(\bar{\boldsymbol{\delta}}'') = \begin{cases} f_N^{(2)}(\bar{\boldsymbol{\delta}}''|\bar{\boldsymbol{\delta}}', h\mathbf{n}^*_\delta) & \text{if } h \text{ is known ,} \\ f_S^{(2)}(\bar{\boldsymbol{\delta}}''|\bar{\boldsymbol{\delta}}', \mathbf{n}^*_\delta/v', \nu') & \text{if } h \text{ is unknown ,} \end{cases} \qquad (5\text{-}132\text{b})$$

$$\mathbf{n}_\delta^{*-1} = (\mathbf{B\,k})[\mathbf{n}'^{-1} - (\mathbf{n}' + \mathbf{n})^{-1}](\mathbf{B\,k})^t\ . \qquad (5\text{-}132\text{c})$$

The problems of evaluating the integrals (5-131) and (5-132) are clearly identical when both densities are of the same type, Normal or Student, and we may therefore take as our task the evaluation of what we shall call the *expected value of information* without specifying whether the information is perfect or comes from a sample:

$$\mathrm{EVI} \equiv \int_{-\infty}^{\infty}\int_{-\infty}^{\infty} \max\{\delta_1, \delta_2, 0\}\ \mathrm{D}(\boldsymbol{\delta})\ d\delta_1\, d\delta_2 \qquad (5\text{-}133\text{a})$$

where $\boldsymbol{\delta}$ is now simply a dummy variable,

$$\mathrm{D}(\boldsymbol{\delta}) = \begin{cases} f_N^{(2)}(\boldsymbol{\delta}|\bar{\boldsymbol{\delta}}', h\mathbf{n}^\square_\delta) & \text{if } h \text{ is known} \\ f_S^{(2)}(\boldsymbol{\delta}|\bar{\boldsymbol{\delta}}', \mathbf{n}^\square_\delta/v', \nu') & \text{if } h \text{ is unknown} \end{cases} \qquad (5\text{-}133\text{b})$$

and $\mathbf{n}^\square_\delta$ represents *either* $\mathbf{n}'_\delta$ or $\mathbf{n}^*_\delta$. Since moreover many of the problems involved in evaluating this integral are the same whether the density is Normal or Student, we shall sometimes avoid repetition by replacing the symbol for the actual density by a symbol which can stand for either density, writing

$$f(\boldsymbol{\delta}|\bar{\boldsymbol{\delta}}', \mathbf{n}^\square_\delta) \equiv \begin{cases} f_N^{(2)}(\boldsymbol{\delta}|\bar{\boldsymbol{\delta}}', h\mathbf{n}^\square_\delta) & \text{if } h \text{ is known ,} \\ f_S^{(2)}(\boldsymbol{\delta}|\bar{\boldsymbol{\delta}}', \mathbf{n}^\square_\delta/v', \nu') & \text{if } h \text{ is unknown .} \end{cases} \qquad (5\text{-}134)$$

In the same spirit we define for notational convenience what we shall call the *scale parameter*

$$\boldsymbol{\sigma}^{\square} \equiv \begin{bmatrix} \sigma_{11}^{\square} & \sigma_{12}^{\square} \\ \sigma_{21}^{\square} & \sigma_{22}^{\square} \end{bmatrix} \equiv \begin{cases} (h\mathbf{n}_{\delta}^{\square})^{-1} & \text{if } h \text{ is known ,} \\ (\mathbf{n}_{\delta}^{\square}/v')^{-1} & \text{if } h \text{ is unknown .} \end{cases} \tag{5-135}$$

(When h is known, $\boldsymbol{\sigma}$ is the variance of $\tilde{\boldsymbol{\delta}}$, but when h is unknown, $\boldsymbol{\sigma}$ is $(\nu - 2)/\nu$ times the variance.)

The geometry of the general problem is depicted in Figure 5.18A, which shows the contours on which the conditional value of information,

$$\text{CVI} \equiv \max \{\delta_1, \delta_2, 0\} \ , \tag{5-136}$$

is constant and the contours on which some particular density function $\text{D}(\boldsymbol{\delta}) \equiv f(\boldsymbol{\delta}|\bar{\boldsymbol{\delta}}', \mathbf{n}_{\delta}^{\square})$ is constant.

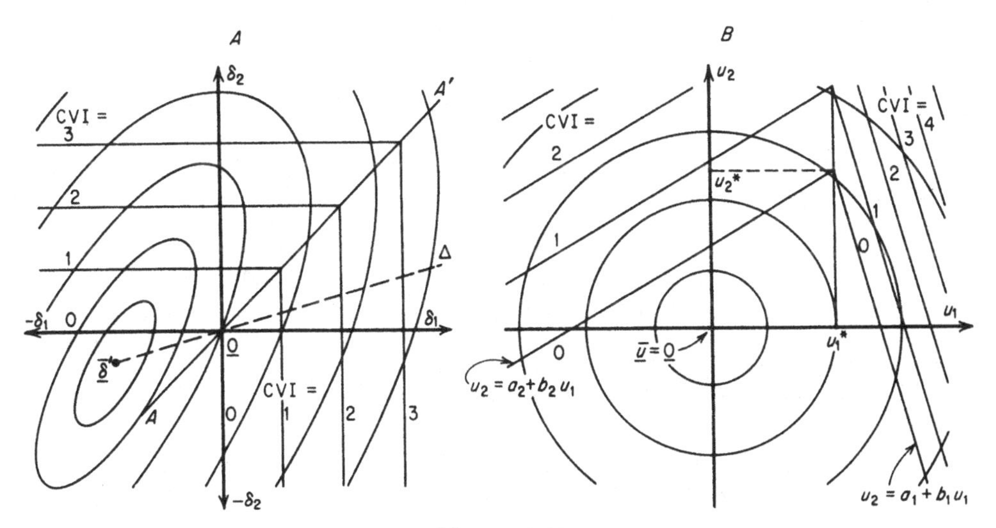

Figure 5.18

Conditional Value of Information and Distributions of Original and Transformed Variables

5.11.2. Transformation to a Unit-Spherical Distribution

The first step towards evaluating the EVI, i.e., towards finding the expectation of the CVI with respect to $\text{D}(\boldsymbol{\delta})$, is to transform Figure 5.18A into Figure 5.18B by a linear transformation

$$\boldsymbol{u} = \mathbf{A}(\boldsymbol{\delta} - \bar{\boldsymbol{\delta}}') \tag{5-137}$$

such that

1. $\tilde{\boldsymbol{u}}$ has mean $\boldsymbol{0}$ and scale parameter $\mathbf{I}$;
2. the line AA' in Figure 5.18A is carried into a line $u_1 = u_1^*$.

This transformation puts the CVI into the special form

$$\text{CVI} = \begin{cases} \max \{0, k(u_2 - [a_2 + b_2 u_1]\} & \text{if } u_1 \le u_1^* \ , \\ \max \{0, k(u_2 - [a_1 + b_1 u_1]\} & \text{if } u_1 \ge u_1^* \ , \end{cases} \tag{5-138}$$

where the constants can be found by computing

$$\alpha = \sqrt{\sigma_{11}^{\square} - 2\sigma_{12}^{\square} + \sigma_{22}^{\square}}\ , \qquad \beta = \sqrt{\sigma_{11}^{\square}\sigma_{22}^{\square} - \sigma_{12}^{\square 2}}\ ,$$

$$k = \beta/\alpha\ , \qquad u_1^* = (\bar{\delta}_2' - \bar{\delta}_1')/\alpha\ , \tag{5-139}$$

$$a_1 = -\bar{\delta}_1'/k\ , \qquad a_2 = -\bar{\delta}_2'/k\ ,$$

$$b_1 = -(\sigma_{11}^{\square} - \sigma_{12}^{\square})/\beta\ , \qquad b_2 = (\sigma_{22}^{\square} - \sigma_{12}^{\square})/\beta\ .$$

The integral (5-133) then becomes

$$\text{EVI} = k \int_{-\infty}^{u_1^*}\left[\int_{a_2+b_2u_1}^{\infty} (u_2 - [a_2 + b_2u_1])\, f^*(u_2)\, du_2\right] f^*(u_1)\, du_1$$

$$+ k \int_{u_1^*}^{\infty}\left[\int_{a_1+b_1u_1}^{\infty} (u_2 - [a_1 + b_1u_1])\, f^*(u_2)\, du_2\right] f^*(u_1)\, du_1 \tag{5-140a}$$

where

$$f^*(u) = \begin{cases} f_{N^*}(u) & \text{if } h \text{ is known}\ , \\ f_{S^*}(u|\nu') & \text{if } h \text{ is unknown}\ . \end{cases} \tag{5-140b}$$

▶ The transformation which takes us from Figure 5.18A to Figure 5.18B can be written (dropping the superscript from $\boldsymbol{\sigma}^{\square}$)

$$\boldsymbol{u} = \mathbf{A}(\boldsymbol{\delta} - \bar{\boldsymbol{\delta}}')\ , \qquad \mathbf{A} \equiv \frac{1}{\alpha\beta}\begin{bmatrix} \beta & -\beta \\ \sigma_{22} - \sigma_{12} & \sigma_{11} - \sigma_{12} \end{bmatrix}.$$

It is obvious from the nature of the transformation that $E(\tilde{\boldsymbol{u}}) = \boldsymbol{0}$, and it is a matter of straightforward algebra to verify that the transformed scale parameter $\mathbf{A}\,\boldsymbol{\sigma}\,\mathbf{A}^t = \mathbf{I}$, thus proving that the transformed densities are as specified by (5-140b).

To verify the transformed CVI function, we first observe that the transformation carries the point (δ_1, δ_2) into the point

$$u_1 = \frac{1}{\alpha}\left[(\delta_1 - \bar{\delta}_1') - (\delta_2 - \bar{\delta}_2')\right]\ ,$$

$$u_2 = \frac{1}{\alpha\beta}\left[(\sigma_{22} - \sigma_{12})(\delta_1 - \bar{\delta}_1') + (\sigma_{11} - \sigma_{12})(\delta_2 - \bar{\delta}_2')\right]\ .$$

It is then again a matter of straightforward algebra to verify that

$$\boldsymbol{\delta} = \bar{\boldsymbol{\delta}}' + \mathbf{A}^{-1}\boldsymbol{u}\ , \qquad \mathbf{A}^{-1} = \frac{1}{\alpha}\begin{bmatrix} \sigma_{11} - \sigma_{12} & \beta \\ -(\sigma_{22} - \sigma_{12}) & \beta \end{bmatrix},$$

and therefore

$$\delta_1 = \bar{\delta}_1' + \frac{1}{\alpha}(\sigma_{11} - \sigma_{12})\, u_1 + \frac{\beta}{\alpha}\, u_2 = k[u_2 - (a_1 + b_1u_1)]\ ,$$

$$\delta_2 = \bar{\delta}_2' - \frac{1}{\alpha}(\sigma_{22} - \sigma_{12})\, u_1 + \frac{\beta}{\alpha}\, u_2 = k[u_2 - (a_2 + b_2u_1)]\ .$$

Comparing the right-hand sides of these two expressions we can verify that the line $\delta_1 = \delta_2$ is transformed into the line $u_1 = u_1^*$; and since

$$b_2 - b_1 = \frac{1}{\beta}(\sigma_{22} - 2\sigma_{12} + \sigma_{11}) > 0$$

this shows that $\delta_1 \le \delta_2$ if and only if $u_1 \le u_1^*$. Formulas (5-138) and hence (5-140a) then follow immediately. ◀

5.11.3. Evaluation of the EVI by Numerical Integration

Numerical integration of the double integrals (5-140) is not at all difficult under either a Normal or Student distribution because the inner integrals are easily evaluated for any given u_1 by means of formulas for linear-loss integrals. For the Normal distribution we have by (11-25b) that (5-140) can be written

$$\text{EVI} = k \int_{-\infty}^{u_1^*} L_{N^*}(a_2 + b_2 u_1)\, f_{N^*}(u_1)\, du_1 + k \int_{u_1^*}^{\infty} L_{N^*}(a_1 + b_1 u_1)\, f_{N^*}(u_1)\, du_1 \qquad \text{(5-141)}$$

and the L_{N^*} function is tabulated in Table II at the end of this monograph. For the Student distribution we have by (11-50b) that (5-140) can be written

$$\text{EVI} = k \int_{-\infty}^{u_1^*} L_{S^*}(a_2 + b_2 u_1|\nu')\, f_{S^*}(u_1|\nu')\, du_1 + k \int_{u_1^*}^{\infty} L_{S^*}(a_1 + b_1 u_1|\nu')\, f_{S^*}(u_1|\nu')\, du_1 \qquad \text{(5-142a)}$$

where the function L_{S^*} is defined in terms of tabulated univariate Student densities and tail areas by

$$L_{S^*}(t|\nu) \equiv \frac{\nu + t^2}{\nu - 1} f_{S^*}(t|\nu) - t\, G_{S^*}(t|\nu) \ . \qquad \text{(5-142b)}$$

The complete double integral can thus be evaluated by choosing a number of equally spaced values u_1, evaluating the inner integral for each by means of the appropriate formula, multiplying each inner integral by the weight $f_{N^*}(u_1)$ or $f_{S^*}(u_1|\nu')$, summing, and dividing by the sum of the weights. A numerical example of this procedure is given in Section 5.11.6.

5.11.4. Evaluation of the EVI by Bivariate Normal Tables When h is Known

When the distribution of $\tilde{u}$ is Normal, the integral (5-141) giving the EVI can also be directly evaluated in terms of tabled functions, specifically the univariate Normal density and cumulative functions and the bivariate Normal integral $V(h, k)$ defined by (8-14); the last named function enters via bivariate "wedge probabilities," written $W_{N^*}(v, \theta)$, which are defined in Section 8.1.2 and which depend on v and θ only through the auxiliary quantities

$$c = v \cos \theta \ , \qquad s \equiv |v \sin \theta| \ , \qquad t = |s/c|$$

To express the results concisely we first define, redefine, or repeat the definition of:

$$\alpha = \sqrt{\sigma_{11}^{\square} - 2\sigma_{12}^{\square} + \sigma_{22}^{\square}} \ , \qquad \rho^2 = \frac{\sigma_{12}^{\square 2}}{\sigma_{11}^{\square}\sigma_{22}^{\square}} \ ,$$

$$\sigma_1^{\square} = \sqrt{\sigma_{11}^{\square}} \ , \qquad \sigma_2^{\square} = \sqrt{\sigma_{22}^{\square}} \ ,$$

$$\zeta_1 = -\bar{\delta}_1'/\sigma_1^{\square} \ , \qquad \zeta_2 = -\bar{\delta}_2'/\sigma_2^{\square} \ ,$$

$$\eta_1 = \frac{\zeta_1 - \rho\zeta_2}{\sqrt{1 - \rho^2}} \ , \qquad \eta_2 = \frac{\zeta_2 - \rho\zeta_1}{\sqrt{1 - \rho^2}} \ ,$$

$$u_1^* = (\bar\delta_2' - \bar\delta_1')/\alpha\ , \qquad u_2^* = (\sigma_2^\square\eta_1 + \sigma_1^\square\eta_2)/\alpha\ , \tag{5-143a}$$

$$c_1 = -\eta_1\ , \qquad s_1 = |\zeta_2|\ ,$$

$$c_2 = -\eta_2\ , \qquad s_2 = |\zeta_1|\ ,$$

$$v = \sqrt{\zeta_1^2 + \eta_2^2} = \sqrt{\zeta_2^2 + \eta_1^2}\ ,$$

$$p' = \frac{\eta_1}{\sigma_1^\square} + \frac{\eta_2}{\sigma_2^\square}\ , \qquad q' = \frac{\zeta_1}{\sigma_2^\square} - \frac{\zeta_2}{\sigma_1^\square}\ ,$$

$$c_3 = v\frac{p'}{\sqrt{p'^2 + q'^2}}\ , \qquad s_3 = v\frac{|q'|}{\sqrt{p'^2 + q'^2}}\ .$$

We then have

$$\text{EVI} = \alpha f_{N^*}(u_1^*)\, G_{N^*}(u_2^*) + \sigma_1^\square f_{N^*}(\zeta_1)\, F_{N^*}(\eta_2) + \sigma_2^\square f_{N^*}(\zeta_2)\, F_{N^*}(\eta_1)$$
$$+ \bar\delta_2' W_{N^*}(v, \theta_1) + \bar\delta_1' W_{N^*}(v, \theta_2) - |\bar\delta_2' - \bar\delta_1'|\ W_{N^*}(v, \theta_3)\ , \tag{5-143b}$$

where $W_{N^*}(v, \theta_i)$ depends only on c_i, s_i, and $t_i = |s_i/c_i|$ and can be evaluated from tables of the bivariate Normal function $V(h, k)$ by use of formulas (8-16).

▶ Before entering on the main proof of (5-143), we clear the ground with two easily verified formulas for integrals. First,

1) $\int G_{N^*}(a + bu) \cdot u f_{N^*}(u)\, du = -G_{N^*}(a + bu) f_{N^*}(u) - b \int f_{N^*}(u) f_{N^*}(a + bu)\, du\ ,$

a simple case of integration by parts. Second,

2) $\int f_{N^*}(a + bu) f_{N^*}(u)\, du = f_{N^*}\left(\frac{a}{\sqrt{1 + b^2}}\right) \frac{1}{\sqrt{1 + b^2}} F_{N^*}\left(\left[u + \frac{ab}{1 + b^2}\right]\sqrt{1 + b^2}\right)\ ,$

as can be shown by completing the square in the exponent on the left and then making the substitution

$$v = \left[u + \frac{ab}{1 + b^2}\right]\sqrt{1 + b^2}\ .$$

Now substituting the definition (11-26) of L_{N^*} in (5-141) we see that we must integrate two expressions of the type

3) $$J \equiv \int [f_{N^*}(a + bu) - (a + bu)\, G_{N^*}(a + bu)]\, f_{N^*}(u)\, du\ ,$$

one over $(-\infty, u_1^*]$ and the other over $[u_1^*, \infty)$. We start by expressing J in two parts,

4) $$J = J_1 - aJ_2\ ,$$

where

5) $$J_1 = \int [f_{N^*}(a + bu) - bu\, G_{N^*}(a + bu)]\, f_{N^*}(u)\, du\ ,$$

6) $$J_2 = \int G_{N^*}(a + bu)\, f_{N^*}(u)\, du\ .$$

The integral J_1 is readily evaluated by substituting (1) in (5) to obtain

7) $$J_1 = b\, G_{N^*}(a + bu)\, f_{N^*}(u) + (1 + b^2) \int f_{N^*}(a + bu)\, f_{N^*}(u)\, du$$

and then substituting (2) in (7) to obtain

$$8)\quad J_1 = b\,G_{N^*}(a + bu)\,f_{N^*}(u) + \sqrt{1+b^2}\,f_{N^*}\left(\frac{a}{\sqrt{1+b^2}}\right) F_{N^*}\left(\left[u + \frac{ab}{1+b^2}\right]\sqrt{1+b^2}\right) .$$

Evaluating this integral with coefficients a_2 and b_2 over $(-\infty, u_1^*]$ and with coefficients a_1 and b_1 over $[u_1^*, \infty)$ and adding the two results we obtain

$$9)\quad b_2\,G_{N^*}(a_2 + b_2u_1^*)\,f_{N^*}(u_1^*) + \sqrt{1+b_2^2}\,f_{N^*}\left(\frac{a_2}{\sqrt{1+b_2^2}}\right) F_{N^*}\left(\left[u_1^* + \frac{a_2b_2}{1+b_2^2}\right]\sqrt{1+b_2^2}\right)$$

$$- b_1\,G_{N^*}(a_1 + b_1u_1^*)\,f_{N^*}(u_1^*) + \sqrt{1+b_1^2}\,f_{N^*}\left(\frac{a_1}{\sqrt{1+b_1^2}}\right) G_{N^*}\left(\left[u_1^* + \frac{a_1b_1}{1+b_1^2}\right]\sqrt{1+b_1^2}\right) .$$

Using (5-139) and (5-143a) to compute

$$a_1 + b_1u_1^* = a_2 + b_2u_1^* = u_2^* , \qquad b_2 - b_1 = \alpha/k ,$$

$$\sqrt{1+b_1^2} = \frac{\alpha}{\beta}\,\sigma_1^{\square} = \sigma_1^{\square}/k , \qquad \sqrt{1+b_2^2} = \frac{\alpha}{\beta}\,\sigma_2^{\square} = \sigma_2^{\square}/k ,$$

$$a_1/\sqrt{1+b_1^2} = \zeta_1 , \qquad a_2/\sqrt{1+b_2^2} = \zeta_2 ,$$

$$\left[u_1^* + \frac{a_1b_1}{1+b_1^2}\right]\sqrt{1+b_1^2} = -\eta_2 , \qquad G_{N^*}(-\eta_2) = F_{N^*}(\eta_2) ,$$

$$\left[u_1^* + \frac{a_2b_2}{1+b_2^2}\right]\sqrt{1+b_2^2} = \eta_1 ,$$

substituting these results in (9), and multiplying the sum by k as called for by (5-141) we obtain the first three terms of (5-143).

We next turn our attention to the two J_2 integrals defined by (6) above:

$$10)\qquad J_2^{(l)} = \int_{-\infty}^{u_1^*} G_{N^*}(a_2 + b_2u_1)\,f_{N^*}(u_1)\,du_1 ,$$

$$11)\qquad J_2^{(r)} = \int_{u_1^*}^{\infty} G_{N^*}(a_1 + b_1u_1)\,f_{N^*}(u_1)\,du_1 .$$

The quantity $J_2^{(l)}$ is clearly the probability that $\tilde{\mathbf{u}}$ lies above the zero loss contour and to the left of the line $u_1 = u_1^*$ in Figure 5.18B and thus corresponds to the probability that $\tilde{\boldsymbol{\delta}}$ lies within the "wedge" $(-\delta_1)OA'$ in Figure 5.18A; and similarly the quantity $J_2^{(r)}$ is the probability that $\tilde{\boldsymbol{\delta}}$ lies within the wedge $AO(-\delta_2)$. Proceeding in the manner described in Section 8.2.2, we express each of these two wedges in Figure 5.18A as a sum or difference of two "canonical" wedges both having one side on the line $\bar{\boldsymbol{\delta}}'O\Delta$ which passes through the mean of the distribution and the common vertex of all the wedges,

$$(-\delta_1)OA' = (-\delta_1)O\Delta - A'O\Delta , \qquad A'O(-\delta_2) = A'O\Delta + \Delta O(-\delta_2) ,$$

and we denote the probabilities of the canonical wedges by

$$12)\qquad \begin{aligned} W_{N^*}(v, \theta_1) &\equiv \mathrm{P}\{(-\delta_1)O\Delta\} , \\ W_{N^*}(v, \theta_2) &\equiv \mathrm{P}\{\Delta O(-\delta_2)\} , \\ W_{N^*}(v, \theta_3) &\equiv \mathrm{P}\{A'O\Delta\} , \end{aligned}$$

so that

$$13)\qquad J_2^{(l)} = W_{N^*}(v, \theta_1) - W_{N^*}(v, \theta_3) , \qquad J_2^{(r)} = W_{N^*}(v, \theta_3) + W_{N^*}(v, \theta_2) .$$

By (5-141) and (4) the contribution of the J_2 integrals to the EVI is

$$k[-a_2J_2^{(l)} - a_1J_2^{(r)}] = \bar{\delta}_2'J_2^{(l)} + \bar{\delta}_1'J_2^{(r)} ,$$

and substituting (13) herein we obtain

14) $$\bar{\delta}_2' W_{N^*}(v, \theta_1) + \bar{\delta}_1' W_{N^*}(v, \theta_2) - (\bar{\delta}_2' - \bar{\delta}_1') W_{N^*}(v, \theta_3) .$$

If the point $\bar{\boldsymbol{\delta}}'$ in Figure 5.18A had been below the line AA' rather than above it, the wedge $A'O\Delta$ would have been added to $(-\delta_1)O\Delta$ and subtracted from $\Delta O(-\delta_2)$ and we would have had instead of (14)

15) $$\bar{\delta}_2' W_{N^*}(v, \theta_1) + \bar{\delta}_1' W_{N^*}(v, \theta_2) - (\bar{\delta}_1' - \bar{\delta}_2') W_{N^*}(v, \theta_3) .$$

It is easily seen, however, that the position of $\bar{\boldsymbol{\delta}}'$ relative to AA' is related to the relative magnitudes of $\bar{\delta}_1'$ and $\bar{\delta}_2'$ in such a way that $(\bar{\delta}_2' - \bar{\delta}_1') \geq 0$ in (14) and $(\bar{\delta}_1' - \bar{\delta}_2') \geq 0$ in (15), so that we may substitute $|\bar{\delta}_2' - \bar{\delta}_1'|$ for either of these coefficients. When we do so, both (14) and (15) become identical to the last three terms of (5-143).

That the three probabilities $W_{N^*}(v, \theta_i)$ can be evaluated by substituting the values of c_i and s_i given by (5-143) in formulas (8-16) follows upon comparison of Figure 5.18A and formulas (5-143a) with Figure 8.4 and formulas (8-23); the quantities p' and q' defined in (5-143a) are $\sqrt{2}$ times the quantities p and q as defined by (8-23) for the wedge $A'O\Delta$. ◀

EVI When $\bar{\delta}_1' = \bar{\delta}_2'$ *and* $\sigma_{11}^{\square} = \sigma_{22}^{\square}$. In some circumstances—particularly when the current "standard" process is deemed superior to both of two equally little known contenders—it may be nearly if not exactly true that $\bar{\delta}_1' = \bar{\delta}_2'$ and $\sigma_{11}^{\square} = \sigma_{22}^{\square}$; and in this case formula (5-143b) becomes very easy to evaluate. Defining

$$\bar{\delta}' \equiv \bar{\delta}_1' = \bar{\delta}_2' , \qquad \sigma^{\square} \equiv \sqrt{\sigma_{11}^{\square}} = \sqrt{\sigma_{22}^{\square}} ,$$

$$\zeta \equiv -\bar{\delta}'/\sigma^{\square} , \qquad \eta \equiv \zeta\left[\frac{\sigma^{\square 2} - \sigma_{12}^{\square}}{\sigma^{\square 2} + \sigma_{12}^{\square}}\right]^{\frac{1}{2}} , \tag{5-144a}$$

we have (as is easily verified)

$$\text{EVI} = 2[\sigma^{\square} f_{N^*}(\zeta)\, F_{N^*}(\eta) + \bar{\delta}' W_{N^*}(v, \theta)] \tag{5-144b}$$

where $W_{N^*}(v, \theta)$ depends only on

$$c = -\eta < 0 , \qquad s = \zeta , \qquad t = \left[\frac{\sigma^{\square 2} + \sigma_{12}^{\square}}{\sigma^{\square 2} - \sigma_{12}^{\square}}\right]^{\frac{1}{2}} . \tag{5-144c}$$

5.11.5. *Bounds on the EVI*

Whereas exact evaluation of the EVI is fairly laborious except for the special case just mentioned and some other special cases which we shall examine presently, upper and lower bounds on the EVI can be found very easily and in some applications these bounds may be such as to eliminate the need for exact evaluation.

If we recall that by definition

$$\text{EVI} = \text{E} \max \{\tilde{\delta}_1, \tilde{\delta}_2, 0\}$$

then it is clear that a *lower* bound on the EVI is provided by

$$\text{EVI} \geq \max [\text{E} \max \{\tilde{\delta}_1, 0\}, \text{E} \max \{\tilde{\delta}_2, 0\}] \tag{5-145}$$

while an *upper* bound is provided by

$$\text{EVI} \leq \text{E} \max \{\tilde{\delta}_1, 0\} + \text{E} \max \{\tilde{\delta}_2, 0\} . \tag{5-146}$$

Evaluation of either of the two expectations involved in these bounds is a simple univariate problem involving only the marginal distribution of $\bar{\delta}_1$ or $\bar{\delta}_2$ as the case may be. *If h is known* and the distribution of $\tilde{\boldsymbol{\delta}}$ is Normal, then by (8-20) and (5-135)

$$D(\delta_i) = f_N(\delta_i|\bar{\delta}'_i, \sigma_{ii}^{\square\,-1})$$

and hence by (11-25b)

$$E \max \{\tilde{\delta}_i, 0\} = \sqrt{\sigma_{ii}^{\square}}\, L_{N^*}(u_i) \ , \qquad u_i = -\bar{\delta}'_i/\sqrt{\sigma_{ii}^{\square}} \ . \tag{5-147}$$

If h is unknown and the distribution of $\tilde{\boldsymbol{\delta}}$ is Student, then by (8-30) and (5-135)

$$D(\delta_i) = f_S(\delta_i|\bar{\delta}'_i, \sigma_{ii}^{\square\,-1}, \nu')$$

and hence by (11-50b)

$$E \max \{0, \tilde{\delta}_i\} = \sqrt{\sigma_{ii}^{\square}}\, L_{S^*}(t_i|\nu') \ , \qquad t_i = -\bar{\delta}'_i/\sqrt{\sigma_{ii}^{\square}} \ . \tag{5-148}$$

5.11.6. *Example*

A chemical manufacturer wishes to choose one of three possible processes for producing a certain product; the criterion of choice is to be expected monetary profit. The chosen process will be definitely used for one year, after which time the entire question will be reexamined; accordingly the manufacturer wishes any investments in fixed assets required by the choice of process to be considered as expense in this year. If process A is used, it will be possible to produce 110 batches during the year; materials and labor will cost \$500 per batch; for process B, the corresponding figures are 100 batches at \$600 per batch; for process C, 90 batches at \$700 per batch. Processes B and C require no equipment other than that already available; but if process A is used, it will be necessary to invest \$7500 in a special mixer. The product sells for \$1 per pound. Ten pilot-plant experiments have been conducted on each of the three processes in order to determine their yields; the means of the indicated yields are 950 lb/batch for process A, 1000 for B, and 1150 for C; and management feels that in comparison with this experimental evidence all other information concerning the yields is of negligible weight. Further experiments can be conducted at a cost of \$100 each for process A, \$110 for B, and \$120 for C. Letting x denote the yield of a single experimental trial multiplied by a factor converting it into an estimate of the yield of a batch of full production size, these experiments have indicated that the standard deviation of $\tilde{x}$ is 50 lbs/batch for any one of the three processes; the experiments are considered to be unbiassed in the sense that the average of $\tilde{x}$ over an infinite number of experiments on any process would be equal to the full-scale mean yield of that process.

Since management wishes to treat its outside or judgmental information as negligible in comparison with the experimental results already obtained, we have as the expected yields per batch of the three processes

$$A: 950 \ , \qquad B: 1000 \ , \qquad C: 1150 \ .$$

From these we can compute the expected profits

$$\begin{aligned} &A: 110(\$950 - \$500) - \$7500 = \$42{,}000 \ , \\ &B: 100(\$1000 - \$600) \qquad\quad\;\; = \;\;40{,}000 \ , \\ &C: 90(\$1150 - \$700) \qquad\quad\;\;\; = \;\;40{,}500 \ . \end{aligned}$$

By our convention (5-61) we must assign the subscript 3 to process A, and therefore we now define

μ_1: yield of process B ,
μ_2: yield of process C ,
μ_3: yield of process A .

We are now ready to formalize the statement of the problem; in so doing we shall express all monetary amounts in units of \$100. The economic structure is described by

$$v = K + \mathbf{k}\,\mu\ , \qquad \delta = \mathbf{B}\,v = \mathbf{B}\,K + \mathbf{B}\,\mathbf{k}\,\mu\ ,$$

where

$$K = \begin{bmatrix} -100 \times 6 \\ -90 \times 7 \\ -110 \times 5 - 75 \end{bmatrix} = \begin{bmatrix} -600 \\ -630 \\ -625 \end{bmatrix}, \qquad \mathbf{k} = \begin{bmatrix} 1.0 & 0 & 0 \\ 0 & .9 & 0 \\ 0 & 0 & 1.1 \end{bmatrix},$$

$$\mathbf{B}\,K = \begin{bmatrix} +25 \\ -5 \end{bmatrix}, \qquad \mathbf{B}\,\mathbf{k} = \begin{bmatrix} 1.0 & 0 & -1.1 \\ 0 & .9 & -1.1 \end{bmatrix}.$$

Regarding the data-generating processes, we may treat as known for all practical purposes

$$h_1 = h_2 = h_3 = h = 1/50^2\ .$$

As regards the prior distribution of $\tilde{\mu}$, we have already seen that

$$m' = \bar{\mu}' = [1000 \quad 1150 \quad 950]^t\ .$$

Since the manufacturer wishes to treat any judgmental information he may have as negligible in comparison with information obtained from the experiments already conducted, the marginal variance of each of the $\tilde{\mu}$s is simply $50^2/10 = 250$ and the covariances are all 0:

$$\mathrm{V}'(\tilde{\mu}) = (h\mathbf{n}')^{-1} = \begin{bmatrix} 250 & 0 & 0 \\ 0 & 250 & 0 \\ 0 & 0 & 250 \end{bmatrix};$$

and therefore

$$\mathbf{n}'^{-1} = \begin{bmatrix} .1 & 0 & 0 \\ 0 & .1 & 0 \\ 0 & 0 & .1 \end{bmatrix}, \qquad \mathbf{n}' = \begin{bmatrix} 10 & 0 & 0 \\ 0 & 10 & 0 \\ 0 & 0 & 10 \end{bmatrix}.$$

From these results we can compute

$$\bar{v}' = K + \mathbf{k}\,\bar{\mu}' = [400 \quad 405 \quad 420]^t$$

(a result already obtained), and

$$\bar{\delta}' = \mathbf{B}\,\bar{v}' = [-20 \quad -15]^t\ ,$$

$$\mathbf{n}_\delta'^{-1} = (\mathbf{B}\,\mathbf{k})\,\mathbf{n}'^{-1}(\mathbf{B}\,\mathbf{k})^t = \begin{bmatrix} .221 & .121 \\ .121 & .202 \end{bmatrix}.$$

Terminal Analysis. If the manufacturer is to choose one of the three processes without further experimentation, he will simply choose process 3 because it maximizes his expected utility.

Bounds on the EVPI. To analyze the expected value of perfect information we first compute

$$\boldsymbol{\sigma}' = (h\mathbf{n}'_\delta)^{-1} = 2500\,\mathbf{n}'^{-1}_\delta = \begin{bmatrix} 552.5 & 302.5 \\ 302.5 & 505.0 \end{bmatrix} .$$

Our next step is to find upper and lower bounds for the EVPI by the method of Section 5.11.5, and to do so we first compute

$$\sqrt{\sigma'_{11}} = 23.51 \ , \qquad \sqrt{\sigma'_{22}} = 22.47 \ ,$$

$$u_1 = 20/23.51 = .8507 \ , \qquad u_2 = 15/22.47 = .6676 \ ,$$

find from Table II at the end of this monograph that

$$L_{N^*}(.8507) = .1099 \ , \qquad L_{N^*}(.6676) = .1509 \ ,$$

and compute

$$\mathrm{E}\max\{\tilde\delta_1, 0\} = 23.51 \times .1099 = 2.584 \ ,$$

$$\mathrm{E}\max\{\tilde\delta_2, 0\} = 22.47 \times .1509 = 3.391 \ ,$$

$$\text{EVPI} \begin{cases} \le 2.584 + 3.391 = 5.975 & (= \$597.50) \\ \ge 3.391 & (= \$339.10) \ . \end{cases}$$

Since a single additional observation on just one of the three processes will cost from \$100 to \$120 depending on the process, it seems obvious that the net gain of further experimentation will almost certainly be negative for all possible experiments; at best it can be a negligible positive amount. The common-sense conclusion is to choose process 3 without further ado; the calculations in the remainder of this analysis are given solely to illustrate computational method.

Before going on to these computational exercises, however, let us pause to contrast our results so far with those which would have been obtained by a traditional test of significance. In a situation where the economics are such that the process with the lowest yield may be the most profitable, it clearly makes no sense to test the null hypothesis that the three *yields* are equal; and we assume therefore that the test called for in this problem is of the null hypothesis that all the *profitabilities* are equal. The sufficient "estimators" are

$$\hat v_1 = -600 + 1.0\,\bar x_1 = 400 \ ,$$

$$\hat v_2 = -630 + .9\,\bar x_2 = 405 \ ,$$

$$\hat v_3 = -625 + 1.1\,\bar x_3 = 420 \ ,$$

and their sampling distributions can be treated as Normal with means v_1, v_2, and v_3 and known variances

$$\sigma_1^2 \equiv \mathrm{V}(\tilde{\hat v}_1|v_1) = 1.0^2 \times 250 = 250.0 \ ,$$

$$\sigma_2^2 \equiv \mathrm{V}(\tilde{\hat v}_2|v_2) = .9^2 \times 250 = 202.5 \ ,$$

$$\sigma_3^2 \equiv \mathrm{V}(\tilde{\hat v}_3|v_3) = 1.1^2 \times 250 = 302.5 \ .$$

The test statistic will be

$$z = \Sigma \frac{1}{\sigma_i^2}(\hat v - \bar v)^2 \ , \qquad \bar v = \frac{1}{\Sigma\,\sigma_i^{-2}}\,\Sigma\,\hat v_i/\sigma_i^2 \ ,$$

which has a chi-square distribution with 2 degrees of freedom; and since its observed value is .8 the experiment is just significant at the level

$$P_{\chi^2}\{\tilde{z} \geq .8|\nu = 2\} = .67 \ .$$

This means, of course, that if the three profitabilities are in fact equal, two samples out of three would show at least as much (mean-square) *in*equality as was actually observed. Traditionally, the decision maker must now decide by the use of judgment whether or not this level of significance is sufficient (numerically small enough) to warrant a terminal decision without further experimentation.

Exact EVPI by Numerical Integration. To evaluate the exact value of perfect information(or expected opportunity loss of an immediate decision to use process 3) by the method of Section 5.11.3, we first compute the constants defined by (5-139):

$$\alpha = \sqrt{552.5 - 2(302.5) + 505.0} = 21.27 \ ,$$

$$\beta = \sqrt{(552.5)(505.0) - (302.5)^2} = 433.0 \ ,$$

$$k = 433.0/21.27 = 20.36 \ , \qquad u_1^* = \frac{(-15) - (-20)}{21.27} = .2351 \ ,$$

$$a_1 = 20/20.36 = .9823 \ , \qquad a_2 = 15/20.36 = .7367 \ ,$$

$$b_1 = \frac{-(552.5 - 302.5)}{433.0} = -.5774 \ , \qquad b_2 = \frac{505.0 - 302.5}{433.0} = .4677 \ .$$

Then dividing the interval $(-3.625 \leq u_1 \leq +3.625)$ into 29 equal intervals centered on $-3.50, -3.25, \cdots, +3.25, +3.50$ we can lay out the computations as shown in Table 5.3 below; recall that for $u_1 \leq u_1^* = .2351$ we compute $a + bu_1$ with coefficients a_2 and b_2 while for $u_1 \geq u_1^*$ we use a_1 and b_1. From the totals in the table we compute

$$\text{weighted mean of } L_{N^*}(a + bu_1) = \frac{.94955}{3.9987} = .2375 \ ,$$

$$\text{EVPI} = .2375\,k = .2375 \times 20.36 = 4.835 \qquad (= \$483.50) \ .$$

It is worth remarking that if we base our computation on only the 7 values $u_1 = -3.00, -2.00, \cdots, +3.00$ we get a weighted mean of .2377, virtually identical to the .2375 obtained with all 29 values.

Table 5.3

Computation of EVPI

u_1	$a + bu_1$	$L_{N^*}(a + bu_1)$	$f_{N^*}(u_1)$	$L_{N^*}f_{N^*}$
-3.50	$-.9002$	1.0006	.0009	.00090054
....	....	....	....	....
0	$+ .7367$	.1342	.3989	.05353238
....	....	....	....	....
$+3.50$	-1.0386	1.1160	.0009	.00100440
total			3.9987	.94955

EVPI Using Bivariate Tables. To evaluate the EVPI by the method of Section 5.11.4, we first compute the first set of constants defined by (5-143a)

$$\alpha = \sqrt{552.5 - 2(302.5) + 505.0} = 21.27 \ ,$$

$$\rho^2 = \frac{(302.5)^2}{(552.5)(505.0)} = .3280 \ , \qquad \rho = .5727 \ , \qquad \sqrt{1-\rho^2} = .8198 \ ,$$

$$\sigma_1' = \sqrt{552.5} = 23.51 \ , \qquad \sigma_2' = \sqrt{505.0} = 22.47 \ ,$$

$$\zeta_1 = 20/23.51 = .8507 \ , \qquad \zeta_2 = 15/22.47 = .6676 \ ,$$

$$\eta_1 = \frac{.8507 - (.5727)(.6676)}{.8198} = .5713 \ , \qquad \eta_2 = \frac{.6676 - (.5727)(.8507)}{.8198} = .2201 \ ,$$

$$u_1^* = \frac{(-15) - (-20)}{21.27} = .2351 \ , \quad u_2^* = \frac{(22.47)(.5713) + (23.51)(.2201)}{21.27} = .0468 \ ,$$

from which we can compute the first three terms of (5-143b)

$$\begin{aligned} A &\equiv \alpha f_{N^*}(u_1^*)\, G_{N^*}(u_2^*) + \sigma_1' f_{N^*}(\zeta_1)\, F_{N^*}(\eta_2) + \sigma_2' f_{N^*}(\zeta_2)\, F_{N^*}(\eta_1) \\ &= (21.27)(.3880)(.1986) + (23.51)(.2778)(.5871) + (22.47)(.3192)(.7161) \\ &= 1.639 + 3.834 + 5.136 = 10.609 \ . \end{aligned}$$

We then compute the remaining constants called for by (5-143a) and (8-16a)

$$c_1 = -.5713 \ , \qquad s_1 = .6676 \ , \qquad 1/t_1 = .5713/.6676 = .8557 \ ,$$

$$c_2 = -.2201 \ , \qquad s_2 = .8507 \ , \qquad 1/t_2 = .2201/.8507 = .2587 \ ,$$

$$v = \sqrt{(.8507)^2 + (.2201)^2} = \sqrt{(.6676)^2 + (.5713)^2} = .8787 \ ,$$

$$p' = \frac{.5713}{23.51} + \frac{.2201}{22.47} = .03409 \ , \qquad q' = \frac{.8507}{22.47} - \frac{.6676}{23.51} = .00946 \ ,$$

$$c_3 > 0 \ , \qquad s_3 = .8787\,\frac{.00946}{.03538} = .2349 \ , \qquad t_3 = .00946/.03409 = .2775 \ ,$$

use (8-16b) to compute

$$\begin{aligned} W_{N^*}(v, \theta_1) &= \tfrac{1}{2}G_{N^*}(.6676) + V(\infty, .8557\,\infty) - V(.6676, .8557\, s_1) \\ &= .1261 + .1125 - .0265 = .2121 \ , \end{aligned}$$

$$\begin{aligned} W_{N^*}(v, \theta_2) &= \tfrac{1}{2}G_{N^*}(.8507) + V(\infty, .2587\,\infty) - V(.8507, .2587\, s_2) \\ &= .0988 + .0402 - .0125 = .1265 \ , \end{aligned}$$

$$\begin{aligned} W_{N^*}(v, \theta_3) &= \tfrac{1}{2}G_{N^*}(.2349) - V(.2775\,\infty, \infty) + V(.2775\, s_3, .2349) \\ &= .2036 - .2070 + .0012 = -.0022 \ . \end{aligned}$$

The obviously incorrect negative result for $W_{N^*}(v, \theta_3)$ is due to the use of linear interpolation λ-wise in the bivariate tables, and following the indication of this result we reduce all the bivariate probabilities to 2 decimal places. We can then compute the last three terms of (5-143b)

$$\begin{aligned} B &\equiv \bar\delta_2' \, W_{N^*}(v, \theta_1) + \bar\delta_1' \, W_{N^*}(v, \theta_2) - |\bar\delta_2' - \bar\delta_1'| \, W_{N^*}(v, \theta_3) \\ &= -15(.21) - 20(.13) - 5(0) = -5.75 \ , \end{aligned}$$

and we then have

$$\text{EVPI} = A + B = 10.61 - 5.75 = 4.86 \qquad (= \$486) .$$

The result is in reasonably good agreement with the $483.50 obtained by numerical integration, which is the more accurate as well as the less laborious method in the general case. (Use of the bivariate tables will, however, be advantageous in the special case to which (5-144) applies.)

Expected Value of Ten Additional Observations on Each Process. To illustrate the computation of EVSI, we shall work out the arbitrarily chosen case where 10 observations are to be taken on each process and therefore (since $h_i/h = 1$ for all i)

$$\mathbf{n} = 10\,\mathbf{I} .$$

The first step is to compute

$$\mathbf{n}'' = \mathbf{n}' + \mathbf{n} = 20\,\mathbf{I} , \qquad \mathbf{n}''^{-1} = .05\,\mathbf{I} ,$$

$$\mathbf{n}'^{-1} - \mathbf{n}''^{-1} = .05\,\mathbf{I} ,$$

$$\mathbf{n}_\delta^{*-1} = (\mathbf{B}\,\mathbf{k})(\mathbf{n}'^{-1} - \mathbf{n}''^{-1})(\mathbf{B}\,\mathbf{k})^t = \begin{bmatrix} .1105 & .0605 \\ .0605 & .1010 \end{bmatrix} .$$

Because all the entries in this matrix are simply half those in $\mathbf{n}_\delta'^{-1}$, $\boldsymbol{\sigma}^*$ for our problem will be simply half $\boldsymbol{\sigma}'$ for the EVPI and the various constants required for numerical integration of the EVSI can be easily obtained from the constants computed above for integration of the EVPI.

$$\alpha = \frac{21.27}{\sqrt{2}} = 15.04 , \qquad \beta = \frac{433.0}{2} = 216.5 ,$$

$$k = 216.5/15.04 = 14.39 , \qquad u_1^* = 5/15.04 = .3324 ,$$

$$a_1 = 20/14.39 = 1.390 , \qquad a_2 = 15/14.39 = 1.042 ,$$

$$b_1 = -125/216.5 = -.578 , \qquad b_2 = 101.2/216.5 = .467 .$$

Using 13 values $u_1 = -3.0, -2.5, \cdots, +3.0$, we then obtain by the method of Table 5.3

$$\text{weighted mean of } L_{N^*}(a + bu_1) = .141$$

$$\text{EVSI} = .141 \times 14.39 = 2.03 \qquad (= \$203.00) ;$$

using only the 7 values $u_1 = -3, -2, \cdots, +3$ we obtain $207. *If observations cost on the average something less than $203/30 ≐ $7 each, rather than $100 or more, a sample of this size would be worth taking;* although of course some other sample might have a still larger net gain.

5.11.7. *EVI When the Prior Expected Utilities are Equal*

When the prior expected utilities of all three acts are equal, $\bar{v}_1' = \bar{v}_2' = \bar{v}_3'$, the prior expected utility differences $\bar{\delta}_1'$ and $\bar{\delta}_2'$ are both zero and very simple formulas can be given for the EVI whether or not h is known. Defining

$$\mathbf{u} \equiv \begin{cases} \mathbf{n}'^{-1} & \text{if EVPI is to be evaluated ,} \\ \mathbf{n}'^{-1} - \mathbf{n}''^{-1} & \text{if EVSI is to be evaluated ,} \end{cases} \qquad \text{(5-149a)}$$

$$
\begin{aligned}
U_{12} &\equiv \sqrt{k_1^2 u_{11} - 2k_1k_2u_{12} + k_2^2 u_{22}} \ , \\
U_{13} &\equiv \sqrt{k_1^2 u_{11} - 2k_1k_3u_{13} + k_3^2 u_{33}} \ , \\
U_{23} &\equiv \sqrt{k_2^2 u_{22} - 2k_2k_3u_{23} + k_3^2 u_{33}} \ , \\
U &\equiv \frac{1}{2\sqrt{2\pi}} (U_{12} + U_{13} + U_{23}) \ ,
\end{aligned}
\tag{5-149b}
$$

we have

$$
\text{EVI} = \begin{cases} U\, h^{-\frac{1}{2}} & \text{if } h \text{ is known} \ , \\ U \dfrac{(\frac{1}{2}\nu' - \frac{3}{2})!}{(\frac{1}{2}\nu' - 1)!} \sqrt{\tfrac{1}{2}\nu' v'} & \text{if } h \text{ is unknown} \ . \end{cases}
\tag{5-150}
$$

If ν' is large enough to justify the use of Stirling's approximation, the last formula reduces to

$$
\text{EVI} \doteq U \left[\frac{\nu'}{\nu' - 2} v' \right]^{\frac{1}{2}} \qquad \text{if } h \text{ is unknown and } \nu' \text{ is large} \ . \tag{5-151}
$$

▶ *First Proof.* Formula (5-150) for h known can be obtained as a special case of (5-143), after which the formula for h unknown can be obtained by substituting $\text{E}(\tilde{h}^{-\frac{1}{2}})$ for $h^{-\frac{1}{2}}$.

1. We first clear the ground by establishing the relation between the quantity U which appears in (5-149b) and the quantities α, $\sigma_1^\square$, and $\sigma_2^\square$ which appear in (5-143a). By (5-135), (5-133b), (5-132c), (5-131c), and (5-149) we have

$$
\boldsymbol{\sigma}^\square = \frac{1}{h} (\mathbf{B\,k})\, \mathbf{u}\, (\mathbf{B\,k})^t \ , \qquad \mathbf{B\,k} = \begin{bmatrix} k_1 & 0 & -k_3 \\ 0 & k_2 & -k_3 \end{bmatrix}
$$

from which we can readily compute

$$
\alpha \equiv \sqrt{\sigma_{11}^\square - 2\sigma_{12}^\square + \sigma_{22}^\square} = U_{12}\, h^{-\frac{1}{2}} \ ,
$$

1)

$$
\sigma_1^\square \equiv \sqrt{\sigma_{11}^\square} = U_{13}\, h^{-\frac{1}{2}} \ , \qquad \sigma_2^\square \equiv \sqrt{\sigma_{22}^\square} = U_{23}\, h^{-\frac{1}{2}} \ .
$$

2. Substituting $\bar{\delta}_1' = \bar{\delta}_2' = 0$ in (5-143a) we obtain

$$
0 = \zeta_1 = \zeta_2 = \eta_1 = \eta_2 = u_1^* = u_2^* = \bar{\delta}_1' = \bar{\delta}_2' \ .
$$

Substituting these values in (5-143b) and recalling that

$$
f_{N^*}(0) = (2\pi)^{-\frac{1}{2}} \ , \qquad F_{N^*}(0) = G_{N^*}(0) = \tfrac{1}{2} \ ,
$$

we obtain

2)

$$
\text{EVI} = \frac{1}{2\sqrt{2\pi}} (\alpha + \sigma_1^\square + \sigma_2^\square) \ ;
$$

and substituting (1) herein we obtain (5-150) for h known.

3. If the marginal distribution of $\tilde{h}$ is gamma-2 with parameter (v', ν'), then by (7-58a) the distribution of $\tilde{h}^{-\frac{1}{2}}$ is inverted-gamma-2 with parameter $(v'^{\frac{1}{2}}, \nu')$ and hence by (7-59a)

$$
E(\tilde{h}^{-\frac{1}{2}}) = \frac{(\frac{1}{2}v' - \frac{3}{2})!}{(\frac{1}{2}\nu' - 1)!} \sqrt{\tfrac{1}{2}\nu' v'} \ .
$$

Substitution of this result for $h^{-\frac{1}{2}}$ in formula (5-150) for the EVI with h known yields the corresponding formula for h unknown. The approximate formula (5-151) is then obtained by defining

$$x \equiv \tfrac{1}{2}\nu' - 1 ,$$

writing

$$\frac{(\frac{1}{2}\nu' - \frac{3}{2})!}{(\frac{1}{2}\nu' - 1)!}\sqrt{\tfrac{1}{2}\nu'} = \frac{(x - \frac{1}{2})!(x + 1)^{\frac{1}{2}}}{x!} ,$$

and then using Legendre's formula and Stirling's approximation

$$(x - \tfrac{1}{2})! = \frac{(2x)!\pi^{\frac{1}{2}}}{2^{2x}\, x!} , \qquad x! \doteq (2\pi)^{\frac{1}{2}}\, x^{x+\frac{1}{2}}\, e^{-x} .$$

Alternate Proof. Formulas (5-150) can also be proved by direct integration of formula (5-140) for the EVI. Again we proceed by first deriving the result for h known, after which the result for h unknown follows by step (3) in our previous proof.

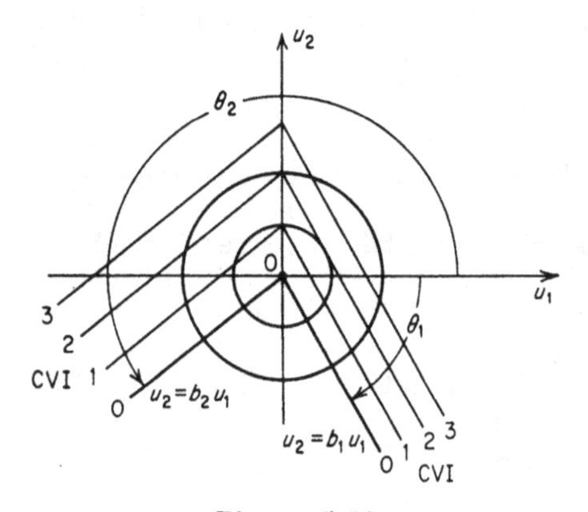

Figure 5.19

Conditional Value of Information and Distribution of $\tilde{u}$ When $\bar{\delta} = 0$

When $\bar{\delta}_1' = \bar{\delta}_2' = 0$, Figure 5.18B assumes the form of Figure 5.19; the constants a_1 and a_2 in (5-140) have the value 0. Changing to polar coordinates ρ and θ defined by

$$u_1 = \rho\cos\theta , \qquad u_2 = \rho\sin\theta ,$$

the integral (5-140) becomes

$$\text{EVI} = \frac{k}{\sqrt{2\pi}}\Big[\int_{\theta_1}^{\pi/2}\int_0^\infty \rho(\sin\theta - b_1\cos\theta)\, f_{N^*}(\rho)\, \rho\, d\rho\, d\theta$$

$$+ \int_{\pi/2}^{\theta_2}\int_0^\infty \rho(\sin\theta - b_2\cos\theta)\, f_{N^*}(\rho)\, \rho\, d\rho\, d\theta\Big] .$$

The integral with respect to ρ is simply one half the unit variance of f_{N^*}, so that

3)
$$\text{EVI} = \frac{k}{2\sqrt{2\pi}}\Big[\int_{\theta_1}^{\pi/2}(\sin\theta - b_1\cos\theta)\, d\theta + \int_{\pi/2}^{\theta_2}(\sin\theta - b_2\cos\theta)\, d\theta\Big]$$

$$= \frac{k}{2\sqrt{2\pi}}\left[\cos\theta_1 - \cos\theta_2 - b_1(1 - \sin\theta_1) + b_2(1 - \sin\theta_2)\right] .$$

By Figure 5.19

$$\cos\theta_1 = (1+b_1^2)^{-\frac{1}{2}}\;, \qquad \cos\theta_2 = -(1+b_2^2)^{-\frac{1}{2}}$$

$$\sin\theta_1 = b_1(1+b_1^2)^{-\frac{1}{2}}\;, \qquad \sin\theta_2 = -b_2(1+b_2^2)^{-\frac{1}{2}}\;;$$

substituting these values in (3) we obtain

$$\text{EVI} = \frac{k}{2\sqrt{2\pi}}[b_2 - b_1 + (1+b_1^2)^{\frac{1}{2}} + (1+b_2^2)^{\frac{1}{2}}]\;;$$

and substituting herein the values of b_1 and b_2 as defined by (5-139) we obtain formula (2) of our previous proof. ◀

A still more special case occurs when the $\tilde{v}$s or $\bar{\tilde{v}}''$s not only have common means but are *independent* and *identically distributed;* when this is true, the EVI can be derived as easily for r processes as for 3 processes. For the moment letting $\tilde{\boldsymbol{v}}$ stand for either the vector of actual utilities or the vector of posterior expected utilities $\bar{\tilde{\boldsymbol{v}}}''$, we define the parameter n_v by writing

$$\text{D}(\boldsymbol{v}) = \begin{cases} f_N(\boldsymbol{v}|\bar{\boldsymbol{v}}', hn_v\,\mathbf{I}) & \text{if } h \text{ is known}\;, \\ f_S(\boldsymbol{v}|\bar{\boldsymbol{v}}', \mathbf{I}n_v/v', \nu') & \text{if } h \text{ is unknown}\;. \end{cases} \tag{5-152}$$

The assumption that $\bar{v}_1' = \bar{v}_2' = \cdots = \bar{v}_r'$ allows us to write

$$\begin{aligned}\text{EVI} &= \text{E}\max\{\tilde{v}_1, \tilde{v}_2, \cdots \tilde{v}_r\} - \bar{v}_r' \\ &= \text{E}\max\{(\tilde{v}_1 - \bar{v}_1'), (\tilde{v}_2 - \bar{v}_2'), \cdots, (\tilde{v}_r - \bar{v}_r')\}\;;\end{aligned}$$

and the additional assumption that the $\tilde{v}$s are independent with equal scale parameters then gives us

$$\text{EVI} = \begin{cases} c_r(hn_v)^{-\frac{1}{2}} & \text{if } h \text{ is known}\;, \quad (5\text{-}153\text{a}) \\ c_r(n_v/v')^{-\frac{1}{2}}\dfrac{(\frac{1}{2}\nu' - \frac{3}{2})!}{(\frac{1}{2}\nu' - 1)!}\sqrt{\tfrac{1}{2}\nu'} & \text{if } h \text{ is unknown}\;, \quad (5\text{-}153\text{b}) \end{cases}$$

where c_r is the expected value of the maximum of r independent unit-Normal random variables. The value of c_r is half the expectation d_r of the *range* of a sample of r independent unit-Normal variables, and d_r is a well tabulated function.†

Observe that if the $\tilde{v}$s are independent and have equal means *a priori*, then even though the prior variances are unequal it will be possible to allocate the sample observations in such a way that the variances of the $\bar{\tilde{v}}''$s *are* equal. When the relative precisions and the costs of sampling from the various processes are the same, such an allocation will be optimal.

▶ Formula (5-153a) for known h was obtained by Dunnett, *Jour. Roy. Stat. Soc.* Series B, 1960, by further specializing some formulas for the case where, in our notation, the prior distribution of $\tilde{\boldsymbol{\mu}}$ is spherical Normal, the matrix $\mathbf{k} = k\,\mathbf{I}$, and the matrix $\mathbf{n} = n\,\mathbf{I}$. If in this case we define

† E.g., *Biometrika Tables for Statisticians*, Table 27 (Page 174), giving d_n to 5 decimal places for $n = 2(1)500(10)1000$.

$$k_1 = k_2 = k_3 \equiv k \ ,$$

$$u_{ij} = \begin{cases} u & \text{if } i = j \ , \\ 0 & \text{if } i \neq j \ , \end{cases}$$

then our formula (5-149b) for U reduces to

$$U = \frac{3k}{2\sqrt{\pi}} \sqrt{u} \ ,$$

and (5-150) with this value of U is equivalent to (5-153) for $r = 3$. ◀

5.11.8. *EVSI When Only One Process is to be Sampled*

If only one of the three processes is to be sampled, then as shown in Section 5.9.5 above the distribution of $\tilde{\boldsymbol{\delta}}''$ is degenerate or singular and the EVSI cannot be obtained by any of the methods we have discussed above. It can be very easily calculated by an appropriate method, however, since the fact that only one process is sampled means that the entire problem is univariate. For convenience of both notation and computation we shall express the analysis in terms of

$$\mathbf{u}' \equiv \mathbf{n}'^{-1} \ . \tag{5-154}$$

If it is the ith process which is to be sampled, then by (5-95) or (5-98) the distribution of

$$\tilde{x} \equiv \tilde{\mu}_i'' - \mu_i' \tag{5-155}$$

is univariate with density

$$\mathrm{D}(x) = \begin{cases} f_N(x|0,\, hn_m^*) & \text{if } h \text{ is known} \ , \\ fs(x|0,\, n_m^*/v',\, \nu') & \text{if } h \text{ is unknown} \ , \end{cases} \tag{5-156a}$$

where

$$n_m^{*-1} = u_{ii}' - (u_{ii}'^{-1} + n_{ii})^{-1} \ , \qquad n_{ii} = n_i(h_i/h) \ . \tag{5-156b}$$

As shown by (5-100), this distribution determines the distribution of $\tilde{\boldsymbol{\delta}}'' = [\tilde{\delta}_1'' \;\; \tilde{\delta}_2'']^t$ through two linear relations

$$\begin{aligned} \tilde{\delta}_1'' &= \bar{\delta}_1' + b_1\tilde{x} \ , \\ \tilde{\delta}_2'' &= \bar{\delta}_2' + b_2\tilde{x} \ , \end{aligned} \tag{5-157}$$

but because of our convention that the subscript $r = 3$ is to be assigned to the process with the greatest prior expected utility we must distinguish two cases in giving formulas for the coefficients b_1 and b_2.

1. If the process whose prior expected utility is greatest is *not* to be sampled, we may assign the subscript 1 to the process which *is* to be sampled. We then have by (5-100) that

$$b_1 = k_1 - k_3 u_{31}' u_{11}'^{-1} \ , \qquad b_2 = (k_2 u_{21}' - k_3 u_{31}') \, u_{11}'^{-1} \ ; \tag{5-158a}$$

and the distribution of $\tilde{x}$ is given by (5-156) with $i = 1$.

2. If the process whose prior expected utility is greatest and which therefore bears the subscript 3 *is* to be sampled, then by (5-100)

$$b_1 = k_1 u'_{13} u'^{-1}_{33} - k_3 \ , \qquad b_2 = k_2 u'_{23} u'^{-1}_{33} - k_3 \ ; \tag{5-158b}$$

and the distribution of $\tilde{x}$ is given by (5-156) with $i = 3$.

In this second case b_1 and b_2 will usually (though not necessarily) be negative, so that $\bar{\delta}_1''$ and $\bar{\delta}_2''$ would behave as functions of x in one or the other of the two

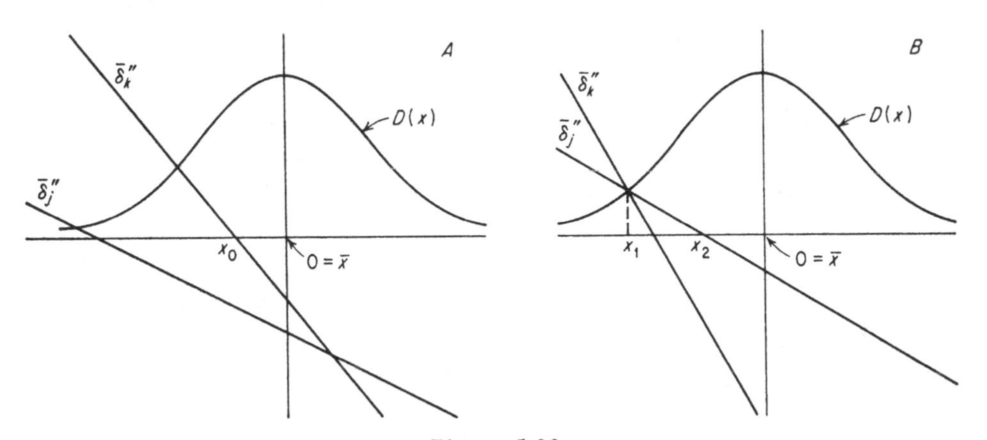

Figure 5.20

CVSI When One of Three Processes is Sampled

ways depicted in Figure 5.20A and B; and in either of these cases the EVSI is easily found by applying the univariate theory of Section 5.3.2. Defining

$$L(x) = \begin{cases} (hn_m^*)^{-\frac{1}{2}} \, L_{N^*}(|x|\sqrt{hn_m^*}) & \text{if } h \text{ is known} \ , \\ (n_m^*/v')^{-\frac{1}{2}} \, L_{S^*}(|x|\sqrt{n_m^*/v'}|\nu') & \text{if } h \text{ is unknown} \ , \end{cases} \tag{5-159}$$

we have by (5-26) and (11-25) or (11-50) for the case of Figure 5.20A

$$v_i^*(e) = |b_k| \, L(x_0) \tag{5-160a}$$

and by (5-25) and (11-25) or (11-50) for the case of Figure 5.20B

$$v_i^*(e) = |b_j| \, L(x_2) + |b_k| \, L(x_1) \ . \tag{5-160b}$$

The reader can readily convince himself that regardless of which process is sampled and regardless of the numerical values of b_1 and b_2 the EVSI will always be given by a formula of one or the other of these two types; recall that by the definition of $\boldsymbol{\delta}$, both $\bar{\delta}_j''$ and $\bar{\delta}_k''$ must be negative when $x = 0$.

5.12. Value of Information When There are More Than Three Independent-Normal Processes

When $r > 3$, we know of no general expression in terms of tabulated functions for the integrals

$$\text{EVI} = \int_{-\infty}^{\infty} \cdots \int_{-\infty}^{\infty} \max\,\{\delta_1, \cdots, \delta_{r-1}, 0\} \, f^{(r-1)}(\boldsymbol{\delta}|\bar{\boldsymbol{\delta}}', \mathbf{n}_\delta^\square) \, d\delta_1 \cdots d\delta_{r-1}$$

where

$$f^{(r-1)}(\boldsymbol{\delta}|\bar{\boldsymbol{\delta}}', \mathbf{n}_\delta^\square) = \begin{cases} f_N^{(r-1)}(\boldsymbol{\delta}|\bar{\boldsymbol{\delta}}', h\,\mathbf{n}_\delta^\square) & \text{if } h \text{ is known ,} \\ f_S^{(r-1)}(\boldsymbol{\delta}|\bar{\boldsymbol{\delta}}', \mathbf{n}_\delta^\square/v', \nu') & \text{if } h \text{ is unknown .} \end{cases}$$

(A formula for one very special case was given as (5-153) above.) The EVI can always be evaluated by numerical methods, however, and we shall now suggest one method of evaluation.

5.12.1. The Nondegenerate Case

When the distribution of $\tilde{\boldsymbol{\delta}}$ is nondegenerate (as it always is when the EVPI is evaluated and as it is when EVSI is evaluated provided that at least $r - 1$ of the r processes are to be sampled), our first step is to find a transformation $\mathbf{A}$ such that

$$\tilde{\boldsymbol{u}} \equiv \mathbf{A}(\tilde{\boldsymbol{\delta}} - \bar{\boldsymbol{\delta}}') \tag{5-161}$$

is standardized Normal or Student; and this can be accomplished by first diagonalizing the symmetric matrix

$$\boldsymbol{\sigma}^\square = \begin{cases} (h\,\mathbf{n}_\delta^\square)^{-1} & \text{if } h \text{ is known ,} \\ (\mathbf{n}_\delta^\square/v')^{-1} & \text{if } h \text{ is unknown ,} \end{cases} \tag{5-162}$$

and then reducing the diagonal matrix to the identity matrix. There already exist programs for digital computers which employ the method of Jacobi and von Neumann for finding all eigenvalues and eigenvectors of a real, symmetric matrix;† the expected operating time is approximately $10(2\nu + \mu)n^3$ where n is the order of the matrix ($= r - 1$ in our application), ν is the computer addition time, and μ is the computer multiplication time. Having found in this way an orthogonal matrix $\mathbf{C}$ such that $\mathbf{C}\,\boldsymbol{\sigma}^\square\,\mathbf{C}^t$ is diagonal, it is easy to find a diagonal matrix $\mathbf{D}$ such that

$$(\mathbf{D}\,\mathbf{C})\,\boldsymbol{\sigma}^\square(\mathbf{D}\,\mathbf{C})^t = \mathbf{I} \ . \tag{5-163}$$

The desired transformation can thus be obtained by taking

$$\mathbf{A} = \mathbf{D}\,\mathbf{C} \ ; \tag{5-164a}$$

and the inverse transformation is then very easy to obtain because

$$\mathbf{A}^{-1} = \mathbf{C}^t\,\mathbf{D}^{-1} \ . \tag{5-164b}$$

The distribution of $\tilde{\boldsymbol{u}}$ as defined by (5-161) is

$$\mathrm{D}(\boldsymbol{u}) = \begin{cases} f_{N*}^{(r-1)}(\boldsymbol{u}) & \text{if } h \text{ is known ,} \\ f_{S*}^{(r-1)}(\boldsymbol{u}|\nu') & \text{if } h \text{ is unknown .} \end{cases} \tag{5-165}$$

There exist many methods for rapid generation on digital computers of $\boldsymbol{u}$ vectors having either of these distributions; and for each $\boldsymbol{u}$ the quantities

$$\boldsymbol{\delta} = \bar{\boldsymbol{\delta}}' + \mathbf{A}^{-1}\boldsymbol{u} \ , \qquad \psi(\boldsymbol{u}) = \max\{\delta_1, \delta_2, \cdots, \delta_{r-1}, 0\} \ , \tag{5-166}$$

can be computed. As the sequence $\boldsymbol{u}^{(1)}, \cdots, \boldsymbol{u}^{(j)}, \cdots$ is generated, the computer needs to retain at the mth stage only the summary statistics

† John Greenstadt, "The determination of the characteristic roots of a matrix by the Jacobi Method," in *Mathematical Methods for Digital Computers* (edited by A. Ralston and H. S. Wilf), John Wiley and Sons, 1960.

$$m\ , \qquad t \equiv \sum_{j=1}^{m} \psi(\mathbf{u}^{(j)})\ , \qquad S \equiv \sum_{j=1}^{m} [\psi(\mathbf{u}^{(j)})]^2\ . \tag{5-167}$$

The EVI is estimated at any stage by

$$\bar{\psi}_m \equiv t/m$$

and the variance of this estimator is estimated by

$$\frac{1}{m(m-1)}\left(S - \frac{t^2}{m}\right) .$$

When for an m greater than some specified m_o the estimated variance is sufficiently small, we take $\bar{\psi}_m$ as the approximation to the EVI.

5.12.2. *The Degenerate Case*

When the EVSI is to be computed for an experiment in which $p < r - 1$ processes are to be sampled, it follows from (5-100) that one can generate $\bar{\boldsymbol{\delta}}''$ vectors by first generating p-dimensional $(\bar{\boldsymbol{\mu}}_i'' - \bar{\boldsymbol{\mu}}_i')$ vectors. From (5-95b) or (5-98b)

$$\mathrm{D}(\bar{\boldsymbol{\mu}}_i'' - \bar{\boldsymbol{\mu}}_i') = \begin{cases} f_N^{(p)}(\bar{\boldsymbol{\mu}}_i'' - \bar{\boldsymbol{\mu}}_i'|\mathbf{0}, h\mathbf{n}_m^*) & \text{if } h \text{ is known}\ , \\ f_S^{(p)}(\bar{\boldsymbol{\mu}}_i'' - \bar{\boldsymbol{\mu}}_i'|\mathbf{0}, \mathbf{n}_m^*/v', \nu') & \text{if } h \text{ is unknown}\ . \end{cases} \tag{5-168}$$

We next find a transformation in p-space to reduce

$$\boldsymbol{\sigma}_m^* = \begin{cases} (h\mathbf{n}_m^*)^{-1} & \text{if } h \text{ is known}\ , \\ (\mathbf{n}_m^*/v')^{-1} & \text{if } h \text{ is unknown} \end{cases} \tag{5-169}$$

to the identity matrix. If we let $\mathbf{A}$ denote this transformation, then

$$\tilde{\boldsymbol{u}} = \mathbf{A}(\tilde{\boldsymbol{\mu}}_i'' - \bar{\boldsymbol{\mu}}_i') \tag{5-170}$$

has a standardized Normal or Student distribution; and by generating a sequence of $\boldsymbol{u}$ vectors we can compute a sequence of vectors

$$\bar{\boldsymbol{\mu}}_i'' - \bar{\boldsymbol{\mu}}_i' = \mathbf{A}^{-1}\,\boldsymbol{u}$$

and hence by (5-100) of vectors

$$\bar{\boldsymbol{\delta}}'' = \bar{\boldsymbol{\delta}}' + \mathbf{B}\,\mathbf{k}\,\mathbf{C}^{(i)}\,\mathbf{A}^{-1}\,\boldsymbol{u}\ . \tag{5-171}$$

The remainder of the procedure is identical to that suggested for the nondegenerate case.

5.12.3. *Choice of the Optimal Experiment*

The real problem of preposterior analysis is of course not simply to compute the EVSI of some given experiment e but to find the *optimal* experiment. Since any e is characterized by an r-tuple $(n_1, \cdots, n_r)$, we may display the dependence of $v^*(e)$ on the n_i by writing $v^*(n_1, n_2, \cdots, n_r)$, and we may view our task as that of finding the maximum (or an approximation to the maximum) of the v^* surface. Since a good deal is known or can be learned about the general nature of this surface, it would seem not unlikely that the maximization problem can be solved by use of search techniques based on those suggested by G. P. Box and others. With the collaboration of Marshall Freimer, we are currently investigating digital computer programming techniques for coupling Monte-Carlo evaluations of $v^*(n_1, n_2, \ldots, n_r)$ with search routines.

CHAPTER 6

Problems in Which the Act and State Spaces Coincide

6.1. Introduction

In the present chapter we take leave of the class of problems characterized by linear terminal utility which we studied in Chapters 5A and 5B and take up another special class of problems characterized essentially by the fact that *the act space coincides with the state space or with some transformation thereof.* Thus in a problem of inventory control where the act space A consists of all possible numbers of units stocked and the state space Θ consists of all possible numbers of units demanded, A and Θ may both be the set of all positive integers or may both be treated approximately as consisting of the positive half of the real line. Or in a more complicated problem of the same sort, the state may be described by an r-tuple $(\theta_1, \cdots, \theta_r)$ where each component describes the quantity demanded by a particular subgroup of customers, but the act space A may still coincide with a transformed state space Ω the generic element of which is $\omega = \theta_1 + \cdots + \theta_r$. The problem of point estimation of an unknown parameter usually falls in this same class because the space of possible estimates $\{a\}$ usually coincides with the space of possible true values $\{\omega\}$.

6.1.1. *Basic Assumptions*

Formally, this chapter will be devoted to the consideration of problems in which

> There exists a mapping W from the state space Θ into a new space Ω, carrying θ into $\omega = W(\theta)$, such that Ω and the act space A coincide.†

The class of problems considered will be further restricted by two assumptions about the utility function $u(\cdot, \cdot, \cdot, \cdot)$ on $E \times Z \times A \times \Theta$. As in Chapters 4 and 5 we shall assume that terminal and sampling utilities are additive; letting c_s denote the negative of the sampling utility, this assumption can be written

$$u(e, z, a, \theta) = u_t(a, \omega) - c_s(e, z) , \qquad \omega = W(\theta) . \tag{6-1}$$

For motivation of this assumption, see Section 4.2. We shall also assume that, for any ω in Ω,

$$u_t(\omega, \omega) \geq u_t(a, \omega) , \qquad \text{all } a \in A = \Omega , \tag{6-2}$$

† Most of the theory developed in this chapter will apply to the even more general class of problems in which the possible states $\{\omega\}$ are a *subset* of the possible acts $\{a\}$, thus including problems in which the act a may fall between two successive possible values of the state ω. We leave the generalization to the reader.

so that by the definition (4-14) of terminal opportunity loss,

$$l_t(\omega, \omega) = 0 \ . \tag{6-3}$$

In words rather than symbols: if the true state is ω, then the best act is ω—the best number of units to stock is the number demanded, the best estimate of ω is ω, and so forth.

6.1.2. Example

As an example of a problem where assessment of the terminal opportunity losses $l_t(a, \omega)$ is particularly straightforward, consider an inventory problem of the kind suggested above. A retailer stocks a product which spoils if it is not sold by the end of the day on which it is stocked; each unit stocked costs the retailer an amount k_o; a unit is sold by him for k_u more than it costs; the utility of money is linear. The retailer must decide on the quantity to stock, which we shall denote by $q = a$; and if we let $d = \omega$ denote the number of units actually demanded by the customers, the *terminal opportunity loss* of q given d is obviously

$$l_t(q, d) = \begin{cases} k_o(q - d) & \text{if} \quad d \leq q \ , \\ k_u(d - q) & \text{if} \quad d \geq q \ . \end{cases}$$

The function $l_t(q, \cdot)$ is graphed against d in Figure 6.1 for $k_o = 1$, $k_u = 4$, and $q = 10$.

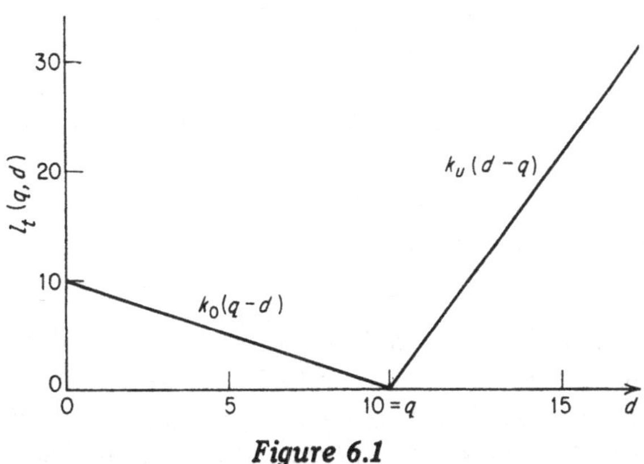

Figure 6.1

Terminal Opportunity Loss of the Act q = 10

6.2. Certainty Equivalents and Point Estimation

In most applied decision problems the number of unknown quantities is so great that the decision maker cannot practically take full formal account of all of the uncertainties that he would like to introduce into his analysis: he cannot assign a proper joint probability distribution to all the unknowns he would like to treat as random variables and then compute exact expected utilities under this distribution. Even in the simplest problems of quality control, the costs of acceptance and rejection (and often the very number of pieces in a lot) may be subject to uncertainty; but only rarely will it be economically sensible to take full formal account of all such uncertainties in selecting a sampling plan.

6.2.1. *Certainty Equivalents*

In some cases a "subsidiary" uncertainty can be disregarded in the analysis without loss because some summary measure of the marginal distribution of the unknown quantity in question is a *certainty equivalent* in the sense that treatment of this summary measure *as if* it were the *true* value of the quantity will lead to exactly the *same course of action* that would be chosen if an analysis were based on the complete distribution. In Section 1.4.1 we have already examined one very important example: substitution of the *expected value* of an *unknown cost* for the complete distribution of the cost will have no effect on the results of analysis whenever (as often happens) the cost enters linearly into the analysis and the distribution of the cost is independent of the other state variables.

We shall now examine the general problem of the finding and use of certainty equivalents—i.e., of summary measures (including but not restricted to means) which in some particular problem or class of problems can be substituted without loss for the complete distribution of some unknown quantity. We begin our investigation of this problem with an artificially simple example in which exact results can be obtained; after this example has fixed the ideas, we shall develop a more pragmatic approach to the general problem.

6.2.2. *Example*

Returning to the problem of inventory control discussed in Section 6.1.2 above, assume that the demand $\tilde{d}$ has a probability measure of known form with parameter σ, denote the *expected* terminal opportunity loss of an act (quantity stocked) q by

$$L_t(q|\sigma) \equiv \mathrm{E}_{d|\sigma}\, l_t(q, \tilde{d}) \ , \tag{6-4}$$

and denote by q_σ an act which minimizes $L_t(\cdot|\sigma)$, i.e., which satisfies

$$L_t(q_\sigma|\sigma) \le L_t(q|\sigma) \ , \qquad \text{all } q \ . \tag{6-5}$$

Consider first what will happen if for any reason whatever the retailer, instead of stocking an amount q_σ which satisfies (6-5), stocks an amount $q_{\hat{\sigma}}$ which *would* satisfy (6-5) *if* the parameter σ had the value $\hat{\sigma}$ rather than its actual value σ. By (6-4) this means that his expected terminal opportunity loss is $L_t(q_{\hat{\sigma}}|\sigma)$ rather than $L_t(q_\sigma|\sigma)$; the decision to stock $q_{\hat{\sigma}}$ instead of q_σ has *increased* his expected opportunity loss by the amount

$$\lambda_t(\hat{\sigma}, \sigma) \equiv L_t(q_{\hat{\sigma}}|\sigma) - L_t(q_\sigma|\sigma) \ . \tag{6-6}$$

Now suppose that the decision maker does not know the true value of the parameter σ but believes that it is close to some number $\hat{\sigma}$ and proposes to act as if this were in fact the true value, i.e., to stock the quantity $q_{\hat{\sigma}}$. By (6-4) and (6-6) his *conditional* expected terminal opportunity loss, *given* any particular true σ, is

$$L_t(q_{\hat{\sigma}}|\sigma) = L_t(q_\sigma|\sigma) + \lambda_t(\hat{\sigma}, \sigma) \ . \tag{6-7a}$$

If he assigns a probability measure to $\tilde{\sigma}$ we can say that his *unconditional* expected terminal opportunity loss is the expectation of (6-7a) with respect to this measure,

$$L_t^*(q_{\hat\sigma}) \equiv \mathrm{E}_\sigma L_t(q_{\hat\sigma}|\tilde\sigma) \ ; \tag{6-7b}$$

and if we define

$$L_t^* \equiv \mathrm{E}_\sigma\, L_t(\tilde q_\sigma|\,\tilde\sigma) \ , \tag{6-8a}$$

$$\Lambda_t(\hat\sigma) \equiv \mathrm{E}_\sigma\, \lambda_t(\hat\sigma, \tilde\sigma) \ , \tag{6-8b}$$

this expectation can be written

$$L_t^*(q_{\hat\sigma}) = L_t^* + \Lambda_t(\hat\sigma) \ . \tag{6-8c}$$

The decision maker's "overall" expected terminal opportunity loss is thus the sum of (1) the terminal opportunity loss L_t^* which he would "expect" to suffer if he were to be told the true σ (but not the true d) before choosing his terminal act, and (2) the additional opportunity loss $\Lambda_t(\hat\sigma)$ which he must expect to suffer because his act will actually be based on the "*estimate*" $\hat\sigma$ rather than on the true value σ. Because the first of these two quantities does not depend on any decision variable, the problem of finding the terminal act q which will minimize the decision maker's expected terminal opportunity loss as defined by (6-7b) is equivalent to the problem of finding the estimate $\hat\sigma^*$ which will minimize the expected "estimation loss" as defined by (6-8b); and once this *optimal* estimate $\hat\sigma^*$ has been found, it will be a certainty equivalent in the terminal action problem of deciding on the actual number q of units that should be stocked. Because the set of possible estimates $\{\hat\sigma\}$ coincides with the set of possible true values $\{\sigma\}$ and the "imputed" loss function λ_t defined by (6-6) has the property $\lambda_t(\sigma, \sigma) = 0$, *this problem falls in exactly the same formal category as the problem of direct terminal action discussed in Section* 6.1.2 *above.*

To obtain a better feeling for the nature of these imputed loss functions λ_t, let us assume by way of example that the demand $\tilde d$ is known to be generated by an Independent Normal process with known mean μ but unknown standard deviation $\tilde\sigma$. Then as shown by (6-42) and (11-24b), the expected terminal opportunity loss defined for any act q by (6-4) is given by the formula

$$L_t(q|\sigma) = (k_u + k_o)\,\sigma\{f_{N^*}(u) + u[k_o^*\,F_{N^*}(u) - k_u^*\,G_{N^*}(u)]\} \tag{6-9a}$$

where

$$u \equiv \frac{q-\mu}{\sigma} \ , \qquad k_o^* = \frac{k_o}{k_u + k_o} \ , \qquad k_u^* = \frac{k_u}{k_u + k_o} \ , \tag{6-9b}$$

and f_{N^*}, F_{N^*}, and G_{N^*} are respectively the unit-Normal density and left and right-tail cumulative functions; and as shown by (6-44) the *optimal* act q_σ defined by (6-5) must satisfy

$$F_{N^*}\left(\frac{q_\sigma - \mu}{\sigma}\right) = k_u^* \ . \tag{6-10}$$

From (6-10) it is easy to calculate the stock $q_{\hat\sigma}$ which will result if any estimate $\hat\sigma$ is treated as if it were the true σ, and from (6-9) it is easy to calculate the *increase* in terminal opportunity loss due to stocking $q_{\hat\sigma}$ rather than q_σ—i.e., the loss $\lambda_t(\hat\sigma, \sigma)$ to be *imputed* to use of the estimate $\hat\sigma$ when σ is true. In Figure 6.2 we graph the function $\lambda_t(\hat\sigma, \cdot)$ for the cost parameters of Section 6.1.2 and the estimate $\hat\sigma = 1$; this graph should be compared with the graph Figure 6.1 of the terminal opportunity loss of the act q itself as a function of the actual demand d.

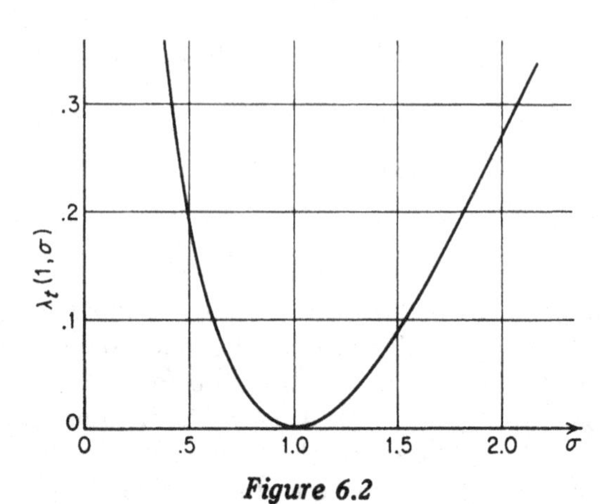

Figure 6.2

Imputed Opportunity Loss of the Estimate $\hat{\sigma} = 1$

6.2.3. *General Theory of Certainty Equivalents*

We now proceed to generalize the theory of certainty equivalents suggested by the example we have just discussed.

Terminal Theory. Let the state θ be expressible as a doublet $(\theta_1, \theta_2) = (\omega, \zeta)$; define u'_t and l'_t by

$$u'_t(a, \omega, \zeta) \equiv u_t(a, \theta) \;; \qquad l'_t(a, \omega, \zeta) \equiv l_t(a, \theta) \;; \tag{6-11a}$$

define U_t and L_t by

$$U_t(a, \omega) \equiv \text{E}_{\zeta|\omega}\, u'_t(a, \omega, \tilde{\zeta}) \;; \qquad L_t(a, \omega) \equiv \text{E}_{\zeta|\omega}\, l'_t(a, \omega, \tilde{\zeta}) \;; \tag{6-11b}$$

define a_ω by

$$U_t(a_\omega, \omega) \geq U_t(a, \omega) \;, \qquad \text{all } a \in A \;, \tag{6-12a}$$

or equivalently by

$$L_t(a_\omega, \omega) \leq L_t(a, \omega) \;, \qquad \text{all } a \in A \;; \tag{6-12b}$$

and define the *imputed loss* of the *estimate* $\hat{\omega}$ to be

$$\lambda_t(\hat{\omega}, \omega) \equiv U_t(a_\omega, \omega) - U_t(a_{\hat{\omega}}, \omega) = L_t(a_{\hat{\omega}}, \omega) - L_t(a_\omega, \omega) \;. \tag{6-13}$$

Then since $\text{E}_\theta \equiv \text{E}_\omega\, \text{E}_{\zeta|\omega}$

$$\text{E}_\theta\, u_t(a_{\hat{\omega}}, \tilde{\theta}) = \text{E}_\omega\, U_t(\tilde{a}_\omega, \tilde{\omega}) - \text{E}_\omega\, \lambda_t(\hat{\omega}, \tilde{\omega}) \tag{6-14a}$$

or

$$\text{E}_\theta\, l_t(a_{\hat{\omega}}, \tilde{\theta}) = \text{E}_\omega\, L_t(\tilde{a}_\omega, \tilde{\omega}) + \text{E}_\omega\, \lambda_t(\hat{\omega}, \tilde{\omega}) \;. \tag{6-14b}$$

Provided that the set $\{a_\omega : \omega \in \Omega\}$ is coextensive with the set $\{a\} = A$, i.e. that every a in A is optimal for some ω in Ω, it follows immediately that

$$\max_a \text{E}_\theta\, u_t(a, \tilde{\theta}) = \text{E}_\omega\, U_t(\tilde{a}_\omega, \tilde{\omega}) - \min_{\hat{\omega}} \text{E}_\omega\, \lambda_t(\hat{\omega}, \tilde{\omega}) \;, \tag{6-15a}$$

and that

$$\min_a \text{E}_\theta\, l_t(a, \tilde{\theta}) = \text{E}_\omega\, L_t(\tilde{a}_\omega, \tilde{\omega}) + \min_{\hat{\omega}} \text{E}_\omega\, \lambda_t(\hat{\omega}, \tilde{\omega}) \;. \tag{6-15b}$$

In words rather than symbols, *minimization of imputed estimation loss as defined by* (6-13) *followed by treatment of the optimal estimate as if it were the true value of the quantity estimated is equivalent to maximization of expected utility or minimization of expected "overall" opportunity loss.*

Preposterior Theory. In general the decision maker is not obliged to proceed immediately to make an estimate $\hat{\omega}$ on the basis of which he will choose his terminal act a but can if he wishes perform an experiment e which will give him additional information about Ω. Formally, consider a set of experiments E with potential outcomes Z such that the conditional measures $P_{z|e}$ on Z depend on $\theta = (\omega, \zeta)$ only through the component ω, and let $\lambda_i^*(e)$ denote the *prior* expected value of the estimation loss due to the estimate $\hat{\omega}$ which will be made *posterior* to e:

$$\lambda_i^*(e) \equiv \mathrm{E}_{z|e} \min_{\hat{\omega}} \mathrm{E}''_{\omega|z} \lambda_t(\hat{\omega}, \tilde{\omega}) \ . \tag{6-16}$$

Then by (4-3) and (6-15a) we have for the expected utility of e

$$\begin{aligned} u^*(e) &= u_i^*(e) + u_s^*(e) \equiv u_i^*(e) - c_s^*(e) \\ &= \mathrm{E}_{z|e} [\mathrm{E}''_{\omega|z} U_t(\tilde{a}_\omega, \tilde{\omega}) - \min_{\hat{\omega}} \mathrm{E}''_{\omega|z} \lambda_t(\hat{\omega}, \tilde{\omega})] - c_s^*(e) \\ &= \mathrm{E}'_\omega U_t(\tilde{a}_\omega, \tilde{\omega}) - [\lambda_i^*(e) + c_s^*(e)] \ ; \end{aligned} \tag{6-17}$$

and since the first term on the right is independent of any decision variable, the optimal e must minimize

$$\lambda^*(e) \equiv \lambda_i^*(e) + c_s^*(e) \ . \tag{6-18}$$

We conclude that

> Any decision problem involving an unknown parameter ω can be decomposed without loss of accuracy into two parts: (1) a first-stage "estimation" problem of deciding how much information should be obtained concerning ω and what estimate $\hat{\omega}$ should be chosen in the light of this information; (2) a second-stage "terminal-action" problem in which a terminal act a is chosen by treating this estimate as if it were the true ω.

It is perhaps worth remarking that this principle extends even to problems in which ω is the *only* unknown quantity involved. Our example of Section 6.1.2, where the only unknown was the actual demand d, can if we like be treated as a problem of "estimating" d or of finding a "certainty equivalent" for d. If we assume that an estimate $\hat{d}$ will be treated as if it were the true demand d, the "imputed" losses $\lambda_t(\hat{d}, d)$ will be simply the "terminal" opportunity losses $l_t(\hat{d}, d)$:

$$\lambda_t(\hat{d}, d) = \begin{cases} k_o(\hat{d} - d) & \text{if} \quad d \le \hat{d} \ , \\ k_u(d - \hat{d}) & \text{if} \quad d \ge \hat{d} \ . \end{cases}$$

Choice of the Quantity to Estimate. In most discussions of point estimation an effort is made to define the properties of a "good"—or even a "best"—estimator without reference to the use to which the estimate will ultimately be put; it is easy to find statements in the literature than an estimate "should" be unbiassed or that the "best" unbiassed estimator of ω is one which minimizes $\mathrm{E}(\tilde{\hat{\omega}} - \omega)^2$ or something of the sort. Now it is perfectly obvious from the role of the imputed losses $\lambda_t(\hat{\omega}, \omega)$ in the above discussion that it is flatly impossible to find *any* one estimator which is "good"—let alone "best"—for *all* applications in any absolute sense, but this discussion has another implication which is less obvious.

It is well known that if $\hat{\sigma}$ is, say, an unbiassed estimate of the standard deviation σ, then $\hat{\sigma}^2$ is *not* an unbiassed estimate of the variance σ^2; if $\tilde{\hat{\omega}}$ is an estimator which minimizes $\max_\omega \mathrm{E}(\tilde{\hat{\omega}} - \omega)^2$, then $f(\tilde{\hat{\omega}})$ does *not* in general minimize $\max_\omega \mathrm{E}[f(\tilde{\hat{\omega}}) - f(\omega)]^2$; and in general, many classical estimation procedures are dependent on the parametrization of the problem. When we think, however, of the way in which a point estimate is actually *used*, we see that if an estimate $\hat{\omega}$ of ω leads to the terminal act $a_{\hat{\omega}}$, an estimate $\hat{\omega}^2$ of ω^2 will lead to this same terminal act $a_{\hat{\omega}}$; and it follows at once that if a particular estimate $\hat{\omega}^*$ of ω minimizes imputed estimation loss, then the estimate $f(\hat{\omega}^*)$ of $f(\omega)$ will also minimize imputed estimation loss. In terms of our example: if $\hat{\sigma}^*$ is the best estimate of the standard deviation of demand $\tilde{d}$ *in the circumstances of that particular decision problem*, then $\hat{\sigma}^{*2}$ is the best estimate of the variance of $\tilde{d}$, and so forth.

6.2.4. *Approximation of* λ_t

Although it is possible in principle to obtain an exact solution of a problem of point estimation by application of the theory which we have just developed, this procedure will rarely if ever be of much practical use. In order to obtain the exact conditional losses $\lambda_t(\hat{\omega}, \omega)$ for the "first-stage" estimation problem, we must first make a complete conditional analysis of the "second-stage" terminal problem for each possible ω; and if we are to carry out this analysis and then take the exact expected values of the derived conditional losses we might just as well eliminate the decomposition and introduce the uncertainty about ω directly into the analysis of the terminal problem itself.

The practical value of the theory developed in the last section lies rather in the fact that, by giving us a *meaningful definition* of the estimation problem, it guides us in the choice of suitable methods for finding *approximate solutions* of the problem. When it is practically impossible to allow for uncertainty about ω in the terminal analysis, it may be possible to calculate $\lambda_t(\hat{\omega}, \omega)$ for a *few* $(\hat{\omega}, \omega)$ pairs, fit some convenient function $\lambda_t'(\cdot\,, \cdot)$ to these calculated values, and then select the estimate which minimizes $\mathrm{E}_\omega\, \lambda_t'(\hat{\omega}, \tilde{\omega})$ or to choose the experiment e which minimizes $\mathrm{E}_{z|e} \min_{\hat{\omega}} \mathrm{E}''_{\omega|z}\, \lambda_t'(\hat{\omega}, \tilde{\omega})$.

In choosing the *class* of functions one member of which will be selected as λ_t' by specification of numerical values for the parameters, the decision maker will of course be guided not only by the values of $\lambda_t(\hat{\omega}, \tilde{\omega})$ which he has actually computed but also by a more general intuitive analysis of the "second-stage" decision problem in which the estimate $\hat{\omega}$ will ultimately be used. Whatever the nature of the second-stage problem, he knows that the imputed loss of a *correct* estimate is zero, $\lambda_t(\omega, \omega) = 0$; and *if* the second-stage problem is an *infinite-action* problem (e.g., a problem like the example of Section 6.1.2 above or a problem of optimal sample size), then it will usually be reasonable to assume that $\lambda_t(\hat{\omega}, \omega)$ increases in some smooth manner with the "error" $(\hat{\omega} - \omega)$ or $(\omega - \hat{\omega})$ in the estimate. When this condition is met, an appropriate approximation function will often be found within one of two particularly tractable classes analyzed in detail in Sections 6.3 through 6.5 below: (1) the *linear* class defined by

$$\lambda_t(\hat{\omega}, \omega) = \begin{cases} k_o(\hat{\omega} - \omega)\, g(\omega) & \text{if} \quad \omega \leq \hat{\omega} \\ k_u(\omega - \hat{\omega})\, g(\omega) & \text{if} \quad \omega \geq \hat{\omega} \end{cases} \tag{6-19}$$

and (2) the *quadratic* class defined by

$$\lambda_t(\hat{\omega}, \omega) = k_t(\hat{\omega} - \omega)^2 g(\omega) \tag{6-20}$$

where $g(\omega)$ may be any function which is "compatible" with the prior density of $\tilde{\omega}$ in a sense to be discussed in Section 6.5.

In choosing between and within these two classes of functions, the decision maker will probably find it easier to think of $\hat{\omega}$ rather than ω as the "variable", letting ω play the role of a "parameter". Thus in the linear class (6-19), k_o represents the loss which will result for each unit by which ω is *over*estimated, while k_u is the loss for each unit by which ω is *under*estimated. The addition of $g(\omega)$ to the loss function makes it possible to find a function which will properly allow for the fact that the seriousness of a given error $(\hat{\omega} - \omega)$ may depend on the true value of ω as well as on the error itself. Thus while

$$\lambda_t(\hat{\mu}, \mu) = k_t(\hat{\mu} - \mu)^2$$

may often be an appropriate loss function for estimation of the mean μ of an Independent Normal process, the loss due to misestimation of the mean p of a Bernoulli process will often be better described by the function

$$\lambda_t(\hat{p}, p) = \left[\frac{\hat{p} - p}{p}\right]^2 ,$$

which looks at the relative rather than the absolute error in $\hat{p}$, or by the more symmetric function

$$\lambda_t(\hat{p}, p) = \frac{(\hat{p} - p)^2}{p(1 - p)} .$$

Choice of the Quantity to Estimate. Although it makes no difference whether we estimate ω or some function of ω when we optimize the estimate with respect to the *exact* loss function λ_t, it does make a considerable difference if we are going to optimize with respect to an approximation λ'_t chosen from a limited class of tractable functions: the class may contain a very good approximation to the true loss function when the problem is parametrized in one way but may contain no even tolerable approximation when the problem is parametrized in another way.

As an example, let us consider once again the problem of estimating the standard deviation of a Normal distribution of demand which was discussed in Section 6.2.2 above. Substituting formula (6-9) for the *terminal* opportunity loss $L_t(q|\sigma)$ in the definition (6-6) of the imputed estimation loss and defining the constant u by

$$F_{N^*}(u) = k_u^*$$

we obtain after a little algebraic reduction

$$\frac{1}{k_u + k_o} \lambda_t(\hat{\sigma}, \sigma) = \sigma\, \phi(\hat{\sigma}/\sigma) \tag{6-21a}$$

where

$$\phi(\hat{\sigma}/\sigma) \equiv f_{N^*}(u\hat{\sigma}/\sigma) - f_{N^*}(u) + (u\hat{\sigma}/\sigma)[k_o^* F_{N^*}(u\hat{\sigma}/\sigma) - k_u^* G_{N^*}(u\hat{\sigma}/\sigma)] . \tag{6-21b}$$

From this it follows that if we define

$$\omega \equiv \sigma^k \tag{6-22}$$

where k is any real number, we can write

$$\frac{1}{k_u + k_o} \lambda_t(\hat{\omega}, \omega) \equiv \omega^{1/k} \phi_k(\hat{\omega}/\omega) \tag{6-23}$$

where ϕ_k is implicitly defined.

If now we wish to *approximate* (6-23) by a quadratic of type (6-20), we will wish to know whether we will do better to take $k = 1$ and $\omega = \sigma$, i.e., to approximate the losses by a function of the form $(\hat{\sigma} - \sigma)^2 g_1(\sigma)$, or to take $k = 2$ and $\omega = \sigma^2 = v$ (say), i.e., to approximate the losses by a function of the form $(\hat{v} - v)^2 g_2(v)$, or possibly to work with some still other power of σ. The question can be settled

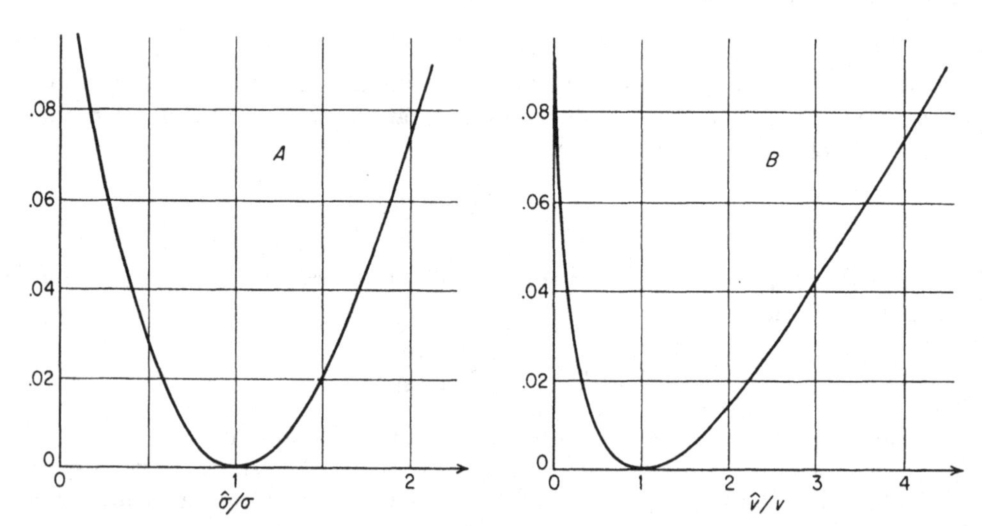

Figure 6.3

Estimation Loss as Function of $\hat{\sigma}/\sigma$ or of $\hat{v}/v$

by examining the functions ϕ_k graphically, as is done in Figure 6.3 for $k = 1$ and 2. This figure makes it clear that the approximation

$$\phi_1(\hat{\sigma}/\sigma) \propto [(\hat{\sigma}/\sigma) - 1]^2 = \frac{(\hat{\sigma} - \sigma)^2}{\sigma^2}$$

will be much better over a wide range than the approximation

$$\phi_2(\hat{v}/v) \propto [(\hat{v}/v) - 1]^2 = \frac{(\hat{v} - v)^2}{v^2} \;;$$

and we conclude that an approximation of the type

$$\lambda_t(\hat{\sigma}, \sigma) \propto \sigma \frac{(\hat{\sigma} - \sigma)^2}{\sigma^2} = (\hat{\sigma} - \sigma)^2 \frac{1}{\sigma}$$

will be much better than one of the type

$$\lambda_t(\hat{v}, v) \propto \sqrt{v}\,\frac{(\hat{v} - v)^2}{v^2} = (\hat{v} - v)^2\, v^{-\frac{3}{2}}\,.$$

In this sense, it is better in this particular example to estimate the standard deviation σ than to estimate the variance v.

6.2.5. *Subjective Evaluation of* λ_t

In many situations the decision maker may wish to base the imputed opportunity losses $\lambda_t(\hat{\omega}, \omega)$ in an "estimation" problem *entirely* on subjective judgment instead of computing exact values $\lambda_t(\hat{\omega}, \omega)$ for selected $(\hat{\omega}, \omega)$ pairs and then "fitting" some function to these computed values. Such a procedure is in perfect accord with the general principles of decision under uncertainty which have been advocated throughout this monograph. We have already seen in Section 1.4.1 that in preparing for the formal analysis of *any* decision problem the decision maker is free to cut the decision tree at any level L that he likes and to assign "subjective" utilities to the branch ends thus created, whether or not he *could* cut the tree and assign subjective utilities farther out and then *compute* the utilities at level L. Again as we pointed out in Section 1.4.1, a really "complete" decision tree would be infinitely complex and therefore totally useless; the proper scope of formal analysis in *any* decision problem is *always* a matter of judgment.

Judgmental evaluation of the imputed-loss function λ_t in an "estimation" problem will be particularly reasonable when previous exact analyses of *complete* decision problems of the type in which the estimate is to be used has provided substantial even though indirect evidence on which to base the assessment of λ_t; and in our opinion it is an important objective of statistical research to make such previous experience generally available. To give a single important example of the kind of numerical research which would be extremely useful, consider the "ultimate" problem of determining optimal sample size for choice between two terminal acts whose utilities depend linearly on the unknown mean of a Normal population whose standard deviation σ_ϵ is also unknown. Because this "ultimate" problem is very laborious to solve as it stands but would be very easy to solve if σ_ϵ were known (as we have seen in Section 5.5 above), one would like to solve the ultimate problem by treating a point estimate $\hat{\sigma}_\epsilon$ *as if* it were the true value σ_ϵ itself. Systematic analysis of the loss function $\lambda_t(\hat{\sigma}_\epsilon, \sigma_\epsilon)$ in problems of this general class would provide a really sound basis both for making an optimal estimate on given information about σ_ϵ and (in situations where information about σ_ϵ can be obtained separately from information about μ, e.g. by means of uniformity trials) for deciding how much information about σ_ϵ should be collected before proceeding to determine the sample size for the "ultimate" decision about μ.

6.2.6. *Rough and Ready Estimation*

Unfortunately there is often a very great difference between what is conceptually desirable and what is mathematically feasible. We have already referred to the difficulty of finding the exact imputed loss function λ_t; but even if we substitute a very well behaved approximation for the exact function it may still be impossible to find the estimate $\hat{\omega}^*$ which minimizes the expected value of the loss because the posterior distribution theory may not be tractable.

In such situations, the best practical solution may well be to exploit the classical methods of point estimation; we give one example to illustrate the point. Suppose that the imputed loss function λ_t is such that the true certainty equivalent is some particular fractile of the posterior distribution of $\tilde{\omega}$. Although we may be unable to find the posterior distribution of $\tilde{\omega}$ based on the entire sample or on a sufficient statistic—i.e., the posterior distribution which represents *all* of our information about Ω—we may know the approximate distribution of the maximum-likelihood "estimator", and if so then from the observed value of this estimator we may be able to find the corresponding posterior distribution of $\tilde{\omega}$. Our actual *estimate*—i.e., the certainty equivalent on which our choice of a terminal act will be based—will then be the appropriate fractile of this posterior distribution, *not* the "maximum-likelihood estimate" itself.

It should also be remarked that when the posterior distribution of $\tilde{\omega}$ is "tight" (relative to the context of the decision problem), it may not be worth the effort to formalize the nature of λ_t or to worry unduly about approximating it, since it may be clear that *any* reasonable summary measure of the posterior distribution of $\tilde{\omega}$ will lead to the choice of a terminal act whose "true" expected utility differs negligibly from the expected utility of the "true" optimal act.

6.2.7. *Multipurpose Estimation*

All of our discussion hitherto has been based on the assumption that some estimate $\hat{\omega}$ of ω is to be used as a certainty equivalent in a single, well defined second-stage action problem. It was on the basis of this assumption that it was possible to introduce at least conceptually the imputed opportunity losses $\lambda_t(\hat{\omega}, \omega)$. We now turn to the case where a given body of information concerning Ω may be used in a number of different second-stage action problems.

Terminal Considerations

In some situations it may be true that even though the imputed loss functions λ_t are not *identical* in the various second-stage action problems, they are nevertheless *similar* in the sense that *the same estimate $\hat{\omega}$ is a certainty equivalent for all the problems.* We cite two examples.

Consider first a two-action problem involving a Bernoulli p and let $u_t(a_1, p) = kp$ and $u_t(a_2, p) = K$; the quantity p might represent a fraction defective, kp the cost of accepting the lot, and K the cost of rejecting the lot. Now let $\tilde{K}$ be unknown and let it play the role of $\tilde{\omega}$; and assume that $\tilde{K}$ and $\tilde{p}$ are independent. If our estimate of K is $\hat{K}$, then

$$\lambda_t(\hat{K}, K) = \begin{cases} 0 & \text{if} \quad K \text{ and } \hat{K} \text{ are on the same side of } k\bar{p} \;, \\ |K - k\bar{p}| & \text{if} \quad K \text{ and } \hat{K} \text{ straddle } k\bar{p} \;, \end{cases}$$

and it is easy to see that for any distribution of $\tilde{K}$, the mean $\bar{K}$ is an optimal estimate $\hat{K}^*$. Observe that $\bar{K}$ *is a certainty equivalent for any problem in which the* *utilities and therefore the imputed losses are of this* form, *regardless of the numerical* values *of the coefficient k and the mean $\bar{p}$ of $\tilde{p}$.* We can think of $\bar{K}$ as being a "broad" rather than a "restricted" certainty equivalent.

As a second example, consider the well known production-smoothing problem which in *very* restricted form can be presented as follows: at time n, the decision maker must choose a production level x_n and demand will be $\tilde{d}_n$; the terminal opportunity loss at time n is of the form

$$a_n(x_n - \tilde{d}_n)^2 + b_n(x_n - x_{n-1})^2 \ .$$

Given the boundary condition $x_0 = c$ and the constants a_n, b_n for $n = 1, 2, \cdots, N$, the problem is to choose x_1 as the first stage of a policy which will minimize the expected sum of the terminal opportunity losses in N periods. It is known† that if the problem is modified by replacing the random variables $\tilde{d}_1, \cdots, \tilde{d}_N$ by their expected values $\bar{d}_1, \cdots, \bar{d}_N$, then the original and modified problems have the same optimal first stage act x_1^*. The means $\bar{d}_1, \cdots, \bar{d}_N$ are certainty equivalents for the joint distribution of $\tilde{d}_1, \cdots, \tilde{d}_N$ *regardless of the numerical values of the parameters* $c, a_1, b_1, a_2, b_2, \cdots, a_N, b_N$.

Usually, however, it will *not* be true that any single estimate $\hat{\omega}$ can serve simultaneously as a certainty equivalent for a number of different second-stage action problems. If the posterior distribution of $\tilde{\omega}$ is "tight," a "reasonable" summary measure like a mean or mode may serve all purposes *reasonably* well; but when this is not the case, one should refrain from making unnecessary compromises. It is much better—and nearly as easy—*to report the posterior distribution of $\tilde{\omega}$ and then to try to find the best possible estimate for each terminal-action problem individually;* cf. Section 3.4.6 on scientific reporting.

Preposterior Considerations

Suppose now that our immediate problem is not to make an estimate or set of estimates of ω for use in a number of terminal-action problems but to choose an experiment e which will give us additional information about Ω. If this information is *ultimately* to be used in a number of different second-stage action problems, then the *immediate* result of the experiment will presumably be a posterior distribution of $\tilde{\omega}$ in one guise or another. But how does this help to decide on sample size? One possible approach is to define some index of "tightness" of the posterior distribution (such as the variance if the distribution is symmetric) and then to decide on a substitution rate between the index of tightness and the cost of sampling. A second possible approach would be to concoct a composite imputed loss structure λ_t *as if* we were eventually going to choose a single estimate $\hat{\omega}$. A standard preposterior analysis can then indicate a "reasonable" sample size, even though after the experiment has been conducted we will actually report the *posterior distribution* of $\tilde{\omega}$ and then use this as a basis for finding a *different* certainty equivalent $\hat{\omega}^*$ for each terminal-action problem that arises.

† See H. Theil, "A note on certainty equivalence in dynamic planning", *Econometrica* 25 (1957) pages 346–349. In this paper Theil refers to an alternate proof given previously by H. Simon.

6.3. Quadratic Terminal Opportunity Loss

In this section and the next we shall examine some of the analytical problems which arise when the conditional *terminal opportunity losses* $l_t(a, \omega)$ are quadratic or linear in $(a - \omega)$; the entire discussion will obviously apply without change to problems in which the conditional *"imputed estimation losses"* $\lambda_t(\hat{\omega}, \omega)$ of an estimate $\hat{\omega}$ are quadratic or linear in $(\hat{\omega} - \omega)$. In the present section we shall consider the quadratic structure

$$l_t(a, \omega) = k_t(a - \omega)^2 \tag{6-24}$$

where k_t is a constant; in Section 6.4 we shall consider the linear structure

$$l_t(a, \omega) = \begin{cases} k_o(a - \omega) & \text{if} \quad \omega \leq a \, , \\ k_u(\omega - a) & \text{if} \quad \omega \geq a \, ; \end{cases} \tag{6-25}$$

and in Section 6.5 we shall see how either of these structures can be made much more flexible without loss of tractability through multiplication by suitably chosen functions of ω.

6.3.1. *Terminal Analysis*

If the conditional terminal opportunity loss of an act a given the state ω is of the simple quadratic form (6-24), then the expected opportunity loss of a under either a prior or a posterior distribution of $\tilde{\omega}$ is

$$\begin{aligned} \mathrm{E}\, l_t(a, \tilde{\omega}) &= k_t\, \mathrm{E}(a - \tilde{\omega})^2 = k_t\, \mathrm{E}(a - \bar{\omega} + \bar{\omega} - \tilde{\omega})^2 \\ &\equiv k_t[(a - \bar{\omega})^2 + \breve{\omega}] \end{aligned} \tag{6-26}$$

where

$$\bar{\omega} \equiv \mathrm{E}(\tilde{\omega}) \, , \qquad \breve{\omega} \equiv \mathrm{V}(\tilde{\omega}) \equiv \mathrm{E}(\tilde{\omega} - \bar{\omega})^2 \, . \tag{6-27}$$

It follows at once that the optimal act is given by

$$a^* = \bar{\omega} \tag{6-28}$$

and that the expected terminal opportunity loss of this optimal act is

$$\mathrm{E}\, l_t(a^*, \tilde{\omega}) = \min_a \mathrm{E}\, l_t(a, \tilde{\omega}) = k_t\, \breve{\omega} \, . \tag{6-29}$$

In Table 6.1 formulas for $\bar{\omega}$ and $\breve{\omega}$ are given for various $\tilde{\omega} = W(\tilde{\theta})$ associated with the univariate data-generating processes listed in Section 3.2.5 above and discussed in Part III below. In each case it is assumed that the distribution of the process parameter $\tilde{\theta}$ is conjugate to the likelihood of a sample.

▶ Except in the case of $\omega = \sigma^2$, all the formulas in the table are derived in Part III; references to the formula in Part III are given in the last column of the table. To derive the formulas for $\mathrm{E}(\tilde{\sigma}^2)$ and $\mathrm{V}(\tilde{\sigma}^2)$, we first observe that if $\tilde{h}$ has the gamma-2 density (7-50) with parameter (v, ν), then by (7-51a) and (7-55) $1/\tilde{h} = \tilde{\sigma}^2$ has the inverted-gamma-1 density (7-54) with parameter $(\frac{1}{2}\nu, \frac{1}{2}\nu v)$. The formulas in the table then follow immediately from (7-56). ◀

Table 6.1

Mean and Variance of $\tilde{\omega}$

Process	Conjugate Distribution	ω	$\bar{\omega} = E(\tilde{\omega})$	$\breve{\omega} = V(\tilde{\omega})$	Reference
Bernoulli: $p^x(1-p)^{1-x}$	beta: $p^{r-1}(1-p)^{n-r-1}$	p	$\frac{r}{n}$	$\frac{r(n-r)}{n^2(n+1)}$	(9-11)
		$\rho = p^{-1}$	$\frac{n-1}{r-1}$	$\frac{(n-1)(n-r)}{(r-1)^2(r-2)}$	(9-15)
Poisson: $e^{-\lambda x}\lambda$	gamma-1: $e^{-\lambda t}\lambda^{r-1}$	λ	$\frac{r}{t}$	$\frac{r}{t^2}$	(10-10)
		$\mu = \lambda^{-1}$	$\frac{t}{r-1}$	$\frac{t^2}{(r-1)^2(r-2)}$	(10-15)
Normal: $e^{-\frac{1}{2}h(x-\mu)^2}h^{\frac{1}{2}}$	h known, $\tilde{\mu}$ Normal: $e^{-\frac{1}{2}hn(\mu-m)^2}$	μ	m	$\frac{1}{hn}$	(11-24)
	μ known, $\tilde{h}$ gamma-2: $e^{-\frac{1}{2}h\nu v}h^{\frac{1}{2}\nu-1}$	h	$\frac{1}{v}$	$\frac{1}{\frac{1}{2}\nu v^2}$	(11-10)
		$\sigma = h^{-\frac{1}{2}}$	$\frac{(\frac{1}{2}\nu-\frac{3}{2})!}{(\frac{1}{2}\nu-1)!}\sqrt{\frac{1}{2}\nu v}$	$\frac{\nu v}{\nu-2} - \bar{\sigma}^2$	(7-59); *cf.* Section 11.1.4
		$\sigma^2 = h^{-1}$	$\frac{\nu v}{\nu-2}$	$\frac{2\nu^2 v^2}{(\nu-2)^2(\nu-4)}$	*see note*
	Normal-gamma: $e^{-\frac{1}{2}hn(\mu-m)^2}h^{\frac{1}{2}} \cdot e^{-\frac{1}{2}h\nu v}h^{\frac{1}{2}\nu-1}$	μ	m	$\frac{v}{n}\frac{\nu}{\nu-2}$	(11-49)
		h, σ, σ^2	*same as for μ known*		

6.3.2. Preposterior Analysis

If we now introduce primes and double primes to distinguish between values prior and posterior to a sample, we have by (6-28) and (6-29) that if the decision maker does not experiment but acts under his prior distribution of $\tilde\omega$, then the optimal act is $\mathrm{E}'(\tilde\omega) \equiv \bar\omega'$ and its expected terminal opportunity loss is $k_t\,\mathrm{V}'(\tilde\omega) \equiv k_t\,\breve\omega'$; whereas if he does perform an experiment e and observes z, then the optimal act is $\mathrm{E}''_{\omega|z}(\tilde\omega) \equiv \bar\omega''_z$ and its expected terminal opportunity loss is

$$k_t\,\mathrm{V}''_{\omega|z}(\tilde\omega) \equiv k_t\,\breve\omega''_z \; . \tag{6-30}$$

When an experiment e is being considered but has not yet been performed, its outcome $\tilde z$ is a random variable and therefore the posterior terminal opportunity loss $k_t\,\tilde{\breve\omega}''$ is in general a random variable. In order to make a preposterior analysis of any e the decision maker must compute the prior expected value of this posterior terminal opportunity loss; and since this expectation is

$$l_t^*(e) = E_{z|e}(k_t\,\tilde{\breve\omega}''_z) = k_t\,\mathrm{E}_{z|e}(\tilde{\breve\omega}''_z) \; , \tag{6-31}$$

this problem is equivalent to the problem of computing the *prior expectation of the posterior variance.*

When the prior variance of the posterior mean is known, the prior expectation of the posterior variance can be easily obtained by using a result derived in Chapter 5 above. Defining

$$\begin{aligned} \bar{\breve\omega}'' &\equiv \mathrm{E}_{z|e}(\tilde{\breve\omega}''_z) \equiv \mathrm{E}_{z|e}\,\mathrm{V}''_{\omega|z}(\tilde\omega) \; , \\ \breve{\bar\omega}'' &\equiv \mathrm{V}_{z|e}(\tilde{\bar\omega}''_z) \equiv \mathrm{V}_{z|e}\,\mathrm{E}''_{\omega|z}(\tilde\omega) \; , \end{aligned} \tag{6-32}$$

we have by (5-28) that for any distributions whatever on $\Omega \times Z$, provided only that the means and variances in question exist,

$$\bar{\breve\omega}'' = \breve\omega' - \breve{\bar\omega}'' \; . \tag{6-33}$$

In Table 6.2 formulas for $\breve{\bar\omega}''$ and $\bar{\breve\omega}''$ in terms of $\breve\omega'$ are given for one or more kinds of experiments which might be conducted in order to obtain information on each of the various ωs appearing in Table 6.1 (page 189). The nature of the experiment is indicated by listing the components of the sufficient statistic with a tilde over those which are not predetermined but left to chance. For the parametrization, see Table 6.1 or Section 3.2.5 above; for formulas for $\breve\omega'$ in terms of the parameters of the prior distribution, see Table 6.1. References in parentheses are to formulas for $\breve{\bar\omega}''$ in Part III; references in square brackets are to derivations of $\bar{\breve\omega}''$ in the notes immediately below.

▶ *Note* 1. In this case $\breve\mu''$ is independent of the sample outcome; its value and the value of $\breve\mu'$ are given by (11-24c) and (11-22).

Note 2. By (11-10b) the prior and posterior variances of $\tilde h$ are

$$\breve h' = \frac{1}{\frac{1}{2}\nu' v'^2} \; , \qquad \tilde{\breve h}'' = \frac{1}{\frac{1}{2}\nu''\,\tilde v''^2} \; .$$

Table 6.2

Expected Posterior Variance of $\tilde{\omega}$

Process	Prior Distribution	Experiment	ω	$\breve{\omega}''/\breve{\omega}'$	$\bar{\breve{\omega}}''/\breve{\omega}'$	Reference
Bernoulli	beta	$\tilde{r}\|n$	p	$\dfrac{n}{n+n'}$	$\dfrac{n'}{n+n'}$	(9-21c)
			$\rho = 1/p$	*see reference*		(9-29)
		$\tilde{n}\|r$	p	*see reference*		(9-38)
			$\rho = 1/p$	$\dfrac{r}{r+r'-1}$	$\dfrac{r'-1}{r+r'-1}$	(9-43c)
Poisson	gamma-1	$\tilde{r}\|t$	λ	$\dfrac{t}{t+t'}$	$\dfrac{t'}{t+t'}$	(10-37c)
			$\mu = 1/\lambda$	*see reference*		(10-45)
		$\tilde{t}\|r$	λ	$\dfrac{r}{r+r'+1}$	$\dfrac{r'+1}{r+r'+1}$	(10-22c)
			$\mu = 1/\lambda$	$\dfrac{r}{r+r'-1}$	$\dfrac{r'-1}{r+r'-1}$	(10-28b)
Normal	h known, $\tilde{\mu}$ Normal	$\tilde{m}\|n$	μ	$\dfrac{n}{n+n'}$	$\dfrac{n'}{n+n'}$	[1]
	μ known, $\tilde{h}$ gamma-2	$\tilde{w}\|\nu$	h	$\dfrac{\nu}{\nu+\nu'+2}$	$\dfrac{\nu'+2}{\nu+\nu'+2}$	[2]
			$\sigma = h^{-\frac{1}{2}}$	*see note*		[3]
			$\sigma^2 = h^{-1}$	$\dfrac{\nu}{\nu+\nu'-2}$	$\dfrac{\nu'-2}{\nu+\nu'-2}$	[4]
	Normal-gamma	$\tilde{m}, \tilde{v}\|n, \nu$	μ	$\dfrac{n}{n+n'}$	$\dfrac{n'}{n+n'}$	(11-66a)
			h, σ, σ^2	*same as for μ known*		—

By (11-15) and (11-61) the density of $\tilde{v}''$ is the same whether or not μ is known, and by the second formula in the proof of (11-65)

$$\mathrm{E}(\tilde{v}''^{-2}) = \left[\frac{\nu' v'}{\nu''}\right]^{-2} \frac{(\frac{1}{2}\nu''-1)!(\frac{1}{2}\nu'+1)!}{(\frac{1}{2}\nu''+1)!(\frac{1}{2}\nu'-1)!} = \frac{\nu''(\nu'+2)}{\nu'(\nu''+2)v'^2} \; .$$

We thus have for the *expected* posterior variance of $\tilde{h}$

$$\bar{\breve{h}}'' = \frac{1}{\frac{1}{2}\nu''} \cdot \frac{\nu''(\nu'+2)}{\nu'(\nu''+2)v'^2} = \frac{\nu'+2}{\nu''+2}\,\breve{h}' \; .$$

Note 3. By Section 11.1.4 and formulas (7-59) the posterior variance of $\tilde{\sigma}$ is

$$\bar{\breve{\sigma}}'' = \left[\frac{1}{\frac{1}{2}\nu'' - 1} - \left(\frac{(\frac{1}{2}\nu'' - \frac{3}{2})!}{(\frac{1}{2}\nu'' - 1)!}\right)^2\right] \tfrac{1}{2}\nu''\tilde{v}'' \;.$$

By (11-15) and (11-61) the density of $\tilde{v}''$ is the same whether or not μ is known, and by (7-27a) or the second formula in the proof of (11-65)

$$\mathrm{E}(\tilde{v}'') = \frac{\frac{1}{2}\nu'(\frac{1}{2}\nu'' - 1)}{\frac{1}{2}\nu''(\frac{1}{2}\nu' - 1)}\, v' \;.$$

Substituting this result for $\tilde{v}''$ in the previous formula we obtain

$$\bar{\breve{\sigma}}'' = \left[\frac{1}{\frac{1}{2}\nu'' - 1} - \left(\frac{(\frac{1}{2}\nu'' - \frac{3}{2})!}{(\frac{1}{2}\nu'' - 1)!}\right)^2\right] \frac{\frac{1}{2}\nu'' - 1}{\frac{1}{2}\nu' - 1}\, \tfrac{1}{2}\nu' v' \;.$$

Note 4. By Table 6.1 (page 189), the prior and posterior variances of $\tilde{\sigma}^2$ are

$$\mathrm{V}'(\tilde{\sigma}^2) = \frac{2\nu'^2\, v'^2}{(\nu' - 2)^2(\nu' - 4)} \;, \qquad \mathrm{V}''(\tilde{\sigma}^2) = \frac{2\nu''^2\, \tilde{v}''^2}{(\nu'' - 2)^2(\nu'' - 4)} \;.$$

By (11-15), (11-61), and the second formula in the proof of (11-65)

$$\mathrm{E}(\tilde{v}''^2) = \frac{\nu'^2\, v'^2}{\nu''^2} \cdot \frac{(\nu'' - 2)(\nu'' - 4)}{(\nu' - 2)(\nu' - 4)} \;.$$

Substituting this result for $\tilde{v}''^2$ in the formula for $\mathrm{V}''(\tilde{\sigma}^2)$ we obtain

$$\mathrm{EV}''(\tilde{\sigma}^2) = \frac{2\nu''^2}{(\nu'' - 2)^2(\nu'' - 4)} \cdot \frac{(\nu'' - 2)(\nu'' - 4)\,\nu'^2}{(\nu' - 2)(\nu' - 4)\,\nu''^2}\, v'^2 = \frac{\nu' - 2}{\nu'' - 2}\, \mathrm{V}'(\tilde{\sigma}^2) \;. \qquad \blacktriangleleft$$

6.3.3. *Optimal Sample Size*

The decision maker's objective is to choose e in such a way as to minimize his expected opportunity loss, which by (4-19), (6-31) and (6-32) is

$$l^*(e) = l_t^*(e) + c_s^*(e) = k_t\, \bar{\breve{\omega}}'' + c_s^*(e) \;. \tag{6-34}$$

We are interested particularly in the case where every e in E is characterized by a real number η (the "sample size") such that the expected sampling cost is a linear function of η,

$$c_s(e_\eta) = K_s + k_s\eta \;.$$

Since by (6-32) the *expected* posterior variance $\bar{\breve{\omega}}''$ is a function of e and therefore of η, we add a subscript η to $\bar{\breve{\omega}}''$ and write (6-34) in the form

$$l^*(e_\eta) = k_t\, \bar{\breve{\omega}}_\eta'' + K_s + k_s\eta \;. \tag{6-36}$$

If now η is continuous or can be treated as continuous for practical purposes, and if $\bar{\breve{\omega}}_\eta''$ has a derivative with respect to η (as it does in all cases listed in Table 6.2 on page 191), then the *optimal* η is either 0 or a solution of the equation

$$\frac{d}{d\eta}\, \bar{\breve{\omega}}_\eta'' = -\frac{1}{k^*} \qquad \text{where} \qquad k^* \equiv \frac{k_t}{k_s} \;. \tag{6-37}$$

For each ω for which a formula for $\bar{\breve{\omega}}''$ was given in Table 6.2, we give in Table 6.3 a formula for the root η° of this equation which, if it is positive, corresponds to the unique local minimum of $l^*(e_\eta)$. If however η° is negative, then the optimal sample size is 0; and even if η° is in fact positive, the optimal sample size *may* be 0: in this case the question must be settled by actually computing $l^*(e_{\eta^\circ})$ and comparing it with $l^*(e_0) = k_t\,\breve{\omega}'$.

Table 6.3

Optimal Sample Size

Process	Prior Distribution	Experiment	ω	S	Optimality Condition: $\sqrt{k^*S} =$	Formula Number
Bernoulli	beta	$\tilde{r}\|n$	p	$\dfrac{r'(n'-r')}{n'(n'+1)}$	$n^\circ + n'$	1
		$\tilde{n}\|r$	$\rho = 1/p$	$\dfrac{(n'-1)(n'-r')}{(r'-1)(r'-2)}$	$r^\circ + r' - 1$	2
Poisson	gamma-1	$\tilde{r}\|t$	λ	$\dfrac{r'}{t'}$	$t^\circ + t'$	3
		$r\|\tilde{t}$	λ	$\dfrac{r'(r'+1)}{t'^2}$	$r^\circ + r' + 1$	4
			$\mu = 1/\lambda$	$\dfrac{t'^2}{(r'-1)(r'-2)}$	$r^\circ + r' - 1$	5
Normal	h known, $\tilde{\mu}$ Normal	$\tilde{m}\|n$	μ	$\dfrac{1}{h}$	$n^\circ + n'$	6
	μ known, $\tilde{h}$ gamma-2	$\tilde{w}\|\nu$	h	$\dfrac{2(\nu'+2)}{\nu' v'^2}$	$\nu^\circ + \nu' + 2$	7
			$\sigma^2 = 1/h$	$\dfrac{2\nu' v'^2}{(\nu'-2)(\nu'-4)}$	$\nu^\circ + \nu' - 2$	8
	Normal-gamma	$\tilde{m}, \tilde{v}\|n, \nu$	μ	$\dfrac{\nu' v'}{\nu'-2}$	$n^\circ + n'$	9
			h, σ^2	*same as for μ known*		10

▶ All the formulas for $\breve{\bar{\omega}}''_\eta$ given in Table 6.2 (page 191) are of the form

$$\breve{\bar{\omega}}''_\eta = \frac{\eta' + c}{\eta + \eta' + c}\,\breve{\omega}'$$

where c is a constant. Differentiating twice with respect to η we obtain

$$\text{(1)} \qquad \frac{d}{d\eta}\breve{\bar{\omega}}''_\eta = -\frac{\eta' + c}{(\eta + \eta' + c)^2}\,\breve{\omega}' ,$$

$$\text{(2)} \qquad \frac{d^2}{d\eta^2}\breve{\bar{\omega}}''_\eta = \frac{2(\eta' + c)}{(\eta + \eta' + c)^3}\,\breve{\omega}' .$$

Substituting (1) in (6-37) we see that $l^*(e_\eta)$ is stationary if η satisfies

$$\eta + \eta' + c = \pm\sqrt{k^*(\eta' + c)\breve{\omega}'} ;$$

and it then follows from (2) that the solution corresponding to a local minimum is the one for which the square root is taken positive. The formulas in Table 6.3 are obtained by substituting in this result the formulas for $\check{\omega}'$ given in Table 6.1 (page 189). ◀

Interpretation of the Formulas for Optimal Sample Size. The role of the factor $k^* = k_t/k_s$ in the sample-size formulas of Table 6.3 is easy to "understand" economically: it says simply that optimal sample size increases as the cost of a discrepancy between a and ω increases and that optimal sample size decreases as the cost of sampling increases. When we seek a similar "explanation" of the factor S, it seems at first sight that a clue is offered by formula 6, for the case where the mean μ of an Independent Normal process is unknown but the process precision h is known. In this case we have by (11-2b) that the factor $S = 1/h$ is equal to the sampling variance $V(\tilde{x}|\mu)$ of a single observation $\tilde{x}$; and it seems reasonable enough that optimal sample size should increase with $V(\tilde{x}|\mu)$ because an increase in $V(\tilde{x}|\mu)$ means a decrease in the amount we can learn about μ from each observation in the sample.

In the other cases in Table 6.3 the variance of a single observation depends on the value of the quantity being estimated, so that the interpretation of the factor S must necessarily be more complicated than in the very simple case of formula 6, but it naturally occurs to us to inquire whether in these other cases the factor S may be related to the *expected* sampling variance $E_\omega V(\tilde{x}|\tilde{\omega})$ of a single observation. The factor S does turn out to be actually *equal* to $E_\omega V(\tilde{x}|\tilde{\omega})$ in several cases, specifically in those covered by formulas 1, 2, and 5; and in those covered by formulas 3 and 8 the factor S turns out to be equal to the expected variance of the statistic $\tilde{r}$ or $\tilde{w}$ in a sample of size $\eta = 1$.

▶ In the case covered by formula 1, we have by (9-1), (7-9c), and (7-22ab)

$$E_p V(\tilde{x}|\tilde{p}) = \int_0^1 p(1-p)\, f_\beta(p|r', n')\, dp = \frac{r'}{n'} - \frac{r'(r'+1)}{n'(n'+1)} = \frac{r'(n'-r)}{n'(n'+1)} = S\ .$$

In the case covered by formula 2, we have by (9-2), (7-12b), and (7-21)

$$\begin{aligned} E_p V(\tilde{y}|\tilde{p}) &= \int_0^1 \frac{1-p}{p^2} f_\beta(p|r', n')\, dp \\ &= \int_0^1 \frac{(n'-1)(n'-2)}{(r'-1)(r'-2)} f_\beta(p|r'-2, n'-2)\, dp - \int_0^1 \frac{n'-1}{r'-1} f_\beta(p|r'-1, n'-1)\, dp \\ &= \frac{(n'-1)(n'-2)}{(r'-1)(r'-2)} - \frac{n'-1}{r'-1} = \frac{(n'-1)(n'-r')}{(r'-1)(r'-2)} = S\ . \end{aligned}$$

In the case covered by formula 3, we have by (10-32), (7-33c), and (7-45a)

$$E_\lambda V(\tilde{x}|\tilde{\lambda}, t=1) = \int_0^\infty \lambda f_{\gamma 1}(\lambda|r', t')\, d\lambda = \frac{r'}{t'} = S\ .$$

In the case covered by formula 5, we have by (10-1), (7-45c), and (7-43)

$$\mathrm{E}_\lambda \mathrm{V}(\tilde{x}|\tilde{\lambda}) = \int_0^\infty \frac{1}{\lambda^2} f_{\gamma 1}(\lambda|r', t')\, d\lambda = \int_0^\infty \frac{t'^2}{(r'-1)(r'-2)} f_{\gamma 1}(\lambda|r'-2, t')\, d\lambda$$

$$= \frac{t'^2}{(r'-1)(r'-2)} = S\ .$$

In the case covered by formula 8, we have by (11-12), (7-52b), and (7-50)

$$\mathrm{E}_h \mathrm{V}(\tilde{w}|\tilde{h}, \nu = 1) = \int_0^\infty \frac{1}{\frac{1}{2}h^2} f_{\gamma 2}(h|v', \nu')\, dh$$

$$= \int_0^\infty 2 \frac{(\frac{1}{2}\nu' v')^2}{(\frac{1}{2}\nu' - 1)(\frac{1}{2}\nu' - 2)} f_{\gamma 2}\left(h \middle| \frac{\nu' v'}{\nu' - 4}, \nu' - 4\right) dh$$

$$= \frac{2\nu' v'^2}{(\nu' - 2)(\nu' - 4)} = S\ .$$ ◀

Unfortunately, however, this line of explanation breaks down totally when we examine the two remaining cases in Table 6.3. In the case covered by formula 4, the factor S is equal to the expectation of the *reciprocal* of the sampling variance of a single observation; and in the case covered by formula 7 the factor S is four times the expectation of the reciprocal of the variance of the statistic $\tilde{w}$ in a sample of size $\nu = 1$. We must therefore confess that we have as yet been unable to attach any clear economic "meaning" to the factor S.

▶ In the case covered by formula 4, we have by (10-1), (7-45a), and (7-45b)

$$\mathrm{E}_\lambda \left[\frac{1}{\mathrm{V}(\tilde{x}|\tilde{\lambda})}\right] = \int_0^\infty \lambda^2 f_{\gamma 1}(\lambda|r', t')\, d\lambda = \frac{r'(r'+1)}{t'^2} = S\ .$$

In the case covered by formula 7, we have by (11-12), (7-52b), and (7-50)

$$\mathrm{E}_h \left[\frac{1}{\mathrm{V}(\tilde{w}|\tilde{h}, \nu = 1)}\right] = \int_0^\infty \tfrac{1}{2}h^2 f_{\gamma 2}(h|v', \nu')\, dh$$

$$= \int_0^\infty \tfrac{1}{2} \frac{\frac{1}{2}\nu'(\frac{1}{2}\nu' + 1)}{(\frac{1}{2}\nu' v')^2} f_{\gamma 2}\left(h \middle| \frac{\nu' v'}{\nu' + 4}, \nu' + 4\right) dh$$

$$= \tfrac{1}{2} \frac{\nu' + 2}{\nu' v'^2} = \tfrac{1}{4} S\ .$$ ◀

6.4. Linear Terminal Opportunity Loss

We next consider the linear loss structure

$$l_t(a, \omega) = \begin{cases} k_o(a - \omega) & \text{if} \quad \omega \le a\ , \\ k_u(\omega - a) & \text{if} \quad \omega \ge a\ , \end{cases} \tag{6-38}$$

where k_o and k_u are positive constants.

6.4.1. *Terminal Analysis*

If the conditional terminal opportunity loss of an act a given ω is of the linear form (6-38), then the expected terminal opportunity loss of a under either a prior or a posterior distribution of $\tilde{\omega}$ is

$$\begin{aligned} \mathrm{E}\, l_t(a, \tilde{\omega}) &= k_o \int_{-\infty}^{a} (a - \omega)\, d\mathrm{P}_\omega + k_u \int_a^{\infty} (\omega - a)\, d\mathrm{P}_\omega \\ &\equiv k_o\, L_\omega^{(l)}(a) + k_u\, L_\omega^{(r)}(a) \ . \end{aligned} \tag{6-39}$$

The "linear-loss integrals" $L^{(l)}$ and $L^{(r)}$ have already been defined in (5-12), and it has already been shown in Table 5.1 (page 99) that they can often be easily evaluated in terms of well tabulated functions when $\tilde{\omega}$ is a reasonably simple function of the state $\tilde{\theta}$ and $\tilde{\theta}$ has one of the conjugate distributions listed in Section 3.2.5 above and discussed in Part III below.

If we let F_ω and G_ω respectively denote the left and right-tail cumulative functions of $\tilde{\omega}$,

$$F_\omega(a) \equiv \mathrm{P}\{\tilde{\omega} \leq a\} \ , \qquad G_\omega(a) \equiv \mathrm{P}\{\tilde{\omega} \geq a\} \ , \tag{6-40}$$

and if we define the "partial expectation" E_α^β of $\tilde{\omega}$ by

$$\mathrm{E}_\alpha^\beta(\tilde{\omega}) \equiv \int_\alpha^\beta \omega\, d\mathrm{P}_\omega \ , \tag{6-41}$$

then (6-39) can be written in the alternative form

$$\mathrm{E}\, l_t(a, \tilde{\omega}) = k_o[a\, F_\omega(a) - \mathrm{E}_{-\infty}^{a}(\tilde{\omega})] + k_u[\mathrm{E}_a^\infty(\tilde{\omega}) - a\, G_\omega(a)] \ . \tag{6-42}$$

If now $\tilde{\omega}$ possesses a density function D_ω and a is continuous or can be treated as such for practical purposes, then the act a^* which minimizes (6-42) can be found by setting the derivative of (6-42) equal to zero: a^* must satisfy

$$\begin{aligned} 0 &= k_o[a^*\mathrm{D}_\omega(a^*) + F_\omega(a^*) - a^*\mathrm{D}_\omega(a^*)] + k_u[-a^*\mathrm{D}_\omega(a^*) + a^*\mathrm{D}_\omega(a^*) - G_\omega(a^*)] \\ &= k_o\, F_\omega(a^*) - k_u\, G_\omega(a^*) \ . \end{aligned} \tag{6-43}$$

That the stationary value of $\mathrm{E}\, l_t(a, \tilde{\omega})$ corresponding to $a = a^*$ is in fact a minimum follows at once from the fact that the right-hand side of (6-43) is an increasing function of a.

To put the optimality condition (6-43) into a more convenient form we substitute $G_\omega(a^*) = 1 - F_\omega(a^*)$, thus obtaining

$$F_\omega(a^*) = \frac{k_u}{k_u + k_o} \ . \tag{6-44}$$

Observe that if the loss structure is symmetric, $k_u = k_o$, then the optimal act is equal to the *median* of the distribution of $\tilde{\omega}$. Observe also that if ϕ is any increasing function of ω and if

$$l_t(a, \omega) = \begin{cases} k_o[\phi(a) - \phi(\omega)] & \text{if} \quad \phi(\omega) \leq \phi(a) \ , \\ k_u[\phi(\omega) - \phi(a)] & \text{if} \quad \phi(\omega) \geq \phi(a) \ , \end{cases}$$

so that the optimal act a^* is given by

$$F_{\phi(\omega)}[\phi(a^*)] = \frac{k_u}{k_u + k_o} \ ,$$

then because

$$F_{\phi(\omega)}[\phi(a)] = F_\omega(a)$$

the optimal act is exactly the same as if the losses were linear in $(a - \omega)$ itself with the *same* coefficients k_o and k_u.

Formulas for the optimal act in terms of standardized, tabulated cumulative functions are given for a number of $\tilde{\omega}$s in the fourth column of Table 6.4, where $F_{\beta*}$ and $G_{\beta*}$ refer to the beta function (7-15), $F_{\gamma*}$ and $G_{\gamma*}$ refer to the gamma function (7-34), F_{N*} refers to the unit-Normal function, and F_{S*} refers to the standardized Student function.

► The optimality condition in row 1 is derived from (9-9), (7-23), and (7-18); row 2 then follows from the definition $\rho = 1/p$. The condition in row 3 is derived from (10-8) and (10-12); row 4 then follows from the definition $\mu = 1/\lambda$. Row 5 is derived from (11-21) and (11-23). Row 6 is derived from (11-8) and (11-11); rows 7 and 8 then follow from the definitions $\sigma = h^{-\frac{1}{2}}$ and $\sigma^2 = 1/h$. Row 9 is derived from (11-47) and (11-48). Row 10 follows from the fact that (11-8) and (11-44) are identical. ◄

When the optimality condition (6-43) is substituted in the general formula (6-42) for the expected terminal opportunity loss we obtain the expected terminal opportunity loss of the *optimal* act a^*:

$$\mathrm{E}\, l_t(a^*, \tilde{\omega}) = k_u\, \mathrm{E}_{a^*}^{\infty}(\tilde{\omega}) - k_o\, \mathrm{E}_{-\infty}^{a^*}(\tilde{\omega}) \ . \tag{6-45}$$

Specific formulas for a number of $\tilde{\omega}$s are given in the fifth and sixth columns of Table 6.4 just below. The formulas in the fifth column are the more convenient for computations, being expressed in terms of the well tabulated binomial (f_b) and Poisson (f_P) mass functions and unit-Normal (f_{N*}) and Student (f_{S*}) density functions. The formulas in the sixth column, on the other hand, show more clearly the essential similarity of all the separate cases: with a single exception, the expected loss is of the form

$$\mathrm{E}\, l_t(a^*, \tilde{\omega}) = (k_u + k_o)\, \mathrm{V}(\tilde{\omega})\, f(a^*)$$

where f is a member of the same family as the conjugate density of $\tilde{\omega}$ but with the parameters possibly augmented or diminished by 1 or 2. A formula of this type can in general be obtained when (and only when) ω is an integral power of the "natural" process parameter θ.

► To derive the formulas for $\mathrm{E}\, l_t(a^*, \tilde{\omega})$ in Table 6.4 we first observe that by the definition (6-41)

$$\mathrm{E}_{a^*}^{\infty}(\tilde{\omega}) = \mathrm{E}(\tilde{\omega}) - \mathrm{E}_{-\infty}^{a^*}(\tilde{\omega}) \equiv \bar{\omega} - \mathrm{E}_{-\infty}^{a^*}(\tilde{\omega})$$

provided that the distribution of $\tilde{\omega}$ is continuous as it is in all the cases in Table 6.4. We then substitute this formula in (6-45) to obtain for the dimensionless terminal opportunity loss of a^*

$$L \equiv \frac{1}{k_u + k_o}\, \mathrm{E}\, l_t(a^*, \tilde{\omega}) = \bar{\omega}\left[\frac{k_u}{k_u + k_o} - \frac{1}{\bar{\omega}}\, \mathrm{E}_{-\infty}^{a^*}(\tilde{\omega})\right] . \tag{1}$$

Table 6.4
Optimal Act and Expected Terminal Opportunity Loss

Process	Conjugate Distribution	ω	Optimal Act: $k_u/(k_u + k_o) =$	$\dfrac{\mathrm{E}\, l_t(a^*, \tilde{\omega})}{k_u + k_o}$	$\dfrac{\mathrm{E}\, l_t(a^*, \tilde{\omega})}{(k_u + k_o)\, \mathrm{V}(\tilde{\omega})}$	Row
Bernoulli	beta	p	$F_{\beta^*}(a^*\|r, n - r)$	$\dfrac{r(n-r)}{n^2} f_b(r\|a^*, n)$	$f_\beta(a^*\|r + 1, n + 2)$	1
		$\rho = 1/p$	$G_{\beta^*}(1/a^*\|r, n - r)$	$\dfrac{n-r}{r-1} f_b\left(r - 1 \middle\| \dfrac{1}{a^*}, n - 1\right)$	$f_{i\beta 1}(a^*\|r - 2, n - 1, 1)$	2
Poisson	gamma-1	λ	$F_{\gamma^*}(a^* t\|r)$	$\dfrac{r}{t} f_P(r\|a^* t)$	$f_{\gamma 1}(a^*\|r + 1, t)$	3
		$\mu = 1/\lambda$	$G_{\gamma^*}(t/a^*\|r)$	$\dfrac{t}{r-1} f_P(r - 1\|t/a^*)$	$f_{i\gamma 1}(a^*\|r - 2, t)$	4
Normal	h known, $\tilde{\mu}$ Normal	μ	$F_{N^*}(u^*)$, $u^* \equiv (a^* - m)\sqrt{hn}$	$(hn)^{-\frac{1}{2}} f_{N^*}(u^*)$	$f_N(a^*\|m, hn)$	5
	μ known, $\tilde{h}$ gamma-2	h	$F_{\gamma^*}(\frac{1}{2}\nu v a^*\|\frac{1}{2}\nu)$	$\dfrac{1}{v} f_P(\frac{1}{2}\nu\|\frac{1}{2}\nu v a^*)$	$f_{\gamma 2}\left(a^* \middle\| \dfrac{\nu v}{\nu + 2}, \nu + 2\right)$	6
		$\sigma = h^{-\frac{1}{2}}$	$G_{\gamma^*}(\frac{1}{2}\nu v/a^{*2}\|\frac{1}{2}\nu)$	*see note*		7
		$\sigma^2 = 1/h$	$G_{\gamma^*}(\frac{1}{2}\nu v/a^*\|\frac{1}{2}\nu)$	$\dfrac{\nu v}{\nu - 2} f_P(\frac{1}{2}\nu - 1\|\frac{1}{2}\nu v/a^*)$	$f_{i\gamma 1}(a^*\|\frac{1}{2}\nu - 2, \frac{1}{2}\nu v)$	8
	Normal-gamma	μ	$F_{S^*}(t^*\|\nu)$, $t^* \equiv (a^* - m)\sqrt{n/v}$	$(n/v)^{-\frac{1}{2}} \dfrac{\nu + t^{*2}}{\nu - 1} f_{S^*}(t^*\|\nu)$	$f_{S^*}\left(a^*\middle\|m, \dfrac{\nu - 2}{\nu} \dfrac{n}{v}, \nu - 2\right)$	9
		h, σ, σ^2		*all formulas same as for μ known*		10

All the derivations except number 7 then follow the same general pattern. By manipulating the expression given in Part III for $E_{-\infty}^{a^*}(\tilde{\omega})$ we find a function ϕ such that

$$\frac{1}{\bar{\omega}} E_{-\infty}^{a^*}(\tilde{\omega}) = F_\omega(a^*) - \phi(a^*, y)$$

where y denotes the parameters of the conjugate density of $\tilde{\omega}$. We then use this result and the optimality condition (6-44)

$$F_\omega(a^*) = \frac{k_u}{k_u + k_o}$$

to put (1) in the form

$$L = \bar{\omega}\, \phi(a^*, y) \ ;$$

and substituting herein the value of $\bar{\omega}$ given in Table 6.1 (page 189) we obtain the formula in column 5 of Table 6.4. The formula in column 6 of Table 6.4 can then be derived by using the formula in Table 6.1 for $\breve{\omega}$ and the formula in Part III for the conjugate density f_ω.

Throughout these derivations the definitions (7-8) of the binomial function f_b and (7-11) of the Pascal function f_{Pa} are to be understood as applying to all real positive r, n (not necessarily integral), while the corresponding cumulative functions G_b and F_{Pa} are to be taken as *defined* in terms of the cumulative beta function by (7-20) and (7-13). The arguments originally given to prove these latter formulas now become proofs that the redefined functions have the properties typified by

$$G_b(r|p, n) = F_{Pa}(n|p, r) \ ,$$
$$F_{Pa}(n + 1|p, r) = F_{Pa}(n|p, r) + f_{Pa}(n + 1|p, r) \ .$$

Similarly the definition (7-32) of the Poisson function f_P is extended to all real positive r, the function G_P is taken as defined in terms of the cumulative gamma function by (7-39), and what was originally a proof of this latter formula becomes a proof that

$$G_P(r|m) = G_P(r + 1|m) + f_P(r|m) \ .$$

Row 1. By (9-11b),

$$\frac{1}{\bar{p}} E_0^{a^*}(\tilde{p}) = F_\beta(a^*|r + 1, n + 1) = 1 - G_\beta(a^*|r + 1, n + 1) \ . \tag{2}$$

By (7-24) and (7-13)

$$\begin{aligned} G_\beta(a^*|r + 1, n + 1) &= G_b(n - r|1 - a^*, n) = F_{Pa}(n|1 - a^*, n - r) \\ &= F_{Pa}(n - 1|1 - a^*, n - r) + f_{Pa}(n|1 - a^*, n - r) \ . \end{aligned} \tag{3}$$

By (7-13) and (7-24) and the optimality condition

$$F_{Pa}(n - 1|1 - a^*, n - r) = G_b(n - r|1 - a^*, n - 1) = G_\beta(a^*|r, n) = 1 - \frac{k_u}{k_u + k_o} \ . \tag{4}$$

Substituting (4) in (3), the result in (2), and this result in (1), we obtain

$$L = \bar{p}\, f_{Pa}(n|1 - a^*, n - r) \ ;$$

and by Table 6.1 and then (7-11) and (7-8)

$$L = \frac{r}{n} f_{Pa}(n|1 - a^*, n - r) = \frac{r(n - r)}{n^2} f_b(r|a^*, n) \ .$$

This is the formula in column 5 of Table 6.4; the formula in column 6 follows by (7-8), (7-21), and Table 6.1.

Row 2. By (9-15b), (7-20), and (7-13)

$$(5)\qquad \frac{1}{\bar{p}}\,\mathrm{E}_1^{a^*}(\tilde{p}) = G_{\beta*}\left(\frac{1}{a^*}\middle|\, r-1, n-r\right) = G_b\left(n-r|1-\frac{1}{a^*}, n-2\right)$$

$$= F_{Pa}\left(n-2|1-\frac{1}{a^*}, n-r\right)$$

$$= F_{Pa}\left(n-1|1-\frac{1}{a^*}, n-r\right) - f_{Pa}\left(n-1|1-\frac{1}{a^*}, n-r\right) .$$

By (7-13), (7-20), and the optimality condition

$$(6)\qquad F_{Pa}\left(n-1|1-\frac{1}{a^*}, n-r\right) = G_b\left(n-r|1-\frac{1}{a^*}, n-1\right)$$

$$= G_{\beta*}(a^*|r, n-r) = \frac{k_u}{k_u+k_o} \cdot$$

Substituting (6) in (5) and the result in (1), we obtain

$$L = \bar{p}\, f_{Pa}\left(n-1|1-\frac{1}{a^*}, n-r\right) ;$$

and by Table 6.1 and then (7-11) and (7-8)

$$L = \frac{n-1}{r-1} f_{Pa}\left(n-1|1-\frac{1}{a^*}, n-r\right) = \frac{n-r}{r-1} f_b\left(r-1\middle|\frac{1}{a^*}, n-1\right) .$$

This is the formula in column 5 of Table 6.4; the formula in column 6 follows by (7-8), (7-25), and Table 6.1.

Row 3. By (10-10b) and (7-39)

$$(7)\qquad \frac{1}{\bar{\lambda}}\,\mathrm{E}_0^{a^*}(\tilde{\lambda}) = F_{\gamma*}(a^*t|r+1) = G_P(r+1|a^*t) = G_P(r|a^*t) - f_P(r|a^*t) .$$

By (7-39) and the optimality condition

$$(8)\qquad G_P(r|a^*t) = F_{\gamma*}(a^*t|r) = \frac{k_u}{k_u+k_o} \cdot$$

Substituting (8) in (7) and the result in (1) and then using Table 6.1 we obtain

$$L = \bar{\lambda}\, f_P(r|a^*t) = \frac{r}{t} f_P(r|a^*t) .$$

This is the formula in column 5 of Table 6.4; the formula in column 6 follows by (7-32), (7-43), and Table 6.1.

Row 4. By (10-15b) and (7-39)

$$(9)\qquad \frac{1}{\bar{\mu}}\,\mathrm{E}_0^{a^*}(\tilde{\mu}) = G_{\gamma*}(t/a^*|r-1) = 1 - G_P(r-1|t/a^*)$$

$$= 1 - G_P(r|t/a^*) - f_P(r-1|t/a^*) .$$

By (7-39) and the optimality condition

$$(10)\qquad 1 - G_P(r|t/a^*) = G_{\gamma*}(t/a^*|r) = \frac{k_u}{k_u+k_o} \cdot$$

Substituting (10) in (9) and the result in (1) and then using Table 6.1 we obtain

$$L = \tilde{\mu}\, f_P(r-1|t/a^*) = \frac{t}{r-1} f_P(r-1|t/a^*) \ .$$

This is the formula in column 5 of Table 6.4; the formula in column 6 follows by (7-32), (7-54), and Table 6.1.

Row 5. By (11-24b) and (11-24a)

$$(11) \qquad \frac{1}{\bar{\mu}} \mathrm{E}^{a^*}_{-\infty}(\tilde{\mu}) = F_{N^*}(u^*) - \frac{1}{\bar{\mu}}(hn)^{-\frac{1}{2}} f_{N^*}(u^*) \ , \qquad u^* \equiv (a^* - m)\sqrt{hn} \ .$$

By the optimality condition

$$(12) \qquad F_{N^*}(u^*) = \frac{k_u}{k_u + k_o} \ .$$

Substituting (12) in (11) and the result in (1) we obtain

$$L = (hn)^{-\frac{1}{2}} f_{N^*}(u^*) \ .$$

This is the formula in column 5 of Table 6.4; the formula in column 6 follows by (7-61), (7-63), and Table 6.1.

Row 6. By (11-8), (7-52a), and (7-39)

$$(13) \qquad (1/\bar{h})\, \mathrm{E}^{a^*}_0(\tilde{h}) = F_{\gamma^*}(\tfrac{1}{2}\nu v a^*|\tfrac{1}{2}\nu + 1) = G_P(\tfrac{1}{2}\nu + 1|\tfrac{1}{2}\nu v a^*)$$
$$= G_P(\tfrac{1}{2}\nu|\tfrac{1}{2}\nu v a^*) - f_P(\tfrac{1}{2}\nu|\tfrac{1}{2}\nu v a^*) \ .$$

By (7-39) and the optimality condition

$$(14) \qquad G_P(\tfrac{1}{2}\nu|\tfrac{1}{2}\nu v a^*) = F_{\gamma^*}(\tfrac{1}{2}\nu v a^*|\tfrac{1}{2}\nu) = \frac{k_u}{k_u + k_o} \ .$$

Substituting (14) in (13) and the result in (1) and then using Table 6.1 we obtain

$$L = \bar{h}\, f_P(\tfrac{1}{2}\nu|\tfrac{1}{2}\nu v a^*) = \frac{1}{v} f_P(\tfrac{1}{2}\nu|\tfrac{1}{2}\nu v a^*) \ .$$

This is the formula in column 5 of Table 6.4; the formula in column 6 follows by (7-32), (7-50), and Table 6.1.

Row 7. By Section 11.1.4 the distribution of $\tilde{\sigma}$ is inverted-gamma-2 with parameter $(v^{\frac{1}{2}}, \nu)$, and by an obvious modification of the proof of formula (7-59a) for $\mathrm{E}(\tilde{\sigma})$ we have

$$\mathrm{E}^{a^*}_0(\tilde{\sigma}) = \bar{\sigma}\, F_{\gamma^*}(\tfrac{1}{2}\nu v/a^{*2}|\tfrac{1}{2}\nu - \tfrac{1}{2}) \ .$$

Substituting this expression in (1) we obtain

$$\frac{1}{k_u + k_o} \mathrm{E}\, l_t(a^*, \tilde{\sigma}) = \bar{\sigma}\left[\frac{k_u}{k_u + k_o} - F_{\gamma^*}(\tfrac{1}{2}\nu v/a^{*2}|\tfrac{1}{2}\nu - \tfrac{1}{2})\right] .$$

It is impossible to obtain a formula of the same type as those in columns 5 and 6 of Table 6.4 because we cannot reduce $F_{\gamma^*}(z|\tfrac{1}{2}\nu - \tfrac{1}{2})$ to an expression of the form $F_{\gamma^*}(z|\tfrac{1}{2}\nu) + \phi(z)$.

Row 8. If $\tilde{h}$ has a gamma-2 density with parameter (v, ν), then by (7-51a) and (7-55) $\tilde{\sigma}^2 = 1/\tilde{h}$ has an inverted-gamma-1 density with parameter $(\tfrac{1}{2}\nu, \tfrac{1}{2}\nu v)$. Then since by (10-13) the distribution of $\tilde{\mu}$ in row 4 of Table 6.4 is inverted-gamma-1 with parameter (r, t), the formulas in row 8 of Table 6.4 can be derived from the formulas in row 4 by substituting $\tfrac{1}{2}\nu$ for r and $\tfrac{1}{2}\nu v$ for t.

Row 9. By (11-49c) and (11-49a)

$$(15) \qquad \frac{1}{\bar{\mu}} \mathrm{E}^{a^*}_{-\infty}(\tilde{\mu}) = F_{S^*}(t^*|\nu) - \frac{1}{\bar{\mu}}(n/v)^{-\frac{1}{2}} \frac{\nu + t^{*2}}{\nu - 1} f_{S^*}(t^*|\nu)$$

where

$$t^* = (a^* - m)\sqrt{n/v} \ .$$

By the optimality condition

$$F_{S^*}(t^*|\nu) = \frac{k_u}{k_u + k_o} \ . \tag{16}$$

Substituting (16) in (15) and the result in (1) we obtain

$$L = (n/v)^{-\frac{1}{2}} \frac{\nu + t^{*2}}{\nu - 1} f_{S^*}(t^*|\nu) \ . \tag{17}$$

This is the formula in column 5 of Table 6.4. To obtain the formula in column 6 we simplify notation by defining

$$z = a^* - m \ , \qquad H = n/v \ ,$$

and then use the definition (7-66) of f_{S^*} to put (17) in the form

$$L = H^{-\frac{1}{2}} \frac{\nu + Hz^2}{\nu - 1} \frac{(\frac{1}{2}\nu - \frac{1}{2})!\nu^{\frac{1}{2}\nu}}{\sqrt{\pi}\,(\frac{1}{2}\nu - 1)!} (\nu + Hz^2)^{-\frac{1}{2}\nu - \frac{1}{2}}$$

$$= \frac{(\frac{1}{2}\nu - \frac{3}{2})!(\nu - 2)^{\frac{1}{2}\nu - 1}}{\sqrt{\pi}\,(\frac{1}{2}\nu - 2)!} \left[\left(\frac{\nu}{\nu - 2}\right)^{\frac{1}{2}\nu - \frac{1}{2}} (\nu + Hz^2)^{-\frac{1}{2}\nu + \frac{1}{2}}\right]\left[\frac{\nu - 2}{\nu} H\right]^{\frac{1}{2}} \frac{(\frac{1}{2}\nu - \frac{1}{2})\,\nu H^{-1}}{(\frac{1}{2}\nu - 1)(\nu - 1)}$$

$$= f_S\left(z|0, \frac{\nu - 2}{\nu} H, \nu - 2\right)\left(\frac{1}{H}\frac{\nu}{\nu - 2}\right) = f_S\left(a^*|m, \frac{\nu - 2}{\nu}\frac{n}{v}, \nu - 2\right)\left(\frac{v}{n}\frac{\nu}{\nu - 2}\right)$$

and by Table 6.1 the last factor is $V(\tilde{\mu})$.

Row 10. This follows from the fact that (11-8) is identical to (11-44). ◀

6.4.2. *Preposterior Analysis*

In order to evaluate an e which is being considered but has not yet been performed, the decision maker must compute

$$l_t^*(e) = E_{z|e} \min_a E''_{\omega|z}\, l_t(a, \tilde{\omega}) = E_{z|e} E''_{\omega|z}\, l_t(\tilde{a}_z^*, \tilde{\omega}) \tag{6-46}$$

where a_z^* is the optimal act for a *given* sample outcome z, i.e. for a given posterior distribution $P''_{\omega|z}$, and $E''_{\omega|z}\, l_t(\tilde{a}_z^*, \tilde{\omega})$ is given by formula (6-45) or, for specific $\tilde{\omega}$s, by Table 6.4 (page 198).

To see the problems involved in the expectation with respect to $\tilde{z}$, suppose for example that $\tilde{\omega}$ is the parameter $\tilde{p}$ of a Bernoulli process and that the e which is being considered consists in taking a predetermined number n of observations, the number $\tilde{r}$ of successes being left to chance. Then by (6-46) and row 1 of Table 6.4

$$l_t^*(e) = (k_u + k_o)\,E\left[\frac{\tilde{r}''(n'' - \tilde{r}'')}{n''^2} f_b(\tilde{r}''|\tilde{a}_r^*, n'')\right] \tag{6-47a}$$

where $\tilde{r}'' = r' + \tilde{r}$, where $\tilde{a}_r^*$ is the very complex function of $\tilde{r}$ defined by the optimality condition

$$F_{\beta^*}(\tilde{a}_r^*|\tilde{r}'', n'' - \tilde{r}'') = \frac{k_u}{k_u + k_o} \ , \tag{6-47b}$$

and where the expectation is with respect to the beta-binomial distribution (9-18) of $\tilde{r}$. It seems difficult if not impossible to obtain an algebraic formula for the expectation of $f_b(\tilde{r}''|\tilde{a}^*_r, n'')$ alone, let alone the product of this random variable and another; and in fact we have not succeeded in finding a formula for $l^*_t(e)$ for any case of this type. Even numerical evaluation with the aid of a high-speed computer seems difficult; the successive values of the beta-binomial mass function of $\tilde{r}$ can be computed by means of simple recursion relations, but determination of the value of a^*_r for each r is not easy.

In some cases, however, the problem of evaluating (6-46) is much simpler because the factor corresponding to $f_b(\tilde{r}''|\tilde{a}^*_r, n'')$ in (6-47a) is a constant independent of the experimental outcome $\tilde{z}$. Suppose for example that $\tilde{\omega}$ is the intensity $\tilde{\lambda}$ of a Poisson process and that e consists in observing the process until the rth event occurs, the amount of "time" $\tilde{t}$ being left to chance. Then by (6-46) and row 3 of Table 6.4 (page 198)

$$l^*_t(e) = (k_u + k_o)\,\mathrm{E}\left[\frac{r''}{\tilde{t}''} f_P(r''|\tilde{a}^*_t\,\tilde{t}'')\right] \tag{6-48a}$$

where $\tilde{t}'' = t' + \tilde{t}$, where $\tilde{a}^*_t$ is defined as a function of $\tilde{t}$ by the optimality condition

$$F_{\gamma^*}(\tilde{a}^*_t\,\tilde{t}''|r'') = \frac{k_u}{k_u + k_o}\,, \tag{6-48b}$$

and where the expectation is with respect to the inverted-beta-2 distribution (10-18) of $\tilde{t}$. Because the value of the *product* $\tilde{a}^*_t\,\tilde{t}''$ is fixed by (6-48b) and r'' is known in advance, the value of the factor $f_P(r''|\tilde{a}^*_t\,\tilde{t}'')$ in (6-48a) is independent of the sample outcome and evaluation of $l^*_t(e)$ requires merely the finding of the expected value of $1/\tilde{t}''$. If on the contrary the experiment consisted in observing the process for a predetermined amount of "time" t and leaving the number of successes $\tilde{r}$ to chance, then

$$l^*_t(e) = (k_u + k_o)\,\mathrm{E}\left[\frac{\tilde{r}''}{t''} f_P(\tilde{r}''|\tilde{a}^*_r\,t'')\right];$$

and since $\tilde{r}''$ is a random variable while the value of $\tilde{a}^*_r\,t''$ is determined by the optimality condition, the factor $f_P(\tilde{r}''|\tilde{a}^*_r\,t'')$ is a random variable and the expectation is very difficult to obtain. We conjecture, however, that the expected terminal opportunity loss with fixed $t = t_o$ is very nearly equal to expected terminal opportunity loss with fixed r such that $\mathrm{E}(\tilde{t}) = t_o$.

In Table 6.5 we show an algebraic formula for $l^*_t(e)$ for every ω and every e for which we have found such a formula; the derivations are given in the following notes.

▶ *Note 1.* In all four cases with reference to this note we can see by comparing the formula for the expected loss in column 5 of Table 6.4 (page 198) with the formula for $\bar{\omega}$ in Table 6.1 (page 189) that the *posterior* expected loss is of the form

$$\frac{1}{k_u + k_o}\,\mathrm{E}''_{\theta|z}\, l_t(a^*, \tilde{\omega}) = \bar{\omega}''_z\, f_P(\eta'' + c|\kappa^*)$$

Table 6.5

Prior Expectation of Posterior Terminal Opportunity Loss

Process	Prior Distribution	Experiment	ω	Definition of κ^*: $\frac{k_u}{k_u + k_o} =$	$\frac{l_t^*(e)}{k_u + k_o}$	Reference Note
Poisson	gamma-1	$\tilde{t}\|r$	λ	$F_{\gamma^*}(\kappa^*\|r'')$	$\frac{r'}{t'} f_P(r''\|\kappa^*)$	1
			$\mu = 1/\lambda$	$G_{\gamma^*}(\kappa^*\|r'')$	$\frac{t'}{r'-1} f_P(r''-1\|\kappa^*)$	1
Normal	h known, $\tilde{\mu}$ Normal	$\tilde{m}\|n$	μ	$F_{N^*}(\kappa^*)$	$(hn'')^{-\frac{1}{2}} f_{N^*}(\kappa^*)$	2
	μ known, $\tilde{h}$ gamma-2	$\tilde{w}\|\nu$	h	$F_{\gamma^*}(\kappa^*\|\frac{1}{2}\nu'')$	$\frac{1}{v'} f_P(\frac{1}{2}\nu''\|\kappa^*)$	1
			$\sigma = h^{-\frac{1}{2}}$		*see note*	3
			$\sigma^2 = 1/h$	$G_{\gamma^*}(\kappa^*\|\frac{1}{2}\nu'')$	$\frac{\nu'}{\nu'-2} v' f_P(\frac{1}{2}\nu''-1\|\kappa^*)$	1
	Normal-gamma	$\tilde{m}, \tilde{v}\|n, \nu$	μ		*see note*	4
			h, σ, σ^2		*same as for μ known*	

where c is a constant (0 in two of the four cases), η'' is predetermined by e, and κ^* is fixed by the optimality condition given in column 4 of Table 6.4. Since $E_{z|e}(\tilde{\bar{\omega}}_z'') = \bar{\omega}'$ by (5-27), we have at once

$$l_t^*(e) = E_{z|e}\left[\tilde{\bar{\omega}}_z'' f_P(\eta'' + c|\kappa^*)\right] = \bar{\omega}' f_P(\eta'' + c|\kappa^*) \ .$$

Note 2. In this case we have by Table 6.4 that the posterior expected opportunity loss is completely independent of the sample outcome; no expectation is required.

Note 3. By the note on row 7 of Table 6.4,

$$\frac{1}{k_u + k_o} l_t^*(e) = \left[\frac{k_u}{k_u + k_o} - F_{\gamma^*}(\kappa^*|\tfrac{1}{2}\nu'' - \tfrac{1}{2})\right] E(\tilde{\bar{\sigma}}'')$$

where κ^* is determined by the optimality condition

$$G_{\gamma^*}(\kappa^*|\tfrac{1}{2}\nu'') = \frac{k_u}{k_u + k_o} \ ;$$

and by (5-27) and Table 6.1

$$E(\tilde{\bar{\sigma}}'') = \bar{\sigma}' = \frac{(\frac{1}{2}\nu' - \frac{3}{2})!}{(\frac{1}{2}\nu' - 1)!} \sqrt{\tfrac{1}{2}\nu' v'} \ .$$

Note 4. By Table 6.4

$$\frac{1}{k_u + k_o} l_t^*(e) = \frac{\nu'' + t^{*2}}{\nu'' - 1} f_{S^*}(t^*|\nu'') \, n''^{-\frac{1}{2}} E(\tilde{v}''^{\frac{1}{2}})$$

where t^* is determined by the optimality condition

$$F_{S^*}(t^*|\nu'') = \frac{k_u}{k_u + k_o} .$$

By the second formula in the proof of (11-65)

$$\text{E}(\tilde{v}''^{\frac{1}{2}}) = \left[\frac{\nu' v'}{\nu''}\right]^{\frac{1}{2}} \frac{(\frac{1}{2}\nu'' - 1)!(\frac{1}{2}\nu' - \frac{3}{2})!}{(\frac{1}{2}\nu'' - \frac{3}{2})!(\frac{1}{2}\nu' - 1)!} ;$$

if ν' is large enough to permit use of Stirling's approximation

$$x! = (2\pi)^{\frac{1}{2}} x^{x+\frac{1}{2}} e^{-x} ,$$

then by a procedure like that used to prove (11-66b) it can be shown that

$$\text{E}(\tilde{v}''^{\frac{1}{2}}) \doteq v'^{\frac{1}{2}} \left[\frac{1 - 2/\nu''}{1 - 2/\nu'}\right]^{\frac{1}{2}} .$$ ◀

6.4.3. *Optimal Sample Size*

The decision maker's objective is to minimize

$$l^*(e) = l_t^*(e) + c_s^*(e)$$

where $c_s^*(e)$ is the (expected) cost of performing the experiment e; we are particularly interested in the case where $c_s^*(e)$ is linear in the "sample size" η and therefore

$$l^*(e_\eta) = l_t^*(e_\eta) + K_s + k_s\eta . \tag{6-49}$$

Of the cases for which a formula for $l_t^*(e)$ has been given in Table 6.5 (page 204) there is only one for which we can obtain an explicit formula for optimal sample size by differentiating (6-49) with respect to η and setting the result equal to zero. This is the case of the Independent Normal process with known precision h, for which n plays the role of η and (6-49) becomes

$$l^*(e_n) = (k_u + k_o)(hn'')^{-\frac{1}{2}} f_{N^*}(\kappa^*) + K_s + k_s n$$

where $n'' = n' + n$. On differentiating with respect to n and setting the result equal to 0 we obtain the condition for a stationary value of $l^*(e_n)$,

$$n' + n = \left[\frac{h^{-\frac{1}{2}}(k_u + k_o) f_{N^*}(\kappa^*)}{2k_s}\right]^{\frac{2}{3}} , \tag{6-50}$$

and it is easy to show that the unique solution of this equation corresponds to a local minimum. If the root is negative, the optimal sample size is 0; if the root is positive, it is certainly the optimal sample size if $K_s = 0$ but if $K_s > 0$ a check must be made to determine whether $l^*(e_n)$ is greater or less than the opportunity loss $l_t^*(e_0)$ of immediate terminal action.

6.5. Modified Linear and Quadratic Loss Structures

We have already remarked in Section 6.2.4 that in some situations the decision maker may feel that for any *given* ω the function $l_t(a, \omega)$ is linear or quadratic in $(a - \omega)$ but that the coefficient or coefficients in this linear or quadratic function depend on the value of ω. We shall now see that it is often possible to allow for this kind of dependence without any loss of mathematical tractability.

Let $l'_t(a, \omega)$ denote *any* "simple" loss function which is analytically tractable, consider the family of "modified" loss functions obtained by multiplying this simple function by some function g of ω,

$$l_t(a, \omega) = l'_t(a, \omega)\, g(\omega) \ ,$$

and let D denote the conjugate density function of $\tilde{\omega}$ and y the parameters of this density, so that the *expected* terminal opportunity loss of a is then

$$\mathrm{E}\, l_t(a, \tilde{\omega}) = \int l'_t(a, \omega)\, g(\omega)\, \mathrm{D}(\omega|y)\, d\omega \ .$$

If now g is such that

$$g(\omega)\, \mathrm{D}(\omega|y) = \mathrm{D}(\omega|y^*) \ ,$$

i.e. if multiplication of D by g changes the parameter of D but leaves its algebraic form unchanged, then the expected terminal opportunity loss of a can be written

$$\mathrm{E}\, l_t(a, \tilde{\omega}) = \int l'_t(a, \omega)\, \mathrm{D}(\omega|y^*)\, d\omega$$

and we see that any analytical results previously obtained for the "simple" loss function l'_t apply unchanged to the "modified" loss function l_t.

As an example, suppose that ω is the parameter p of a Bernoulli process, that the act a is an estimate $\hat{p}$ of p, and that the decision maker feels that although for any given p the imputed loss of an error $(\hat{p} - p)$ is quadratic in $(\hat{p} - p)$, it is really the relative rather than the absolute error which counts; his loss function is

$$l_t(\hat{p}, p) = k_t\left[\frac{\hat{p} - p}{p}\right]^2 = k_t(\hat{p} - p)^2\, p^{-2} \ ,$$

so that $g(p) = p^{-2}$. If the density of $\tilde{p}$ is beta with parameter $y = (r, n)$, then

$$\begin{aligned}
\mathrm{E}\, l_t(\hat{p}, \tilde{p}) &= \int k_t(\hat{p} - p)^2\, p^{-2} f_\beta(p|r, n)\, dp \\
&= \int k_t(\hat{p} - p)^2\, p^{-2} \frac{(n-1)!}{(r-1)!(n-r-1)!}\, p^{r-1}(1-p)^{n-r-1}\, dp \\
&= \int k_t(\hat{p} - p)^2 \frac{(n-1)(n-2)}{(r-1)(r-2)} \frac{(n-3)!}{(r-3)!(n-r-1)!}\, p^{r-3}(1-p)^{n-r-1}\, dp \\
&= \int \left[\frac{(n-1)(n-2)}{(r-1)(r-2)}\, k_t\right] (\hat{p} - p)^2 f_\beta(p|r-2, n-2)\, dp \ .
\end{aligned}$$

The problem is thus identical to a problem with "simple" quadratic loss in which the loss constant k_t and the parameter (r, n)—but not the algebraic form—of the density of $\tilde{p}$ have been suitably modified.

Following this example the reader can easily find the class of functions g which is compatible with a density D of a given family. As one illustration, if the density of $\tilde{p}$ is beta,

$$\mathrm{D}(p) \propto p^{r-1}(1-p)^{n-r-1} \ ,$$

we may take

$$g(p) = p^\alpha(1-p)^\beta \ , \qquad \begin{aligned} &\alpha > -r \\ &\beta > -(n-r) \ ; \end{aligned}$$

as another, if the density of $\tilde{\mu}$ is Normal,

$$D(\mu) \propto e^{-\frac{1}{2}hn(\mu-m)^2} ,$$

we may take

$$g(\mu) \propto e^{\alpha(\mu-\beta)^2} , \qquad \alpha < \tfrac{1}{2}hn ;$$

as a third, if the density of $\tilde{\sigma}$ is inverted-gamma-2,

$$D(\sigma) \propto e^{-\frac{1}{2}\nu s^2/\sigma^2} \sigma^{-\nu-1} ,$$

we may take

$$g(\sigma) \propto e^{\alpha/\sigma^2} \sigma^{\beta} , \qquad \begin{array}{l} \alpha < \frac{1}{2}\nu s^2 , \\ \beta < \nu . \end{array}$$

In all these cases, the class g is rich enough to allow the decision maker very considerable flexibility in approximating his "true" imputed loss function.

PART THREE

DISTRIBUTION THEORY

CHAPTER 7

Univariate Normalized Mass and Density Functions

7.0. Introduction

7.0.1. Normalized Mass and Density Functions

A real-valued scalar function f defined on the real line will be called a *normalized mass function* for the denumerable set X if it has the properties

$$a) \begin{cases} f(x) \geq 0 \; , & x \in X \; , \\ f(x) = 0 \; , & x \notin X \; , \end{cases} \qquad b) \sum f(x) = 1 \tag{7-1a}$$

where the summation is over all $x \in X$.

A real-valued scalar function f defined on the real line will be called a *normalized density function* if it has the properties

$$a) \; f(x) \geq 0 \; , \qquad -\infty < x < \infty \; , \qquad b) \int_{-\infty}^{\infty} f(x)\, dx = 1 \; . \tag{7-1b}$$

In addition to $f(x)$ the symbol $\mathrm{D}(x)$ will often be used to denote "the density of the random variable $\tilde{x}$ at the point x".

7.0.2. Cumulative Functions

If f is a normalized mass or density function, the function F defined by

$$F(a) \equiv \sum_{x \leq a} f(x) \qquad \text{or} \qquad \int_{x \leq a} f(x)\, dx \tag{7-2a}$$

will be called the *left-tail cumulative* function; the function G defined by

$$G(a) \equiv \sum_{x \geq a} f(x) \qquad \text{or} \qquad \int_{x \geq a} f(x)\, dx \tag{7-2b}$$

will be called the *right-tail cumulative* function. Notice that in the discrete case

$$F(a) + G(a) = 1 + f(a) \; . \tag{7-2c}$$

7.0.3. Moments

The *first moment* or *mean* of a normalized mass or density function f is defined to be

$$\mu_1 \equiv \sum x\, f(x) \qquad \text{or} \qquad \int x\, f(x)\, dx \tag{7-3a}$$

where the signs $\sum$ and $\int$ without indicated limits denote summation or integration over the entire domain of f. The *incomplete first moment* of f is defined to be

$$\mu_1(x) \equiv \sum_{-\infty}^{x} z\, f(z) \qquad \text{or} \qquad \int_{-\infty}^{x} z\, f(z)\, dz \;. \tag{7-3b}$$

For $k \neq 1$, the k*th moment about the origin* is defined to be

$$\mu_k' \equiv \sum x^k f(x) \qquad \text{or} \qquad \int x^k f(x)\, dx \tag{7-3c}$$

while the k*th moment about the mean* is defined to be

$$\mu_k \equiv \sum (x - \mu_1)^k f(x) \qquad \text{or} \qquad \int (x - \mu_1)^k f(x)\, dx \;. \tag{7-3d}$$

When the sum or integral defining a moment does not converge, we shall say that the moment does not exist.

In computing second moments about the mean we shall often make use of the fact that

$$\mu_2 = \mu_2' - \mu_1^2 \;. \tag{7-4}$$

► This relation is proved in the discrete case by writing

$$\mu_2 \equiv \sum (x - \mu_1)^2 f(x) = \sum x^2 f(x) - 2\mu_1 \sum x\, f(x) + \mu_1^2 \sum f(x)$$
$$= \sum x^2 f(x) - \mu_1^2 \sum f(x) \equiv \mu_2' - \mu_1^2 \;.$$

The proof in the continuous case is similar. ◄

7.0.4. *Expectations and Variances*

When a normalized mass or density function f is interpreted as the mass or density function of a *random variable* $\tilde{x}$, the first moment as defined by (7-3a) is by definition the *expectation* of the random variable,

$$E(\tilde{x}) \equiv \bar{x} \equiv \mu_1 \;; \tag{7-5a}$$

and by the same definition of expectation we have for the moments about the origin defined by (7-3c)

$$E(\tilde{x}^k) \equiv \mu_k' \;. \tag{7-5b}$$

The *partial expectation* of $\tilde{x}$ is defined in terms of the incomplete first moment (7-3b) by

$$E_{-\infty}^{x}(\tilde{x}) \equiv \mu_1(x) \;. \tag{7-5c}$$

The second moment about the mean is by definition the *variance* of $\tilde{x}$:

$$V(\tilde{x}) \equiv \breve{x} \equiv \mu_2 \;. \tag{7-5d}$$

7.0.5. *Integrand Transformations*

If f is a normalized density function defined on X and if X is mapped into Y by a one-to-one differentiable transformation h,

$$y = h(x) \;, \qquad x = h^{-1}(y) \;, \tag{7-6a}$$

the density f^* on Y must have the property

$$\int_{y_1}^{y_2} f^*(y)\,dy = \int_{x_1}^{x_2} f(x)\,dx\ , \qquad \begin{cases} y_1 = \min\ \{h(x_1), h(x_2)\}\ , \\ y_2 = \max\ \{h(x_1), h(x_2)\}\ ; \end{cases}$$

and for this to be true f^* must be such that

$$f^*(y) = f[h^{-1}(y)]\ |dx/dy|\ . \tag{7-6b}$$

Thus if

$$f(x) = e^{-x}\ , \qquad 0 \le x < \infty\ ,$$

and

$$y = h(x) = x^2\ , \qquad x = y^{\frac{1}{2}}\ ,$$

the transformed density is

$$f^*(y) = e^{-\sqrt{y}} \cdot \tfrac{1}{2}y^{-\frac{1}{2}}\ .$$

Such transformations of normalized density functions will be called *integrand transformations* to remind the reader that substitution of a new variable must be accompanied by multiplication by the absolute value of the *Jacobian* dx/dy; and because we shall use Jacobians only in connection with integrand transformations, we shall ordinarily take the words "absolute value" as understood and speak simply of multiplication by "the Jacobian".

7.0.6. *Effect of Linear Transformations on Moments*

In the applications we shall frequently require the moments of the density of a random variable $\tilde{y}$ which is a linear function of another random variable $\tilde{x}$ whose moments are known. We give the relations between the two sets of moments in the more flexible random-variable notation of Section 7.0.4 rather than in the moment notation of Section 7.0.3. Letting

$$\tilde{y} = a + b\tilde{x} \tag{7-7a}$$

and letting F denote the left-tail cumulative function of the density of $\tilde{x}$ we have

$$\begin{aligned} \mathrm{E}(\tilde{y}) &\equiv \bar{y} = a + b\ \mathrm{E}(\tilde{x}) \equiv a + b\ \bar{x}\ , \\ \mathrm{E}^{y}_{-\infty}(\tilde{y}) &= a\ F(x) + b\ \mathrm{E}^{x}_{-\infty}(\tilde{x})\ , \qquad x = \frac{y-a}{b}\ , \quad b \ge 0\ , \\ \mathrm{E}(\tilde{y}^2) &= a^2 + 2ab\ \mathrm{E}(\tilde{x}) + b^2\ \mathrm{E}(\tilde{x}^2)\ , \\ \mathrm{V}(\tilde{y}) &\equiv \breve{y} = b^2\ \mathrm{V}(\tilde{x}) \equiv b^2\ \breve{x}\ . \end{aligned} \tag{7-7b}$$

The proofs are obvious.

A. *Natural Univariate Mass and Density Functions*

7.1. Binomial Function

The binomial normalized mass function is defined by

$$f_b(r|p, n) \equiv \frac{n!}{r!(n-r)!}\, p^r(1-p)^{n-r}\ , \qquad \begin{aligned} &0 < p < 1\ , \\ &n, r = 0, 1, 2, \cdots\ , \\ &n \ge r\ . \end{aligned} \tag{7-8}$$

The first two moments of f_b are

$$\mu_1 = np \; , \qquad \mu_1(r) = np\, F_b(r-1|p, n-1) \; , \tag{7-9a}$$
$$\mu_2' = np + n(n-1)\, p^2 \; , \tag{7-9b}$$
$$\mu_2 = np(1-p) \; . \tag{7-9c}$$

The mass function is tabulated in

Tables of the Binomial Probability Distribution, National Bureau of Standards, Applied Mathematics Series 6, U. S. Government Printing Office, Washington, 1950: Table 1 is a 7-place table of $f_b(r|p, n)$ for

$$n = 1(1)49, \qquad p = .01(.01).50, \qquad r = 1(1)n \; .$$

H. G. Romig, *50–100 Binomial Tables*, Wiley, New York, 1953: gives $f_b(r|p, n)$ to 6 places for

$$n = 50(5)100 \; , \qquad p = .01(.01).50 \; , \qquad r = 1(1)n \; .$$

Both these tables also give values of the cumulative function, but the following tables of the cumulative function are more extensive:

Tables of the Cumulative Binomial Probabilities, Ordnance Corps Pamphlet ORDP-1, 1952, distributed by U. S. Dept. of Commerce, Office of Technical Services, Washington: a 7-place table of $P(c, n, p) = G_b(c|p, n)$ for

$$n = 1(1)150, \qquad p = .01(.01).50, \qquad c = 1(1)n \; .$$

Tables of the Cumulative Binomial Probability Distribution, Annals of the Computation Laboratory XXXV, Harvard U. Press, Cambridge, Mass., 1955: a 5-place table of $E(n, r, p) = G_b(r|p, n)$ for

$$n = 1(1)50(2)100(10)500(50)1000 \; , \qquad r = 1(1)n \; ,$$
$$p = .01(.01).50 \; , \qquad \tfrac{1}{12}(\tfrac{1}{12})\tfrac{6}{12} \; , \qquad \tfrac{1}{16}(\tfrac{1}{16})\tfrac{8}{16} \; .$$

Values of the cumulative function can also be obtained from Pearson's tables of the beta function $I_x(p, q)$ discussed in Section 7.3.1 below. The relations

$$\begin{aligned} G_b(r|p, n) &= I_p(r, n-r+1) \; , \qquad r \geq \frac{n+1}{2} \; , \qquad r \leq 50 \; , \\ &= 1 - I_{1-p}(n-r+1, r) \; , \qquad r \leq \frac{n+1}{2} \; , \qquad n - r \leq 49 \; , \end{aligned} \tag{7-10}$$

lead directly to tabled entries.

▶ That f_b is a proper normalized mass function is proved by observing (a) that it is obviously nonnegative over its entire domain and (b) that its complete sum

$$F_b(n|p, n) \equiv \sum_{r=0}^{n} f_b(r|p, n) = \sum_{r=0}^{n} \frac{n!}{r!(n-r)!}\, p^r (1-p)^{n-r}$$

is an expansion of $(p + [1-p])^n = 1^n$ and therefore has the value 1.

Formulas (7-9a) for the complete and incomplete first moments are proved by starting from the definition

$$\mu_1(r) \equiv \sum_{j=0}^{r} j\, f_b(j|p, n) \ .$$

Substituting its definition (7-8) for f_b and suppressing the term for $j = 0$ because its value is 0 we may write

$$\mu_1(r) = np \sum_{j=1}^{r} \frac{(n-1)!}{(j-1)!(n-j)!} p^{j-1}(1-p)^{n-j} \ ;$$

and defining $i = j - 1$ this becomes

$$\mu_1(r) = np \sum_{i=0}^{r-1} \frac{(n-1)!}{i!(n-1-i)!} p^i(1-p)^{n-1-i} = np\, F_b(r-1|p, n-1) \ ,$$

which becomes np when $r = n$.

For the second moment about the origin (7-9b), we have by definition

$$\mu_2' \equiv \sum_{r=0}^{n} r^2 f_b(r|p, n) = \sum_{r=0}^{n} (r + r[r-1])\, f_b(r|p, n) \ .$$

Dropping terms of 0 value this can be written

$$\mu_2' = \mu_1 + \sum_{r=2}^{n} r(r-1)\, f_b(r|p, n)$$

$$= \mu_1 + n(n-1)\, p^2 \sum_{r=2}^{n} \frac{(n-2)!}{(r-2)!(n-r)!} p^{r-2}(1-p)^{n-r} \ :$$

and defining $j = r - 2$ this becomes

$$\mu_2' = \mu_1 + n(n-1)\, p^2 \sum_{j=0}^{n-2} f_b(j|p, n-2) = np + n(n-1)p^2 \ .$$

The second central moment (7-9c) then follows from (7-4):

$$\mu_2 = \mu_2' - \mu_1^2 = np + n(n-1)\, p^2 - (np)^2 = np(1-p) \ .$$

The relation (7-10) with the beta function follows from formulas (7-20), (7-18), and (7-19) which are proved in Section 7.3 below. ◀

7.2. Pascal Function

The Pascal normalized mass function is defined by

$$f_{Pa}(n|p, r) \equiv \frac{(n-1)!}{(r-1)!(n-r)!} p^r(1-p)^{n-r} \ , \qquad \begin{array}{l} 0 < p < 1 \ , \\ n, r = 1, 2, \cdots, \\ n \geq r \ . \end{array} \qquad (7\text{-}11)$$

The first two moments of f_{Pa} are

$$\mu_1 = \frac{r}{p} \ , \qquad \mu_1(n) = \frac{r}{p} F_{Pa}(n+1|p, r+1) \ , \qquad (7\text{-}12a)$$

$$\mu_2 = \frac{r(1-p)}{p^2} \; . \tag{7-12b}$$

The Pascal function is related to the binomial function and to the beta function discussed in Section 7.3.1 below by

$$F_{Pa}(n|p, r) = G_b(r|p, n) = F_{\beta^*}(p|r, n - r + 1) \; . \tag{7-13}$$

The Pascal function is essentially identical to the negative-binomial function defined by (7-73) below, the relation being

$$f_{Pa}(n|p, r) \equiv f_{nb}(n - r|1 - p, r) \; . \tag{7-14}$$

Notice, however, that the Pascal function is defined only for integral n and r whereas the negative-binomial function will be defined also for nonintegral n and r (although only for integral $n - r$).

▶ The fact that the Pascal function is a proper normalized mass function follows via (7-14) from the fact proved in Section 7.10 below that the negative-binomial function is a proper normalized mass function.

Formula (7-12a) is proved by defining $i = j + 1$ and writing

$$\mu_1(n) = \sum_{j=r}^{n} j\, f_{Pa}(j|p, r) = \frac{r}{p} \sum_{j=r}^{n} \frac{j!}{r!(j-r)!} p^{r+1}(1-p)^{j-r}$$

$$= \frac{r}{p} \sum_{i=r+1}^{n+1} f_{Pa}(i|p, r+1) = \frac{r}{p} F_{Pa}(n+1|p, r+1) \; .$$

Formulas (7-12b) and (7-13) follow from the corresponding negative-binomial formulas (7-74b) and (7-75) and the relation (7-14) between the negative-binomial and Pascal functions. ◀

7.3. Beta Functions

7.3.1. *Standardized Beta Function*

The standardized beta normalized density function is defined by

$$f_{\beta^*}(z|p, q) \equiv \frac{1}{B(p, q)} z^{p-1}(1-z)^{q-1} \; , \qquad \begin{array}{l} 0 \le z \le 1 \; , \\ p, q > 0 \; , \end{array} \tag{7-15}$$

where $B(p, q)$ is the complete beta function

$$B(p, q) = \frac{(p-1)!(q-1)!}{(p+q-1)!} \tag{7-16}$$

discussed in books on advanced calculus. The first two moments of f_{β^*} are

$$\mu_1 = \frac{p}{p+q} \; , \qquad \mu_1(z) = \frac{p}{p+q} F_{\beta^*}(z|p+1, q) \; , \tag{7-17a}$$

$$\mu_2' = \frac{p(p+1)}{(p+q)(p+q+1)} \; , \tag{7-17b}$$

$$\mu_2 = \frac{pq}{(p+q)^2(p+q+1)} \; . \tag{7-17c}$$

The cumulative function has been tabulated by

K. Pearson, *Tables of the Incomplete Beta-Function*, Biometrika, London, 1934: a 7-place table of

$$I_x(p, q) \equiv F_{\beta^*}(x|p, q) \ , \qquad p, q = .5(.5)10(1)50 \ , \qquad p \geq q \ , \qquad x = .01(.01)1 \ . \tag{7-18}$$

For $p < q$, it is necessary to use the relation

$$G_{\beta^*}(x|p, q) = I_{1-x}(q, p) \ . \tag{7-19}$$

When $p, q \geq 1$, values of the cumulative beta function can also be obtained from tables of the cumulative binomial function $G_b(p|r, n)$ by use of

$$\begin{aligned} F_{\beta^*}(x|p, q) &= G_b(p|x, p + q - 1) \ , \qquad & x \leq \tfrac{1}{2} \ , \\ G_{\beta^*}(x|p, q) &= G_b(q|1 - x, p + q - 1) \ , \qquad & x \geq \tfrac{1}{2} \end{aligned} \tag{7-20}$$

because these relations hold for all p, q for which G_b is defined.

▶ That f_{β^*} is a proper normalized density function follows from the facts (*a*) that it is obviously nonnegative over its entire domain and (*b*) that the integral

$$\int_0^1 z^{p-1}(1 - z)^{q-1} \, dz \equiv B(p, q) \equiv B(q, p)$$

converges for all $p, q > 0$ as proved in books on advanced calculus.

To prove formulas (7-17) for the moments we write

$$\mu_1(z) \equiv \int_0^z y \, f_{\beta^*}(y|p, q) \, dy \equiv \int_0^z y \frac{(p + q - 1)!}{(p - 1)!(q - 1)!} y^{p-1}(1 - y)^{q-1} \, dy$$

$$= \frac{p}{p + q} \int_0^z \frac{(p + q)!}{p!(q - 1)!} y^p(1 - y)^{q-1} \, dy = \frac{p}{p + q} F_{\beta^*}(z|p + 1, q) \ ;$$

$$\mu_2' \equiv \int_0^1 z^2 f_{\beta^*}(z|p, q) \, dz = \frac{p(p + 1)}{(p + q)(p + q + 1)} \int_0^1 f_{\beta^*}(z|p + 2, q) \, dz = \frac{p(p + 1)}{(p + q)(p + q + 1)} \ ;$$

$$\mu_2 = \mu_2' - \mu_1^2 = \frac{p(p + 1)}{(p + q)(p + q + 1)} - \left[\frac{p}{p + q}\right]^2 = \frac{pq}{(p + q)^2(p + q + 1)} \ .$$

To prove (7-19), observe that by (7-18) and (7-15)

$$I_{1-x}(q, p) \equiv \int_0^{1-x} \frac{1}{B(q, p)} z^{q-1}(1 - z)^{p-1} \, dz$$

and substitute y for $1 - z$ and $B(p, q)$ for $B(q, p)$ to obtain

$$I_{1-x}(q, p) = \int_x^1 \frac{1}{B(p, q)} y^{p-1}(1 - y)^{q-1} \, dy = \int_x^1 f_{\beta^*}(y|p, q) \, dy \ .$$

To prove (7-20), write

$$F_{\beta^*}(x|p, q) \equiv \int_0^x f_{\beta^*}(z|p, q) \, dz = \int_0^x \frac{(p + q - 1)!}{(p - 1)!(q - 1)!} (1 - z)^{q-1} \cdot z^{p-1} \, dz$$

and integrate by parts to obtain (provided that $p, q > 0$)

$$\frac{(p+q-1)!}{(p-1)!(q-1)!}(1-z)^{q-1}\cdot\frac{z^p}{p}\Big|_0^x+\int_0^x\frac{z^p}{p}\cdot\frac{(p+q-1)!}{(p-1)!(q-2)!}(1-z)^{q-2}\,dz$$

$$=\frac{(p+q-1)!}{p!(q-1)!}x^p(1-x)^{q-1}+\int_0^x f_{\beta^*}(z|p+1,q-1)\,dz\ .$$

Provided that q is integral, iteration of the integration by parts yields finally

$$F_{\beta^*}(x|p,q)=\sum_{j=p}^{p+q-1}\frac{(p+q-1)!}{j!(p+q-1-j)!}x^j(1-x)^{p+q-1-j}\ ;$$

and provided that p is also integral, this is $G_b(p|x, p+q-1)$. ◀

7.3.2. Beta Function in Alternate Notation

In addition to the "standardized" beta density function f_{β^*} defined by (7-15) it will be convenient to define what we shall call simply "the" beta normalized density function:

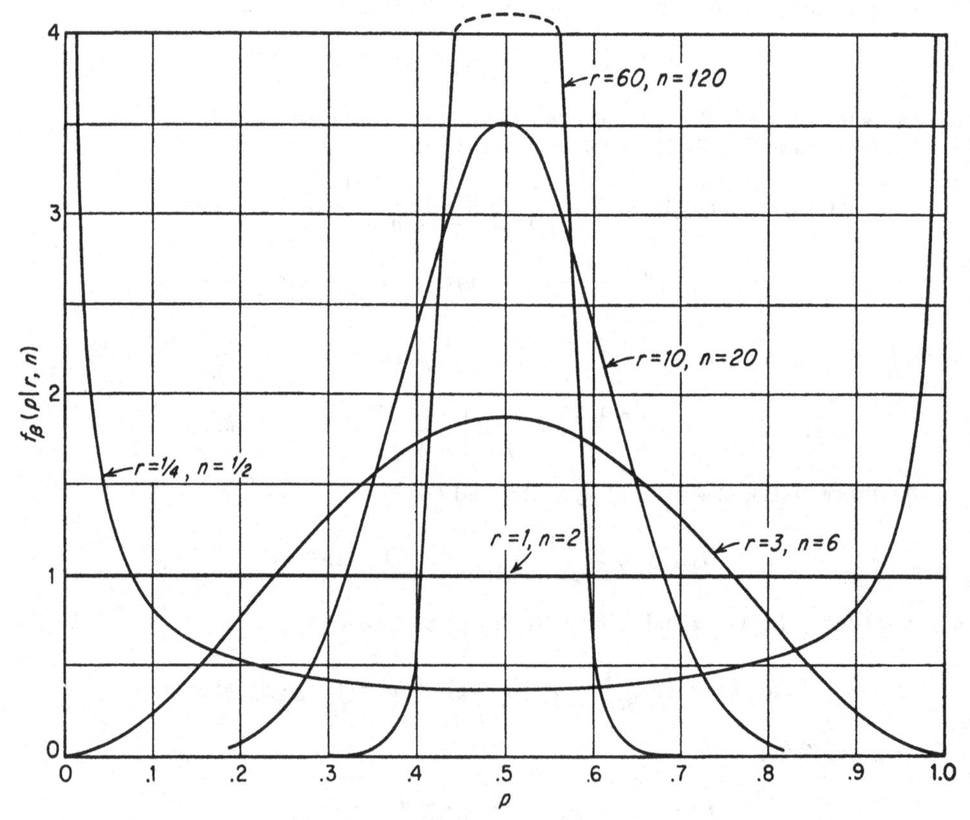

Figure 7.1

Beta Densities, $\mu_1 = .5$

$$f_\beta(z|r, n) \equiv \frac{1}{B(r, n-r)} z^{r-1}(1-z)^{n-r-1} , \qquad \begin{array}{l} 0 \le z \le 1 , \\ n > r > 0 . \end{array} \qquad (7\text{-}21)$$

Graphs of the function for selected values of the parameters are shown in Figures 7.1 and 7.2.

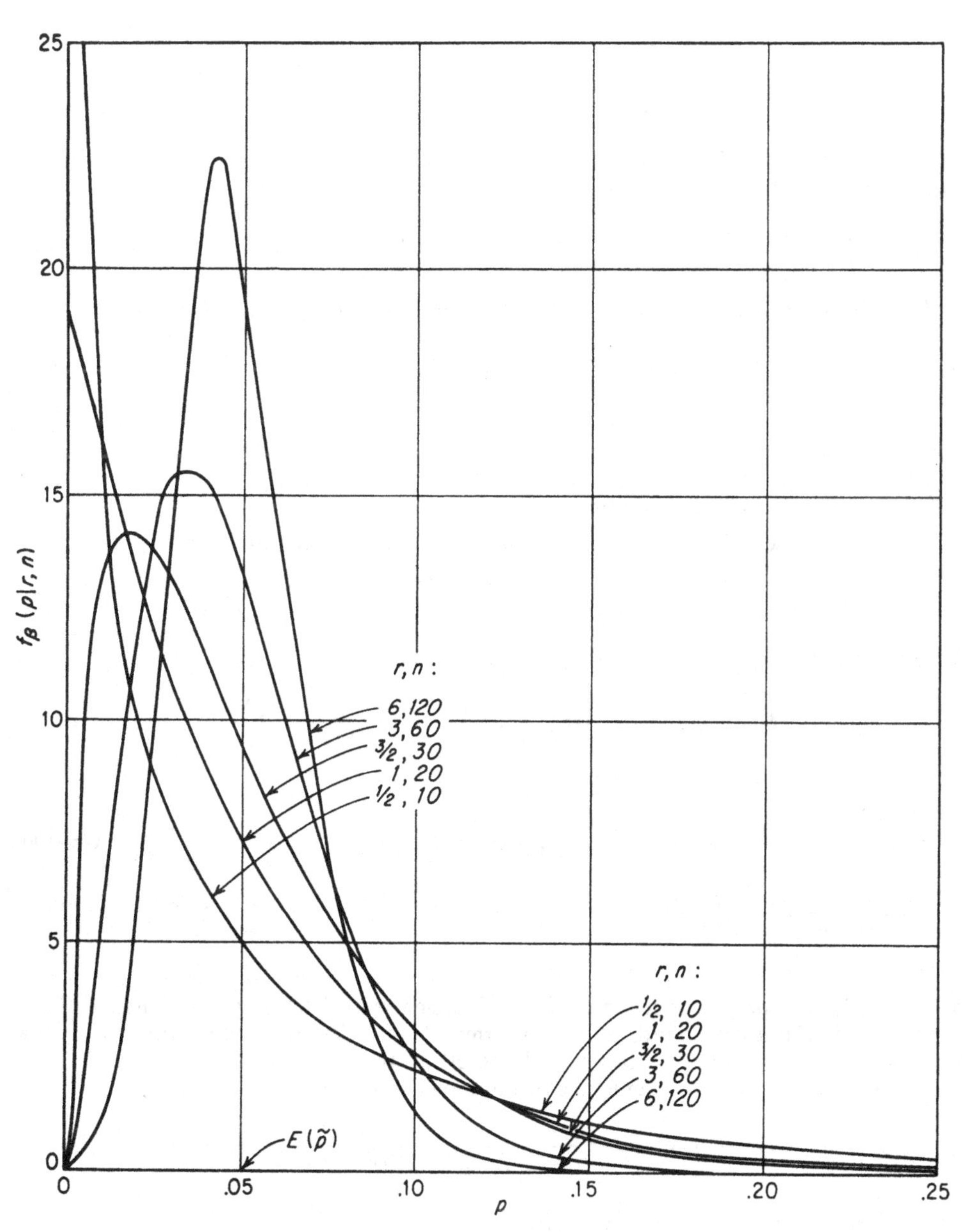

Figure 7.2

Beta Densities, $\mu_1 = .05$

This function is of course identical in essence to f_{β^*}; we have simply redefined the parameters. The moments are therefore given directly by (7-17) as

$$\mu_1 = \frac{r}{n} , \qquad \mu_1(z) = \frac{r}{n} F_\beta(z|r+1, n+1) , \tag{7-22a}$$

$$\mu_2' = \frac{r(r+1)}{n(n+1)} , \tag{7-22b}$$

$$\mu_2 = \frac{r(n-r)}{n^2(n+1)} ; \tag{7-22c}$$

and it follows from (7-18) to (7-20) that values of the cumulative function can be obtained from Pearson's tables of the beta function $I_z(p, q)$ by use of

$$\begin{aligned} F_\beta(z|r, n) &= I_z(r, n-r) , & r &\geq \tfrac{1}{2}n , & r &\leq 50 , \\ G_\beta(z|r, n) &= I_{1-z}(n-r, r) , & r &\leq \tfrac{1}{2}n , & n-r &\leq 50 , \end{aligned} \tag{7-23}$$

or from tables of the cumulative binomial function $G_b(r|p, n)$ by use of

$$\begin{aligned} F_\beta(z|r, n) &= G_b(r|z, n-1) , & z &\leq \tfrac{1}{2} , \\ G_\beta(z|r, n) &= G_b(n-r|1-z, n-1) , & z &\geq \tfrac{1}{2} . \end{aligned} \tag{7-24}$$

7.4. Inverted Beta Functions

7.4.1. *Inverted-beta-1 Function*

The inverted-beta-1 normalized density function is defined by

$$f_{i\beta 1}(y|r, n, b) \equiv \frac{1}{B(r, n-r)} \frac{(y-b)^{n-r-1} b^r}{y^n} , \qquad \begin{aligned} &0 \leq b \leq y < \infty , \\ &n > r > 0 . \end{aligned} \tag{7-25}$$

It is related to the beta functions (7-21) and (7-15) by

$$F_{i\beta 1}(y|r, n, b) = G_\beta\left(\frac{b}{y}\middle| r, n\right) = G_{\beta^*}\left(\frac{b}{y}\middle| r, n-r\right) . \tag{7-26}$$

The first two moments of $f_{i\beta 1}$ are

$$\mu_1 = b\frac{n-1}{r-1} , \qquad \mu_1(y) = b\frac{n-1}{r-1} G_\beta(b/y|r-1, n-1) , \qquad r > 1 , \tag{7-27a}$$

$$\mu_2 = b^2 \frac{(n-1)(n-r)}{(r-1)^2(r-2)} , \qquad r > 2 . \tag{7-27b}$$

▶ That $f_{i\beta 1}$ is a proper normalized density function related to f_β by (7-26) follows from the fact that it is simply an integrand transformation of f_β derived by substituting b/y for z in the definition (7-21) and multiplying the result by the Jacobian $|dz/dy| = b/y^2$.

The formulas (7-27) for the moments are found as follows.

$$\begin{aligned} \mu_1(y) &\equiv \int_b^y z\, f_{i\beta 1}(z|r, n, b)\, dz = b\frac{n-1}{r-1} \int_b^y \frac{(n-2)!}{(r-2)!(n-r-1)!} \frac{(z-b)^{n-r-1} b^{r-1}}{z^{n-1}} dz \\ &= b\frac{n-1}{r-1} F_{i\beta 1}(y|r-1, n-1, b) \end{aligned}$$

from which (7-27a) follows by (7-26).

$$\mu_2' \equiv \int_b^\infty z^2 f_{i\beta 1}(z|r, n, b)\, dz = b^2 \frac{(n-1)(n-2)}{(r-1)(r-2)} \int_b^\infty f_{i\beta 1}(z|r-2, n-2, b)\, dz \;.$$

$$\mu_2 = \mu_2' - \mu_1^2 = b^2 \frac{n-1}{r-1}\left[\frac{n-2}{r-2} - \frac{n-1}{r-1}\right] = b^2 \frac{(n-1)(n-r)}{(r-1)^2(r-2)} \;. \qquad ◀$$

7.4.2. Inverted-beta-2 Function

The inverted-beta-2 normalized density function is defined by

$$f_{i\beta 2}(y|p, q, b) \equiv \frac{1}{B(p, q)} \frac{y^{p-1}\, b^q}{(y+b)^{p+q}} \;, \qquad \begin{matrix} 0 \le y < \infty \;, \\ p, q, b > 0 \;. \end{matrix} \tag{7-28}$$

It is related to the standardized beta function (7-15) by

$$F_{i\beta 2}(y|p, q, b) = F_{\beta^*}\left(\frac{y}{y+b}\middle| p, q\right) \;. \tag{7-29}$$

The first two moments of $f_{i\beta 2}$ are

$$\mu_1 = \frac{p}{q-1}\, b \;, \qquad \mu_1(y) = \frac{p}{q-1}\, b\, F_{\beta^*}\left(\frac{y}{y+b}\middle| p+1, q-1\right) \;, \qquad q > 1 \;, \tag{7-30a}$$

$$\mu_2 = \frac{p(p+q-1)}{(q-1)^2(q-2)}\, b^2 \;, \qquad q > 2 \;. \tag{7-30b}$$

The proofs of these formulas are similar to the proofs of the corresponding formulas for $f_{i\beta 1}$.

7.4.3. F Function

The special case of the inverted-beta-2 density with parameters

$$p = \tfrac{1}{2}\nu_1 \;, \qquad q = \tfrac{1}{2}\nu_2 \;, \qquad b = \nu_2/\nu_1 \;,$$

is known as the F density:

$$f_F(y|\nu_1, \nu_2) \equiv f_{i\beta 2}(y|\tfrac{1}{2}\nu_1, \tfrac{1}{2}\nu_2, \nu_2/\nu_1) \;. \tag{7-31}$$

7.5. Poisson Function

The Poisson normalized mass function is defined by

$$f_P(r|m) \equiv \frac{e^{-m}\, m^r}{r!} \;, \qquad \begin{matrix} r = 0, 1, 2, \cdots \;, \\ m > 0 \;. \end{matrix} \tag{7-32}$$

Its first two moments are

$$\mu_1 = m \;, \qquad \mu_1(r) = m\, F_P(r-1|m) \;, \tag{7-33a}$$

$$\mu_2' = m(m+1) \;, \tag{7-33b}$$

$$\mu_2 = m \;. \tag{7-33c}$$

The function is tabulated in

E. C. Molina, *Poisson's Exponential Binomial Limit,* Van Nostrand, New York 1942: contains 6-place tables of both $f_P(x|a)$ (Table I) and $P(c, a) = G_P(c|a)$ (Table II) for

$$a = .001(.001).01(.01).3(.1)15(1)100 \; , \qquad x, c = 0(1)\infty \; .$$

Biometrika Tables for Statisticians, I: Table 39 (pages 194–202) gives $f_P(i|m)$ to 6 places for

$$m = .1(.1)15.0 \qquad i = 0(1)\infty \; ;$$

Table 7 (pages 122–129) gives $F_P(c - 1|m)$ to 5 places for

$$m = .0005(.0005).005(.005).05(.05)1(.1)5(.25)10(.5)20(1)60 \; , \qquad c = 1(1)35 \; .$$

T. Kitagawa, *Tables of Poisson Distribution*, Baifukan, Tokyo, 1952: gives $f_P(x|m)$ for $x = 0(1)\infty$ and

$m = .001(.001)1.000$, 8 places (Table 1) ,
$m = 1.01(.01)5.00$, 8 places (Table 2) ,
$m = 5.01(.01)10.00$, 7 places (Table 3) .

► That (7-32) is a proper normalized mass function follows from the facts (*a*) that it is obviously everywhere positive and (*b*) that, as is proved in books on calculus,

$$\sum_{r=0}^{\infty} \frac{m^r}{r!} = e^m \; .$$

Formulas (7-33) are proved in much the same way as (7-9):

$$\mu_1(r) \equiv \sum_{j=0}^{r} j \frac{e^{-m} m^j}{j!} = m \sum_{j=1}^{r} \frac{e^{-m} m^{j-1}}{(j-1)!} = m \sum_{i=0}^{r-1} \frac{e^{-m} m^i}{i!} = m \, F_P(r - 1|m) \; .$$

$$\mu_2' \equiv \sum_{r=0}^{\infty} (r + r[r - 1]) \frac{e^{-m} m^r}{r!} = \mu_1 + m^2 \sum_{r=2}^{\infty} \frac{e^{-m} m^{r-2}}{(r-2)!} = m + m^2 \; .$$

$$\mu_2 = \mu_2' - \mu_1^2 = m + m^2 - m^2 = m \; .$$ ◄

7.6. Gamma Functions

7.6.1. *Standardized Gamma Function*

The standardized gamma normalized density function is defined by

$$f_{\gamma^*}(z|r) \equiv \frac{e^{-z} z^{r-1}}{(r-1)!} \; , \qquad z \geq 0 \; , \quad r > 0 \; , \tag{7-34}$$

where

$$(r - 1)! \equiv \Gamma(r) \tag{7-35}$$

is the complete gamma function or generalized factorial discussed in books on advanced calculus. The first two moments of f_{γ^*} are

$$\mu_1 = r \; , \qquad \mu_1(z) = r \, F_{\gamma^*}(z|r + 1) \; , \tag{7-36a}$$

$$\mu_2' = r(r + 1) \; , \tag{7-36b}$$

$$\mu_2 = r \; . \tag{7-36c}$$

The cumulative function is tabulated in

K. Pearson, *Tables of the Incomplete Γ-Function*, Biometrika, London, 1922: contains 7-place tables of

$$I(u, p) \equiv F_{\gamma^*}(u\sqrt{p+1}|p+1) , \tag{7-37}$$
$$p = -.95(.05)0 \text{ (Table II)}, 0(.1)5(.2)50 \text{ (Table I)} ,$$
$$u = 0(.1)\infty .$$

The relation inverse to (7-37) is

$$F_{\gamma^*}(z|r) = I\left(\frac{z}{\sqrt{r}}, r-1\right), \qquad 0 < r \leq 51 . \tag{7-38}$$

Values of the cumulative gamma function may also be obtained from tables of the cumulative Poisson function by use of the relation

$$F_{\gamma^*}(z|r) = G_P(r|z) \tag{7-39}$$

which holds for all values of r, z for which G_P is defined. When used in this way, the Molina Table II cited in Section 7.5 above is equivalent to a table of

$$F_{\gamma^*}(a|c) , \qquad \begin{aligned} c &= 0(1)\infty , \\ a &= .001(.001).01(.01).3(.1)15(1)100 , \end{aligned}$$

while the *Biometrika* Table 7 also cited in Section 7.5 is equivalent to a table of

$$G_{\gamma^*}(m|c) , \qquad \begin{aligned} c &= \tfrac{1}{2}(\tfrac{1}{2})15(1)35 , \\ m &= .0005(.0005).005(.005).05(.05)1(.1)5(.25)10(.5)20(1)60. \end{aligned}$$

Beyond the range of the available tables, values of the cumulative function may be obtained from tables of the standardized Normal function by use of one of three approximations which we list in order of increasing accuracy:

$$F_{\gamma^*}(z|r) \doteq F_{N^*}(u) \begin{cases} u = \dfrac{z-r}{\sqrt{r}} , & \text{direct} , \\ u = \sqrt{4z} - \sqrt{4r-1} , & \text{Fisher} , \\ u = 3\sqrt{r}\left[\sqrt[3]{\dfrac{z}{r}} - \dfrac{9r-1}{9r}\right] , & \text{Wilson-Hilferty} . \end{cases} \tag{7-40}$$

For a discussion of the accuracy of these approximations, see M. G. Kendall and A. Stuart, *The Advanced Theory of Statistics*, Vol. I, London, Griffin, 1958, pages 371–374.

▶ That f_{γ^*} as defined in (7-34) is a proper normalized density function follows from the fact proved in books on advanced calculus that the integral

$$\int_0^\infty e^{-z} z^{r-1}\, dz \equiv \Gamma(r) \equiv (r-1)!$$

converges for $r > 0$.

Formulas (7-36) for the moments are proved in much the same way as (7-17):

$$\mu_1(z) \equiv \int_0^z y\, f_{\gamma^*}(y|r)\, dy = \int_0^z y\, \frac{e^{-y}\, y^{r-1}}{(r-1)!}\, dy = r\, F_{\gamma^*}(z|r+1) .$$

$$\mu_2' \equiv \int_0^\infty z^2 f_{\gamma^*}(z|r)\,dz = r(r+1)\int_0^\infty f_{\gamma^*}(z|r+2)\,dz = r(r+1)\ .$$
$$\mu_2 = \mu_2' - \mu_1'^2 = r(r+1) - r^2 = r\ .$$

The proof of (7-38) is obvious.
To prove (7-39) we write

$$1 - F_{\gamma^*}(z|r) = \int_z^\infty \frac{t^{r-1}}{(r-1)!} e^{-t}\,dt$$

and integrate by parts to obtain (provided that $r > 0$)

$$-\frac{t^{r-1}}{(r-1)!}e^{-t}\Big|_z^\infty + \int_z^\infty e^{-t}\frac{t^{r-2}}{(r-2)!}\,dt = \frac{e^{-z}z^{r-1}}{(r-1)!} + \int_z^\infty \frac{t^{r-2}}{(r-2)!}e^{-t}\,dt.$$

Provided that r is integral, iteration of the integration by parts yields finally

$$1 - F_{\gamma^*}(z|r) = \sum_{j=0}^{r-1}\frac{e^{-z}z^j}{j!} \equiv 1 - G_P(r|z)\ ,$$

from which (7-39) follows immediately. ◀

The *convolution* g of any n density functions $f_1, \cdots f_n$, denoted by

$$g = f_1 * f_2 * \cdots * f_n\ , \tag{7-41a}$$

is defined by

$$g(z) = \int_{R_z} f_1(x_1)\, f_2(x_2) \cdots f_n(x_n)\, dA \tag{7-41b}$$

where R_z is the $(n-1)$-dimensional hyperplane

$$x_1 + x_2 + \cdots + x_n = z\ . \tag{7-41c}$$

In the applications we shall require the following theorem on convolutions of gamma densities:

Theorem: The convolution of n standardized gamma densities with parameters $r_1, \cdots, r_i, \cdots, r_n$, is a standardized gamma density with parameter $\Sigma\, r_i$:

$$f_{\gamma^*}(\cdot|r_1) * f_{\gamma^*}(\cdot|r_2) * \cdots * f_{\gamma^*}(\cdot|r_n) = f_{\gamma^*}(\cdot|\Sigma\, r_i)\ . \tag{7-42}$$

▶ We first show that the convolution of two standardized gamma densities,

$$f_{\gamma^*}(x|a) = \frac{e^{-x}x^{a-1}}{(a-1)!}\ , \qquad f_{\gamma^*}(y|b) = \frac{e^{-y}y^{b-1}}{(b-1)!}\ ,$$

is a standardized gamma density

$$f_{\gamma^*}(z|a+b) = \frac{e^{-z}z^{a+b-1}}{(a+b-1)!}\ .$$

To show this, let g be the convolution, which since x and y are necessarily positive is

$$g(z) = \int_0^z f_{\gamma^*}(z-x|b)\, f_{\gamma^*}(x|a)\,dx = \int_0^z \frac{e^{-(z-x)}(z-x)^{b-1}}{(b-1)!}\,\frac{e^{-x}x^{a-1}}{(a-1)!}\,dx\ .$$

Substituting herein

$$x = zu\ , \qquad dx = z\,du\ ,$$

we obtain

$$g(z) = \frac{e^{-z}\, z^{a+b-1}}{(a+b-1)!} \int_0^1 \frac{(a+b-1)!}{(a-1)!(b-1)!} u^{a-1}(1-u)^{b-1}\, du$$

$$= f_{\gamma*}(z|a+b)\, F_{\beta*}(1|a, b) = f_{\gamma*}(z|a+b) \ .$$

The extension of this result to the convolution of n densities is an obvious induction. ◀

7.6.2. *Gamma-1 Function*

The gamma-1 normalized density function is defined by

$$f_{\gamma 1}(z|r, y) \equiv \frac{e^{-yz}(yz)^{r-1}}{(r-1)!}\, y \ , \qquad \begin{array}{l} z \ge 0 \ , \\ r, y > 0 \ . \end{array} \tag{7-43}$$

It is related to the standardized gamma function (7-34) by

$$F_{\gamma 1}(z|r, y) = F_{\gamma*}(yz|r) \tag{7-44a}$$

and thus to the gamma function tabulated by Pearson and to the cumulative Poisson function by

$$F_{\gamma 1}(z|r, y) = I\left(\frac{yz}{\sqrt{r}}, r-1\right) = G_P(r|yz) \ . \tag{7-44b}$$

Its first two moments are

$$\mu_1 = \frac{r}{y} \ , \qquad \mu_1(z) = \frac{r}{y} F_{\gamma*}(yz|r+1) \tag{7-45a}$$

$$\mu_2' = \frac{r(r+1)}{y^2} \ , \tag{7-45b}$$

$$\mu_2 = \frac{r}{y^2} \ . \tag{7-45c}$$

Graphs of the function are shown in Figure 7.3 for selected values of the parameter r, the parameter y being set equal to r so as to make the mean $\mu_1 = 1$.

The *convolution* of n gamma-1 densities with possibly differing parameters r_i but with the *same scale parameter* y is a gamma-1 density with parameters $\Sigma\, r_i$ and y:

$$f_{\gamma 1}(\cdot|r_1, y) * \cdots * f_{\gamma 1}(\cdot|r_i, y) * \cdots * f_{\gamma 1}(\cdot|r_n, y) = f_{\gamma 1}(\cdot|\Sigma\, r_i, y) \ . \tag{7-46}$$

▶ All results expressed by (7-43) through (7-45) follow immediately from the fact that $f_{\gamma 1}$ is an obvious integrand transformation of $f_{\gamma*}$ as defined by (7-34). The proof of (7-46) is very similar to the proof of (7-42). ◀

7.6.3. *Chi-square Function*

The special case of the gamma-1 density with parameters $r = \frac{1}{2}\nu$, $y = \frac{1}{2}$, is known as the *chi-square* density:

$$f_{\chi^2}(z|\nu) \equiv f_{\gamma 1}(z|\tfrac{1}{2}\nu, \tfrac{1}{2}) \ , \qquad \begin{array}{l} z \ge 0 \ , \\ \nu > 0 \ . \end{array} \tag{7-47}$$

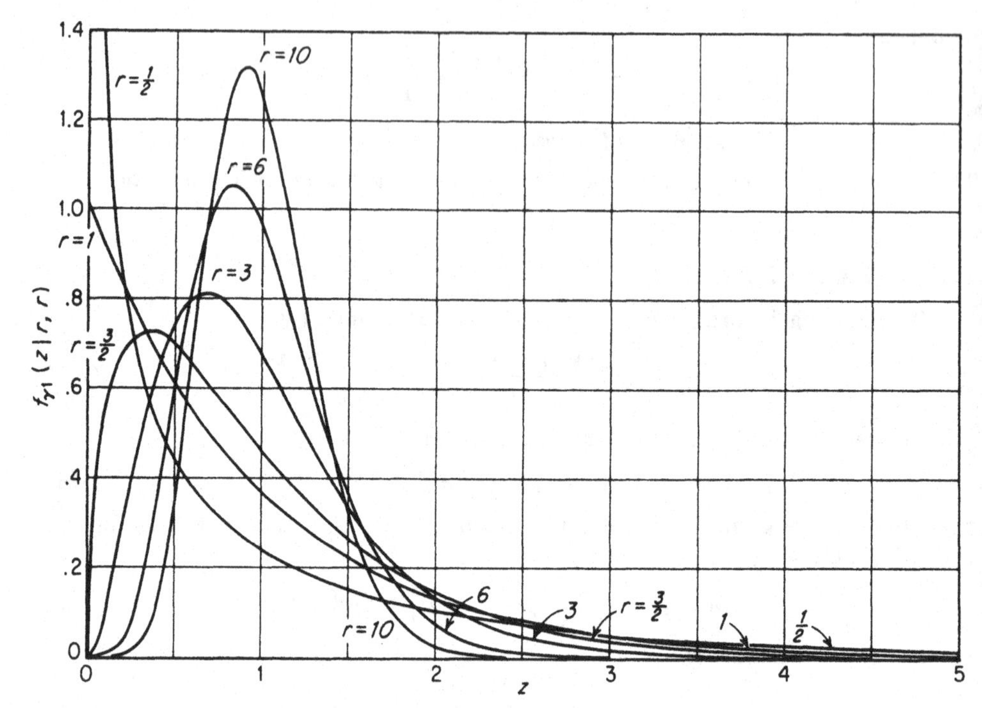

Figure 7.3

Gamma-1 Densities, $\mu_1 = 1$

It follows from (7-44) that

$$F_{\chi^2}(z|\nu) = I\left(\frac{\frac{1}{2}z}{\sqrt{\frac{1}{2}\nu}}\,,\,\tfrac{1}{2}\nu - 1\right) = G_P(\tfrac{1}{2}\nu|\tfrac{1}{2}z) \tag{7-48}$$

and from (7-45) that

$$\mu_1 = \nu\ , \qquad \mu_2 = 2\nu\ . \tag{7-49}$$

7.6.4. Gamma-2 Function

The gamma-2 normalized density function is defined by

$$f_{\gamma 2}(z|y, \nu) \equiv \frac{e^{-\frac{1}{2}\nu y z}(\frac{1}{2}\nu y z)^{\frac{1}{2}\nu - 1}}{(\frac{1}{2}\nu - 1)!}\,\tfrac{1}{2}\nu y\ , \qquad \begin{array}{l} z \geq 0\ , \\ \nu, y > 0\ . \end{array} \tag{7-50}$$

It is related to the standardized gamma function (7-34) by

$$F_{\gamma 2}(z|y, \nu) = F_{\gamma^*}(\tfrac{1}{2}\nu y z|\tfrac{1}{2}\nu) \tag{7-51a}$$

and thus to the gamma function tabulated by Pearson and to the cumulative Poisson function by

$$F_{\gamma 2}(z|y, \nu) = I(yz\sqrt{\tfrac{1}{2}\nu}, \tfrac{1}{2}\nu - 1) = G_P(\tfrac{1}{2}\nu|\tfrac{1}{2}\nu y z)\ . \tag{7-51b}$$

Its first two moments are

$$\mu_1 = \frac{1}{y}\ , \qquad \mu_1(z) = \frac{1}{y} F_{\gamma*}(\tfrac{1}{2}\nu y z|\tfrac{1}{2}\nu + 1)\ , \tag{7-52a}$$

$$\mu_2 = \frac{1}{\frac{1}{2}\nu y^2}\ . \tag{7-52b}$$

The curves in Figure 7.3 can be interpreted as graphs of $f_{\gamma 2}$ for $y = 1$ and selected values of the parameter $\nu = 2r$.

The convolution of n gamma-2 densities with the *same parameters* y *and* ν is a gamma-2 density with parameters y/n and $n\nu$:

$$f_{\gamma 2}(\cdot|y, \nu) * \cdots * f_{\gamma 2}(\cdot|y, \nu) = f_{\gamma 2}\left(\cdot \left|\frac{y}{n}, n\nu\right.\right)\ . \tag{7-53}$$

▶ All results expressed by (7-50) through (7-52) follow immediately from the fact that $f_{\gamma 2}$ is an obvious integrand transformation of $f_{\gamma*}$ as defined by (7-34). The proof of (7-53) is very similar to the proof of (7-42). ◀

7.7. Inverted Gamma Functions

7.7.1. Inverted-gamma-1 Function

The inverted-gamma-1 normalized density function is defined by

$$f_{i\gamma 1}(t|r, y) \equiv \frac{e^{-y/t}(y/t)^{r+1}}{(r-1)!}\frac{1}{y}\ , \qquad \begin{array}{l} t \geq 0\ , \\ r, y > 0\ . \end{array} \tag{7-54}$$

It is related to the gamma-1 and standardized gamma functions by

$$G_{i\gamma 1}(t|r, y) = F_{\gamma 1}(1/t|r, y) = F_{\gamma*}(y/t|r)\ . \tag{7-55}$$

Its first two moments are

$$\mu_1 = \frac{y}{r-1}\ , \qquad \mu_1(t) = \frac{y}{r-1} G_{\gamma*}(y/t|r-1)\ , \qquad r > 1\ , \tag{7-56a}$$

$$\mu_2 = \frac{y^2}{(r-1)^2(r-2)}\ , \qquad r > 2\ . \tag{7-56b}$$

This function is also known as Pearson's Type V.

▶ That $f_{i\gamma 1}$ as defined by (7-54) is a proper normalized density function follows from the fact that it is simply an integrand transformation of the gamma density $f_{\gamma 1}(z|r, y)$, derived by substituting $1/t = z$ in the definition (7-43) of $f_{\gamma 1}$ and multiplying by the Jacobian $|dz/dt| = 1/t^2$.

To prove formulas (7-56) we write

$$\mu_1(t) \equiv \int_0^t u\, f_{i\gamma 1}(u|r, y)\, du = \frac{y}{r-1}\int_0^t \frac{e^{-y/u}(y/u)^r}{(r-2)!}\frac{1}{y}\, du = \frac{y}{r-1} F_{i\gamma 1}(t|r-1, y)$$

and then use (7-55).

$$\mu_2' \equiv \int_0^\infty t^2 f_{i\gamma 1}(t|r, y)\, dt = \frac{y^2}{(r-1)(r-2)} \int_0^\infty \frac{e^{-y/t}(y/t)^{r-1}}{(r-3)!} \frac{1}{y}\, dt = \frac{y^2}{(r-1)(r-2)} \cdot$$

$$\mu_2 = \mu_2' - \mu_1^2 = \frac{y^2}{r-1}\left[\frac{1}{r-2} - \frac{1}{r-1}\right] = \frac{y^2}{(r-1)^2(r-2)} \cdot$$ ◀

7.7.2. *Inverted-gamma-2 Function*

The inverted-gamma-2 normalized density function is defined by

$$f_{i\gamma 2}(\sigma|s, \nu) \equiv \frac{2\, e^{-\frac{1}{2}\nu s^2/\sigma^2}(\frac{1}{2}\nu s^2/\sigma^2)^{\frac{1}{2}\nu + \frac{1}{2}}}{(\frac{1}{2}\nu - 1)!(\frac{1}{2}\nu s^2)^{\frac{1}{2}}} , \qquad \begin{array}{l} \sigma \geq 0 , \\ s, \nu > 0 . \end{array} \tag{7-57}$$

It is related to the gamma-2 and standardized gamma functions by

$$G_{i\gamma 2}(\sigma|s, \nu) = F_{\gamma 2}(1/\sigma^2|s^2, \nu) = F_{\gamma^*}(\tfrac{1}{2}\nu s^2/\sigma^2|\tfrac{1}{2}\nu) \tag{7-58a}$$

and thus to the gamma function (7-37) tabulated by Pearson and to the cumulative Poisson function by

$$G_{i\gamma 2}(\sigma|s, \nu) = I\left(\frac{s^2}{\sigma^2}\sqrt{\tfrac{1}{2}\nu}, \tfrac{1}{2}\nu - 1\right) = G_P(\tfrac{1}{2}\nu|\tfrac{1}{2}\nu s^2/\sigma^2) . \tag{7-58b}$$

Its first two moments are

$$\mu_1 = s\sqrt{\tfrac{1}{2}\nu}\, \frac{(\frac{1}{2}\nu - \frac{3}{2})!}{(\frac{1}{2}\nu - 1)!} , \qquad \nu > 1 , \tag{7-59a}$$

$$\mu_2 = s^2 \frac{\nu}{\nu - 2} - \mu_1^2 , \qquad \nu > 2 . \tag{7-59b}$$

The mode is at

$$\sigma_{\text{mod}} = s\sqrt{\frac{\nu}{\nu + 1}} . \tag{7-60}$$

The function is graphed in Figure 7.4 for $s = 1$ and selected values of the parameter ν. The ratio of each of the quartiles to s is graphed as a function of ν in Figure 7.5A; certain ratios between quartiles are graphed as a function of ν in Figure 7.5B.

▶ That $f_{i\gamma 2}$ as defined by (7-57) is a proper normalized density function related to the gamma-2 function by (7-58a) follows from the fact that it is simply an integrand transformation of $f_{\gamma 2}(z|s^2, \nu)$ as defined by (7-50), obtained by substituting $1/\sigma^2$ for z in the definition and multiplying by the Jacobian $|dz/d\sigma| = 2/\sigma^3$.

To prove formulas (7-59) we first make an integrand transformation of $f_{i\gamma 2}$ by substituting $\sigma = \sqrt{\frac{1}{2}\nu s^2/u}$ in (7-57) and multiplying by the Jacobian $|d\sigma/du| = (\frac{1}{2}\nu s^2)^{\frac{1}{2}}\, \frac{1}{2}u^{-\frac{3}{2}}$, thus obtaining the standardized gamma density

$$f_{\gamma^*}(u|\tfrac{1}{2}\nu) = \frac{e^{-u}\, u^{\frac{1}{2}\nu - 1}}{(\frac{1}{2}\nu - 1)!} \cdot$$

We then have

$$\mu_1 \equiv \int_0^\infty \sigma\, f_{i\gamma 2}(\sigma|s, \nu)\, d\sigma = \int_0^\infty \sqrt{\tfrac{1}{2}\nu s^2/u}\, f_{\gamma^*}(u|\tfrac{1}{2}\nu)\, du$$

$$= \sqrt{\tfrac{1}{2}\nu s^2}\,\frac{(\frac{1}{2}\nu - \frac{1}{2} - 1)!}{(\frac{1}{2}\nu - 1)!}\int_0^\infty f_{\gamma*}(u|\tfrac{1}{2}\nu - \tfrac{1}{2})\,du \;;$$

$$\mu_2' \equiv \int_0^\infty \sigma^2 f_{i\gamma 2}(\sigma|s,\nu)\,d\sigma = \int_0^\infty (\tfrac{1}{2}\nu s^2/u)\,f_{\gamma*}(u|\tfrac{1}{2}\nu)\,du$$

$$= \tfrac{1}{2}\nu s^2\,\frac{(\frac{1}{2}\nu - 1 - 1)!}{(\frac{1}{2}\nu - 1)!}\int_0^\infty f_{\gamma*}(u|\tfrac{1}{2}\nu - 1)\,du = s^2\,\frac{\frac{1}{2}\nu}{\frac{1}{2}\nu - 1} = s^2\,\frac{\nu}{\nu - 2}\;.$$

Formula (7-60) for the mode is obtained by differentiating the right-hand side of (7-57) with respect to σ and setting the result equal to 0. ◀

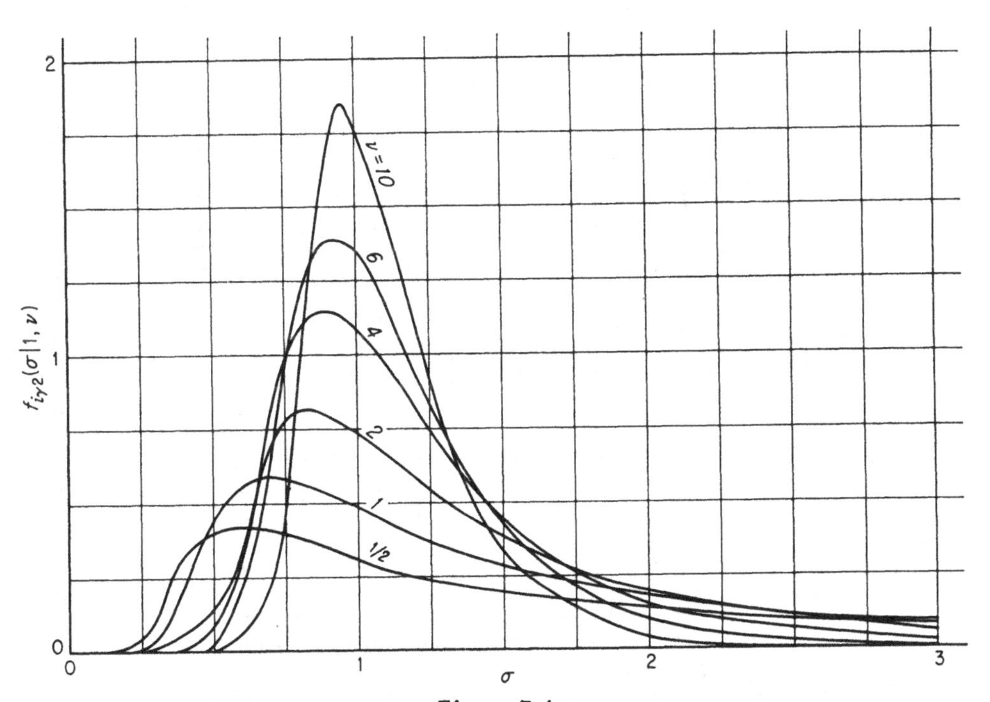

Figure 7.4

Inverted-gamma-2 Densities, s = 1

7.8. Normal Functions

7.8.1. Standardized Normal Function

The standardized or unit Normal normalized density function is defined by

$$f_{N*}(u) \equiv \frac{1}{\sqrt{2\pi}}\,e^{-\frac{1}{2}u^2}\,, \qquad -\infty < u < \infty\;. \tag{7-61}$$

Its first two moments are

$$\mu_1 = 0\,, \qquad \mu_1(u) = -f_{N*}(u)\,, \tag{7-62a}$$
$$\mu_2 = 1\;. \tag{7-62b}$$

The density and cumulative functions are very extensively tabulated in

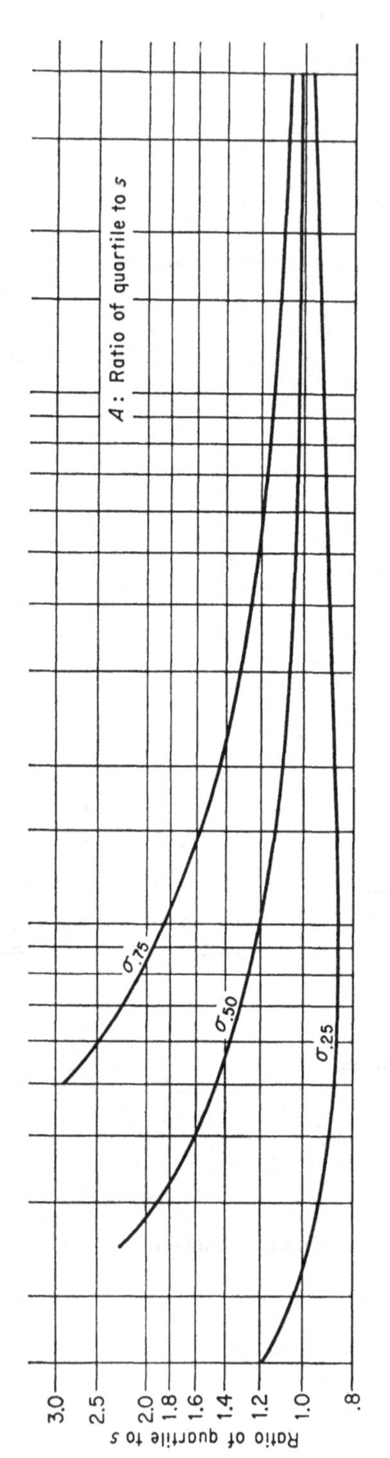

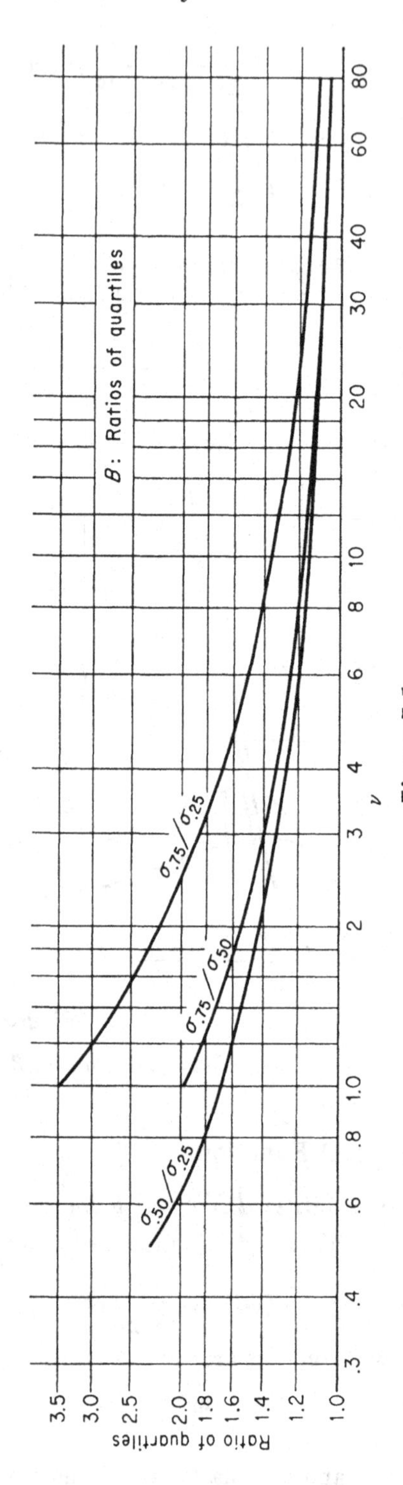

Figure 7.5

Quartiles of the Cumulative Inverted-gamma-2 Function

Tables of the Normal Probability Functions, National Bureau of Standards Applied Mathematics Series 23, U. S. Dept. of Commerce, Washington, 1953: contain 15-place tables of $f_{N*}(x)$ and $1 - 2G_{N*}(x)$ for

$$x = 0(.0001)1(.001)7.8 \ .$$

For a discussion of approximations to the cumulative function see J. T. Chu, "On Bounds for the Normal Integral," *Biometrika* 42 (1955) pages 263–265. The function

$$Hh_n(x) \equiv \int_x^\infty Hh_{n-1}(u)\, du$$

where

$$Hh_0(x) \equiv \int_x^\infty \sqrt{2\pi}\, f_{N*}(u)\, du$$

is tabulated to 10 decimal places in

British Association Mathematical Tables, Vol. I, 3rd edition, Cambridge University Press, 1951, Table XV (Pages 60–67) for

$$n = -7(1)21 \ , \qquad x = -7.0(.1)6.6 \ .$$

We shall have occasion to make use of the function

$$L_{N*}(u) \equiv f_{N*}(u) - u\, G_{N*}(u) = (2\pi)^{-\frac{1}{2}}\, Hh_1(u)$$

which is also tabulated (to 3 or 4 significant figures) in Table II at the end of this monograph for

$$u = 0(.01)4.00 \ ;$$

values for negative u can be found by means of the easily verified relation

$$L_{N*}(-u) = u + L_{N*}(u) \ .$$

► That (7-61) is a proper normalized density function is proved by (*a*) observing that it is obviously everywhere positive and (*b*) showing that its complete integral has the value 1. The latter step is equivalent to showing that

$$\int_{-\infty}^{\infty} e^{-\frac{1}{2}u^2}\, du = \sqrt{2\pi} \ ,$$

and this is accomplished by substituting

$$v = \tfrac{1}{2}u^2 \ , \qquad du = \frac{dv}{\sqrt{2v}} \ ,$$

to obtain

$$2\int_0^\infty \frac{1}{\sqrt{2}}\, e^{-v}\, v^{-\frac{1}{2}}\, dv = \sqrt{2}(-\tfrac{1}{2})!\int_0^\infty \frac{e^{-v}\, v^{\frac{1}{2}-1}}{(\frac{1}{2}-1)!}\, dv \ .$$

By (7-34), the integral on the right is $F_{\gamma*}(\infty\,|\,\frac{1}{2}) = 1$; and it is shown in books on advanced calculus that $(-\frac{1}{2})! \equiv \Gamma(\frac{1}{2}) = \sqrt{\pi}$.

Formulas (7-62) for the moments are proved as follows:

$$\mu_1(u) \equiv \int_{-\infty}^{u} z\, \frac{1}{\sqrt{2\pi}}\, e^{-\frac{1}{2}z^2}\, dz = \frac{-1}{\sqrt{2\pi}}\, e^{-\frac{1}{2}z^2}\Big|_{-\infty}^{u} = -f_{N*}(u)$$

$$\mu_2' = \int_{-\infty}^{\infty} u^2 \frac{1}{\sqrt{2\pi}} e^{-\frac{1}{2}u^2} du = \int_{-\infty}^{\infty} u \cdot \frac{1}{\sqrt{2\pi}} e^{-\frac{1}{2}u^2} u\, du ,$$

$$= -u \cdot \frac{1}{\sqrt{2\pi}} e^{-\frac{1}{2}u^2} \Big|_{-\infty}^{\infty} + \int_{-\infty}^{\infty} \frac{1}{\sqrt{2\pi}} e^{-\frac{1}{2}u^2} du = 0 + 1 = 1 .$$

$$\mu_2 = \mu_2' - \mu_1^2 = 1 - 0 = 1 .$$

◀

7.8.2. General Normal Function

The general or nonstandardized Normal normalized density function is defined by

$$f_N(z|m, H) \equiv \frac{1}{\sqrt{2\pi}} e^{-\frac{1}{2}H(z-m)^2} \sqrt{H} . \tag{7-63}$$

It is related to the standardized density (7-61) by

$$F_N(z|m, H) = F_{N*}([z - m] \sqrt{H}) . \tag{7-64}$$

The first two complete moments of f_N are

$$\mu_1 = m , \tag{7-65a}$$

$$\mu_2 = \frac{1}{H} . \tag{7-65b}$$

The incomplete first moment is most usefully expressed as

$$\mu_1(z) = m\, F_{N*}(u) - \frac{1}{\sqrt{H}} f_{N*}(u) , \qquad u = (z - m) \sqrt{H} . \tag{7-65c}$$

▶ That f_N as defined by (7-63) is a proper normalized density function follows from the obvious fact that it results from the integrand transformation $u = (z - m) \sqrt{H}$ of the standardized density (7-61).

Formulas (7-65) for the moments of f_N follow trivially from (7-62) and (7-7). ◀

B. Compound Univariate Mass and Density Functions

7.9. Student Functions

7.9.1. Standardized Student Function

The standardized Student normalized density function is defined as a gamma mixture of Normal density functions by

$$\begin{aligned} f_{S*}(t|\nu) &\equiv \int_0^\infty f_N(t|0, h)\, f_{\gamma 2}(h|1, \nu)\, dh \\ &= \frac{\nu^{\frac{1}{2}\nu}}{B(\frac{1}{2}, \frac{1}{2}\nu)} (\nu + t^2)^{-\frac{1}{2}(\nu+1)} . \end{aligned} \qquad \begin{aligned} -\infty < t < \infty , \\ \nu > 0 , \end{aligned} \tag{7-66}$$

The first two moments of f_{S*} are

$$\mu_1 = 0 \ , \qquad \mu_1(t) = -\frac{\nu + t^2}{\nu - 1} f_{S^*}(t|\nu) \ , \qquad \nu > 1 \ , \tag{7-67a}$$

$$\mu_2 = \frac{\nu}{\nu - 2} \ , \qquad \nu > 2 \ . \tag{7-67b}$$

The density function was tabulated by P. V. Sukhatme in *Sankhyā*, 1938; this table, which is reproduced as Table I at the end of this monograph, gives $f_{S^*}(t|n)$ to 6 places for

$$n = 1(1)10,\ 12,\ 15,\ 20,\ 24,\ 30,\ 60,\ \infty \ ; \qquad t = .05(.05)7.25 \ .$$

The cumulative function is tabled in

Biometrika Tables for Statisticians I: Table 9 (pages 132–134) gives $P(t, \nu) = F_{S^*}(t|\nu)$ to 5 places for

$$\nu = 1(1)20 \ , \qquad t = 0(.1)4(.2)8 \ ;$$
$$\nu = 20(1)24,\ 30,\ 40,\ 60,\ 120,\ \infty \ , \qquad t = 0(.05)2(.1)4,\ 5 \ .$$

Values of the cumulative function can also be obtained from tables of the cumulative beta function by use of the relation

$$F_{S^*}(-t|\nu) = G_{S^*}(t|\nu) = \tfrac{1}{2} F_{\beta^*}\left(\frac{\nu}{\nu + t^2}\middle| \tfrac{1}{2}\nu, \tfrac{1}{2}\right) \ , \qquad t > 0 \ . \tag{7-68}$$

For approximations to the cumulative function in terms of the Normal cumulative function, see D. L. Wallace, *Annals of Mathematical Statistics*, 30 (1959) pages 1124–1127.

▶ To evaluate the integral (7-66) we substitute for f_N and $f_{\gamma 2}$ their definitions (7-63) and (7-50), thus obtaining

$$f_{S^*}(t|\nu) = \int_0^\infty (2\pi)^{-\frac{1}{2}} e^{-\frac{1}{2}ht^2} h^{\frac{1}{2}} \cdot \frac{e^{-\frac{1}{2}\nu h}(\frac{1}{2}\nu h)^{\frac{1}{2}\nu - 1}}{(\frac{1}{2}\nu - 1)!} \tfrac{1}{2}\nu \, dh$$

$$= \frac{\nu^{\frac{1}{2}\nu}(\frac{1}{2}\nu - \frac{1}{2})!}{\pi^{\frac{1}{2}}(\frac{1}{2}\nu - 1)!} \frac{1}{(\nu + t^2)^{\frac{1}{2}\nu + \frac{1}{2}}} \int_0^\infty \frac{e^{-\frac{1}{2}h(\nu + t^2)}[\frac{1}{2}(\nu + t^2)h]^{\frac{1}{2}\nu - \frac{1}{2}}}{(\frac{1}{2}\nu - \frac{1}{2})!} \tfrac{1}{2}(\nu + t^2) \, dh \ .$$

The integral on the right is $F_{\gamma 1}(\infty|\frac{1}{2}\nu + \frac{1}{2}, \frac{1}{2}[\nu + t^2]) = 1$ and since (as is proved in books on advanced calculus)

$$\pi^{\frac{1}{2}} = \Gamma(\tfrac{1}{2}) \equiv (-\tfrac{1}{2})! \ , \qquad \frac{(\frac{1}{2}\nu - \frac{1}{2})!}{(-\frac{1}{2})!(\frac{1}{2}\nu - 1)!} = \frac{\Gamma(\frac{1}{2}\nu + \frac{1}{2})}{\Gamma(\frac{1}{2})\Gamma(\frac{1}{2}\nu)} = \frac{1}{B(\frac{1}{2}, \frac{1}{2}\nu)} \ ,$$

the factors preceding this integral may be written as in (7-66).

To show that f_{S^*} as defined by (7-66) is a proper normalized density function we first observe that it is everywhere positive because f_N and $f_{\gamma 2}$ are proper density functions and are therefore everywhere positive. That the complete integral exists for $\nu > 0$ is proved by the fact that $(\nu + t^2)^{-\frac{1}{2}(\nu+1)}$ is of order $t^{-\nu-1}$ at infinity; that the integral has the value 1 can then be shown most conveniently by using the integral definition of f_{S^*} and then reversing the order of integration:

$$\int_{-\infty}^{\infty} f_{S^*}(t|\nu) \, dt = \int_{-\infty}^{\infty} \int_0^\infty f_N(t|0, h) \, f_{\gamma 2}(h|1, \nu) \, dh \, dt$$

$$= \int_0^\infty \left[\int_{-\infty}^{\infty} f_N(t|0, h) \, dt\right] f_{\gamma 2}(h|1, \nu) \, dh = \int_0^\infty f_{\gamma 2}(h|1, \nu) \, dh = 1 \ .$$

To find the complete moments we first observe that the kth moment exists if $\nu > k$ because $t^k(\nu + t^2)^{-\frac{1}{2}(\nu+1)}$ is of order $t^{k-\nu-1}$ at infinity. Provided that this condition is met, it is apparent by symmetry that all complete odd moments about the origin are 0 and hence that all complete odd moments about the mean are 0.

Complete even moments about the mean are most conveniently evaluated by using the integral definition of f_{S^*}, reversing the order of integration, and then using formulas (7-65) for the moments of f_N. Using μ_k^* to denote the kth moment of f_N we thus obtain

$$\mu_k \equiv \int_{-\infty}^{\infty} (t - \mu_1)^k f_{S^*}(t|\nu)\, dt = \int_{-\infty}^{\infty} t^k f_{S^*}(t|\nu)\, dt$$

$$= \int_{-\infty}^{\infty} \int_0^{\infty} t^k f_N(t|0, h)\, f_{\gamma 2}(h|1, \nu)\, dh\, dt = \int_0^{\infty} \left[\int_{-\infty}^{\infty} t^k f_N(t|0, h)\, dt\right] f_{\gamma 2}(h|1, \nu)\, dh$$

$$= \int_0^{\infty} \mu_k^* f_{\gamma 2}(h|1, \nu)\, dh = \int_0^{\infty} \mu_k^* \frac{e^{-\frac{1}{2}\nu h}(\frac{1}{2}\nu h)^{\frac{1}{2}\nu - 1}}{(\frac{1}{2}\nu - 1)!} \tfrac{1}{2}\nu\, dh\ .$$

Substituting on the right

$$u = \tfrac{1}{2}\nu h\ , \qquad du = \tfrac{1}{2}\nu\, dh\ ,$$

we have finally

$$\mu_k = \int_0^{\infty} \mu_k^* \frac{e^{-u}\, u^{\frac{1}{2}\nu - 1}}{(\frac{1}{2}\nu - 1)!}\, du = \int_0^{\infty} \mu_k^* f_{\gamma^*}(u|\tfrac{1}{2}\nu)\, du\ .$$

We now apply this result to prove formula (7-67b) for the second moment about the mean. By (7-65b) we have

$$\mu_2^* = h^{-1} = \tfrac{1}{2}\nu u^{-1}\ ,$$

and therefore

$$\mu_2 = \tfrac{1}{2}\nu \int_0^{\infty} u^{-1} f_{\gamma^*}(u|\tfrac{1}{2}\nu)\, du = \tfrac{1}{2}\nu \frac{(\frac{1}{2}\nu - 2)!}{(\frac{1}{2}\nu - 1)!} \int_0^{\infty} f_{\gamma^*}(u|\tfrac{1}{2}\nu - 1)\, du = \frac{\nu}{\nu - 2}\ .$$

Formula (7-67a) for the incomplete first moment is given by

$$\mu_1(t) \equiv \int_{-\infty}^{t} z\, f_{S^*}(z|\nu)\, dz = \frac{\nu^{\frac{1}{2}\nu}}{B(\frac{1}{2}, \frac{1}{2}\nu)} \int_{-\infty}^{t} (\nu + z^2)^{-\frac{1}{2}\nu - \frac{1}{2}}\, z\, dz$$

$$= \frac{\nu^{\frac{1}{2}\nu}}{B(\frac{1}{2}, \frac{1}{2}\nu)} \frac{(\nu + z^2)^{-\frac{1}{2}\nu + \frac{1}{2}}}{2(-\frac{1}{2}\nu + \frac{1}{2})}\Bigg|_{-\infty}^{t} = -\frac{\nu + t^2}{\nu - 1} f_{S^*}(t|\nu)$$

provided that $\nu > 1$.

The relation (7-68) is proved by writing

$$G_{S^*}(t|\nu) \equiv \int_t^{\infty} \frac{\nu^{\frac{1}{2}\nu}}{B(\frac{1}{2}, \frac{1}{2}\nu)} (\nu + u^2)^{-\frac{1}{2}\nu - \frac{1}{2}}\, du\ , \qquad t > 0\ ,$$

and then substituting

$$u = \left[\frac{\nu z}{1 - z}\right]^{\frac{1}{2}}, \qquad du = \frac{1}{2}\left[\frac{\nu z}{1 - z}\right]^{-\frac{1}{2}} \frac{\nu}{(1 - z)^2}\, dz\ ,$$

to obtain

$$\int_{z^*}^{1} \frac{1}{2} \frac{1}{B(\frac{1}{2}, \frac{1}{2}\nu)} z^{\frac{1}{2} - 1}(1 - z)^{\frac{1}{2}\nu - 1}\, dz$$

where

$$z^* \equiv \frac{t^2}{\nu + t^2}\ .$$

This integral is $\frac{1}{2}G_{\beta^*}(z^*|\frac{1}{2}, \frac{1}{2}\nu)$ by the definition (7-15); and (7-68) then follows by (7-18) and (7-19). ◀

7.9.2. *General Student Function*

The general or nonstandardized Student normalized density function is defined by

$$f_S(z|m, H, \nu) \equiv \int_0^\infty f_N(z|m, hH)\, f_{\gamma 2}(h|1, \nu)\, dh$$

$$= \frac{\nu^{\frac{1}{2}\nu}}{B(\frac{1}{2}, \frac{1}{2}\nu)} [\nu + H(z - m)^2]^{-\frac{1}{2}(\nu+1)} \sqrt{H} \;, \qquad \begin{matrix} -\infty < z < \infty \\ H, \nu > 0 \;. \end{matrix} \tag{7-69}$$

It is related to the standardized density (7-66) by

$$F_S(z|m, H, \nu) = F_{S^*}([z - m]\sqrt{H}|\nu) \;. \tag{7-70}$$

The first two complete moments of f_S are

$$\mu_1 = m \;, \qquad \nu > 1 \;, \tag{7-71a}$$

$$\mu_2 = \frac{1}{H}\frac{\nu}{\nu - 2} \;, \qquad \nu > 2 \;. \tag{7-71b}$$

The incomplete first moment for $\nu > 1$ is most usefully expressed as

$$\mu_1(z) = m\, F_{S^*}(t|\nu) - \frac{1}{\sqrt{H}}\frac{\nu + t^2}{\nu - 1} f_{S^*}(t|\nu) \;, \qquad t \equiv (z - m)\sqrt{H} \;. \tag{7-71c}$$

The following formula will be required in the applications:

$$\int_0^\infty f_N(z|m, hn)\, f_{\gamma 2}(h|v, \nu)\, dh = f_S(z|m, n/v, \nu) \;. \tag{7-72}$$

▶ To evaluate the integral (7-69) we substitute for f_N and $f_{\gamma 2}$ their definitions (7-63) and (7-50), thus obtaining

$$f_S(z) = \int_0^\infty (2\pi)^{-\frac{1}{2}} e^{-\frac{1}{2}hH(z-m)^2} h^{\frac{1}{2}} H^{\frac{1}{2}} \cdot \frac{e^{-\frac{1}{2}\nu h}(\frac{1}{2}\nu h)^{\frac{1}{2}\nu - 1}}{(\frac{1}{2}\nu - 1)!} \tfrac{1}{2}\nu\, dh$$

$$= \frac{H^{\frac{1}{2}}(\frac{1}{2}\nu)^{\frac{1}{2}\nu}}{(\frac{1}{2}\nu - 1)!(2\pi)^{\frac{1}{2}}} \int_0^\infty e^{-\frac{1}{2}h(H[z-m]^2+\nu)} h^{\frac{1}{2}\nu + \frac{1}{2} - 1}\, dh \;.$$

Substituting herein

$$u = \tfrac{1}{2}h(H[z - m]^2 + \nu)$$

we obtain

$$f_S(z) = \frac{H^{\frac{1}{2}}(\frac{1}{2}\nu)^{\frac{1}{2}\nu}(\frac{1}{2}\nu + \frac{1}{2} - 1)!}{(2\pi)^{\frac{1}{2}}(\frac{1}{2}\nu - 1)!} [\tfrac{1}{2}(\nu + H[z - m]^2)]^{-\frac{1}{2}\nu - \frac{1}{2}} \int_0^\infty \frac{e^{-u} u^{\frac{1}{2}\nu + \frac{1}{2} - 1}}{(\frac{1}{2}\nu + \frac{1}{2} - 1)!} du \;;$$

and since the integral is $F_{\gamma^*}(\infty|\frac{1}{2}\nu + \frac{1}{2}) = 1$, this expression can be reduced to (7-69).

To prove the relation (7-70) with the standardized function, substitute $t = (z - m)\sqrt{H}$ in the right-hand member of (7-69) and multiply by the Jacobian $|dz/dt| = 1/\sqrt{H}$.

Formulas (7-71) follow from formulas (7-67) for the standardized moments via the relation (7-70) and (7-7).

To prove (7-72), substitute $h = y/v$ in the definitions of f_N and $f_{\gamma 2}$, multiply by the Jacobian $|dh/dy| = 1/v$ to obtain

$$\int_0^\infty f_N(z|m, hn)\, f_{\gamma 2}(h|v, \nu)\, dh = \int_0^\infty f_N(z|m, yn/v)\, f_{\gamma 2}(y|1, \nu)\, dy \; ,$$

and apply (7-69). ◀

7.10. Negative-Binomial Function

The negative-binomial normalized mass function is defined by

$$f_{nb}(r|p, r') \equiv \int_0^\infty f_P(r|pz)\, f_{\gamma 1}(z|r', 1-p)\, dz$$

$$= \frac{(r+r'-1)!}{r!(r'-1)!} p^r (1-p)^{r'} \; , \qquad \begin{array}{l} r = 0, 1, 2 \cdots , \\ r' > 0 \; , \\ 0 < p < 1 \; . \end{array} \tag{7-73}$$

The first two complete moments are

$$\mu_1 = r' \frac{p}{1-p} \; , \tag{7-74a}$$

$$\mu_2 = r' \frac{p}{(1-p)^2} \; . \tag{7-74b}$$

The negative-binomial function is related to the beta and binomial functions by

$$G_{nb}(r|p, r') = F_{\beta *}(p|r, r') = G_b(r|p, r+r'-1) \; . \tag{7-75}$$

▶ That (7-73) is a proper normalized mass function is proved by (*a*) observing that the integrand and therefore the integral is everywhere positive and (*b*) showing that the complete sum of f_{nb} has the value 1. The latter step is accomplished by interchanging summation and integration:

$$\sum_{r=0}^\infty \int_0^\infty f_P(r|pz)\, f_{\gamma 1}(z|r', 1-p)\, dz = \int_0^\infty \left[\sum_{r=0}^\infty f_P(r|pz)\right] f_{\gamma 1}(z|r', 1-p)\, dz$$

$$= \int_0^\infty f_{\gamma 1}(z|r', 1-p)\, dz = 1 \; .$$

The formula on the right of (7-73) is obtained by substituting formulas (7-32) and (7-43) in the integral, factoring,

$$\int_0^\infty \frac{e^{-pz}(pz)^r}{r!} \frac{e^{-z(1-p)}(z[1-p])^{r'-1}}{(r'-1)!} (1-p)\, dz$$

$$= \frac{(r+r'-1)!}{r!(r'-1)!} p^r (1-p)^{r'} \int_0^\infty \frac{e^{-z} z^{r+r'-1}}{(r+r'-1)!} dz$$

and observing that the integral on the right is $F_{\gamma *}(\infty|r+r'-1) = 1$.

To prove (7-74a) we again interchange summation and integration and in addition make use of formulas (7-33a) for the mean of f_P and (7-45a) for the mean of $f_{\gamma 1}$:

$$\sum_{r=0}^\infty r \int_0^\infty f_P(r|pz)\, f_{\gamma 1}(z|r', 1-p)\, dz = \int_0^\infty \left[\sum_{r=0}^\infty r\, f_P(r|pz)\right] f_{\gamma 1}(z|r', 1-p)\, dz$$

$$= \int_0^\infty pz\, f_{\gamma 1}(z|r', 1-p)\, dz = \frac{p}{1-p} r' \; .$$

The second moment about the origin is similarly obtained by use of (7-33b) and (7-45ab):

$$\sum_{r=0}^{\infty} r^2 \int_0^\infty f_P(r|pz)\, f_{\gamma 1}(z|r', 1-p)\, dz = \int_0^\infty \left[\sum_{r=0}^{\infty} r^2 f_P(r|pz)\right] f_{\gamma 1}(z|r', 1-p)\, dz$$

$$= \int_0^\infty ([pz]^2 + pz)\, f_{\gamma 1}(z|r', 1-p)\, dz = \frac{p^2}{(1-p)^2} r'(r'+1) + \frac{p}{1-p} r' .$$

The second moment about the mean is then

$$\frac{p^2 r'(r'+1)}{(1-p)^2} + \frac{pr'}{1-p} - \left[\frac{pr'}{1-p}\right]^2 = r' \frac{p}{(1-p)^2} .$$

To prove (7-75) we use (7-73) and (7-39) and then (7-34) and (7-43) to write

$$G_{nb}(r|p, r') = \int_0^\infty G_P(r|pz)\, f_{\gamma 1}(z|r', 1-p)\, dz = \int_0^\infty F_{\gamma*}(pz|r)\, f_{\gamma 1}(z|r', 1-p)\, dz$$

$$= \int_0^\infty \left[\int_0^{pz} \frac{e^{-t}\, t^{r-1}}{(r-1)!}\, dt\right] \frac{e^{-(1-p)z}([1-p]z)^{r'-1}}{(r'-1)!} (1-p)\, dz .$$

Substituting herein

$$t = uv , \qquad z = \frac{v(1-u)}{1-p} , \qquad dt\, dz = \frac{v}{1-p}\, du\, dv ,$$

we obtain

$$G_{nb}(r|p, r') = \frac{(r+r'-1)!}{(r-1)!(r'-1)!} \int_0^p \left[\int_0^\infty \frac{e^{-v}\, v^{r+r'-1}}{(r+r'-1)!}\, dv\right] u^{r-1}(1-u)^{r'-1}\, du .$$

The integral with respect to v is $F_{\gamma*}(\infty|r+r') = 1$, and therefore

$$G_{nb}(r|p, r') = \int_0^p \frac{1}{B(r, r')} u^{r-1}(1-u)^{r'-1}\, du \equiv F_{\beta*}(p|r, r') .$$

The relation to G_b then follows from (7-20). ◀

7.11. Beta-binomial and Beta-Pascal Functions

The *beta-binomial* normalized mass function is defined by

$$f_{\beta b}(r|r', n', n) \equiv \int_0^1 f_b(r|p, n)\, f_\beta(p|r', n')\, dp \qquad r = 0, 1, 2, \cdots ,$$

$$= \frac{(r+r'-1)!(n+n'-r-r'-1)!\, n!(n'-1)!}{r!(r'-1)!(n-r)!(n'-r'-1)!(n+n'-1)!} , \qquad n = 1, 2, \cdots , \quad n \ge r , \quad n' > r' > 0 . \tag{7-76}$$

The first two complete moments are

$$\mu_1 = n \frac{r'}{n'} , \tag{7-77a}$$

$$\mu_2 = n(n+n') \frac{r'(n'-r')}{n'^2(n'+1)} . \tag{7-77b}$$

Notice that the function which assigns equal mass to $r = 0, 1, \cdots, n$ is a special case of the beta-binomial function:

$$f_{\beta b}(r|1, 2, n) = \frac{1}{n+1} .$$

The *beta-Pascal* normalized mass function is defined by

$$f_{\beta Pa}(n|r', n', r) \equiv \int_0^1 f_{Pa}(n|p, r)\, f_\beta(p|r', n')\, dp$$

$$= \frac{(r+r'-1)!(n+n'-r-r'-1)!(n-1)!(n'-1)!}{(r-1)!(r'-1)!(n-r)!(n'-r'-1)!(n+n'-1)!}\,, \qquad \begin{aligned} &n, r = 1, 2, \cdots, \\ &n \geq r\,, \\ &n' > r' > 0\,. \end{aligned} \tag{7-78}$$

The first two complete moments are

$$\mu_1 = r\,\frac{n'-1}{r'-1}\,, \tag{7-79a}$$

$$\mu_2 = r(r+r'-1)\,\frac{(n'-1)(n'-r')}{(r'-1)^2(r'-2)}\,. \tag{7-79b}$$

The beta-binomial and beta-Pascal cumulative functions are related by

$$F_{\beta Pa}(n|r', n', r) = G_{\beta b}(r|r', n', n)\,. \tag{7-80}$$

▶ The proofs that (7-76) and (7-78) are proper normalized mass functions are similar to the proof that (7-73) is a proper normalized mass function; and the proofs of the moment formulas (7-77) and (7-79) are similar to the proofs of (7-74).

The relation (7-80) between the cumulative beta-binomial and beta-Pascal functions follows directly from the corresponding relation (7-13) between the binomial and Pascal functions:

$$F_{\beta Pa}(n|r', n', r) \equiv \int_0^1 F_{Pa}(n|p, r)\, f_\beta(p|r', n')\, dp$$

$$= \int_0^1 G_b(r|p, n)\, f_\beta(p|r', n')\, dp \equiv G_{\beta b}(r|r', n', n)\,. \quad ◀$$

7.11.1. *Relations with the Hypergeometric Function*

An interesting and useful relation exists between the beta-binomial and beta-Pascal functions and the hypergeometric function defined by

$$f_h(s|S, F, \nu) \equiv \frac{\nu!(S+F-\nu)!\,S!\,F!}{s!(\nu-s)!(S-s)!(F-\nu+s)!(S+F)!}\,, \qquad \begin{aligned} &r, S, F, \nu = 0, 1, 2, \cdots \\ &S, \nu \geq s \\ &F \geq \nu - s\,. \end{aligned} \tag{7-81}$$

When the parameters r' and n' of the beta-binomial or beta-Pascal functions are integral, these functions are related to the hypergeometric function by

$$F_{\beta Pa}(n|r', n', r) = G_{\beta b}(r|r', n', n) = G_h(r|n, n'-1, r+r'-1)\,. \tag{7-82}$$

▶ The relation (7-82) can be proved through an interpretation in terms of probabilities. Writing (7-81) in the form

$$f_h(s|S, F, \nu) = \frac{C_s^S\, C_{\nu-s}^F}{C_\nu^{S+F}}\,,$$

we see that $f_h(s|S, F, \nu)$ gives the probability that there will be exactly s successes in a sample of ν items drawn in simple sampling without replacement from a finite population containing S successes and F failures, since C_s^S is the number of ways in which s successes can be chosen among S successes, $C^F_{\nu-s}$ is the number of ways in which $\nu - s$ failures can be chosen among F failures, and C^{S+F}_ν is the number of ways in which ν items can be chosen among $S + F$ items.

The probability that there will be exactly s successes *before the* f*th failure* in this kind of sampling will be called a "negative-hypergeometric" probability by analogy with one interpretation of the negative-binomial function. This probability is equal to the product of (1) the probability of exactly s successes in the first $s + f - 1$ drawings, times (2) the conditional probability, given the event just described, of a failure on the next draw. Using (7-81) we therefore define

$$f_{nh}(s|S, F, f) = \frac{(s + f - 1)!(S + F - s - f + 1)!\,S!\,F!}{s!(f - 1)!(S - s)!(F - f + 1)!(S + F)!} \cdot \frac{F - f + 1}{S + F - s - f + 1} \cdot$$

We next observe that, since there will be *at least* r successes before the fth failure if and only if there are at least r successes in the first $r + f - 1$ drawings, the *cumulative* hypergeometric and negative-hypergeometric functions are related by

$$G_{nh}(r|S, F, f) = G_h(r|S, F, r + f - 1) \ .$$

Now finally, writing (7-76) in the form

$$f_{\beta b}(r|r', n', n) \equiv \frac{(r + r' - 1)!(n + n' - r - r')!\,n!(n' - 1)!}{r!(r' - 1)!(n - r)!(n' - r')!(n + n' - 1)!} \cdot \frac{n' - r'}{n + n' - r - r'}$$

and comparing this with the definition of f_{nh} given just above, we see that

$$f_{\beta b}(r|r', n', n) = f_{nh}(r|n, n' - 1, r')$$

and therefore that

$$G_{\beta b}(r|r', n', n) = G_{nh}(r|n, n' - 1, r') = G_h(r|n, n' - 1, r + r' - 1)$$

as asserted by (7-82). The relation between G_h and $F_{\beta Pa}$ also asserted by (7-82) then follows from (7-80). ◀

7.11.2. *Computation of the Cumulative Beta-binomial and Beta-Pascal Functions*

The cumulative functions $F_{\beta b}$, $G_{\beta b}$, $F_{\beta Pa}$, and $G_{\beta Pa}$ have not been tabulated. Any one of the four can be evaluated by term-by-term computation of the values of $f_{\beta b}$ or $f_{\beta Pa}$ included in it or—in the light of (7-80)—in any one of the other three.

Term-by-term computations will be simplified by the use of one of the recursion relations

$$\begin{aligned} f_{\beta b}(r|r', n', n) &= \frac{(r' - 1 + r)(n + 1 - r)}{r(n + n' - r' - r)} f_{\beta b}(r - 1|r', n', n) \\ &= \frac{(1 + r)(n + n' - r' - 1 - r)}{(r' + r)(n - r)} f_{\beta b}(r + 1|r', n', n) \ , \\ f_{\beta Pa}(n|r', n', r) &= \frac{(n - 1)(n + n' - r - r' - 1)}{(n + n' - 1)(n - r)} f_{\beta Pa}(n - 1|r', n', r) \\ &= \frac{(n + n')(n + 1 - r)}{n(n + n' - r - r')} f_{\beta Pa}(n + 1|r', n', r) \ . \end{aligned} \qquad (7\text{-}83)$$

As a base for the recursion, one term must be evaluated by use of the complete formula (7-76) for $f_{\beta b}$ or (7-78) for $f_{\beta Pa}$ and this is usually most easily accomplished by use of tables of $n!$ or of $\Gamma(n) = (n-1)!$ The most complete tabulation is found in

H. T. Davis, *Tables of the Higher Mathematical Functions*, Principia, Bloomington, 1933, vol. I: gives to 10 or more decimal places

$\Gamma(x)$:	$x = 1.0000(.0001)1.1000$	(Table 1) ,
	$1.100(.001)2.000$	(Table 2) ;
$\log_{10}\Gamma(x)$:	$x = 1.0000(.0001)1.1000$	(Table 1) ,
	$1.100(.001)2.000$	(Table 2) ,
	$2.00(.01)11.00$	(Table 3) ,
	$11.0(.1)101.0$	(Table 4) .

For many purposes a more convenient table is

J. Brownlee, *Tracts for Computers* (ed. K. Pearson) no. IX, Cambridge University Press, 1923: gives $\log\Gamma(x)$ to 7 decimal places for

$$x = 1.00(.01)50.99.$$

Computation of the beta-binomial function is still more easily accomplished by rewriting (7-76) in the form

$$f_{\beta b}(r|r', n', n) = \frac{C_r^{r+r'-1}\, C_{n-r}^{n+n'-r-r'-1}}{C_n^{n+n'-1}} \tag{7-84}$$

and using tables of log combinatorials—e.g., J. W. Glover, *Tables of Applied Mathematics*, George Wahr, Ann Arbor, 1923. The corresponding simplification for the beta-Pascal function is to write (7-78) in the form

$$f_{\beta Pa}(n|r', n', r) = \frac{B(n, n')}{(n-r)\, B(r, r')\, B(n-r, n'-r')} \tag{7-85}$$

and obtain the values of the complete beta functions from tables—e.g., Pearson's *Tables of the Incomplete Beta-Function.*

In evaluating $F_{\beta b}$ it is tempting to choose

$$f_{\beta b}(0|r', n', n) = \frac{(n+n'-r'-1)!(n'-1)!}{(n'-r'-1)!(n+n'-1)!}$$

as the first term to evaluate because this term contains only 4 factorials whereas the general term (7-76) contains 9; and for the same reason it is tempting to start with $f_{\beta b}(n)$ in the case of $G_{\beta b}$ or to start with $f_{\beta Pa}(r)$ in the case of $F_{\beta Pa}$. It must be remembered, however, that such extreme-tail values will usually contribute an extremely small fraction of the total value of the cumulative function, so that starting in this way makes it necessary to carry a very large number of decimals.

When r' is integral and smaller than either r or $n - r$, evaluation of the cumulative beta-binomial or beta-Pascal function will be simplified by using (7-82), since term-by-term computation of G_h involves only r' terms of type (7-81) whereas direct evaluation of $F_{\beta b}$ involves r terms, $F_{\beta Pa}$ or $G_{\beta Pa}$ involves $n - r$ terms, and $G_{\beta Pa}$ involves (in principle) an infinite number of terms.

The relation (7-82) also enables us to find beta and binomial approximations to the beta-binomial and beta-Pascal cumulative functions and determine the conditions under which these approximations will be accurate. Provided that the conditions

$$r \ll n \;, \qquad r' \ll n' \;, \qquad r + r' \ll \max\{n, n'\} \tag{7-86}$$

are *all* met, very good accuracy will be obtained from the binomial approximations

$$\begin{aligned} F_{\beta Pa}(n|r', n', r) &= G_{\beta b}(r|r', n', n) \\ &\doteq G_b\left(r \left| \frac{n}{n + n' - 1}, r + r' - 1\right.\right), \qquad n \leq n' - 1 \;, \\ &\doteq 1 - G_b\left(r' \left| \frac{n' - 1}{n + n' - 1}, r + r' - 1\right.\right), \qquad n \geq n' - 1 \;, \end{aligned} \tag{7-87}$$

or the equivalent beta approximations

$$\begin{aligned} F_{\beta Pa}(n|r', n', r) &= G_{\beta b}(r|r', n', n) \\ &\doteq F_{\beta *}\left(\left.\frac{n}{n + n' - 1}\right| r, r'\right), \qquad r \geq r' + 1 \;, \\ &\doteq 1 - F_{\beta *}\left(\left.\frac{n' - 1}{n + n' - 1}\right| r', r\right), \qquad r \leq r' + 1 \;. \end{aligned} \tag{7-88}$$

► The conditions (7-86) for the validity of the binomial and beta approximations derive from the fact that the binomial approximation to an individual value of the hypergeometric mass function,

$$f_h(s|S, F, \nu) \doteq f_b\left(s \left| \frac{S}{S + F}, \nu\right.\right),$$

will be good if $s \ll S$ *and* $\nu - s \ll F$. All the values included in $F_h(r)$ will therefore be good if $r \ll S$ and $\nu \ll F$; all the values included in $G_h(r)$ will be good if $\nu \ll S$ and $\nu - r \ll F$. The approximations to *both* F_h and G_h will be good if these conditions are met for *either* one. It then follows from (7-82) that the binomial or beta approximation to the beta-binomial and beta-Pascal cumulative functions will be good if either (1) $r \ll n$ and $r + r' \ll n' - 1$ or (2) $r + r' \ll n$ and $r' \ll n' - 1$. At least one of these two pairs of conditions will be satisfied if, and neither will be satisfied unless, $r \ll n$ *and* $r' \ll n'$ *and* $r + r' \ll \max\{n, n'\}$. ◄

CHAPTER 8

Multivariate Normalized Density Functions

8.0. Introduction

8.0.1. Matrix and Vector Notation

Matrices will be denoted by **boldface roman** letters, either upper or lower-case, or by boldface vertical Greek letters, e.g. $\boldsymbol{\alpha}$, $\boldsymbol{\beta}$, $\boldsymbol{\gamma}$, $\boldsymbol{\epsilon}$, $\boldsymbol{\zeta}$, $\boldsymbol{\theta}$, $\boldsymbol{\mu}$, $\boldsymbol{\nu}$, $\boldsymbol{\xi}$, $\boldsymbol{\sigma}$. Column vectors will be denoted by ***boldface italic*** letters, either upper or lower case, or by boldface Porson Greek letters, e.g. $\boldsymbol{\alpha}$, $\boldsymbol{\beta}$, $\boldsymbol{\gamma}$, $\boldsymbol{\epsilon}$, $\boldsymbol{\zeta}$, $\boldsymbol{\theta}$, $\boldsymbol{\mu}$, $\boldsymbol{\nu}$, $\boldsymbol{\xi}$, $\boldsymbol{\sigma}$. Thus

$$\mathbf{a} \equiv \begin{bmatrix} a_{11} & a_{12} & \cdots & a_{1s} \\ a_{21} & a_{22} & \cdots & a_{2s} \\ \cdot & \cdot & \cdot & \cdot \\ \cdot & \cdot & \cdot & \cdot \\ \cdot & \cdot & \cdot & \cdot \\ a_{r1} & a_{r2} & \cdots & a_{rs} \end{bmatrix}, \qquad \boldsymbol{a} \equiv \begin{bmatrix} a_1 \\ a_2 \\ \cdot \\ \cdot \\ \cdot \\ a_r \end{bmatrix}.$$

Transposed matrices and vectors will be denoted by a superscript t; thus

$$\mathbf{a}^t \equiv \begin{bmatrix} a_{11} & a_{21} & \cdots & a_{r1} \\ a_{12} & a_{22} & \cdots & a_{r2} \\ \cdot & \cdot & \cdot & \cdot \\ \cdot & \cdot & \cdot & \cdot \\ \cdot & \cdot & \cdot & \cdot \\ a_{1s} & a_{2s} & \cdots & a_{rs} \end{bmatrix}, \qquad \boldsymbol{a}^t \equiv [a_1 \quad a_2 \quad \cdots \quad a_r] \;.$$

A matrix or vector whose elements are random variables will be identified by a tilde. Thus

$$\tilde{\mathbf{a}} \equiv \begin{bmatrix} \tilde{a}_{11} & \tilde{a}_{12} & \cdots & \tilde{a}_{1s} \\ \tilde{a}_{21} & \tilde{a}_{22} & \cdots & \tilde{a}_{2s} \\ \cdot & \cdot & \cdot & \cdot \\ \cdot & \cdot & \cdot & \cdot \\ \cdot & \cdot & \cdot & \cdot \\ \tilde{a}_{r1} & \tilde{a}_{r2} & \cdots & \tilde{a}_{rs} \end{bmatrix}, \qquad \tilde{\boldsymbol{a}}^t \equiv [\tilde{a}_1 \quad \tilde{a}_2 \quad \cdots \quad \tilde{a}_r] \;.$$

The *determinant* of the matrix $\mathbf{A}$ will be denoted by $|\mathbf{A}|$.

8.0.2. Inverses of Matrices

If a matrix $\mathbf{A}$ is partitioned into submatrices $\mathbf{A}_{ij}$, the symbol $\mathbf{A}_{ij}^{-1}$ will be used to denote the inverse of $\mathbf{A}_{ij}$; observe that it does *not* denote the ijth element of $\mathbf{A}^{-1}$:

$$\begin{aligned} \mathbf{A}_{ij}^{-1} &\equiv (\mathbf{A}_{ij})^{-1} \\ &\neq (\mathbf{A}^{-1})_{ij} \quad \text{except by coincidence.} \end{aligned}$$

The following easily verified theorems on inverses will be useful in the applications.

1. If both $\mathbf{A}$ and $\mathbf{B}$ are nonsingular $r \times r$ matrices, then

$$(\mathbf{A} + \mathbf{B})^{-1} = \mathbf{B}^{-1}(\mathbf{B}^{-1} + \mathbf{A}^{-1})^{-1}\mathbf{A}^{-1} = \mathbf{A}^{-1}(\mathbf{A}^{-1} + \mathbf{B}^{-1})^{-1}\mathbf{B}^{-1} . \tag{8-1}$$

2. If $\mathbf{A}$ is a partitioned matrix of the form

$$\mathbf{A} = \begin{bmatrix} \mathbf{A}_{11} & \mathbf{A}_{12} \\ \mathbf{0} & \mathbf{A}_{22} \end{bmatrix} , \tag{8-2a}$$

then

$$\mathbf{A}^{-1} = \begin{bmatrix} \mathbf{A}_{11}^{-1} & -\mathbf{A}_{11}^{-1}\mathbf{A}_{12}\mathbf{A}_{22}^{-1} \\ \mathbf{0} & \mathbf{A}_{22}^{-1} \end{bmatrix} . \tag{8-2b}$$

3. If $\mathbf{A}$ and $\mathbf{B}$ are two conformably partitioned matrices such that

$$\begin{bmatrix} \mathbf{A}_{11} & \mathbf{A}_{12} \\ \mathbf{A}_{21} & \mathbf{A}_{22} \end{bmatrix} \cdot \begin{bmatrix} \mathbf{B}_{11} & \mathbf{B}_{12} \\ \mathbf{B}_{21} & \mathbf{B}_{22} \end{bmatrix} = \begin{bmatrix} \mathbf{I} & \mathbf{0} \\ \mathbf{0} & \mathbf{I} \end{bmatrix} \tag{8-3a}$$

and if $\mathbf{A}_{11}$ and $\mathbf{B}_{22}$ are nonsingular, then

$$\mathbf{A}_{11}^{-1} = \mathbf{B}_{11} - \mathbf{B}_{12}\mathbf{B}_{22}^{-1}\mathbf{B}_{21} , \tag{8-3b}$$

$$\mathbf{A}_{11}^{-1}\mathbf{A}_{12} = -\mathbf{B}_{12}\mathbf{B}_{22}^{-1} . \tag{8-3c}$$

▶ The hypothesis (8-3a) implies directly that

$$\mathbf{A}_{11}\mathbf{B}_{11} + \mathbf{A}_{12}\mathbf{B}_{21} = \mathbf{I} , \qquad \mathbf{A}_{11}\mathbf{B}_{12} + \mathbf{A}_{12}\mathbf{B}_{22} = \mathbf{0} .$$

Premultiplying the second equation of this pair by $\mathbf{A}_{11}^{-1}$ and postmultiplying by $\mathbf{B}_{22}^{-1}$ we get (8-3c). Postmultiplying this same second equation by $\mathbf{B}_{22}^{-1}\mathbf{B}_{21}$ and subtracting it from the first we get

$$\mathbf{A}_{11}(\mathbf{B}_{11} - \mathbf{B}_{12}\mathbf{B}_{22}^{-1}\mathbf{B}_{21}) = \mathbf{I} ;$$

and premultiplying this result by $\mathbf{A}_{11}^{-1}$ we obtain (8-3b). ◀

8.0.3. Positive-definite and Positive-semidefinite Matrices

Let $\mathbf{A}$ be an $r \times r$ matrix and let $\boldsymbol{x}$ be an $r \times 1$ vector. Then $\mathbf{A}$ is said to be *positive-semidefinite* if

$$\boldsymbol{x}^t \mathbf{A} \boldsymbol{x} \geq 0 , \qquad \text{all } \boldsymbol{x} ; \tag{8-4a}$$

and $\mathbf{A}$ is said to be *positive-definite* if

$$\boldsymbol{x}^t \mathbf{A} \boldsymbol{x} > 0 , \qquad \text{all } \boldsymbol{x} \neq \boldsymbol{0} . \tag{8-4b}$$

We shall use the abbreviation PDS for "positive-definite and symmetric."

The following theorems on positive-definite and positive-semidefinite matrices will be required in the applications; proofs can be found in the literature.

1. If $\mathbf{A}$ is any matrix, then $\mathbf{A}^t\mathbf{A}$ is positive-semidefinite symmetric.
2. If $\mathbf{B}$ is positive-semidefinite and nonsingular, then $\mathbf{B}$ is positive-definite.
3. If $\mathbf{A}$ is an $s \times r$ matrix of rank $r \leq s$, then $\mathbf{A}^t\mathbf{A}$ is positive-definite symmetric.

4. If $\mathbf{C}$ is a positive-definite $r \times r$ matrix and if the $p \times r$ matrix $\mathbf{B}$ is of rank $p \leq r$, then $\mathbf{B}\,\mathbf{C}\,\mathbf{B}^t$ is positive-definite.

5. If $\mathbf{C}$ is positive-definite, then $\mathbf{C}^{-1}$ is positive-definite.

6. If $\mathbf{C}$ is positive-definite and $\mathbf{D}$ is formed by deleting any number of columns of $\mathbf{C}$ together with the corresponding rows, $\mathbf{D}$ is positive-definite.

7. If $\mathbf{C}$ is positive definite, there exist nonsingular matrices $\mathbf{B}$ and $\mathbf{U} = (\mathbf{B}^t)^{-1}$ such that

$$\mathbf{B}\,\mathbf{C}\,\mathbf{B}^t = \mathbf{I} \; , \qquad \mathbf{U}^t\,\mathbf{I}\,\mathbf{U} = \mathbf{C} \; .$$

8. If the $r \times r$ matrix $\mathbf{D}$ is positive-definite and the $r \times r$ matrix $\mathbf{S}$ is positive-semidefinite, then there exists a matrix $\mathbf{T}$ such that $\mathbf{T}\,\mathbf{D}\,\mathbf{T}^t$ is the identity matrix and $\mathbf{T}\,\mathbf{S}\,\mathbf{T}^t$ is a diagonal matrix with all elements nonnegative and with strictly positive elements equal in number to the rank of $\mathbf{S}$.

8.0.4. Projections

If $\mathbf{A}$ is an $n \times r$ matrix of rank $r \leq n$, its columns considered as vectors span an r-dimensional subspace in Euclidean n-space. Let $\boldsymbol{v}$ be any vector in this n-space. Then the projection of $\boldsymbol{v}$ on the column space of $\mathbf{A}$ is

$$\mathbf{B}\,\boldsymbol{v} \qquad \text{where} \qquad \mathbf{B} \equiv \mathbf{A}(\mathbf{A}^t\,\mathbf{A})^{-1}\,\mathbf{A}^t \; . \tag{8-5a}$$

If we define

$$\boldsymbol{b} \equiv (\mathbf{A}^t\,\mathbf{A})^{-1}\,\mathbf{A}^t\,\boldsymbol{v} \; , \tag{8-5b}$$

the projection can be written

$$\mathbf{B}\,\boldsymbol{v} = \mathbf{A}\,\boldsymbol{b} \tag{8-5c}$$

and the r-tuple $(b_1, \cdots, b_r)$ can be interpreted as the weights in a linear combination of the columns of $\mathbf{A}$.

▶ Decomposing $\boldsymbol{v}$ into two components

$$\boldsymbol{v} = \mathbf{B}\,\boldsymbol{v} + (\boldsymbol{v} - \mathbf{B}\,\boldsymbol{v}) \; ,$$

we observe that whereas $\mathbf{B}\,\boldsymbol{v} = \mathbf{A}\,\boldsymbol{b}$ is a linear combination of the columns of $\mathbf{A}$ and therefore lies wholly *within* the column space of $\mathbf{A}$, the fact that

$$\mathbf{A}^t(\boldsymbol{v} - \mathbf{B}\,\boldsymbol{v}) \equiv \mathbf{A}^t\,\boldsymbol{v} - \mathbf{A}^t\mathbf{B}\,\boldsymbol{v} = \mathbf{A}^t\,\boldsymbol{v} - \mathbf{A}^t\,\boldsymbol{v} = \boldsymbol{0}$$

implies that $(\boldsymbol{v} - \mathbf{B}\,\boldsymbol{v})$ is *orthogonal* to every column of $\mathbf{A}$ and therefore to the column space of $\mathbf{A}$. It then follows by definition that $\mathbf{B}\,\boldsymbol{v} = \mathbf{A}\,\boldsymbol{b}$ is the projection of $\boldsymbol{v}$ on the column space of $\mathbf{A}$. ◀

8.0.5. Notation for Multivariate Densities and Integrals

The definition of a multivariate density function will in general depend on the number r of the dimensions of the Euclidean space $R^{(r)}$ on which the density is defined, and therefore we shall sometimes denote a density by $f^{(r)}(\boldsymbol{z}|\boldsymbol{\theta})$. When no confusion can result, we shall simplify printing by suppressing the index (r) and writing $f(\boldsymbol{z}|\boldsymbol{\theta})$.

Let $f(\boldsymbol{z})$ be a scalar function of an $r \times 1$ vector $\boldsymbol{z}$. We define

$$\int_a^b f(z)\,dz \equiv \int_{a_1}^{b_1} \cdots \int_{a_i}^{b_i} \cdots \int_{a_r}^{b_r} f(z_1, \cdots, z_r)\,\Pi\,dz_i \ .$$

When ∞ appears as a limit of integration, it is to be understood as the vector $[\infty\ \infty\ \cdots\ \infty]^t$.

In an *integrand transformation* (cf. Section 7.0.5) of a multivariate density which substitutes the vector variable y for the vector variable z we shall use $|dz/dy|$ to denote the absolute value of the Jacobian of the transformation, i.e., the absolute value of the determinant whose ijth element is $\partial z_i/\partial y_j$.

The integral

$$\int_a^b y(z)\,f(z)\,dz$$

where y is a vector, is to be understood as the vector whose ith element is

$$\int_a^b y_i(z)\,f(z)\,dz \ ;$$

and similarly

$$\int_a^b \mathbf{y}(z)\,f(z)\,dz \ ,$$

where $\mathbf{y}$ is a matrix, stands for a matrix whose ijth element is

$$\int_a^b y_{ij}(z)\,f(z)\,dz \ .$$

8.0.6. Moments; Expectations and Variances

The *first moment* or *mean* of a multivariate normalized density function $f^{(r)}$ is defined to be the $r \times 1$ vector

$$\mu_1 \equiv \int z\,f^{(r)}(z)\,dz \tag{8-6a}$$

where the integration is to be carried out over the entire domain of $f^{(r)}$. Similarly the *second moment about the mean* of $f^{(r)}$ is defined to be the $r \times r$ matrix

$$\boldsymbol{\mu}_2 \equiv \int (z - \mu_1)(z - \mu_1)^t\,f^{(r)}(z)\,dz \ . \tag{8-6b}$$

When the integral defining a moment does not converge, we shall say that the moment does not exist.

When a normalized density function $f^{(r)}$ is interpreted as the density function of an $r \times 1$ vector *random variable* $\tilde{z}$, the first moment as defined by (8-6a) is also by definition the *expectation* of the random variable,

$$\mathrm{E}(\tilde{z}) \equiv \bar{z} \equiv \mu_1 \ . \tag{8-7a}$$

Similarly the second moment about the mean as defined by (8-6b) is by definition the *variance* of $\tilde{z}$:

$$\mathrm{V}(\tilde{z}) \equiv \check{\mathbf{z}} \equiv \boldsymbol{\mu}_2 \ . \tag{8-7b}$$

The moments of the density of a random variable $\tilde{y}$ which is a *linear function* of a random variable $\tilde{z}$ are easily expressed in terms of the moments of the density of $\tilde{z}$. Letting

$$\tilde{y} = a + \mathbf{B}\,\tilde{z} \tag{8-8a}$$

we have

$$\mathrm{E}(\tilde{y}) \equiv \bar{y} = a + \mathbf{B}\,\bar{z}\ , \qquad (8\text{-}8b)$$

$$\mathrm{V}(\tilde{y}) \equiv \check{y} = \mathbf{B}\,\check{z}\,\mathbf{B}^t\ . \qquad (8\text{-}8c)$$

▶ Letting f denote the density of $\tilde{z}$ and g the density of $\tilde{y}$ we have by (8-6)

$$\bar{y} \equiv \int y\, g(y)\, dy = \int (a + \mathbf{B}\, z)\, f(z)\, dz = a \int f(z)\, dz + \mathbf{B} \int z\, f(z)\, dz \equiv a + \mathbf{B}\,\bar{z}\ ;$$

$$\check{y} = \int (y - \bar{y})(y - \bar{y})^t\, g(y)\, dy = \int \mathbf{B}(z - \bar{z})(z - \bar{z})^t\, \mathbf{B}^t f(z)\, dz = \mathbf{B}\,\check{z}\,\mathbf{B}^t\ . \qquad ◀$$

8.1. Unit-Spherical Normal Function

The nondegenerate unit-spherical Normal density function is defined on Euclidean r-space $R^{(r)}$ by

$$f_{N*}^{(r)}(u) \equiv \Pi([2\pi]^{-\frac{1}{2}}\, e^{-\frac{1}{2}u_i^2}) = (2\pi)^{-\frac{1}{2}r}\, e^{-\frac{1}{2}\Sigma u_i^2}$$

$$= (2\pi)^{-\frac{1}{2}r}\, e^{-\frac{1}{2}u^t \mathbf{I} u}\ , \qquad -\infty < u < \infty\ , \qquad (8\text{-}9)$$

where u is an $r \times 1$ vector and $\mathbf{I}$ is the $r \times r$ identity matrix. The first two moments of f_{N*} are

$$\mu_1 = 0\ , \qquad (8\text{-}10a)$$

$$\boldsymbol{\mu}_2 = \mathbf{I}\ . \qquad (8\text{-}10b)$$

▶ That f_{N*} is a proper normalized density function follows from the facts (a) that it is obviously everywhere nonnegative and (b) that its integral over its entire domain has the value 1. The truth of the latter assertion is obvious if we write the density in the product form of (8-9) and observe that each factor in the product is an independent univariate Normal density with complete integral equal to 1.

To prove (8–10a) we observe that the ith component of μ_1 is

$$\int_{-\infty}^{\infty} \cdots \int_{-\infty}^{\infty} u_i (2\pi)^{-\frac{1}{2}r} \exp\left(-\tfrac{1}{2}\,\Sigma\, u_i^2\right) \Pi\, du_i$$

$$= \int_{-\infty}^{\infty} u_i (2\pi)^{-\frac{1}{2}}\, e^{-\frac{1}{2}u_i^2}\, du_i \int_{-\infty}^{\infty} \cdots \int_{-\infty}^{\infty} (2\pi)^{-\frac{1}{2}(r-1)} \exp\left(-\tfrac{1}{2}\,\Sigma_{j\neq i}\, u_j^2\right) \Pi_{j\neq i}\, du_j$$

$$= \int_{-\infty}^{\infty} u_i (2\pi)^{-\frac{1}{2}u_i^2}\, du_i\ ;$$

and this is 0 by the univariate result (7-62a).

To prove (8-10b) we observe first that the iith component of $\boldsymbol{\mu}_2$ can be similarly reduced to

$$\int_{-\infty}^{\infty} u_i^2 (2\pi)^{-\frac{1}{2}}\, e^{-\frac{1}{2}u_i^2}\, du_i\ ,$$

and this is 1 by the univariate result (7-62b). In the same way the ijth component can be reduced to

$$\int_{-\infty}^{\infty} u_i (2\pi)^{-\frac{1}{2}}\, e^{-\frac{1}{2}u_i^2}\, du_i \int_{-\infty}^{\infty} u_j (2\pi)^{-\frac{1}{2}}\, e^{-\frac{1}{2}u_j^2}\, du_j\ ;$$

and the iterated integral is 0 because either of the two single integrals is 0 by (7-62a). ◀

8.1.1. Conditional and Marginal Densities

Suppose that the r components of $\boldsymbol{u}$ are numbered in any order and then partitioned into the first l and the remaining $r - l$ components,

$$\boldsymbol{u} \equiv [u_1 \cdots u_l \vdots u_{l+1} \cdots u_r]^t \ ; \tag{8-11}$$

and define

$$\boldsymbol{u}_1 \equiv [u_1 \cdots u_l]^t \ , \qquad \boldsymbol{u}_2 \equiv [u_{l+1} \cdots u_r]^t \ , \tag{8-12}$$

in Euclidean spaces $R_1^{(l)}$ and $R_2^{(r-l)}$ respectively. It is apparent that (8-9) can be factored as

$$f_{N^*}^{(l)}(\boldsymbol{u}_1) \cdot f_{N^*}^{(r-l)}(\boldsymbol{u}_2) \ . \tag{8-13}$$

From this it follows immediately that the *conditional* density on R_1 given a particular $\boldsymbol{u}_2$ in R_2 is nondegenerate unit-spherical Normal; and since the integral of $f_{N^*}(\boldsymbol{u}_2)$ over its entire domain in R_2 has the value 1, it also follows that the *marginal* density on R_1 is nondegenerate unit-spherical Normal.

8.1.2. Tables

The integral

$$V(h, k) = \int_0^h \int_0^{ku_1/h} f_{N^*}^{(2)}(\boldsymbol{u}) \, du_2 \, du_1 \tag{8-14}$$

of the unit-spherical bivariate Normal density is tabulated in

> *Tables of the Bivariate Normal Distribution Function and Related Functions*, National Bureau of Standards, Applied Mathematics Series 50, U. S. Government Printing Office, Washington, 1959. These tables include (among others):
>
> Table III: $V(h, \lambda h)$ to 7 places for $h = 0(.01)4(.02)4.6(.1)5.6, \infty$;
> $\lambda = .1(.1)1$;
>
> Table IV: $V(\lambda h, h)$ to 7 places for $h = 0(.01)4(.02)5.6, \infty$;
> $\lambda = .1(.1)1$;
>
> Table V: $\dfrac{1}{2\pi}$ arc sin r to 8 places for $r = 0(.01)1$.

The functions $V(h, \lambda h)$ and $V(\lambda h, h)$, which are tabulated with arguments λ and h rather than λh and h, give the masses over triangles like those shown in Figure 8.1A; when $h = \infty$, the triangles are infinite wedges like those shown in Figure 8.1B. To obtain values of either function for $\lambda > 1$, we can make use of the fact obvious from the geometry that

$$V(h, ah) = V\left(\frac{1}{a}h, h\right) \ ; \qquad V(ah, h) = V\left(h, \frac{1}{a}h\right) \cdot \tag{8-15}$$

The mass $W_{N^*}(v, \theta)$ over infinite wedges avb with included angle θ and one side extending *along an axis* from $+v$ to $+\infty$, like those shown in Figure 8.2, can be found from NBS Table III or IV in conjunction with a table of the right-tail univariate function G_{N^*} by computing

$$c = v \cos\theta \ , \qquad s = |v \sin\theta| \ , \qquad t = |s/c| \ , \tag{8-16a}$$

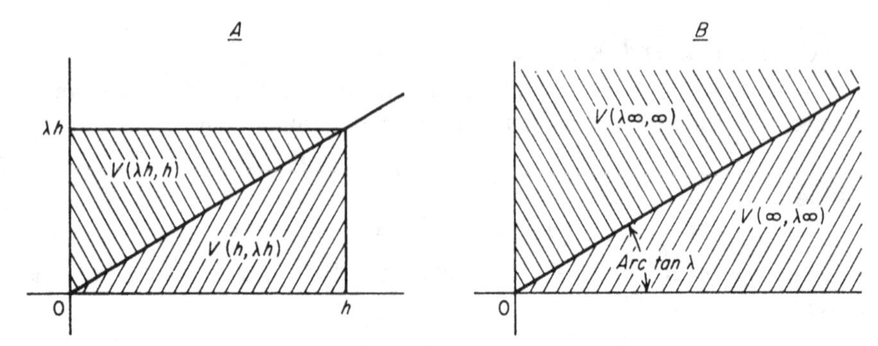

Figure 8.1

NBS Tables III and IV

and then applying the appropriate one of the formulas below.

0° $\leq |\theta| \leq$ *45°*: $(c > 0, t \leq 1)$

$$W_{N*}(v, \theta) = \tfrac{1}{2}G_{N*}(s) - V(t\,\infty, \infty) + V(ts, s)$$

45° $\leq |\theta| \leq$ *90°*: $(c \geq 0, t \geq 1)$

$$W_{N*}(v, \theta) = \tfrac{1}{2}G_{N*}(s) - V\left(\infty, \frac{1}{t}\infty\right) + V\left(s, \frac{1}{t}s\right)$$

90° $\leq |\theta| \leq$ *135°*: $(c \leq 0, t \geq 1)$ (8-16b)

$$W_{N*}(v, \theta) = \tfrac{1}{2}G_{N*}(s) + V\left(\infty, \frac{1}{t}\infty\right) - V\left(s, \frac{1}{t}s\right)$$

135° $\leq |\theta| \leq$ *180°*: $(c < 0, t \leq 1)$

$$W_{N*}(v, \theta) = \tfrac{1}{2}G_{N*}(s) + V(t\,\infty, \infty) - V(ts, s)\ .$$

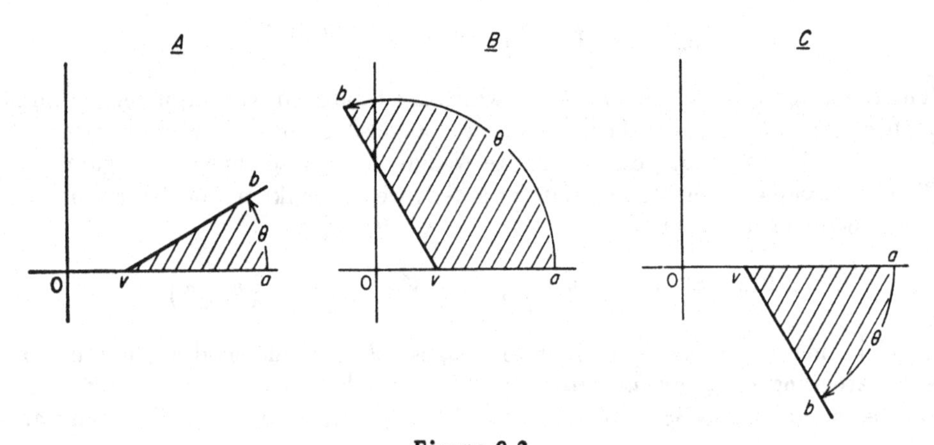

Figure 8.2

Standardized Bivariate Normal Wedges

▶ Formulas (8-16b) can all be proved by elementary trigonometry; we give the proof of the formula for $0° \leq \theta \leq 45°$ as an example. In Figure 8.3 we take axes Ox_1 and Ox_2 respectively perpendicular and parallel to the side vb of the wedge. The distances s and c in the figure and their ratio $t = s/c$ clearly correspond to the definitions (8-16a). We then

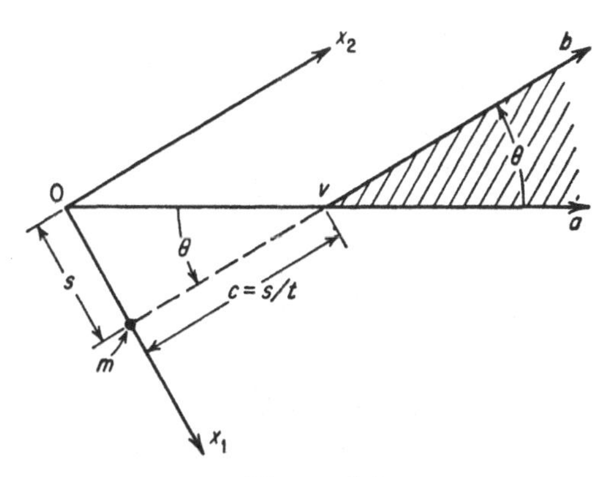

Figure 8.3

express the mass over the wedge as the mass over x_1mb less the mass over x_1mva. The mass over x_1mb is evaluated by observing that it corresponds in terms of probabilities to

$$P\{\tilde{x}_1 > s, \tilde{x}_2 > 0\} = \tfrac{1}{2}G_{N^*}(s) .$$

The mass over x_1mva is evaluated by first expressing it as the mass over x_1Oa less the mass over mOv, observing that by Figure 8.1 this difference is

$$V\left(\infty, \frac{1}{t}\infty\right) - V\left(s, \frac{1}{t}s\right) ,$$

and then $\left(\text{because } \frac{1}{t} > 1\right)$ using (8-15) to rewrite this last expression as

$$V(t\,\infty, \infty) - V(ts, s) .$$ ◀

8.2. General Normal Function

The general nondegenerate Normal density function is defined on Euclidean r-space by

$$f_N^{(r)}(z|m, \mathbf{H}) \equiv (2\pi)^{-\frac{1}{2}r} e^{-\frac{1}{2}(z-m)^t \mathbf{H}(z-m)} |\mathbf{H}|^{\frac{1}{2}} , \qquad \begin{array}{l} -\infty < z < \infty , \\ -\infty < m < \infty , \\ \mathbf{H} \text{ is PDS} , \end{array} \qquad (8\text{-}17)$$

where z and m are $r \times 1$ and $\mathbf{H}$ is $r \times r$. The first two moments of the function are

$$\mu_1 = m , \qquad (8\text{-}18a)$$
$$\mu_2 = \mathbf{H}^{-1} . \qquad (8\text{-}18b)$$

▶ That $f_N^{(r)}$ is a proper normalized density function follows from the fact that it is simply an integrand transformation of the unit-spherical density $f_{N^*}^{(r)}$. Since $\mathbf{H}$ is positive definite, there exists by Theorem 7 of Section 8.0.3 a nonsingular matrix $\mathbf{U}$ such that

$$\mathbf{U}^t\,\mathbf{I}\,\mathbf{U} = \mathbf{H} \ .$$

The general density $f_N^{(r)}$ can therefore be derived from the unit-spherical density $f_{N*}^{(r)}$ by substituting

$$\mathbf{U}(z - m) \qquad \text{for} \qquad u$$

in (8-9) and multiplying by the Jacobian

$$|du/dz| = |\mathbf{U}| = |\mathbf{U}^t\,\mathbf{I}\,\mathbf{U}|^{\frac{1}{2}} = |\mathbf{H}|^{\frac{1}{2}} \ .$$

Formulas (8-18) for the moments are derived from the unit-spherical moments (8-10) by writing

$$z = m + \mathbf{U}^{-1}\,u$$

and using (8-8):

$$\mu_1 = m + \mathbf{U}^{-1}\,0 = m \ ,$$

$$\boldsymbol{\mu}_2 = \mathbf{U}^{-1}\,\mathbf{I}(\mathbf{U}^{-1})^t = \mathbf{U}^{-1}\,\mathbf{I}(\mathbf{U}^t)^{-1} = (\mathbf{U}^t\,\mathbf{I}\,\mathbf{U})^{-1} = \mathbf{H}^{-1} \ .$$ ◀

8.2.1. *Conditional and Marginal Densities*

Consider a nondegenerate Normal density $f_N^{(r)}(z|m, \mathbf{H})$ on $R^{(r)}$, partition the space $R^{(r)}$ into subspaces $R_1^{(q)}$ and $R_2^{(r-q)}$, and partition the vectors z and m and the matrices $\mathbf{H}$ and $\mathbf{V} \equiv \mathbf{H}^{-1}$ correspondingly:

$$z = [z_1 \cdots z_q \vdots z_{q+1} \cdots z_r]^t = [z_1^t \quad z_2^t]^t \ ,$$
$$m = [m_1 \cdots m_q \vdots m_{q+1} \cdots m_r]^t = [m_1^t \quad m_2^t]^t \ ,$$

$$\mathbf{H} = \begin{bmatrix} \mathbf{H}_{11} & \mathbf{H}_{12} \\ \mathbf{H}_{21} & \mathbf{H}_{22} \end{bmatrix} \qquad \text{where} \qquad \mathbf{H}_{11} \text{ is } q \times q \ , \tag{8-19}$$

$$\mathbf{V} = \begin{bmatrix} \mathbf{V}_{11} & \mathbf{V}_{12} \\ \mathbf{V}_{21} & \mathbf{V}_{22} \end{bmatrix} \qquad \text{where} \qquad \mathbf{V}_{11} \text{ is } q \times q \ .$$

Then the *marginal* density on R_1 is

$$f_N^{(q)}(z_1|m_1, \mathbf{V}_{11}^{-1}) = f_N^{(q)}(z_1|m_1, \mathbf{H}_{11} - \mathbf{H}_{12}\,\mathbf{H}_{22}^{-1}\,\mathbf{H}_{21}) \tag{8-20}$$

while the *conditional* density on R_1 given z_2 in R_2 is

$$f_N^{(q)}(z_1|m_1 - \mathbf{H}_{11}^{-1}\,\mathbf{H}_{12}[z_2 - m_2], \mathbf{H}_{11})$$
$$= f_N^{(q)}(z_1|m_1 + \mathbf{V}_{12}\,\mathbf{V}_{22}^{-1}[z_2 - m_2], [\mathbf{V}_{11} - \mathbf{V}_{12}\,\mathbf{V}_{22}^{-1}\,\mathbf{V}_{21}]^{-1}) \ . \tag{8-21}$$

It will be of interest in the applications to observe that, because of its appearance in (8-21), $\mathbf{H}_{11}$ can be interpreted as the precision of our information about z_1 when we know the value of $\tilde{z}_2$; and because of its appearance in (8-20), $\mathbf{V}_{11}^{-1} = \mathbf{H}_{11} - \mathbf{H}_{12}\,\mathbf{H}_{22}^{-1}\,\mathbf{H}_{21}$ can be interpreted as the precision of our information about z_1 when we do *not* know the value of $\tilde{z}_2$. Knowledge of z_2 contributes $\mathbf{H}_{12}\,\mathbf{H}_{22}^{-1}\,\mathbf{H}_{21}$ to the precision of our information about z_1.

▶ To prove (8-20) and (8-21) we first simplify notation by defining

$$\begin{bmatrix} \zeta_1 \\ \zeta_2 \end{bmatrix} \equiv \begin{bmatrix} z_1 - m_1 \\ z_2 - m_2 \end{bmatrix} = z - m \ .$$

Substituting the partitioned forms of z and $\mathbf{H}$ in the density formula (8-17) we obtain a constant multiplied by $e^{-\frac{1}{2}S}$ where

$$(1) \qquad \begin{aligned} S &\equiv \zeta_1^t \mathbf{H}_{11} \zeta_1 + \zeta_1^t \mathbf{H}_{12} \zeta_2 + \zeta_2^t \zeta_1 + \mathbf{H}_{21} \zeta_1 + \zeta_2^t \mathbf{H}_{22} \zeta_2 \\ &= (\zeta_1 + \mathbf{H}_{11}^{-1} \mathbf{H}_{12} \zeta_2)^t \mathbf{H}_{11}(\zeta_1 + \mathbf{H}_{11}^{-1} \mathbf{H}_{12} \zeta_2) + \zeta_2^t(\mathbf{H}_{22} - \mathbf{H}_{21} \mathbf{H}_{11}^{-1} \mathbf{H}_{12}) \zeta_2 \ . \end{aligned}$$

Since the second of the two terms on the right does not contain ζ_1, the *conditional* density on R_1 depends on the first term alone. The matrix $\mathbf{H}_{11}$ in this term is symmetric because $\mathbf{H}$ is symmetric and it is positive-definite by Theorem 6 in Section 8.0.3; and it then follows from the form of the exponent that the conditional density on R_1 is Normal with second parameter $\mathbf{H}_{11}$ as given by (8-21). The value of the first parameter is obvious as soon as we write

$$\zeta_1 + \mathbf{H}_{11}^{-1} \mathbf{H}_{12} \zeta_2 = z_1 - (m_1 - \mathbf{H}_{11}^{-1} \mathbf{H}_{12}[z_2 - m_2]) \ ;$$

the alternative form of the parameters given in (8-21) then follows from (8-3c).

Integrating out ζ_1 from the right-hand side of (1) we are left with the second term as the exponent of the *marginal* density on R_2; and interchanging subscripts 1 and 2 we have the exponent of the marginal density on R_1:

$$\zeta_1^t(\mathbf{H}_{11} - \mathbf{H}_{12} \mathbf{H}_{22}^{-1} \mathbf{H}_{21}) \zeta_1 \ .$$

Using (8-3b) we can write this as

$$(z_1 - m_1)^t \mathbf{V}_{11}^{-1}(z_1 - m_1) \ .$$

$\mathbf{V}_{11}$ is PDS for the same reason as $\mathbf{H}_{11}$; its inverse is therefore PDS by Theorem 5 of Section 8.0.3; and it follows that the marginal density on R_1 is Normal with parameters m_1 and $\mathbf{V}_{11}^{-1}$. ◀

8.2.2. Tables.

The cumulative unit-elliptical bivariate function

$$L(h, k, r) \equiv \int_h^\infty \int_k^\infty f_N^{(2)}(z|0, \mathbf{H}) \, dz_2 \, dz_1 \ , \qquad \mathbf{H}^{-1} = \begin{bmatrix} 1 & r \\ r & 1 \end{bmatrix} , \qquad (8\text{-}22)$$

has been tabled in the National Bureau of Standards publication cited in Section 8.1.2. These tables include (among others)

Table I: $L(h, k, r)$ to 6 places for $h, k = 0(.1)4$; $r = 0(.05).95(.01)1$;
Table II: $L(h, k, -r)$ to 7 places for $h, k = 0(.1)4$; $r = 0(.05).95(.01)1$.

For methods of evaluating cumulative Normal functions of dimensionality greater than 2, see P. Ihm, *Sankhyā* 21 (1959) pages 363–366, and S. John, *ibid.* pages 367–370, both of whom give references to the earlier literature.

Some of the applications discussed in this monograph require the mass over bivariate wedges S_1VS_2 like the shaded areas in Figure 8.4. When VS_1 is parallel to one coordinate axis *and* VS_2 is parallel to the other, the mass over S_1VS_2 can be found most conveniently from NBS Tables I and II cited just above, but in all other cases the Tables III and IV discussed in Section 8.1.2 are more convenient. To solve such problems by the use of these latter tables, we first decompose the wedge into the sum (Figure 8.4A), difference (Figure 8.4B), or complement (Figure 8.4C) of two wedges AVS_1 and AVS_2 having a common side formed by the projection of a ray from the mean M of the density through the vertex V of the original wedge S_1VS_2. We then make a change of variables

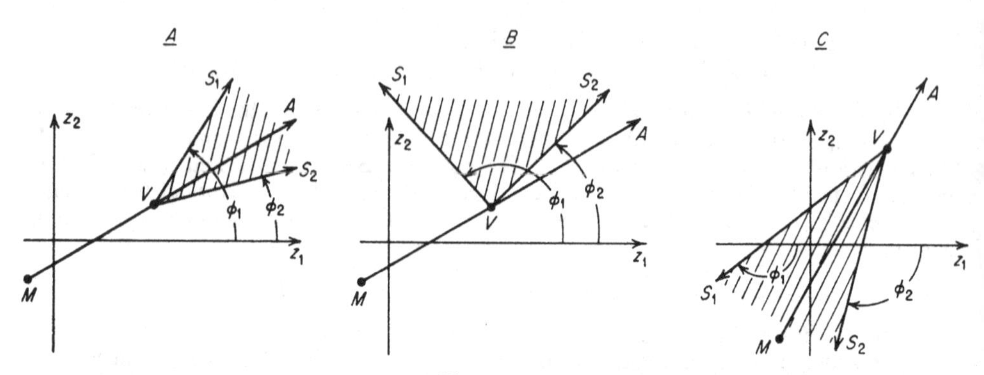

Figure 8.4

General Bivariate Normal Wedges

$\boldsymbol{u} = \mathbf{A}(\boldsymbol{z} - \bar{\boldsymbol{z}})$ such that the density is rendered unit-spherical and the point V is carried into a point $(v, 0)$ on the positive u_1 axis.

The wedges AVS_1 and AVS_2 are thus transformed into wedges of the type illustrated in Figure 8.2 and the masses over each of the transformed wedges can then be found by application of formulas (8-16). These formulas have as arguments the quantities v and θ which describe the transformed wedge, and to express their values in terms of the geometry of the original wedge AVS_1 or AVS_2 we first define

$\boldsymbol{z}$: coordinates of V in the original variables,

$\boldsymbol{\phi}$: angle between the *directed* line VS and the *positive* z_1 axis (cf. Figure 8.4),

$\bar{\boldsymbol{z}}, \boldsymbol{\sigma}$: mean and second central moment of the density in the original variables,

$$\sigma_1 = \sqrt{\sigma_{11}} \, , \qquad \sigma_2 = \sqrt{\sigma_{22}} \, , \qquad \rho = \frac{\sigma_{12}}{\sqrt{\sigma_{11}\sigma_{22}}} \, ,$$

$$\zeta_1 = \frac{z_1 - \bar{z}_1}{\sigma_1} \, , \qquad \zeta_2 = \frac{z_2 - \bar{z}_2}{\sigma_2} \, , \tag{8-23a}$$

$$\eta_1 = \frac{\zeta_1 - \rho\zeta_2}{\sqrt{1 - \rho^2}} \, , \qquad \eta_2 = \frac{\zeta_2 - \rho\zeta_1}{\sqrt{1 - \rho^2}} \, .$$

We can then compute

$$v = \sqrt{\zeta_1^2 + \eta_2^2} = \sqrt{\zeta_2^2 + \eta_1^2} \, ,$$

$$p = \eta_1 \frac{\cos\phi}{\sigma_1} + \eta_2 \frac{\sin\phi}{\sigma_2} \, , \qquad q = \zeta_1 \frac{\sin\phi}{\sigma_2} - \zeta_2 \frac{\cos\phi}{\sigma_1} \, , \tag{8-23b}$$

$$\cos\theta = \frac{p}{\sqrt{p^2 + q^2}} \, , \qquad \sin\theta = \frac{q}{\sqrt{p^2 + q^2}} \, .$$

In the special case where VS is parallel to one or the other of the original coordinate axes, i.e., where either $\sin\phi$ or $\cos\phi$ vanishes, formulas (8-23b) reduce to

$$\begin{array}{ll} \underline{\sin\phi = 0} & \underline{\cos\phi = 0} \\ v\cos\theta = \eta_1 \cos\phi \, , & v\cos\theta = \eta_2 \sin\phi \, , \\ v\sin\theta = -\zeta_2 \cos\phi \, , & v\sin\theta = \zeta_1 \sin\phi \, . \end{array} \tag{8-23b'}$$

▶ Consider the change of variables

$$u = \mathbf{A}(z - \bar{z})$$

where

$$\mathbf{A} \equiv (\zeta_1^2 - 2\rho\zeta_1\zeta_2 + \zeta_2^2)^{-\frac{1}{2}} \begin{bmatrix} \eta_1/\sigma_1 & \eta_2/\sigma_2 \\ -\zeta_2/\sigma_1 & \zeta_1/\sigma_2 \end{bmatrix} .$$

1. That the transformed density is in fact unit-spherical can be verified by the use of Corollary 2.1′ in Section 8.2.3 below, which gives for the parameters of the transformed density

$$-\mathbf{A}\,\bar{z} + \mathbf{A}\,\bar{z} = 0 ,$$

$$[\mathbf{A}\,\boldsymbol{\sigma}\,\mathbf{A}^t]^{-1} = \left\{\mathbf{A}\begin{bmatrix} \sigma_1^2 & \rho\sigma_1\sigma_2 \\ \rho\sigma_1\sigma_2 & \sigma_2^2 \end{bmatrix}\mathbf{A}^t\right\}^{-1} = \mathbf{I} .$$

2. To verify that the vertex $V = z$ is transformed into a point on the positive u_1 axis at a distance from the origin given by the formula for v in (8-23b), let u denote the transformation of z and compute

$$u = \mathbf{A}(z - \bar{z}) = \mathbf{A}\begin{bmatrix} \zeta_1\sigma_1 \\ \zeta_2\sigma_2 \end{bmatrix} = \begin{bmatrix} v \\ 0 \end{bmatrix} .$$

3. To verify the formulas for $\cos\theta$ and $\sin\theta$ in (8-23b), make use of the fact that the transformation leaves parallel lines parallel and consider its effect on a line segment joining the mean $\bar{z}$ of the original density to the point $(\bar{z}_1 + \cos\phi, \bar{z}_2 + \sin\phi)$. The point $\bar{z}$ goes into the point 0 while the point $(\bar{z}_1 + \cos\phi, \bar{z}_2 + \sin\phi)$ goes into

$$\mathbf{A}\begin{bmatrix} \cos\phi \\ \\ \sin\phi \end{bmatrix} = \text{scalar} \times \begin{bmatrix} \eta_1 \dfrac{\cos\phi}{\sigma_1} + \eta_2 \dfrac{\sin\phi}{\sigma_2} \\ \\ -\zeta_2 \dfrac{\cos\phi}{\sigma_1} + \zeta_1 \dfrac{\sin\phi}{\sigma_2} \end{bmatrix} ,$$

from which the formulas for $\cos\theta$ and $\sin\theta$ follow immediately.

4. Verification of (8-23b′) is a matter of simple algebra. ◀

8.2.3. *Linear Combinations of Normal Random Variables*

In the applications we shall require the following theorems concerning linear combinations of random variables whose joint distribution has a density given by the (multivariate) Normal function. These theorems will be restated without reference to random variables before they are proved.

Theorem 1. Let $\tilde{u}_1, \cdots, \tilde{u}_s$ be independent scalar random variables each having a unit Normal distribution, so that

$$\mathrm{D}(u) = f_{N^*}^{(s)}(u) \qquad \text{where} \qquad u \text{ is } s \times 1 ; \tag{8-24a}$$

and let $\tilde{z}_1, \cdots, \tilde{z}_r$ be $r \leq s$ linearly independent linear combinations of the $\tilde{u}$s:

$$\tilde{z} = a + \mathbf{b}\,\tilde{u} \qquad \text{where} \qquad \begin{array}{l} a \text{ is } r \times 1 , \\ \mathbf{b} \text{ is } r \times s \text{ of rank } r \leq s . \end{array} \tag{8-24b}$$

Then the joint distribution of the $\tilde{z}$s is nondegenerate Normal with mean a and variance $\mathbf{b}\,\mathbf{I}\,\mathbf{b}^t$:

$$\mathrm{D}(z) = f_N^{(r)}(z|a, [\mathbf{b}\,\mathbf{I}\,\mathbf{b}^t]^{-1}) . \tag{8-24c}$$

Theorem 2. Let $\tilde{z}_1, \cdots, \tilde{z}_r$ be scalar random variables whose joint distribution is nondegenerate Normal, so that

$$D(z) = f_N^{(r)}(z|m, \mathbf{H}) \qquad \text{where} \qquad z \text{ is } r \times 1 \; . \tag{8-25a}$$

Then it is always possible to express the r $\tilde{z}$s as linear combinations of r independent $\tilde{u}$s each having a unit Normal distribution; or in other words, an $r \times r$ matrix $\mathbf{U}$ can always be found such that

$$u = \mathbf{U}(z - m) \; , \qquad \text{which is} \qquad r \times 1 \; , \tag{8-25b}$$

has the density

$$D(u) = f_{N*}^{(r)}(u) \; . \tag{8-25c}$$

The most important results of the two preceding theorems are the two following corollaries.

Corollary 2.1. Let $\tilde{z}_1, \cdots, \tilde{z}_r$ be scalar random variables whose joint distribution is nondegenerate Normal with mean m and variance $\mathbf{H}^{-1}$, so that

$$D(z) = f_N^{(r)}(z|m, \mathbf{H}) \qquad \text{where} \qquad z \text{ is } r \times 1 \; ; \tag{8-26a}$$

and let $\tilde{y}_1, \cdots, \tilde{y}_q$ be $q \leq r$ linearly independent linear combinations of the $\tilde{z}$s:

$$\tilde{y} = c + \mathbf{d}\,\tilde{z} \qquad \text{where} \qquad \begin{array}{l} c \text{ is } q \times 1 \; , \\ \mathbf{d} \text{ is } q \times r \text{ of rank } q \leq r \; . \end{array} \tag{8-26b}$$

Then the joint distribution of the $\tilde{y}$s is nondegenerate Normal with mean $c + \mathbf{d}\, m$ and variance $\mathbf{d}\,\mathbf{H}^{-1}\,\mathbf{d}^t$:

$$D(y) = f_N^{(q)}(y|c + \mathbf{d}\, m, [\mathbf{d}\,\mathbf{H}^{-1}\,\mathbf{d}^t]^{-1}) \; . \tag{8-26c}$$

Corollary 2.2. Let $\tilde{\mathbf{z}}^{(1)}, \cdots, \tilde{\mathbf{z}}^{(p)}, \cdots, \tilde{\mathbf{z}}^{(n)}$ be independent $r \times 1$ random variables each of which has a nondegenerate Normal distribution; and let $\tilde{\mathbf{y}}$ be an $r \times 1$ linear combination of the $\tilde{\mathbf{z}}$s,

$$\tilde{\mathbf{y}} = \alpha + \Sigma\, \beta_p\, \tilde{\mathbf{z}}^{(p)} \; . \tag{8-27a}$$

Then the distribution of $\tilde{\mathbf{y}}$ is nondegenerate Normal with mean and variance

$$\begin{aligned} E(\tilde{\mathbf{y}}) &= \alpha + \Sigma\, \beta_p\, \bar{\mathbf{z}}^{(p)} \; , \\ V(\tilde{\mathbf{y}}) &= \Sigma\, \beta_p^2\, \check{\mathbf{z}}^{(p)} \end{aligned} \tag{8-27b}$$

where $\check{\mathbf{z}}^{(p)} \equiv V(\tilde{\mathbf{z}}^{(p)})$.

▶ We shall now restate and prove these theorems without reference to random variables.

Theorem 1′. If the mass distributed on $R^{(s)}$ by the density function $f_{N*}^{(s)}$ is mapped onto $R^{(r)}$, $r \leq s$, by

$$z = a + \mathbf{b}\, u \qquad \text{where} \qquad \begin{array}{l} a \text{ is } r \times 1 \\ \mathbf{b} \text{ is } r \times s \text{ of rank } r \leq s \; , \end{array}$$

the density on $R^{(r)}$ is $f_N^{(r)}(\cdot|a, [\mathbf{b}\,\mathbf{I}\,\mathbf{b}^t]^{-1})$.

Assume first that $r = s$. Since $\mathbf{b}$ is of full rank, it has an inverse, and therefore we may write

$$u = \mathbf{b}^{-1}(z - a) \; .$$

Substituting this result for u in the formula for $f_{N*}^{(s)}(u)$ and multiplying by the Jacobian

$$|du/dz| = |\mathbf{b}^{-1}|$$

we obtain

$$(2\pi)^{-\frac{1}{2}s}\, e^{-\frac{1}{2}(\mathbf{b}^{-1}[z-a])^t\,\mathbf{I}\,(\mathbf{b}^{-1}[z-a])}\,|\mathbf{b}^{-1}| \;.$$

That this result is equivalent to $f_N^{(r)}(z|a, [\mathbf{b\,I\,b}^t]^{-1})$ is clear as soon as we observe that

$$(\mathbf{b}^{-1}[z-a])^t\,\mathbf{I}\,(\mathbf{b}^{-1}[z-a]) = (z-a)^t(\mathbf{b}^{-1})^t\,\mathbf{I}\,\mathbf{b}^{-1}(z-a) = (z-a)^t(\mathbf{b\,I\,b}^t)^{-1}(z-a) \;,$$
$$|\mathbf{b}^{-1}| = |(\mathbf{b}^{-1})^t\,\mathbf{I}\,\mathbf{b}^{-1}|^{\frac{1}{2}} = |(\mathbf{b\,I\,b}^t)^{-1}|^{\frac{1}{2}} \;.$$

That $(\mathbf{b\,I\,b}^t)^{-1}$ is positive definite as required by (8-17) follows from Theorems 4 and 5 of Section 8.0.3.

Next consider the case where $\mathbf{b}$ is $r \times s$ of rank $r < s$. We can always augment $\mathbf{b}$ by adding $s - r$ rows which are (*a*) orthogonal to the original r rows and to each other and (*b*) of unit norm, thus creating an $s \times s$ matrix $\mathbf{b}^*$ such that

$$\mathbf{b}^*\,\mathbf{I}\,\mathbf{b}^{*t} = \begin{bmatrix} \mathbf{b\,I\,b}^t & \mathbf{0} \\ \mathbf{0} & \mathbf{I} \end{bmatrix} \qquad \text{is of rank } s \;.$$

Then defining

$$a^* = [a_1 \cdots a_r \,\vdots\, 0 \cdots 0]^t \;, \qquad z^* = a^* + \mathbf{b}^*\,u = [z_1 \cdots z_r \,\vdots\, z_{r+1}^{(0)} \cdots z_s^{(0)}]^t \;,$$

and proceeding exactly as we did when we assumed $\mathbf{b}$ itself to be $s \times s$ of rank s, we can show that the density on the transformed $R^{(s)}$ is

$$f_N(z|a^*, [\mathbf{b}^*\,\mathbf{I}\,\mathbf{b}^{*t}]^{-1}) = (2\pi)^{-\frac{1}{2}s}\, e^{-\frac{1}{2}(z^*-a^*)^t(\mathbf{b}^*\mathbf{I}\mathbf{b}^{*t})^{-1}(z^*-a^*)}\, |(\mathbf{b}^*\,\mathbf{I}\,\mathbf{b}^{*t})^{-1}|^{\frac{1}{2}} \;.$$

Now partitioning $(z^* - a^*)$ and $\mathbf{b}^*\,\mathbf{I}\,\mathbf{b}^{*t}$ as shown above, we see that

$$|\mathbf{b}^*\,\mathbf{I}\,\mathbf{b}^{*t}| = |\mathbf{b\,I\,b}^t| \;, \qquad \text{so that} \qquad |(\mathbf{b}^*\,\mathbf{I}\,\mathbf{b}^{*t})^{-1}| = |(\mathbf{b\,I\,b}^t)^{-1}| \;,$$
$$(z^* - a^*)^t(\mathbf{b}^*\,\mathbf{I}\,\mathbf{b}^{*t})^{-1}(z^* - a^*) = (z-a)^t(\mathbf{b\,I\,b}^t)^{-1}(z-a) + z^{(0)t}\,\mathbf{I}\,z^{(0)} \;;$$

and integrating out $z^{(0)}$ we obtain $f_N^{(r)}(z|a, [\mathbf{b\,I\,b}^t]^{-1})$.

Theorem 2′. Any Normal density $f_N^{(r)}(\cdot|m, \mathbf{H})$ can be considered as a linear integrand transformation

$$z = m + \mathbf{U}^{-1}u \;, \qquad \mathbf{U}^t\,\mathbf{I}\,\mathbf{U} = \mathbf{H} \;,$$

of a unit-spherical Normal density $f_{N*}^{(r)}$.

This theorem has already been proved in the course of proving that $f_N^{(r)}$ as defined by (8-17) is a proper density function.

Corollary 2.1′. If the mass distributed on $R^{(r)}$ by the density function $f_N^{(r)}(\cdot|m, \mathbf{H})$ is mapped onto $R^{(q)}$, $q \le r$, by

$$y = \mathbf{c} + \mathbf{d}\,z \qquad \text{where} \qquad \begin{array}{l} \mathbf{c} \text{ is } q \times 1, \\ \mathbf{d} \text{ is } q \times r \text{ of rank } q \;, \end{array}$$

the density on $R^{(q)}$ is $f_N^{(q)}(\cdot|\mathbf{c} + \mathbf{d}\,m, [\mathbf{d\,H}^{-1}\,\mathbf{d}^t]^{-1})$.

By Theorem 2′, $f_N^{(r)}$ can be expressed as a linear integrand transformation

$$z = m + \mathbf{U}^{-1}u \;, \qquad \mathbf{U}^t\,\mathbf{I}\,\mathbf{U} = \mathbf{H} \;,$$

of $f_{N*}^{(r)}$; and consequently the mapping in which we are interested can be expressed as a linear mapping

$$y = \mathbf{c} + \mathbf{d}\,z = (\mathbf{c} + \mathbf{d}\,m) + (\mathbf{d\,U}^{-1})\,u$$

of $f_{N*}^{(r)}$ onto $R^{(q)}$. It then follows by Theorem 1′ that the density on $R^{(q)}$ is Normal with parameters $\mathbf{c} + \mathbf{d}\,m$ and

$$[(\mathbf{dU}^{-1})\,\mathbf{I}(\mathbf{d\,U}^{-1})^t]^{-1} = [\mathbf{d}(\mathbf{U}^t\,\mathbf{I}\,\mathbf{U})^{-1}\,\mathbf{d}^t]^{-1} = [\mathbf{d\,H}^{-1}\,\mathbf{d}^t]^{-1} \;.$$

Corollary 2.2'. Let n nondegenerate Normal densities with moments $\mu_1^{(1)}, \cdots, \mu_1^{(n)}$ and $\mu_2^{(1)}, \cdots, \mu_2^{(n)}$ be independently defined on subspaces $R_1^{(r)}, \cdots, R_n^{(r)}$; let the density on the product space $R^{(rn)}$ be the product of these densities; and let the density on $R^{(rn)}$ be mapped onto $R^{(r)}$ by

$$y = \alpha + [\mathbf{B}^{(1)} \cdots \mathbf{B}^{(p)} \cdots \mathbf{B}^{(n)}] \begin{bmatrix} z^{(1)} \\ \cdot \\ \cdot \\ \cdot \\ z^{(p)} \\ \cdot \\ \cdot \\ \cdot \\ z^{(n)} \end{bmatrix} ,$$

where α is $r \times 1$,

$$\mathbf{B}^{(p)} \equiv \beta_p \mathbf{I} \qquad \text{is} \qquad r \times r ,$$

and at least one $\beta_p \neq 0$. Then the density on $R^{(r)}$ is $f_N^{(r)}$ with parameters $\alpha + \Sigma\, \beta_p \mu_1^{(p)}$ and $(\Sigma\, \beta_p^2 \mu_2^{(p)})^{-1}$.

Since at least one $\beta_p \neq 0$, the rank of $\mathbf{B} \equiv [\mathbf{B}^{(1)} \cdots \mathbf{B}^{(n)}]$ is r, and therefore the Normality of the density on $R^{(r)}$ follows immediately from Corollary 2.1'. By (8-18a) the first parameters of the n original densities are equal to their first moments, and therefore the first parameter of the mapped density is given by Corollary 2.1' as

$$\alpha + \mathbf{B}[(\mu_1^{(1)})^t \cdots (\mu_1^{(n)})^t]^t = \alpha + \Sigma\, \beta_p \mu_1^{(p)} .$$

Similarly the second parameters of the n original densities are equal by (8-18b) to the inverses of their second moments. Since the n original densities are independent, the second moment or inverse of the second parameter of the density on $R^{(rn)}$ is

$$\mathbf{M}_2 = \begin{bmatrix} \mu_2^{(1)} & & & & & & 0 \\ & \cdot & & & & & \\ & & \cdot & & & & \\ & & & \mu_2^{(p)} & & & \\ & & & & \cdot & & \\ 0 & & & & & \cdot & \\ & & & & & & \mu_2^{(n)} \end{bmatrix} ,$$

and the second parameter of the mapped density is therefore given by Corollary 2.2' as

$$[\mathbf{B}\,\mathbf{M}_2\,\mathbf{B}^t]^{-1} = [\Sigma\, \beta_p^2 \mu_2^{(p)}]^{-1}$$ ◀

8.3. Student Function

The nondegenerate Student density function is defined on $R^{(r)}$ by

$$f_S^{(r)}(z|m, \mathbf{H}, \nu) \equiv \int_0^\infty f_N^{(r)}(z|m, h\mathbf{H})\, f_{\gamma 2}(h|1, \nu)\, dh \qquad \begin{array}{l} -\infty < z < \infty , \\ -\infty < m < \infty , \\ \nu > 0 , \\ \mathbf{H} \text{ is } PDS , \end{array}$$

$$= \frac{\nu^{\frac{1}{2}\nu}(\frac{1}{2}\nu + \frac{1}{2}r - 1)!}{\pi^{\frac{1}{2}r}\, (\frac{1}{2}\nu - 1)!} [\nu + (z - m)^t \mathbf{H}(z - m)]^{-\frac{1}{2}\nu - \frac{1}{2}r} |\mathbf{H}|^{\frac{1}{2}} . \quad \text{(8-28)}$$

The first two moments are

$$\mu_1 = m \ , \qquad \nu > 1 \ , \tag{8-29a}$$

$$\mu_2 = \mathbf{H}^{-1} \frac{\nu}{\nu - 2} \ , \qquad \nu > 2 \ . \tag{8-29b}$$

On the evaluation of the cumulative function when $r = 2$, see C. W. Dunnett and M. Sobel, *Biometrika* 41 (1954) pages 153-169.

▶ That f_S as defined by (8-28) is everywhere positive is clear from the fact that both f_N and $f_{\gamma 2}$ are everywhere positive. That the complete integral of f_S has the value 1 can therefore be proved by reversing the order of integration. Dropping parameters irrelevant to the argument we have

$$\int_{-\infty}^{\infty} f_S(z)\, dz = \int_{-\infty}^{\infty} \int_0^{\infty} f_N(z|h)\, f_{\gamma 2}(h)\, dh\, dz = \int_0^{\infty} \left[\int_{-\infty}^{\infty} f_N(z|h)\, dz\right] f_{\gamma 2}(h)\, dh$$
$$= \int_0^{\infty} f_{\gamma 2}(h)\, dh = 1 \ .$$

It follows that f_S is a proper normalized density function.

To evaluate the integral (8-28) we substitute for f_N and $f_{\gamma 2}$ their definitions (8-17) and (7-50) obtaining

$$f_S(z) = \int_0^{\infty} (2\pi)^{-\frac{1}{2}r}\, e^{-\frac{1}{2}h(z-\mathbf{m})^t \mathbf{H}(z-\mathbf{m})}\, h^{\frac{1}{2}r}\, |\mathbf{H}|^{\frac{1}{2}}\, \frac{e^{-\frac{1}{2}\nu h}(\frac{1}{2}\nu h)^{\frac{1}{2}\nu - 1}}{(\frac{1}{2}\nu - 1)!}\, \tfrac{1}{2}\nu\, dh$$
$$= \frac{|\mathbf{H}|^{\frac{1}{2}}(\frac{1}{2}\nu)^{\frac{1}{2}\nu}}{(\frac{1}{2}\nu - 1)!(2\pi)^{\frac{1}{2}r}} \int_0^{\infty} e^{-\frac{1}{2}h([z-\mathbf{m}]^t \mathbf{H}[z - m] + \nu)}\, h^{\frac{1}{2}\nu + \frac{1}{2}r - 1}\, dh \ .$$

Substituting herein

$$u = \tfrac{1}{2}\, h([z - m]^t \mathbf{H}[z - m] + \nu) \ ,$$

we obtain

$$f_S(z) = \frac{|\mathbf{H}|^{\frac{1}{2}}(\frac{1}{2}\nu)^{\frac{1}{2}\nu}(\frac{1}{2}\nu + \frac{1}{2}r - 1)!}{(2\pi)^{\frac{1}{2}r}(\frac{1}{2}\nu - 1)!}\, [\tfrac{1}{2}(\nu + [z - m]^t \mathbf{H}[z - m])]^{-\frac{1}{2}\nu - \frac{1}{2}r}$$
$$\times \int_0^{\infty} \frac{e^{-u}\, u^{\frac{1}{2}\nu + \frac{1}{2}r - 1}}{(\frac{1}{2}\nu + \frac{1}{2}r - 1)!}\, du \ .$$

The integral is $F_{\gamma *}(\infty\,|\frac{1}{2}\nu + \frac{1}{2}r) = 1$, and the factor preceding the integral can easily be reduced to (8-28).

To prove formula (8-29a) for the first moment we first observe that the moment exists if $\nu > 1$ because

$$z_i[\nu + (z - m)^t \mathbf{H}(z - m)]^{-\frac{1}{2}\nu - \frac{1}{2}r}$$

is of order $(-r - \nu + 1)$ at infinity. We can then obtain its value by using the integral definition of f_S, reversing the order of integration, and using formula (8-18a) for the first moment of f_N:

$$\mu_1 \equiv \int_{-\infty}^{\infty} \int_0^{\infty} z\, f_N^{(r)}(z|m, h\mathbf{H})\, f_{\gamma 2}(h|1, \nu)\, dh\, dz$$
$$= \int_0^{\infty} \left[\int_{-\infty}^{\infty} z\, f_N^{(r)}(z|m, h\mathbf{H})\, dz\right] f_{\gamma 2}(h|1, \nu)\, dh$$
$$= \int_0^{\infty} m\, f_{\gamma 2}(h|1, \nu)\, dh = m \ .$$

To prove formula (8-29b) for the second moment we proceed similarly. The moment exists for $\nu > 2$ because

$$(z_i - m_i)(z_j - m_j)[\nu + (z - m)^t \mathbf{H}(z - m)]^{-\frac{1}{2}\nu - \frac{1}{2}r}$$

is of order $(-r - \nu + 2)$ at infinity; and we can then calculate using (8-18b)

$$\begin{aligned}\boldsymbol{\mu}_2 &\equiv \int_{-\infty}^{\infty}\int_0^{\infty} (z - m)(z - m)^t f_N^{(r)}(z|m, h\mathbf{H})\, f_{\gamma 2}(h|1, \nu)\, dh\, dz \\ &= \int_0^{\infty}\left[\int_{-\infty}^{\infty} (z - m)(z - m)^t f_N^{(r)}(z|m, h\mathbf{H})\, dz\right] f_{\gamma 2}(h|1, \nu)\, dh \\ &= \int_0^{\infty} \mathbf{H}^{-1} h^{-1} f_{\gamma 2}(h|1, \nu)\, dh \\ &= \mathbf{H}^{-1}\frac{\frac{1}{2}\nu}{\frac{1}{2}\nu - 1}\int_0^{\infty}\frac{e^{-\frac{1}{2}\nu h}(\frac{1}{2}\nu h)^{\frac{1}{2}\nu - 2}}{(\frac{1}{2}\nu - 2)!}\,\tfrac{1}{2}\nu\, dh = \mathbf{H}^{-1}\frac{\nu}{\nu - 2}\,.\end{aligned}$$ ◀

8.3.1. *Conditional and Marginal Densities*

Consider a nondegenerate Student density $f_S^{(r)}(z|m, \mathbf{H}, \nu)$ defined on $R^{(r)}$; partition the space $R^{(r)}$ into subspaces $R_1^{(q)}$ and $R_2^{(r-q)}$; and partition the vectors z and m and the matrices $\mathbf{H}$ and $\mathbf{V} \equiv \mathbf{H}^{-1}$ correspondingly as shown in (8-19). Then the *marginal* density on R_1 is

$$f_S^{(q)}(z_1|m_1, \mathbf{V}_{11}^{-1}, \nu) = f_S^{(q)}(z_1|m_1, \mathbf{H}_{11} - \mathbf{H}_{12}\,\mathbf{H}_{22}^{-1}\,\mathbf{H}_{21}, \nu) \tag{8-30}$$

while the *conditional* density on R_1 given z_2 in R_2 is

$$\begin{aligned}&f_S^{(q)}(z_1|m_1 - \mathbf{H}_{11}^{-1}\,\mathbf{H}_{12}[z_2 - m_2], \mathbf{H}_{11}, \nu) \\ &\qquad = f_S^{(q)}(z_1|m_1 + \mathbf{V}_{12}\,\mathbf{V}_{22}^{-1}[z_2 - m_2], [\mathbf{V}_{11} - \mathbf{V}_{12}\,\mathbf{V}_{22}^{-1}\,\mathbf{V}_{21}]^{-1}, \nu)\,.\end{aligned} \tag{8-31}$$

▶ By the definition (8-28) of $f_S^{(r)}$, the marginal density on $R_1^{(q)}$, $q < r$, is

$$\int_{-\infty}^{\infty}\int_0^{\infty} f_N^{(r)}(z|m, h\mathbf{H})\, f_{\gamma 2}(h|1, \nu)\, dh\, dz_2$$

where the integral with respect to z_2 is evaluated over R_2. Reversing the order of integration and using (8-20) we obtain

$$\int_0^{\infty}\left[\int_{-\infty}^{\infty} f_N^{(r)}(z|m, h\mathbf{H})\, dz_2\right] f_{\gamma 2}(h|1, \nu)\, dh = \int_0^{\infty} f_N^{(q)}(z|m_1, h\mathbf{V}_{11}^{-1})\, f_{\gamma 2}(h|1, \nu)\, dh\ ;$$

and by (8-28) the integral on the right is (8-30).
The proof of (8-31) is similar. ◀

8.3.2. *Linear Combinations of Student Random Variables*

In the applications we shall require the following theorems concerning linear combinations of random variables whose joint distribution has a density given by the (multivariate) Student function.

Theorem 1. Let $\tilde{z}_1, \cdots, \tilde{z}_r$ be scalar random variables whose joint distribution is nondegenerate Student with parameters m, $\mathbf{H}$, and ν, so that

$$\mathrm{D}(z) = f_S^{(r)}(z|m, \mathbf{H}, \nu) \qquad \text{where} \qquad z \text{ is } r \times 1\ ; \tag{8-32a}$$

and let $\tilde{y}_1, \cdots, \tilde{y}_q$ be $q \leq r$ linearly independent linear combinations of the $\tilde{z}$s:

$$\tilde{y} = c + d\,\tilde{z} \qquad \text{where} \qquad \begin{array}{l} c \text{ is } q \times 1 \;, \\ d \text{ is } q \times r \text{ of rank } q \;. \end{array} \tag{8-32b}$$

Then the joint distribution of the $\tilde{y}$s is nondegenerate Student with density

$$D(y) = f_S^{(q)}(y|c + d\,m, [d\,H^{-1}\,d^t]^{-1}, \nu) \;. \tag{8-32c}$$

Theorem 2. Let $\tilde{z}^{(1)}, \cdots, \tilde{z}^{(p)}, \cdots, \tilde{z}^{(n)}$ be $r \times 1$ random variables which conditionally on given h have independent Normal densities of the form $f_N^{(r)}(z^{(p)}|\bar{z}^{(p)}, hV_p^{-1})$; let $\tilde{h}$ have a gamma-2 distribution with density $f_{\gamma 2}(h|1, \nu)$; and let $\tilde{y}$ be an $r \times 1$ linear combination of the $\tilde{z}$s,

$$\tilde{y} = \alpha + \Sigma\,\beta_p\,\tilde{z}^{(p)} \;. \tag{8-33a}$$

Then the distribution of $\tilde{y}$ is nondegenerate Student with density

$$D(y) = f_S^{(q)}(y|\alpha + \Sigma\,\beta_p\,\bar{z}^{(p)}, [\Sigma\,\beta_p^2\,V_p]^{-1}, \nu) \;. \tag{8-33b}$$

▶ Theorems 1 and 2 follow from Corollaries 2.1 and 2.2 of Section 8.2.3 by a line of argument identical to the one used to prove (8-30): first obtain the Normal density conditional on h and then integrate with respect to h. ◀

8.4. Inverted-Student Function

The nondegenerate inverted-Student density function is defined on $R^{(r)}$ by

$$f_{iS}^{(r)}(y|m, H, \nu) \equiv \frac{(\frac{1}{2}[\nu + r] - 1)!}{\nu^{\frac{1}{2}(\nu+r)-1}\,\pi^{\frac{1}{2}r}(\frac{1}{2}\nu - 1)!}\,[\nu - (y - m)^t\,H(y - m)]^{\frac{1}{2}\nu-1}\,|H|^{\frac{1}{2}} \tag{8-34a}$$

where

$$-\infty < m < \infty \;, \qquad \nu > 0 \;, \qquad H \text{ is PDS} \;,$$
$$(y - m)^t\,H(y - m) \le \nu \;. \tag{8-34b}$$

It is related to the Student function (8-28) by

$$F_{iS}(y|m, H, \nu) = F_S(z|m, H, \nu) \tag{8-35a}$$

where z is defined by

$$z - m = \frac{y - m}{\sqrt{1 - (y - m)^t(H/\nu)(y - m)}} \;. \tag{8-35b}$$

▶ To prove the relation (8-35) and thus to show that f_{iS} as defined by (8-34) is a proper normalized density function we shall prove that (8-34) results from substituting new variables y defined by (8-35b) for the variables z in the definition (8-28) of the Student function f_S. Since we obtain (8-34) immediately upon substitution of (8-35b) in (8-28) and multiplication by the Jacobian

$$|dz/dy| = [1 - (y - m)^t(H/\nu)(y - m)]^{-\frac{1}{2}r-1} \;,$$

all that we really need to prove is that this is in fact the Jacobian of the transformation; and to prove this, we break the transformation down into three steps. Observing that because H is positive-definite and ν is positive there exists by Theorem 7 of Section 8.0.3 a nonsingular matrix Q such that

$$\mathbf{Q}^t\,\mathbf{Q} = \mathbf{H}/\nu \ ,$$

we define

$$z - m = \mathbf{Q}^{-1} u \ , \qquad u = \frac{v}{\sqrt{1 - v^t v}} \ , \qquad v = \mathbf{Q}(y - m) \ ,$$

so that the Jacobian becomes

$$|dz/dy| = |\mathbf{Q}^{-1}|\,|du/dv|\,|\mathbf{Q}| = |du/dv| \ .$$

The transformation from v to u maps a sphere of radius $\rho = \sqrt{v^t v}$ into a sphere of radius $\rho/\sqrt{1 - \rho^2}$, and because of the symmetry we can derive the Jacobian by comparing the ratio of the volume element $\Pi\, dv_i$ at the particular point $v = (0, 0, \cdots, 0, \rho)$ to the volume element $\Pi\, du_i$ at the corresponding point $u = (0, 0, \cdots, 0, \rho/\sqrt{1 - \rho^2})$. Both elements are rectangular prisms with 1-dimensional altitudes along the rth coordinate axis and $(r - 1)$-dimensional bases tangent to the surface of the sphere and perpendicular to the rth axis. The volume $\Pi\, dv_i$ is thus

$$dV_v = d\rho \cdot (dv_1\, dv_2 \cdots dv_{r-1})$$

while the volume $\Pi\, du_i$ is

$$dV_u = d\left[\frac{\rho}{\sqrt{1 - \rho^2}}\right] \cdot \left[\frac{dv_1}{\sqrt{1 - \rho^2}}\,\frac{dv_2}{\sqrt{1 - \rho^2}} \cdots \frac{dv_{r-1}}{\sqrt{1 - \rho^2}}\right]$$
$$= (1 - \rho^2)^{-\frac{3}{2}}(1 - \rho^2)^{-\frac{1}{2}(r-1)}\, dV_v = (1 - v^t v)^{-\frac{1}{2}r-1}\, dV_v \ .$$

Substituting $v = \mathbf{Q}(y - m)$ in this result we obtain

$$|dz/dy| = |du/dv| = \frac{dV_u}{dV_v} = [1 - (y - m)^t(\mathbf{H}/\nu)(y - m)]^{-\frac{1}{2}r-1}$$

as was to be proved. ◀

CHAPTER 9

Bernoulli Process

9.1. Prior and Posterior Analysis

9.1.1. *Definition of a Bernoulli Process*

A Bernoulli process can be defined as a process generating independent random variables $\tilde{x}_1, \cdots, \tilde{x}_i, \cdots$ with identical binomial mass functions

$$f_b(x|p, 1) = p^x(1-p)^{1-x} , \qquad \begin{array}{l} x = 0, 1 , \\ 0 < p < 1 . \end{array} \tag{9-1}$$

As is customary, we shall refer to the observation of a single random variable $\tilde{x}$ as a trial, to the observation of a variable with value 1 as a success, and to the observation of a variable with value 0 as a failure.

Alternatively, a Bernoulli process can be defined as a process generating independent random variables $\tilde{y}_1, \cdots, \tilde{y}_i, \cdots$ with identical Pascal mass functions

$$f_{Pa}(y|p, 1) = (1-p)^{y-1}\, p , \qquad \begin{array}{l} y = 1, 2, \cdots , \\ 0 < p < 1 . \end{array} \tag{9-2}$$

When the process is regarded in this way, it is more naturally characterized by the parameter

$$\rho \equiv \frac{1}{p} \tag{9-3}$$

than by p itself because, by (7-12a),

$$E(\tilde{y}|p) = \frac{1}{p} = \rho . \tag{9-4}$$

In the applications, y is the length of the "interval" between two adjacent successes, as measured by the number of "trials" occurring *after* one success up to *and including* the trial on which the next success occurs. The quantity $\rho = 1/p$ can be interpreted as the mean number of trials per success.

A unique and very important property of a Bernoulli process is the fact that

$$P\{\tilde{y} > y + \eta | \tilde{y} > \eta\} = P\{\tilde{y} > y\} . \tag{9-5}$$

If y is interpreted as the number of trials from one success to the next, this says that the probability that more than y *additional* trials will be required to obtain the next success is independent of the number η of trials that have already been made since the last success. This implies in turn that the distribution of the number of trials required to obtain a success starting from any arbitrarily chosen point in a sequence of trials is the same as the distribution of the number of trials required starting immediately after a success.

▶ To show that the independence of past history asserted by (9-5) follows from (9-2), we write

$$P\{\tilde{y} > \eta + y | \tilde{y} > \eta\} = \frac{P\{\tilde{y} > \eta + y\}}{P(\tilde{y} > \eta\}} = \frac{\sum_{i=1}^{\infty} (1-p)^{\eta+y+i-1}p}{\sum_{j=1}^{\infty} (1-p)^{\eta+j-1}p}$$

$$= \frac{\sum_{i=1}^{\infty} (1-p)^{\eta+y+i-1}p}{(1-p)^{\eta}} = \sum_{i=1}^{\infty} (1-p)^{y+i-1}p$$

$$= P\{\tilde{y} > y\} \ .$$

To show that this independence of past history is a *unique* property of the Bernoulli process among all processes generating integral-valued random variables $\tilde{y} \geq 1$, we define

$$H(y) \equiv P\{\tilde{y} > y\}$$

and write (9-5) in the form

$$\frac{H(y+\eta)}{H(\eta)} = H(y) \ .$$

Letting $\eta = 1$ we get the recursion relation

$$H(y+1) = H(y)\, H(1)$$

from which we derive

$$H(y) = [H(1)]^y \ .$$

Now defining p by

$$H(1) = 1 - p \ ,$$

we obtain (9-2) by writing

$$P\{\tilde{y} = y\} = H(y-1) - H(y) = [H(1)]^{y-1}[1 - H(1)] = (1-p)^{y-1}p \ . \quad ◀$$

9.1.2. *Likelihood of a Sample*

The conditional probability, given p, that a Bernoulli process will *generate* r successes and $n - r$ failures in some specific order such as $ssfsf \cdots$ is the product of the conditional probabilities of the outcomes of the independent trials, which by (9-1) is

$$\Pi(p^{x_i}[1-p]^{1-x_i}) = p^r(1-p)^{n-r} \ . \tag{9-6}$$

If in addition the stopping process is noninformative in the sense of Section 2.3, the expression (9-6) is the kernel of the *likelihood of the sample.* It is clear that in this case all the information in the sample is conveyed by the statistic (r, n): this statistic is *sufficient* when the stopping process is noninformative.

It is worth emphasizing that we get an identical result by computing the probability that the process will generate r intervals of lengths $y_1, \cdots, y_r$ followed by one interval of length greater than y_{r+1}. The probabilities of the r complete intervals are given by (9-2); the probability of the incomplete interval is

$$P\{\tilde{y} > y|p\} = \Sigma_{j=y+1}^{\infty}(1-p)^{j-1}\,p = (1-p)^y \; ; \tag{9-7}$$

and therefore the joint probability is

$$\Pi_{i=1}^{r}\,[(1-p)^{y_i-1}\,p](1-p)^{y_{r+1}} = p^r(1-p)^{\Sigma y_i - r} \; , \tag{9-8}$$

which reduces to (9-6) on substitution of $n = \Sigma_{i=1}^{r+1} y_i$.

9.1.3. Conjugate Distribution of $\tilde{p}$

When the parameter of a Bernoulli process is treated as a random variable $\tilde{p}$, the most convenient distribution of $\tilde{p}$—the natural conjugate of (9-6)—is the beta distribution defined by (7-21):

$$f_\beta(p|r, n) \propto p^{r-1}(1-p)^{n-r-1} \; . \tag{9-9}$$

If such a distribution with parameter (r', n') has been assigned to $\tilde{p}$ and if then a sample from the process yields a sufficient statistic (r, n), the posterior distribution of $\tilde{p}$ will be beta with parameters

$$r'' = r' + r \; , \qquad n'' = n' + n \; . \tag{9-10}$$

▶ By Bayes' theorem in the form (2-6) the posterior density is proportional to the product of the kernel (9-9) of the prior density and the kernel (9-6) of the sample likelihood:

$$D''(p|r', n'; r, n) \propto p^{r'-1}(1-p)^{n'-r'-1} \cdot p^r(1-p)^{n-r} = p^{r''-1}(1-p)^{n''-r''-1} \; .$$

Comparing this result with (9-9) we see that the density is beta. ◀

The mean, partial expectation, and variance of the distribution (9-9) of $\tilde{p}$ are, by (7-22),

$$E(\tilde{p}|r, n) \equiv \bar{p} = \frac{r}{n} \; , \tag{9-11a}$$

$$E_0^p(\tilde{p}|r, n) = \bar{p}\, F_\beta(p|r+1, n+1) \; , \tag{9-11b}$$

$$V(\tilde{p}|r, n) = \frac{r(n-r)}{n^2(n+1)} = \frac{\bar{p}(1-\bar{p})}{n+1} \; . \tag{9-11c}$$

From (9-11b) it follows that the left- and right-hand linear-loss integrals defined by (5-12) are

$$L_l(p) \equiv \int_0^p (p-z)\, f_\beta(z|r, n)\, dz = p\, F_\beta(p|r, n) - \bar{p}\, F_\beta(p|r+1, n+1) \; ; \tag{9-12a}$$

$$L_r(p) \equiv \int_p^1 (z-p)\, f_\beta(z|r, n)\, dz = \bar{p}\, G_\beta(p|r+1, n+1) - p\, G_\beta(p|r, n) \; . \tag{9-12b}$$

For tables of the cumulative beta function, see Section 7.3.

Limiting Behavior of the Prior Distribution. As the parameters r and n of (9-9) both approach zero in such a way that the ratio $r/n = \bar{p}$ remains fixed, a fraction $\bar{p}$ of the total probability becomes more and more concentrated toward $p = 1$, the remainder toward $p = 0$; the variance approaches $\bar{p}(1 - \bar{p})$.

▶ The behavior of the variance is given by (9-11c). To show the behavior of the density as $n, r \to 0$ with $r/n = \bar{p}$ fixed, write the density (9-9) in the form

$$\frac{(n-1)!}{(n\bar{p}-1)!(n[1-\bar{p}]-1)!}\,p^{n\bar{p}-1}(1-p)^{n(1-\bar{p})-1}$$
$$= \frac{(n\bar{p})(n[1-\bar{p}])}{n}\cdot\frac{n!}{(n\bar{p})!(n[1-\bar{p}])!}\cdot\frac{1}{p^{1-n\bar{p}}(1-p)^{1-n(1-\bar{p})}}\ .$$

As $n \to 0$, the first factor on the right approaches 0, the second approaches 1, and the third approaches $\dfrac{1}{p(1-p)}$, so that the entire expression approaches 0 except at $p = 0$ and 1. The sharing of the total probability between $p = 0$ and $p = 1$ is then clear from the fact that the mean remains fixed at $\bar{p}$. ◀

9.1.4. *Conjugate Distribution of $1/\tilde{p} = \tilde{\rho}$*

In problems where utility is linear in $1/p$ rather than in p, we shall need the same kinds of information concerning the distribution of $\tilde{\rho} = 1/\tilde{p}$ that we have just obtained for the distribution of $\tilde{p}$. Substituting $1/\rho$ for p in the beta density (9-9) and multiplying by the Jacobian we find that the induced density of $\tilde{\rho}$ is inverted-beta-1 as defined by (7-25):

$$f_{i\beta 1}(\rho|r, n, 1) \propto \frac{(\rho-1)^{n-r-1}}{\rho^n}\ . \tag{9-13}$$

Cumulative probabilities may be obtained from tables of the standardized beta or binomial functions by use of the relations

$$G_{i\beta 1}(\rho|r, n, 1) = F_{\beta^*}(1/\rho|r, n-r) = G_b(r|1/\rho, n-1)\ , \tag{9-14}$$

which follow from (7-26), (7-23), and (7-24). The mean, partial expectation, and variance of $\tilde{\rho}$ are, by (7-27),

$$\text{E}(\tilde{\rho}|r, n) \equiv \bar{\rho} = \frac{n-1}{r-1}\ , \qquad r > 1\ , \tag{9-15a}$$

$$\text{E}_1^{\rho}(\tilde{\rho}|r, n) = \bar{\rho}\, G_{\beta^*}(1/\rho|r-1, n-r)\ , \qquad r > 1\ , \tag{9-15b}$$

$$\text{V}(\tilde{\rho}|r, n) = \frac{(n-1)(n-r)}{(r-1)^2(r-2)} = \frac{\bar{\rho}(\bar{\rho}-1)}{r-2}\ , \qquad r > 2\ . \tag{9-15c}$$

It follows from (9-14) and (9-15b) that the linear-loss integrals are

$$L_l(\rho) \equiv \int_1^{\rho} (\rho - z)\, f_{i\beta 1}(z|r, n, 1)\, dz$$
$$= \rho\, G_{\beta^*}(1/\rho|r, n-r) - \bar{\rho}\, G_{\beta^*}(1/\rho|r-1, n-r)\ , \tag{9-16a}$$

$$L_r(\rho) \equiv \int_{\rho}^{\infty} (z - \rho)\, f_{i\beta 1}(z|r, n, 1)\, dz$$
$$= \bar{\rho}\, F_{\beta^*}(1/\rho|r-1, n-r) - \rho\, F_{\beta^*}(1/\rho|r, n-r)\ . \tag{9-16b}$$

9.2. Sampling Distributions and Preposterior Analysis: Binomial Sampling

9.2.1. Definition of Binomial Sampling

If a sample is taken from a Bernoulli process in such a way that the value of the statistic n is predetermined, the value of the statistic $\tilde{r}$ being left to chance, the sampling will be called binomial. In other words, binomial sampling consists in counting the number of successes which occur in a specified number of trials.

We assume throughout Section 9.2 that a binomial sample with fixed n is to be taken from a Bernoulli process whose parameter $\tilde{p}$ has a beta distribution with parameter (r', n').

9.2.2. Conditional Distribution of $(\tilde{r}|p)$

The conditional probability, given p, that a sample of n trials from a Bernoulli process will yield r successes in some particular order was shown in (9-6) to be $p^r(1-p)^{n-r}$; and it is easy to see that there are $\dfrac{n!}{r!(n-r)!}$ orders in which the r successes and $n - r$ failures can occur. Since the orders are mutually exclusive, the conditional probability of r successes in n trials regardless of order is the product of these two factors, and it follows that the conditional probability of $(r|p)$ is given by the binomial mass function defined in (7-8).

$$P\{r|p, n\} = \frac{n!}{r!(n-r)!}\, p^r(1-p)^{n-r} \equiv f_b(r|p, n) \ . \tag{9-17}$$

9.2.3. Unconditional Distribution of $\tilde{r}$

If a sample of size n is drawn by binomial sampling from a Bernoulli process whose parameter is a random variable $\tilde{p}$ having the beta density (9-9) with parameter (r', n'), then by the definition (7-76) of the beta-binomial mass function the unconditional distribution of $\tilde{r}$ is

$$P\{r|r', n'; n\} = \int_0^1 f_b(r|p, n)\, f_\beta(p|r', n')\, dp \equiv f_{\beta b}(r|r', n', n) \ . \tag{9-18}$$

9.2.4. Distribution of $\tilde{\bar{p}}''$

If the prior distribution of $\tilde{p}$ is beta with parameter (r', n') and if a sample then yields a sufficient statistic (r, n), formulas (9-11a) and (9-10) show that the mean of the posterior distribution of $\tilde{p}$ will be

$$\bar{p}'' = \frac{r''}{n''} = \frac{r + r'}{n + n'} \ . \tag{9-19}$$

It follows that when the value of the $\tilde{r}$ to be obtained by binomial sampling is unknown and $\tilde{\bar{p}}''$ is therefore a random variable, the distribution of $\tilde{\bar{p}}''$ can be obtained from the unconditional distribution (9-18) of $\tilde{r}$. Solving (9-19) for r and substituting in (9-18) we have

$$P\{\bar{p}''|r', n'; n\} = f_{\beta b}(n''\bar{p}'' - r'|r', n', n) . \tag{9-20}$$

The mean, partial expectation, and variance of this distribution of $\tilde{\bar{p}}''$ are

$$E(\tilde{\bar{p}}''|r', n'; n) = \frac{r'}{n'} = E(\tilde{p}|r', n') \equiv \bar{p}' , \tag{9-21a}$$

$$E_0^p(\tilde{\bar{p}}''|r', n'; n) = \bar{p}' F_{\beta b}(pn'' - r'|r' + 1, n' + 1, n) , \tag{9-21b}$$

$$V(\tilde{\bar{p}}''|r', n'; n) = \frac{n}{n + n'} \frac{r'(n' - r')}{n'^2(n' + 1)} = \frac{n}{n + n'} V(\tilde{p}|r', n') . \tag{9-21c}$$

It follows at once from (9-20) and (9-21b) that the linear-loss integrals under this distribution of $\tilde{\bar{p}}''$ are given by

$$\begin{aligned} L_l(p) &\equiv \sum_{z \le p} (p - z) P\{\tilde{\bar{p}}'' = z|r', n'; n\} \\ &= p F_{\beta b}(pn'' - r'|r', n', n) - \bar{p}' F_{\beta b}(pn'' - r'|r' + 1, n' + 1, n) , \end{aligned} \tag{9-22a}$$

$$\begin{aligned} L_r(p) &\equiv \sum_{z \ge p} (z - p) P\{\tilde{\bar{p}}'' = z|r', n'; n\} \\ &= \bar{p}' G_{\beta b}(pn'' - r'|r' + 1, n' + 1, n) - p G_{\beta b}(pn'' - r'|r', n', n) . \end{aligned} \tag{9-22b}$$

▶ Formula (9-21a) for the mean of $\tilde{\bar{p}}''$ follows immediately from (5-27).
To prove (9-21b) we write the definition

$$E_0^p(\tilde{\bar{p}}'') \equiv \sum_{\bar{p}'' \le p} \bar{p}'' P\{\bar{p}''\}$$

and then substitute (9-19) for $\bar{p}''$ and replace the mass function $P\{\bar{p}''\}$ by the mass function (9-18) of $\tilde{r}$. Using formula (7-76) for this latter mass function we thus obtain

$$E_0^p(\tilde{\bar{p}}'') = \sum_{r \le r_p} \frac{r + r'}{n + n'} \cdot \frac{(r + r' - 1)!(n + n' - r - r' - 1)!\, n!\, (n' - 1)!}{r!(r' - 1)!(n - r)!(n' - r' - 1)!(n + n' - 1)!}$$

where by (9-19)

$$r_p = pn'' - r' .$$

Defining $r^* = r' + 1$ and $n^* = n' + 1$, the sum on the right can be written

$$\frac{r'}{n'} \sum_{r \le r_p} \frac{(r + r^* - 1)!(n + n^* - r - r^* - 1)!n!\, (n^* - 1)!}{r!(r^* - 1)!(n - r)!(n^* - r^* - 1)!(n + n^* - 1)!} = \bar{p}' \sum_{r \le r_p} f_{\beta b}(r|r^*, n^*, n) ,$$

as was to be shown.

To prove formula (9-21c) for the variance, observe that, by (9-19),

$$\tilde{\bar{p}}'' = \tilde{r}/n'' + \text{constant}$$

is a linear function of $\tilde{r}$, so that

$$V(\tilde{\bar{p}}'') = \frac{1}{n''^2} V(\tilde{r}|r', n'; n) .$$

Formula (9-21c) is obtained by substituting herein the value of $V(\tilde{r})$ as given by (7-77b):

$$V(\tilde{r}|r', n'; n) = n(n + n') \frac{r'(n' - r')}{n'^2(n' + 1)} .$$ ◀

Approximations. In principle, computation of a cumulative probability such as $P\{\tilde{\bar{p}}'' < p|r', n'; n\}$ under the distribution defined by (9-20) or evaluation of a linear-loss integral as given by (9-22) involves computation of a beta-binomial cumulative probability in some one of the ways discussed in Section 7.11.2; and when n is small and the distribution of $\tilde{r}$ and therefore of $\tilde{\bar{p}}''$ is very discrete, this is the only method which will give good results. Even in this case, however, exact term-by-term computation will rarely be necessary because (as is pointed out in Section 7.11.2) a binomial or beta approximation to $F_{\beta b}$ will give good results if $r' \ll n'$, $r \ll n$, and $r + r' \ll \max\{n, n'\}$.

As n increases, the distribution of $\tilde{\bar{p}}''$ becomes more nearly continuous and approaches the beta prior distribution of $\tilde{p}$, so that a direct beta approximation to the distribution of $\tilde{\bar{p}}''$ itself—*not* a Normal approximation—will give good results. It follows from (9-21) and (9-11) that a beta distribution with parameters r^*, n^* defined by

$$n^* + 1 = \frac{n + n'}{n}(n' + 1) , \qquad r^* = \frac{n^*}{n'} r' , \tag{9-23}$$

will have the same mean and variance as the exact distribution of $\tilde{\bar{p}}''$. Using these parameter values, cumulative probabilities may be obtained by use of formula (7-23) or (7-24) and linear-loss integrals may be evaluated by use of formulas (9-12).

9.2.5. *Distribution of $\tilde{\bar{\rho}}''$*

By (9-15a) and (9-10) the posterior expectation of $\tilde{\rho}$ will be

$$\bar{\rho}'' = \frac{n'' - 1}{r + r' - 1} \cdot \tag{9-24}$$

It follows that when $\tilde{r}$ and therefore $\tilde{\bar{\rho}}''$ are random variables, the distribution of $\tilde{\bar{\rho}}''$ can be obtained from the unconditional distribution (9-18) of $\tilde{r}$. Solving (9-24) for r and remembering that $\bar{\rho}''$ is a decreasing function of r we have

$$P\{\tilde{\bar{\rho}}'' \le \rho|r', n'; n\} = G_{\beta b}(r_\rho|r', n', n) \tag{9-25}$$

where

$$r_\rho \equiv \frac{n'' - 1}{\rho} - (r' - 1) . \tag{9-26}$$

The mean and partial expectation of the distribution of $\tilde{\bar{\rho}}''$ defined by (9-25) are

$$E(\tilde{\bar{\rho}}''|r', n'; n) = \frac{n' - 1}{r' - 1} = E(\tilde{\rho}|r', n') \equiv \bar{\rho}' , \tag{9-27a}$$

$$E_0^\rho(\tilde{\bar{\rho}}''|r', n'; n) = \bar{\rho}' \, G_{\beta b}(r_\rho|r' - 1, n' - 1, n) . \tag{9-27b}$$

It follows immediately from (9-25) and (9-27b) that the linear-loss integrals under this distribution of $\tilde{\bar{\rho}}''$ are given by

$$\begin{aligned} L_l(\rho) &\equiv \sum_{z \le \rho} (\rho - z) \, P\{\tilde{\bar{\rho}}'' = z|r', n'; n\} \\ &= \rho \, G_{\beta b}(r_\rho|r', n', n) - \bar{\rho}' \, G_{\beta b}(r_\rho|r' - 1, n' - 1, n) ; \end{aligned} \tag{9-28a}$$

$$L_r(\rho) \equiv \sum_{z \geq \rho} (z - \rho) \, \mathrm{P}\{\tilde{\bar{p}}'' = z | r', n'; n\}$$

$$= \bar{p}' \, F_{\beta b}(r_\rho | r' - 1, n' - 1, n) - \rho \, F_{\beta b}(r_\rho | r', n', n) \; . \qquad \text{(9-28b)}$$

▶ The proof of formulas (9-27) is very similar to the proof of (9-21ab), the main difference being that in determining the region of summation for (9-27b) we must remember that $\bar{p}''$ is a decreasing function of r:

$$\mathrm{E}_0^\rho(\tilde{\bar{p}}'') = \sum_{r \geq r_\rho} \frac{n' + n - 1}{r' + r - 1} f_{\beta b}(r | r', n', n) = \sum_{r \geq r_\rho} \frac{n' - 1}{r' - 1} f_{\beta b}(r | r' - 1, n' - 1, n)$$

where r_ρ is defined by (9-26). ◀

No closed formula for the variance of $\tilde{\bar{p}}''$ or for its second moment about the origin can be found, but the latter quantity is approximated by

$$S(J) \equiv \frac{(n'' - 1)(n' - 1)(n' - 2)}{(n'' - 2)(r' - 1)(r' - 2)}$$

$$\times \left[1 - \sum_{j=2}^{J} \frac{(j - 1)!(-1)^j \, (n' - 3) \cdots (n' - 1 - j)}{(n'' - 3) \cdots (n'' - 1 - j)(r' - 3) \cdots (r' - 1 - j)}\right] \cdot \qquad \text{(9-29a)}$$

The sequence $S(J)$ considered as a function of J does not converge, but as long as $J < r' - 1$ it gives values which are alternately above and below the true value of $\mathrm{E}(\tilde{\bar{p}}''^2)$. The error is therefore less than the last term included, and two or three terms will give a very good approximation if $\bar{p}'$ is reasonably small and n is reasonably large. If for example $n = 10$ and if $r' = 12$, $n' = 15$ (implying $\bar{p}' = 1.273$, $\sqrt{\mathrm{V}(\tilde{p})} = .186$), the series within the brackets of (9-29a) is $1 - .061 + .008 - \cdots$. Notice, however, that at least *two* terms *must* be used, since the first term by itself is simply $(n'' - 1)/(n'' - 2)$ times the second moment of $\tilde{p}$ itself about the origin. The variance can then be approximated by use of

$$\mathrm{V}(\tilde{\bar{p}}'') = \mathrm{E}(\tilde{\bar{p}}''^2) - [\mathrm{E}(\tilde{\bar{p}}'')]^2 \doteq S(J) - \left[\frac{n' - 1}{r' - 1}\right]^2 \cdot \qquad \text{(9-29b)}$$

▶ To show that the quantity $S(J)$ defined by (9-29a) has the properties described in the text we shall prove that for any J such that $1 \leq J < r' - 1$,

$$S(J) \equiv \mathrm{E}\left(\tilde{\bar{p}}''^2\left[1 + \frac{J!(-1)^{J-1}}{(\tilde{r} + r' - 2) \cdots (\tilde{r} + r' - 1 - J)}\right]\right) ,$$

since it is immediately apparent that if this is true, then $S(1)$, $S(2)$, $\cdots$, $S(J)$ will be alternately greater than and less than $\mathrm{E}(\tilde{\bar{p}}''^2)$.

We start by rewriting the definition (9-29a) of $S(J)$ in the form

$$S(J) = (n'' - 1)^2 \sum_{j=1}^{J} \frac{(j - 1)!(-1)^{j-1}}{(n'' - 1) \cdots (n'' - 1 - j)} \cdot \frac{(n' - 1) \cdots (n' - 1 - j)}{(r' - 1) \cdots (r' - 1 - j)} \, .$$

We next observe that we are entitled to write

$$S(J) = (n'' - 1)^2 \sum_{j=1}^{J} \frac{(j-1)!(-1)^{j-1}}{(n''-1)\cdots(n''-1-j)} \cdot \mathrm{E}\left[\frac{(n''-1)\cdots(n''-1-j)}{(\tilde{r}''-1)\cdots(\tilde{r}''-1-j)}\right]$$

because

$$\mathrm{E}\left[\frac{(n''-1)\cdots(n''-1-j)}{(\tilde{r}''-1)\cdots(\tilde{r}''-1-j)}\right] \equiv \sum_{r=0}^{n} \frac{(n+n'-1)\cdots(n+n'-1-j)}{(r+r'-1)\cdots(r+r'-1-j)} f_{\beta b}(r|r', n', n)$$

$$= \frac{(n'-1)\cdots(n'-1-j)}{(r'-1)\cdots(r'-1-j)} \sum_{r=0}^{n} f_{\beta b}(r|r'-j-1, n'-j-1, n)$$

$$= \frac{(n'-1)\cdots(n'-1-j)}{(r'-1)\cdots(r'-1-j)} .$$

If now we reverse the order of expectation and summation in our previous expression for $S(J)$ and then cancel out the factors involving n'', we obtain

$$S(J) = (n''-1)^2 \cdot \mathrm{E}\left[\sum_{j=1}^{J} \frac{(j-1)!(-1)^{j-1}}{(\tilde{r}''-1)\cdots(\tilde{r}''-1-j)}\right] .$$

We can easily verify that the jth term in the summand is equal to

$$-\frac{1}{\tilde{r}''-1} \Delta \frac{(j-1)!(-1)^{j-1}}{(\tilde{r}''-1)\cdots(\tilde{r}''-j)} ,$$

where Δ is the first-difference operator defined by $\Delta\, a_j = a_{j+1} - a_j$. Then since

$$\Sigma_{j=1}^{J} \Delta\, a_j = a_{J+1} - a_1 ,$$

we may write

$$S(J) = (n''-1)^2\, \mathrm{E}\left(\frac{1}{\tilde{r}''-1}\left[\frac{1}{\tilde{r}''-1} - \frac{J!(-1)^J}{(\tilde{r}''-1)\cdots(\tilde{r}''-J-1)}\right]\right)$$

$$= \mathrm{E}\left(\left[\frac{n''-1}{\tilde{r}''-1}\right]^2 \left[1 + \frac{J!(-1)^{J-1}}{(\tilde{r}''-2)\cdots(\tilde{r}''-J-1)}\right]\right) .$$

The proof is completed by recalling that

$$\tilde{\tilde{\rho}}'' = \frac{n''-1}{\tilde{r}''-1} .$$

◀

Approximations. In principle, computation of a cumulative probability such as $\mathrm{P}\{\tilde{\tilde{\rho}}'' < \rho | r', n'; n\}$ under the distribution defined by (9-25) or evaluation of a linear-loss integral as given by (9-28) involves computation of a beta-binomial cumulative probability in some one of the ways discussed in Section 7.11.2, a beta or binomial approximation to the beta-binomial giving good results when $r' \ll n'$, $r \ll n$, and $r + r' \ll \max\{n, n'\}$.

As n increases, the distribution of $\tilde{\tilde{\rho}}''$ becomes more nearly continuous and approaches the inverted-beta-1 prior distribution of $\tilde{\rho}$, so that a direct inverted-beta-1 approximation to the distribution of $\tilde{\tilde{\rho}}''$ itself—not a Normal approximation—will give good results. It follows from (9-15) and (9-27a) that an inverted-beta-1 distribution with parameters

$$\begin{aligned} r^* &= \frac{\mathrm{E}(\tilde{\bar{p}}'')[\mathrm{E}(\tilde{\bar{p}}'') - 1)]}{\mathrm{V}(\tilde{\bar{p}}'')} + 2 = \frac{\bar{p}'(\bar{p}' - 1)}{\mathrm{V}(\tilde{\bar{p}}'')} + 2 \ , \\ n^* &= (r^* - 1)\ \mathrm{E}(\tilde{\bar{p}}'') + 1 = (r^* - 1)\bar{p}' + 1 \ , \\ b^* &= 1 \ , \end{aligned} \tag{9-30}$$

would have the same mean and variance as the exact distribution of $\tilde{\bar{p}}''$. Once (9-30) has been evaluated approximately by use of (9-29), tail probabilities may be computed by use of (9-14) and linear-loss integrals may be evaluated by use of (9-16).

9.3. Sampling Distributions and Preposterior Analysis: Pascal Sampling

9.3.1. Definition of Pascal Sampling

If a sample is taken from a Bernoulli process in such a way that the value of the statistic r is predetermined, the value of the statistic $\tilde{n}$ being left to chance, the sampling will be called Pascal. In other words, Pascal sampling consists in counting the number of trials required to obtain a specified number of successes.

We assume throughout Section 9.3 that a Pascal sample with fixed r is to be taken from a Bernoulli process whose parameter $\tilde{p}$ has a beta distribution with parameter (r', n').

9.3.2. Conditional Distribution of $(\tilde{n}|p)$.

To find the conditional probability, given p, that n trials will be required to secure r successes, we start from the fact that the probability that the process will generate r successes and $n - r$ failures in any one specified order is $p^r(1 - p)^{n-r}$, just as in the case of binomial sampling with n fixed rather than r. The number of possible orders is different, however: we are specifying that the nth trial is a success, and therefore samples with the same n can differ only as regards the arrangement of the first $r - 1$ successes within the first $n - 1$ trials. Accordingly the conditional probability, given r and p, that the rth success will occur on the nth trial is given by the Pascal mass function (7-11):

$$\mathrm{P}\{n|p, r\} = \frac{(n - 1)!}{(r - 1)!(n - r)!}\, p^r(1 - p)^{n-r} \equiv f_{Pa}(n|p, r) \ . \tag{9-31}$$

9.3.3. Unconditional Distribution of $\tilde{n}$

If a sample of size r is drawn by Pascal sampling from a Bernoulli process whose parameter is a random variable $\tilde{p}$ having the beta density (9-9) with parameter (r', n'), the unconditional distribution of $\tilde{n}$ will be beta-Pascal by the definition (7-78) of that distribution:

$$\mathrm{P}\{n|r', n'; r\} = \int_0^1 f_{Pa}(n|p, r)\, f_\beta(p|r', n')\, dp \equiv f_{\beta Pa}(n|r', n', r) \ . \tag{9-32}$$

9.3.4. Distribution of $\tilde{\bar{p}}''$

By (9-11a) and (9-10) the expectation of $\tilde{p}$ posterior to observing the statistic (r, n) in Pascal sampling is

$$\bar{p}'' = \frac{r''}{n''} = \frac{r + r'}{n + n'} \tag{9-33}$$

just as if the statistic had been obtained in binomial sampling. Before the value of $\tilde{n}$ is known, this formula permits us to obtain the distribution of $\tilde{\bar{p}}''$ from the unconditional distribution (9-32) of $\tilde{n}$. Solving (9-33) for n and remembering that $\bar{p}''$ is a decreasing function of n, we have

$$P\{\tilde{\bar{p}}'' \le p|r', n', r\} = G_{\beta Pa}(n_p|r', n', r) \ . \tag{9-34}$$

where

$$n_p = \frac{r''}{p} - n' \ . \tag{9-35}$$

The mean and partial expectation of this distribution of $\tilde{\bar{p}}''$ are

$$E(\tilde{\bar{p}}''|r', n'; r) = \frac{r'}{n'} = E(\tilde{p}|r', n') \equiv \bar{p}' \ , \tag{9-36a}$$

$$E_0^p(\tilde{\bar{p}}''|r', n'; r) = \bar{p}' \, G_{\beta Pa}(n_p|r' + 1, n' + 1, r) \ . \tag{9-36b}$$

It follows immediately from (9-34) and (9-36b) that the linear-loss integrals under this distribution of $\tilde{\bar{p}}''$ are given by

$$\begin{aligned} L_l(p) &\equiv \sum_{z \le p} (p - z) \, P\{\tilde{\bar{p}}'' = z|r', n'; r\} \\ &= p \, G_{\beta Pa}(n_p|r', n', r) - \bar{p}' \, G_{\beta Pa}(n_p|r' + 1, n' + 1, r) \ , \end{aligned} \tag{9-37a}$$

$$\begin{aligned} L_r(p) &\equiv \sum_{z \ge p} (z - p) \, P\{\tilde{\bar{p}}'' = z|r', n'; r\} \\ &= \bar{p}' \, F_{\beta Pa}(n_p|r' + 1, n' + 1, r) - p \, F_{\beta Pa}(n_p|r', n', r) \ . \end{aligned} \tag{9-37b}$$

▶ The proof of formulas (9-36) is very similar to the proof of (9-21ab), the main difference being that we must remember that $\bar{p}''$ is a decreasing function of n:

$$E_0^p(\tilde{\bar{p}}'') = \sum_{n \ge n_p} \frac{r' + r}{n' + n} f_{\beta Pa}(n|r', n', r) = \sum_{n \ge n_p} \frac{r'}{n'} f_{\beta Pa}(n|r' + 1, n' + 1, r) \ ,$$

where n_p is defined by (9-35). ◀

No closed formula for the variance of $\tilde{\bar{p}}''$ or for its second moment about the origin can be found, but the latter quantity can be expressed in terms of an infinite series:

$$E(\tilde{\bar{p}}''^2|r', n'; r) = \frac{r''}{r'' + 1} \cdot \frac{r'(r' + 1)}{n'(n' + 1)} \times \left[1 + \sum_{j=2}^{\infty} \frac{(j - 1)!(r' + 2) \cdots (r' + j)}{(r'' + 2) \cdots (r'' + j)(n' + 2) \cdots (n' + j)}\right] \cdot \tag{9-38a}$$

The variance can then be found by computing

$$V(\tilde{\bar{p}}''|r', n'; r) = E(\tilde{\bar{p}}''^2|r', n'; r) - [E(\tilde{\bar{p}}''|r', n'; r)]^2 \ . \tag{9-38b}$$

▶ To show that the series (9-38a) in fact converges to $E(\tilde{\tilde{p}}''^2)$, we first give it the noncommittal name S and rewrite it in the form

$$S \equiv r''^2 \sum_{j=1}^{\infty} \frac{(j-1)!}{r'' \cdots (r''+j)} \cdot \frac{r' \cdots (r'+j)}{n' \cdots (n'+j)} \cdot$$

We next observe that we are entitled to write

$$S = r''^2 \sum_{j=1}^{\infty} \frac{(j-1)!}{r'' \cdots (r''+j)} \mathrm{E}\left[\frac{r'' \cdots (r''+j)}{\tilde{n}'' \cdots (\tilde{n}''+j)}\right]$$

because

$$\mathrm{E}\left[\frac{r'' \cdots (r''+j)}{\tilde{n}'' \cdots (\tilde{n}''+j)}\right] \equiv \sum_{n=r}^{\infty} \frac{(r+r') \cdots (r+r'+j)}{(n+n') \cdots (n+n'+j)} f_{\beta Pa}(n|r', n', r)$$

$$= \frac{r' \cdots (r'+j)}{n' \cdots (n'+j)} \sum_{n=r}^{\infty} f_{\beta Pa}(n|r'+j+1, n'+j+1, r) = \frac{r' \cdots (r'+j)}{n' \cdots (n'+j)} \cdot$$

If now we reverse the order of expectation and summation in our last expression for S and then cancel out the factors involving r'', we obtain

$$S = r''^2 \mathrm{E} \sum_{j=1}^{\infty} \frac{(j-1)!}{\tilde{n}'' \cdots (\tilde{n}''+j)} \cdot$$

We can easily verify that the jth term in the summand is equal to

$$-\frac{1}{\tilde{n}''} \Delta \frac{(j-1)!}{\tilde{n}'' \cdots (\tilde{n}''+j-1)} ,$$

where Δ is the first-difference operator defined by $\Delta\, a_j = a_{j+1} - a_j$. Then since

$$\sum_{j=1}^{J} \Delta\, a_j = a_{J+1} - a_1 ,$$

we may write

$$S = r''^2 \mathrm{E} \lim_{J\to\infty} \left(\frac{1}{\tilde{n}''}\left[\frac{1}{\tilde{n}''} - \frac{J!}{\tilde{n}'' \cdots (\tilde{n}''+J)}\right]\right) \cdot$$

This limit will be

$$r''^2 \mathrm{E}(1/\tilde{n}''^2) = \mathrm{E}(r''^2/\tilde{n}''^2) = \mathrm{E}(\tilde{\tilde{p}}''^2)$$

as asserted by (9-38a) provided that the limit of

$$R(J) \equiv \frac{J!}{n'' \cdots (n''+J)}$$

is 0 for all possible n''. To prove this, we first observe that because the parameters n' and r are restricted to $n' > 0$ and $r \geq 1$, and because the variable n is restricted to $n \geq r$, therefore $n'' \equiv n' + n$ is restricted to $n'' > 1$. But then

$$R(J) \leq \frac{J!}{1 \cdot 2 \cdots J \cdot (1 + J)} = \frac{1}{1 + J}$$

and therefore

$$\lim_{J \to \infty} R(J) = 0$$

as was to be proved.

[Note: In the first printing there is a slight error in the proof of the above theorem. We hope it is correct this time.] ◀

Approximations. In principle, computation of a cumulative probability such as $P\{\tilde{\bar p}'' < p|r', n'; r\}$ under the distribution defined by (9-34) or evaluation of a linear-loss integral as given by (9-37) involves computation of a beta-Pascal cumulative probability in one of the ways discussed in Section 7.11.2, a beta or binomial approximation to the beta-Pascal giving good results when $r' \ll n'$, $r \ll n$, and $r + r' \ll \max\{n, n'\}$.

As r increases, the distribution of $\tilde{\bar p}''$ becomes more nearly continuous and approaches the beta prior distribution (9-9) of $\tilde p$, so that a direct beta approximation to the distribution of $\tilde{\bar p}''$ itself—*not* a Normal approximation—will give good results. It follows from (9-11) and (9-36a) that a beta distribution with parameters

$$n^* = \frac{\mathrm{E}(\tilde{\bar p}'')[1 - \mathrm{E}(\tilde{\bar p}'')]}{\mathrm{V}(\tilde{\bar p}'')} - 1 = \frac{\bar p'(1 - \bar p')}{\mathrm{V}(\tilde{\bar p}'')} - 1 , \tag{9-39}$$

$$r^* = n^* \, \mathrm{E}(\tilde{\bar p}'') = n^* \, \bar p' ,$$

would have the same mean and variance as the exact distribution of $\tilde{\bar p}''$. Once (9-39) has been evaluated approximately by use of (9-38), tail probabilities may be computed by use of (7-23) or (7-24) and linear-loss integrals may be evaluated by use of (9-12).

9.3.5. *Distribution of $\tilde{\bar\rho}''$*

If a sample drawn by Pascal sampling from a Bernoulli process yields a statistic (r, n), the posterior expectation of $\tilde\rho = 1/\tilde p$ will be, by (9-15a) and (9-10),

$$\bar\rho'' = \frac{n' + n - 1}{r'' - 1} , \tag{9-40}$$

just as if the same statistic had been obtained in binomial sampling with fixed n instead of fixed r. The distribution of $\tilde{\bar\rho}''$ can be obtained from the unconditional distribution (9-32) of $\tilde n$ by solving (9-40) for n:

$$\mathrm{P}\{\bar\rho''|r', n'; r\} = f_{\beta Pa}(n_\rho|r', n', r) \tag{9-41}$$

where

$$n_\rho = \rho(r'' - 1) - (n' - 1) \ . \tag{9-42}$$

The mean, partial expectation, and variance of the distribution of $\tilde{\bar{p}}''$ defined by (9-41) are

$$E(\tilde{\bar{p}}''|r', n'; r) = \frac{n' - 1}{r' - 1} = E(\tilde{\bar{p}}|r', n') \equiv \bar{p}' \ , \tag{9-43a}$$

$$E_0^\rho(\tilde{\bar{p}}''|r', n'; r) = \bar{p}' \, F_{\beta Pa}(n_\rho|r' - 1, n' - 1, r) \ , \tag{9-43b}$$

$$V(\tilde{\bar{p}}''|r', n'; r) = \frac{r}{r + r' - 1} \frac{(n' - 1)(n' - r')}{(r' - 1)^2(r' - 2)} = \frac{r}{r + r' - 1} V(\tilde{\bar{p}}|r', n') \ ; \tag{9-43c}$$

the proofs are very similar to the proofs of (9-21). From (9-41) and (9-43b) it follows immediately that the linear-loss integrals under this distribution of $\tilde{\bar{p}}''$ are given by

$$\begin{aligned} L_l(\rho) &\equiv \sum_{z \le \rho} (\rho - z) \, P\{\tilde{\bar{p}}'' = z|r', n'; r\} \\ &= \rho \, F_{\beta Pa}(n_\rho|r', n', r) - \bar{p}' \, F_{\beta Pa}(n_\rho|r' - 1, n' - 1, r) \ , \end{aligned} \tag{9-44a}$$

$$\begin{aligned} L_r(\rho) &\equiv \sum_{z \ge \rho} (z - \rho) \, P\{\tilde{\bar{p}}'' = z|r', n'; r\} \\ &= \bar{p}' \, G_{\beta Pa}(n_\rho|r' - 1, n' - 1, r) - \rho \, G_{\beta Pa}(n_\rho|r', n', r) \ . \end{aligned} \tag{9-44b}$$

Approximations. In principle, computation of a cumulative probability such as $P\{\tilde{\bar{p}}'' < \rho|r', n'; r\}$ under the distribution (9-41) or evaluation of a linear-loss integral as given by (9-44) involves computation of a beta-Pascal cumulative probability by one of the methods discussed in Section 7.11.2, a beta or binomial approximation to the beta-Pascal giving good results if $r' \ll n$, $r \ll n$, and $r + r' \ll \max\{n, n'\}$.

As r increases, the distribution of $\tilde{\bar{p}}''$ becomes more nearly continuous and approaches the inverted-beta-1 prior distribution (9-13) of $\tilde{\bar{p}}$, so that a direct inverted-beta-1 approximation to the distribution of $\tilde{\bar{p}}''$ itself—*not* a Normal approximation—will give good results. It follows from (9-43) and (9-15) that an inverted-beta-1 distribution with parameters r^*, n^*, b^* defined by

$$r^* - 2 = \frac{r + r' - 1}{r}(r' - 2) \ , \qquad n^* = \frac{r^*}{r'} n' \ , \qquad b^* = 1 \ , \tag{9-45}$$

will have the same mean and variance as the exact distribution of $\tilde{\bar{p}}''$. Using these parameter values, cumulative probabilities may be obtained by use of formula (9-14) and linear-loss integrals may be evaluated by use of (9-16).

CHAPTER 10

Poisson Process

10.1. Prior and Posterior Analysis

10.1.1. Definition of a Poisson Process

A Poisson process can be defined as a process generating independent random variables $\tilde{x}_1, \cdots, \tilde{x}_i, \cdots$ with identical gamma-1 densities

$$f_{\gamma 1}(x|1, \lambda) = e^{-\lambda x}\,\lambda\ , \qquad \begin{array}{l} 0 \le x < \infty\ , \\ \lambda > 0\ . \end{array} \tag{10-1}$$

The parameter λ will be called the *intensity* of the process.

In the applications, x will measure the extent of an *interval* between two adjacent *events* which themselves have no extent. Thus, for example, x may be the time between two successive instantaneous events or the distance between two points. The parameter λ can then be interpreted as the mean number of events per unit of time or space.

A unique and very important property of a Poisson process is the fact that

$$P\{\tilde{x} > \xi + x|\tilde{x} > \xi\} = P\{\tilde{x} > x\}\ . \tag{10-2}$$

If x is interpreted as the time between two events, this says that the conditional probability that more than x *additional* "minutes" will elapse before the next event is independent of the number ξ of minutes that have already elapsed since the last event. This implies in turn that the distribution of the number of minutes from any arbitrarily chosen point of time to the next event is the same as the distribution of the number of minutes between two adjacent events.

▶ To show that the independence of past history asserted by (10-2) follows from (10-1), we write

$$P\{\tilde{x} > \xi + x|\tilde{x} > \xi\} = \frac{\int_{\xi+x}^{\infty} e^{-\lambda u}\,\lambda\,du}{\int_{\xi}^{\infty} e^{-\lambda u}\,\lambda\,du} = \frac{e^{-\lambda(\xi+x)}}{e^{-\lambda\xi}} = e^{-\lambda x} = P\{\tilde{x} > x\}\ .$$

To show that this independence of past history is a *unique* property of the Poisson process among all processes generating continuous random variables $\tilde{x} \ge 0$, we let $G(x)$ denote an as yet undetermined right-tail cumulative function of which it is required that

$$\frac{G(\xi + x)}{G(\xi)} = G(x)\ , \qquad G(0) = 1\ , \qquad G(\infty) = 0\ .$$

If we define

$$h(a) \equiv \log G(a) \ ,$$

these conditions imply that

$$h(\xi + x) = h(x) + h(\xi)$$

for all positive x and ξ. It then follows from the lemma below that h must be of the form

$$h(x) = -\lambda x$$

for some $\lambda \geq 0$. Consequently

$$G(x) = e^{-\lambda x} \ ,$$

and since $G(\infty) = 0$, the parameter $\lambda > 0$. The result (10-1) then follows from the fact that the density is

$$-\frac{d}{dx} G(x) = \lambda \, e^{-\lambda x} \ .$$

Lemma: Let h be a real-valued function defined on the interval $[0, \infty)$ such that

(1) $$h(\xi + x) = h(\xi) + h(x) \ , \qquad \text{all } \xi, x \ ,$$

(2) $$h \text{ is monotonic nonincreasing in } x \ .$$

Then h is of the form

$$h(x) = -\lambda x \qquad \text{for some } \lambda \geq 0$$

Proof: Using (1) to write for m integral

$$h(mx) = h[(m - 1)x + x] = h[(m - 1)x] + h(x)$$

we get

(3) $$h(mx) = m \, h(x) \ .$$

Using (3) to write for n integral

$$h(x) = h\left[n\left(\frac{1}{n} x\right)\right] = n \, h\left(\frac{1}{n} x\right)$$

we get

(4) $$h\left(\frac{1}{n} x\right) = \frac{1}{n} h(x) \ .$$

From (3) and (4) we have for m, n integral

(5) $$h\left(\frac{m}{n} x\right) = \frac{m}{n} h(x) \ .$$

Letting $x = 1$ and $h(1) = -\lambda$ we get from (5) that for any positive rational y

(6) $$h(y) = -\lambda y \ .$$

Since h is nonincreasing by hypothesis, $\lambda \geq 0$; and since h is monotonic, (6) is true for all positive real y. ◀

10.1.2. Likelihood of a Sample

The likelihood that the process defined by (10-1) will generate an interval of length x is given directly by (10-1) as

$$e^{-\lambda x} \lambda \ . \tag{10-3}$$

The conditional probability, given λ, that the process will generate an interval of length $\tilde{x} > x$ is

$$\int_x^\infty e^{-\lambda u}\,\lambda\,du = e^{-\lambda x} . \tag{10-4}$$

The joint likelihood that the process will *generate* r successive intervals of lengths $x_1, \cdots, x_r$ followed by an interval of length $\tilde{x} > x_{r+1}$ is thus

$$\Pi_{i=1}^r(e^{-\lambda x_i}\,\lambda)\,e^{-\lambda x_{r+1}} = e^{-\lambda\Sigma_{i=1}^{r+1} x_i}\,\lambda^r ; \tag{10-5}$$

and if the stopping process is noninformative in the sense of Section 2.3, this is the kernel of the *likelihood of the sample* consisting of the r complete and one incomplete intervals. If we define

$$t \equiv \Sigma_{i=1}^{r+1} x_i , \tag{10-6}$$

the kernel (10-5) can be written

$$e^{-\lambda t}\,\lambda^r , \tag{10-7}$$

so that it is clear that the statistic (r, t) is sufficient when the stopping process is noninformative.

10.1.3. Conjugate Distribution of $\tilde{\lambda}$

When the parameter of a Poisson process is a random variable $\tilde{\lambda}$, the most convenient distribution of $\tilde{\lambda}$—the natural conjugate of (10-7)—is the gamma-1 distribution defined by (7-43):

$$f_{\gamma 1}(\lambda|r, t) \propto e^{-\lambda t}\,\lambda^{r-1} . \tag{10-8}$$

If such a distribution with parameter (r', t') has been assigned to $\tilde{\lambda}$ and if then a sample from the process yields a sufficient statistic (r, t), the posterior distribution of $\tilde{\lambda}$ will be gamma-1 with parameters

$$r'' = r' + r , \qquad t'' = t' + t . \tag{10-9}$$

The proof is similar to the proof of (9-10).

The mean, partial expectation, and variance of the distribution (10-8) of $\tilde{\lambda}$ are, by (7-45),

$$E(\tilde{\lambda}|r, t) \equiv \bar{\lambda} = \frac{r}{t} , \tag{10-10a}$$

$$E_0^\lambda(\tilde{\lambda}|r, t) = \bar{\lambda}\,F_{\gamma^*}(\lambda t|r + 1) , \tag{10-10b}$$

$$V(\tilde{\lambda}|r, t) = \frac{r}{t^2} . \tag{10-10c}$$

It follows immediately from (10-10b) that the linear-loss integrals are

$$L_l(\lambda) \equiv \int_0^\lambda (\lambda - z)\,f_{\gamma 1}(z|r, t)\,dz = \lambda\,F_{\gamma^*}(\lambda t|r) - \bar{\lambda}\,F_{\gamma^*}(\lambda t|r + 1) , \tag{10-11a}$$

$$L_r(\lambda) \equiv \int_\lambda^\infty (z - \lambda)\,f_{\gamma 1}(z|r, t)\,dz = \bar{\lambda}\,G_{\gamma^*}(\lambda t|r + 1) - \lambda\,G_{\gamma^*}(\lambda t|r) . \tag{10-11b}$$

As shown in (7-44b), cumulative probabilities may be obtained from Pearson's tables of the gamma function $I(u, p)$ or from tables of the cumulative Poisson function by use of the relation

$$F_{\gamma 1}(\lambda|r; t) = F_{\gamma^*}(\lambda t|r) = I\left(\frac{t\lambda}{\sqrt{r}}, r-1\right) = G_P(r|t\lambda) \ . \tag{10-12}$$

Limiting Behavior of the Prior Distribution. As the parameters r and t of (10-8) both approach zero in such a way that the ratio $r/t = \bar{\lambda}$ remains fixed, the probability becomes more and more concentrated toward $\lambda = 0$ but the variance becomes infinite. The mean and variance of the limiting distribution—a mass of 1 at $\lambda = 0$—are *not* equal to the limits of the mean and variance.

▶ The behavior of the variance follows immediately from (10-10c). To show that the distribution approaches a mass of 1 located at $\lambda = 0$—i.e., to prove that if r/t is held equal to $\bar{\lambda}$ then

$$\lim_{r,t\to 0} F_{\gamma 1}(\delta|r, t) = 1$$

for any $\delta > 0$ however small—, we first substitute $r/\bar{\lambda}$ for t in the definition (7-43) of $f_{\gamma 1}$ and write

$$F_{\gamma 1}(\delta|r, r/\bar{\lambda}) = \int_0^\delta \frac{e^{-r\lambda/\bar{\lambda}}(r\lambda/\bar{\lambda})^{r-1}}{(r-1)!}(r/\bar{\lambda})\, d\lambda \ .$$

We then substitute

$$u = \lambda/\bar{\lambda}\ , \qquad \delta^* = \delta/\bar{\lambda}\ ,$$

and write

$$F_{\gamma 1}(\delta|r, r/\bar{\lambda}) = \int_0^{\delta^*} \frac{e^{-ru}(ru)^{r-1}}{(r-1)!}\, r\, du = \frac{r^r}{r!}\int_0^{\delta^*} e^{-ru}\, r\, u^{r-1}\, du$$

$$> \frac{r^r}{r!} e^{-r\delta^*} \int_0^{\delta^*} r\, u^{r-1}\, du = \frac{r^r\, e^{-r\delta^*}\, \delta^{*r}}{r!} \ .$$

For any $\delta^* = \delta/\bar{\lambda} > 0$, however small, the limit as $r \to 0$ of each of the four factors in the expression on the right is 1, and therefore the limit of the whole expression is 1. Since on the other hand $F_{\gamma 1}(\delta) < F_{\gamma 1}(\infty) = 1$ for all $r > 0$, it follows that as $r \to 0$ the limit of $F_{\gamma 1}(\delta)$ is 1. ◀

10.1.4. Conjugate Distribution of $1/\tilde{\lambda} = \tilde{\mu}$

In problems where the utility is linear in $1/\lambda$ rather than in λ, we shall need the same kinds of information about the distribution of $\tilde{\mu} = 1/\tilde{\lambda}$ that we have just obtained for the distribution of $\tilde{\lambda}$. Substituting $1/\mu$ for λ in the gamma-1 density (10-8) and multiplying by the Jacobian we find that the induced density of $\tilde{\mu}$ is inverted-gamma-1 as defined by (7-54):

$$f_{i\gamma 1}(\mu|r, t) \propto e^{-t/\mu}\, \mu^{-r-1} \ . \tag{10-13}$$

Cumulative probabilities may be obtained from Pearson's tables (cf. Section 7.6.1) of the gamma function $I(u, p)$ or from tables of the cumulative Poisson function by use of the relations

$$G_{i\gamma 1}(\mu|r, t) = F_{\gamma^*}(t/\mu|r) = I\left(\frac{t/\mu}{\sqrt{r}}, r-1\right) = G_P(r|t/\mu) \tag{10-14}$$

which follow immediately from (7-55), (7-38), and (7-39). The mean, partial expectation, and variance of $\tilde{\mu}$ are, by (7-56),

$$\mathrm{E}(\tilde{\mu}|r, t) \equiv \bar{\mu} = \frac{t}{r-1} , \qquad r > 1 , \qquad (10\text{-}15a)$$

$$\mathrm{E}_0^{\mu}(\tilde{\mu}|r, t) = \bar{\mu}\, G_{\gamma^*}(t/\mu|r-1) , \qquad r > 1 , \qquad (10\text{-}15b)$$

$$\mathrm{V}(\tilde{\mu}|r, t) = \frac{t^2}{(r-1)^2(r-2)} , \qquad r > 2 . \qquad (10\text{-}15c)$$

It follows from (10-14) and (10-15b) that the linear-loss integrals are

$$L_l(\mu) \equiv \int_0^{\mu} (\mu - z)\, f_{i\gamma 1}(z|r, t)\, dz = \mu\, G_{\gamma^*}(t/\mu|r) - \bar{\mu}\, G_{\gamma^*}(t/\mu|r-1) , \quad (10\text{-}16a)$$

$$L_r(\mu) \equiv \int_{\mu}^{\infty} (z - \mu)\, f_{i\gamma 1}(z|r, t)\, dz = \bar{\mu}\, F_{\gamma^*}(t/\mu|r-1) - \mu\, F_{\gamma^*}(t/\mu|r) . \quad (10\text{-}16b)$$

10.2. Sampling Distributions and Preposterior Analysis: Gamma Sampling

10.2.1. Definition of Gamma Sampling

If a sample is taken from a Poisson process in such a way that the value of the statistic r is predetermined, the value of the statistic $\tilde{t}$ being left to chance, the sampling will be called gamma. In other words, gamma sampling consists in observing the process until the rth event occurs and measuring the time, length, or other dimension which "elapses" up to this point.

We assume throughout Section 10.2 that a gamma sample with fixed r is to be taken from a Poisson process whose intensity $\tilde{\lambda}$ has a gamma-1 distribution with parameter (r', t').

10.2.2. Conditional Distribution of $(\tilde{t}\,|\lambda)$

The density of the individual random variables $\tilde{x}$ generated by a Poisson process is by definition (10-1) the special case $f_{\gamma 1}(x|1, \lambda)$ of the gamma-1 density defined by (7-43). The conditional probability, given λ, that the sum of r such variables will have a total value t is given by the convolution of the r individual densities, and we saw in (7-46) that this convolution is a gamma-1 density with parameters r and λ. We thus have for the conditional density of $(\tilde{t}\,|r, \lambda)$

$$\mathrm{D}(t|r, \lambda) = f_{\gamma 1}(t|r, \lambda) . \qquad (10\text{-}17)$$

10.2.3. Unconditional Distribution of $\tilde{t}$

If a sample of size r is drawn by gamma sampling from a Poisson process whose intensity is a random variable $\tilde{\lambda}$ having the gamma-1 density (10-8) with parameter (r', t'), the unconditional distribution of $\tilde{t}$ is inverted-beta-2 as defined by (7-28):

$$\mathrm{D}(t|r', t'; r) = \int_0^{\infty} f_{\gamma 1}(t|r, \lambda)\, f_{\gamma 1}(\lambda|r', t')\, d\lambda = f_{i\beta 2}(t|r, r', t') . \qquad (10\text{-}18)$$

▶ To prove (10-18) we replace $f_{\gamma 1}(t)$ and $f_{\gamma 1}(\lambda)$ in the integrand by their definitions (7-43) and write

$$\int_0^\infty \frac{e^{-\lambda t}(\lambda t)^{r-1}}{(r-1)!}\lambda \cdot \frac{e^{-\lambda t'}(\lambda t')^{r'-1}}{(r'-1)!} t'\,d\lambda$$

$$= \frac{(r+r'-1)!}{(r-1)!(r'-1)!}\frac{t^{r-1}t'^{r'}}{(t+t')^{r+r'}}\int_0^\infty \frac{e^{-\lambda(t+t')}(\lambda[t+t'])^{r+r'-1}}{(r+r'-1)!}(t+t')\,d\lambda .$$

The integral on the right is $F_{\gamma 1}(\infty|r+r', t+t') = 1$, while by (7-28) the factors preceding the integral sign are $f_{i\beta 2}(t|r, r', t')$. ◀

10.2.4. Distribution of $\tilde{\bar{\lambda}}''$

If the prior distribution of $\tilde{\lambda}$ is gamma-1 with parameter (r', t') and if a sample then yields a sufficient statistic (r, t), formulas (10-10a) and (10-9) show that the mean of the posterior distribution of $\tilde{\lambda}$ will be

$$\bar{\lambda}'' = \frac{r''}{t''} = \frac{r+r'}{t+t'} \cdot \tag{10-19}$$

It follows that when the value of the $\tilde{t}$ to be obtained by gamma sampling is still unknown and $\tilde{\bar{\lambda}}''$ is therefore a random variable, the distribution of $\tilde{\bar{\lambda}}''$ can be obtained from the unconditional distribution (10-18) of $\tilde{t}$. Solving (10-19) for t and remembering that $\bar{\lambda}''$ is a decreasing function of t we have

$$P\{\tilde{\bar{\lambda}}'' < \lambda|r', t'; r\} = G_{i\beta 2}\left(\frac{r''}{\lambda} - t'|r, r', t'\right) \cdot \tag{10-20}$$

It now follows from (7-29), (7-19), and (7-20) that cumulative probabilities may be obtained from tables of the standardized beta function or tables of the binomial function by use of the relation

$$P\{\tilde{\bar{\lambda}}'' < \lambda|r', t'; r\} = F_{\beta*}(\lambda t'/r''|r', r) = G_b(r'|\lambda t'/r'', r''-1) . \tag{10-21}$$

The mean, partial expectation, and variance of this distribution of $\tilde{\bar{\lambda}}''$ are

$$E(\tilde{\bar{\lambda}}''|r', t'; r) = \frac{r'}{t'} = E(\tilde{\lambda}|r', t') \equiv \bar{\lambda}' , \tag{10-22a}$$

$$E_0^\lambda(\tilde{\bar{\lambda}}''|r', t'; r) = \bar{\lambda}' F_{\beta*}(\lambda t'/r''|r'+1, r) , \tag{10-22b}$$

$$V(\tilde{\bar{\lambda}}''|r', t'; r) = \frac{r}{r+r'+1}\cdot\frac{r'}{t'^2} = \frac{r}{r+r'+1}V(\tilde{\lambda}|r', t') . \tag{10-22c}$$

It follows at once from (10-21) and (10-22b) that the linear-loss integrals under this distribution of $\tilde{\bar{\lambda}}''$ are given by

$$L_l(\lambda) \equiv \int_0^\lambda (\lambda - \bar{\lambda}'')\,D(\bar{\lambda}'')\,d\bar{\lambda}''$$
$$= \lambda F_{\beta*}(\lambda t'/r''|r', r) - \bar{\lambda}' F_{\beta*}(\lambda t'/r''|r'+1, r) , \tag{10-23a}$$

$$L_r(\lambda) \equiv \int_\lambda^\infty (\bar{\lambda}'' - \lambda)\,D(\bar{\lambda}'')\,d\bar{\lambda}''$$
$$= \bar{\lambda}' G_{\beta*}(\lambda t'/r''|r'+1, r) - \lambda G_{\beta*}(\lambda t'/r''|r', r) . \tag{10-23b}$$

▶ Formula (10-22a) for the mean of $\tilde{\bar{\lambda}}''$ follows immediately from (5-27).

To prove formula (10-22b) we write the definition

$$E_0^\lambda(\tilde{\bar{\lambda}}'') \equiv \int_0^\lambda \bar{\lambda}'' \, D(\bar{\lambda}'') \, d\bar{\lambda}''$$

and then substitute (10-19) for $\bar{\lambda}''$ and replace the density of $\tilde{\bar{\lambda}}''$ by the density (10-18) of $\tilde{t}$. Using formula (7-28) for this latter density we thus obtain, remembering that $\bar{\lambda}''$ is a decreasing function of t,

$$E_0^\lambda(\tilde{\bar{\lambda}}'') = \int_{t^*}^\infty \frac{r+r'}{t+t'} \frac{(r+r'-1)!}{(r-1)!(r'-1)!} \frac{t^{r-1} t'^{r'}}{(t+t')^{r+r'}} dt$$

$$= \frac{r'}{t'} \int_{t^*}^\infty \frac{(r+r')!}{(r-1)!\,r'!} \frac{t^{r-1} t'^{r'+1}}{(t+t')^{r+r'+1}} dt = \bar{\lambda}' \, G_{i\beta 2}(t^*|r, r'+1, t')$$

where by (10-19)

$$t^* = \frac{r''}{\lambda} - t' \; .$$

Formula (10-22b) then follows by (10-20) and (10-21).

To prove formula (10-22c) for the variance we first find the second moment about the origin; and to do so we again substitute for $\bar{\lambda}''$ in terms of t:

$$E(\tilde{\bar{\lambda}}''^2) \equiv \int_0^\infty \bar{\lambda}''^2 \, D(\bar{\lambda}'') \, d\bar{\lambda}'' = \int_0^\infty \left[\frac{r+r'}{t+t'}\right]^2 f_{i\beta 2}(t|r, r', t') \, dt$$

$$= \frac{(r+r')r'(r'+1)}{(r+r'+1)t'^2} \int_0^\infty f_{i\beta 2}(t|r, r'+2, t') \, dt = \frac{(r+r')r'(r'+1)}{(r+r'+1)\, t'^2} \; .$$

We then obtain the variance by use of the usual formula

$$\mu_2 = \mu_2' - \mu_1^2 = \frac{(r+r')r'(r'+1)}{(r+r'+1)\, t'^2} - \left[\frac{r'}{t'}\right]^2 = \frac{r}{r+r'+1} \frac{r'}{t'^2} \; .$$ ◀

Approximations. As r increases, the distribution of $\tilde{\bar{\lambda}}''$ approaches the gamma-1 prior distribution of $\tilde{\lambda}$, so that when r is large a gamma—*not* a Normal—approximation to the distribution of $\tilde{\bar{\lambda}}''$ will give good results. It follows from (10-22) and (10-10) that a gamma-1 distribution with parameters

$$r^* = \frac{r''+1}{r} r' \; , \qquad t^* = \frac{r''+1}{r} t' \; , \tag{10-24}$$

will have the same mean and variance as the exact distribution (10-20) of $\tilde{\bar{\lambda}}''$. Using these parameter values, cumulative probabilities may be obtained by use of formula (10-12) and linear-loss integrals may be evaluated by use of formulas (10-11).

10.2.5. Distribution of $\tilde{\bar{\mu}}''$

By (10-15a) and (10-9) the posterior expectation of $\tilde{\mu}$ will be

$$\bar{\mu}'' = \frac{t''}{r''-1} = \frac{t'+t}{r''-1} \; , \tag{10-25}$$

so that when $\tilde{t}$ and therefore $\tilde{\mu}''$ are random variables the distribution of $\tilde{\mu}''$ can be obtained from the unconditional distribution (10-18) of $\tilde{t}$. Solving (10-25) for t we have

$$P\{\tilde{\mu}'' < \mu | r', t'; r\} = F_{i\beta 2}(t_\mu | r, r', t') \tag{10-26a}$$

where

$$t_\mu = \mu(r'' - 1) - t' \ . \tag{10-26b}$$

It follows from (7-29) and (7-20) that cumulative probabilities may be obtained from tables of the standardized beta function or the binomial function by use of the relations

$$P\{\tilde{\mu}'' < \mu | r', t'; r\} = F_{\beta*}(\pi_\mu | r, r') = G_b(r | \pi_\mu, r + r' - 1) \tag{10-27a}$$

where

$$\pi_\mu = \frac{t_\mu}{t_\mu + t'} = 1 - \frac{t'}{\mu(r'' - 1)} \ . \tag{10-27b}$$

The mean and variance of this distribution of $\tilde{\mu}''$ are

$$E(\tilde{\mu}'' | r', t'; r) = \frac{t'}{r' - 1} = E(\tilde{\mu} | r', t') \equiv \bar{\mu}' \ , \tag{10-28a}$$

$$V(\tilde{\mu}'' | r', t'; r) = \frac{r}{r'' - 1} \frac{t'^2}{(r' - 1)^2 (r' - 2)} = \frac{r}{r'' - 1} V(\tilde{\mu} | r', t') \ . \tag{10-28b}$$

The partial expectation is

$$E_0^\mu(\tilde{\mu}'' | r', t'; r) = \bar{\mu}' F_{\beta*}(\pi_\mu | r, r' - 1) \tag{10-28c}$$

where π_μ is defined by (10-27b). It follows from (10-27) and (10-28c) that the linear-loss integrals are

$$\begin{aligned} L_l(\mu) &\equiv \int_0^\mu (\mu - \bar{\mu}'') D(\bar{\mu}'') \, d\bar{\mu}'' \\ &= \mu F_{\beta*}(\pi_\mu | r, r') - \bar{\mu}' F_{\beta*}(\pi_\mu | r, r' - 1) \ , \end{aligned} \tag{10-29a}$$

$$\begin{aligned} L_r(\mu) &\equiv \int_\mu^\infty (\bar{\mu}'' - \mu) D(\bar{\mu}'') \, d\bar{\mu}'' \\ &= \bar{\mu}' G_{\beta*}(\pi_\mu | r, r' - 1) - \mu G_{\beta*}(\pi_\mu | r, r') \ . \end{aligned} \tag{10-29b}$$

▶ Formula (10-28a) follows directly from (5-27).

To prove formula (10-28b) we use the general formula for the effect of a linear transformation on moments. Since

$$\tilde{\mu}'' = \frac{t'}{r'' - 1} + \frac{\tilde{t}}{r'' - 1} \ ,$$

we have by (7-7)

$$V(\tilde{\mu}'') = \frac{1}{(r'' - 1)^2} V(\tilde{t}) \ ;$$

and formula (10-28b) results when we use (10-18) and (7-30b) to substitute herein

$$V(\tilde{t}) = \frac{r(r + r' - 1) \, t'^2}{(r' - 1)^2 (r' - 2)} \ .$$

Formula (10-28c) is most easily derived by writing the definition

$$E_0^{\mu}(\tilde{\mu}'') \equiv \int_0^{\mu} \bar{\mu}''\, D(\bar{\mu}'')\, d\bar{\mu}''$$

and then substituting (10-25) for $\bar{\mu}''$ and replacing the density of $\tilde{\bar{\mu}}''$ by the density (10-18) of $\tilde{t}$:

$$E_0^{\mu}(\tilde{\bar{\mu}}'') = \int_0^{t_\mu} \frac{t'+t}{r''-1} f_{i\beta 2}(t|r, r', t')\, dt = \frac{t'}{r'-1} \int_0^{t_\mu} f_{i\beta 2}(t|r, r'-1, t')\, dt \ .$$

The result (10-28c) follows by (7-29). ◀

Approximations. As r increases, the distribution of $\tilde{\bar{\mu}}''$ approaches the inverted-gamma-1 prior distribution of $\tilde{\mu}$, so that when r is large an inverted-gamma-1 —*not* a Normal—approximation to the distribution of $\tilde{\bar{\mu}}''$ will give good results. It follows from (10-28) and (10-15) that an inverted-gamma-1 distribution with parameters r^* and t^* determined by

$$r^* - 2 = (r'-2)\,\frac{r+r'-1}{r} \ , \qquad t^* = \frac{r^*-1}{r'-1}\,t' \ , \tag{10-30}$$

will have the same mean and variance as the exact distribution (10-26) of $\tilde{\bar{\mu}}''$. Using these parameter values, cumulative probabilities may be obtained by use of (10-14) and linear-loss integrals may be evaluated by use of (10-16).

10.3. Sampling Distributions and Preposterior Analysis: Poisson Sampling

10.3.1. Definition of Poisson Sampling

If a sample is taken from a Poisson process in such a way that the value of the statistic t is predetermined, the value of $\tilde{r}$ being left to chance, the sampling will be called Poisson. In other words, Poisson sampling consists of observing the process over a predetermined amount of time, length or other dimension, and counting the number of events which occur.

We assume throughout Section 10.3 that a Poisson sample with fixed t is to be taken from a Poisson process whose intensity $\tilde{\lambda}$ has a gamma-1 distribution with parameter (r', t').

10.3.2. Conditional Distribution of $(\tilde{r}|\lambda)$

Since there will be r or more "events" in t "minutes" if and only if the number of minutes preceding the rth event is t or less, the distribution of $(\tilde{r}|t, \lambda)$ can be obtained immediately from the distribution (10-17) of $(\tilde{t}|r, \lambda)$:

$$P\{\tilde{r} \geq r|t, \lambda\} = P\{\tilde{t} \leq t|r, \lambda\} = F_{\gamma 1}(t|r, \lambda) \ . \tag{10-31}$$

It then follows by (7-44b) that the distribution of $\tilde{r}$ is given by the Poisson mass function (7-32):

$$P\{r|t, \lambda\} = f_P(r|\lambda t) \ . \tag{10-32}$$

10.3.3. Unconditional Distribution of $\tilde{r}$

If a sample of size t is drawn by Poisson sampling from a Poisson process whose intensity is a random variable $\tilde{\lambda}$ having the gamma-1 density (10-8) with parameter (r', t'), the unconditional distribution of $\tilde{r}$ is negative-binomial as defined by (7-73):

$$P\{r|r', t'; t\} = \int_0^\infty f_P(r|\lambda t)\, f_{\gamma 1}(\lambda|r', t')\, d\lambda \equiv f_{nb}\left(r\left|\frac{t}{t+t'}, r'\right.\right) \cdot \qquad (10\text{-}33)$$

▶ Substituting for f_P and $f_{\gamma 1}$ in the integral (10-33) their definitions (7-32) and (7-43) we obtain

$$P\{r\} = \int_0^\infty \frac{e^{-\lambda t}(\lambda t)^r}{r!} \frac{e^{-\lambda t'}(\lambda t')^{r'-1}}{(r'-1)!}\, t'\, d\lambda$$

$$= \frac{(r+r'-1)!}{r!(r'-1)!} \frac{t^r t'^{r'}}{(t+t')^{r+r'}} \int_0^\infty \frac{e^{-\lambda(t+t')}(\lambda[t+t'])^{r+r'-1}}{(r+r'-1)!} (t+t')\cdot d\lambda$$

$$= f_{nb}\left(r\left|\frac{t}{t+t'}, r'\right.\right) F_{\gamma 1}(\infty|r+r', t+t')$$

and the second factor on the right is 1. ◀

10.3.4. Distribution of $\bar{\lambda}''$

If the prior distribution of $\tilde{\lambda}$ is gamma-1 with parameter (r', t') and if a sample then yields a sufficient statistic (r, t), formulas (10-10a) and (10-9) show as we have already seen that the mean of the posterior distribution of $\tilde{\lambda}$ will be

$$\bar{\lambda}'' = \frac{r''}{t''} = \frac{r+r'}{t+t'} \cdot \qquad (10\text{-}34)$$

It follows that when the value of the $\tilde{r}$ to be obtained by Poisson sampling is still unknown and $\bar{\lambda}''$ is therefore a random variable, the distribution of $\tilde{\bar{\lambda}}''$ can be obtained from the unconditional distribution (10-33) of $\tilde{r}$. Solving (10-34) for r and substituting in (10-33) we have

$$P\{\tilde{\bar{\lambda}}'' = \lambda|r', t'; t\} = f_{nb}\left(r_\lambda\left|\frac{t}{t+t'}, r'\right.\right), \qquad r_\lambda = \lambda t'' - r' . \qquad (10\text{-}35)$$

Then recalling that the argument of f_{nb} is integer-valued although its parameters are continuous we have by (7-75) that cumulative probabilities under the distribution of $\tilde{\bar{\lambda}}''$ may be obtained from tables of the standardized beta function by use of the relations

$$P\{\tilde{\bar{\lambda}}'' \le \lambda|r', t'; t\} = F_{nb}\left(r_\lambda\left|\frac{t}{t+t'}, r'\right.\right) = G_{\beta^*}\left(\frac{t}{t+t'}\left|r^- + 1, r'\right.\right),$$
$$P\{\tilde{\bar{\lambda}}'' \ge \lambda|r', t'; t\} = G_{nb}\left(r_\lambda\left|\frac{t}{t+t'}, r'\right.\right) = F_{\beta^*}\left(\frac{t}{t+t'}\left|r^+, r'\right.\right), \qquad (10\text{-}36a)$$

where

$$r^- \equiv \text{greatest integer} \le \lambda t'' - r' ,$$
$$r^+ \equiv \text{least integer} \ge \lambda t'' - r' . \qquad (10\text{-}36b)$$

The mean, partial expectation, and variance of this distribution of $\tilde{\bar{\lambda}}''$ are

$$\mathrm{E}(\tilde{\bar{\lambda}}''|r', t'; t) = \frac{r'}{t'} = \mathrm{E}(\tilde{\lambda}|r', t') \equiv \bar{\lambda}' \ , \tag{10-37a}$$

$$\mathrm{E}_0^\lambda(\tilde{\bar{\lambda}}''|r', t'; t) = \bar{\lambda}' \, G_{\beta^*}\left(\frac{t}{t+t'}\middle| r^- + 1, r' + 1\right) , \tag{10-37b}$$

$$\mathrm{V}(\tilde{\bar{\lambda}}''|r', t'; t) = \frac{t}{t+t'} \frac{r'}{t'^2} = \frac{t}{t+t'} \mathrm{V}(\tilde{\lambda}|r', t') \ . \tag{10-37c}$$

It follows at once from (10-36) and (10-37b) that the linear-loss integrals under this distribution of $\tilde{\bar{\lambda}}''$ are given by

$$\begin{aligned} L_l(\lambda) &\equiv \sum_{z \le \lambda} (\lambda - z) \, \mathrm{P}\{\tilde{\bar{\lambda}}'' = z|r', t'; t\} \\ &= \lambda \, G_{\beta^*}\left(\frac{t}{t+t'}\middle| r^- + 1, r'\right) - \bar{\lambda}' \, G_{\beta^*}\left(\frac{t}{t+t'}\middle| r^- + 1, r' + 1\right) , \qquad (10\text{-}38a) \\ L_r(\lambda) &\equiv \sum_{z \ge \lambda} (z - \lambda) \, \mathrm{P}\{\tilde{\bar{\lambda}}'' = z|r', t'; t\} \\ &= \bar{\lambda}' \, F_{\beta^*}\left(\frac{t}{t+t'}\middle| r^+, r' + 1\right) - \lambda \, F_{\beta^*}\left(\frac{t}{t+t'}\middle| r^+, r'\right) . \qquad (10\text{-}38b) \end{aligned}$$

▶ Formula (10-37a) follows immediately from (5-27).

To prove formula (10-37b) we write the definition

$$\mathrm{E}_0^\lambda(\tilde{\bar{\lambda}}'') \equiv \sum_{\bar{\lambda}'' \le \lambda} \bar{\lambda}'' \, \mathrm{P}\{\bar{\lambda}''\}$$

and then substitute (10-34) for $\bar{\lambda}''$ and replace the probability of $\bar{\lambda}''$ by the probability (10-33) of r. Using formula (7-73) for this latter probability we thus obtain

$$\begin{aligned} \mathrm{E}_0^\lambda(\tilde{\bar{\lambda}}'') &= \sum_{r \le r^-} \frac{r + r'}{t + t'} \frac{(r + r' - 1)!}{r!(r' - 1)!} \frac{t^r t'^{r'}}{(t + t')^{r+r'}} \\ &= \frac{r'}{t'} \sum_{r \le r^-} \frac{(r + r')!}{r! r'!} \frac{t^r t'^{r'+1}}{(t + t')^{r+r'+1}} = \bar{\lambda}' \, F_{nb}\left(r^-\middle| \frac{t}{t+t'}, r' + 1\right) . \end{aligned}$$

Formula (10-37b) then follows by (10-36).

To prove formula (10-37c) for the variance we make use of the fact that, by (10-34),

$$\tilde{\bar{\lambda}}'' = \frac{\tilde{r}}{t''} + \text{constant}$$

is a linear function of $\tilde{r}$, so that

$$\mathrm{V}(\tilde{\bar{\lambda}}'') = \frac{1}{t''^2} \mathrm{V}(\tilde{r}|r', t'; t) \ .$$

Formula (10-37c) is obtained by substituting herein from (7-74b)

$$\mathrm{V}(\tilde{r}|r', t'; t) = r' \frac{t/(t + t')}{[t'/(t + t')]^2} = r' \frac{t/t''}{(t'/t'')^2} = \frac{r' t'' t}{t'^2} \ . \qquad ◀$$

Approximations. As r increases, the distribution of $\tilde{\bar{\lambda}}''$ approaches the gamma-1 prior distribution of $\tilde{\lambda}$, so that when r is large a gamma—*not* a Normal—approximation to the distribution of $\tilde{\bar{\lambda}}''$ will give good results. It follows from (10-37) and (10-10) that a gamma-1 distribution with parameters

$$r^* = \frac{t''}{t} r' , \qquad t^* = \frac{t''}{t} t' , \tag{10-39}$$

will have the same mean and variance as the exact distribution (10-35) of $\tilde{\bar{\lambda}}''$. Using these parameter values, cumulative probabilities may be obtained by use of formula (10-12) and linear-loss integrals may be evaluated by use of formulas (10-11).

10.3.5. Distribution of $\tilde{\bar{\mu}}''$

If a sample drawn by Poisson sampling with fixed t from a Poisson process yields a statistic (r, t), the posterior expectation of $\tilde{\mu} = 1/\tilde{\lambda}$ will be, by (10-15a) and (10-9),

$$\bar{\mu}'' = \frac{t''}{r + r' - 1} , \tag{10-40}$$

just as if the same statistic had been obtained in gamma sampling with fixed r rather than fixed t. The distribution of $\tilde{\bar{\mu}}''$ can be obtained from the unconditional distribution (10-33) of $\tilde{r}$ by solving (10-40) for r:

$$P\{\tilde{\bar{\mu}}'' = \mu | r', t'; t\} = f_{nb}\left(r_\mu \middle| \frac{t}{t + t'}, r'\right) , \qquad r_\mu = \frac{t''}{\mu} - (r' - 1) . \tag{10-41}$$

Since $\bar{\mu}''$ is a decreasing function of r and r is integer-valued, we have by (7-75) that

$$\begin{aligned} P\{\tilde{\bar{\mu}}'' \le \mu | r', t'; t\} &= G_{nb}\left(r_\mu \middle| \frac{t}{t + t'}, r'\right) = F_{\beta^*}\left(\frac{t}{t + t'} \middle| r^+, r'\right) , \\ P\{\tilde{\bar{\mu}}'' \ge \mu | r', t'; t\} &= F_{nb}\left(r_\mu \middle| \frac{t}{t + t'}, r'\right) = G_{\beta^*}\left(\frac{t}{t + t'} \middle| r^- + 1, r'\right) , \end{aligned} \tag{10-42a}$$

where

$$\begin{aligned} r^- &\equiv \text{greatest integer} \le \frac{t''}{\mu} - (r' - 1) , \\ r^+ &\equiv \text{least integer} \ge \frac{t''}{\mu} - (r' - 1) . \end{aligned} \tag{10-42b}$$

The mean and partial expectation of this distribution of $\tilde{\bar{\mu}}''$ are

$$E(\tilde{\bar{\mu}}'' | r', t'; t) = \frac{t'}{r' - 1} = E(\tilde{\mu} | r', t') \equiv \bar{\mu}' , \tag{10-43a}$$

$$E_0^\mu(\tilde{\bar{\mu}}'' | r', t'; t) = \bar{\mu}' F_{\beta^*}\left(\frac{t}{t + t'} \middle| r^+, r' - 1\right) . \tag{10-43b}$$

It follows immediately from (10-42) and (10-43) that the linear-loss integrals under this distribution of $\tilde{\bar{\mu}}''$ are

$$L_l(\mu) \equiv \sum_{z \leq \mu} (\mu - z)\, \mathrm{P}\{\tilde{\mu}'' = z\}$$

$$= \mu\, F_{\beta^*}\left(\frac{t}{t+t'}\middle| r^+, r'\right) - \bar{\mu}'\, F_{\beta^*}\left(\frac{t}{t+t'}\middle| r^+, r'-1\right), \tag{10-44a}$$

$$L_r(\mu) \equiv \sum_{z \geq \mu} (z - \mu)\, \mathrm{P}\{\tilde{\mu}'' = z\}$$

$$= \bar{\mu}'\, G_{\beta^*}\left(\frac{t}{t+t'}\middle| r^- + 1, r'-1\right) - \mu\, G_{\beta^*}\left(\frac{t}{t+t'}\middle| r^- + 1, r'\right) \cdot \tag{10-44b}$$

▶ Formula (10-43a) follows immediately from (5-27).

To prove formula (10-43b) for the incomplete expectation, we write the definition

$$\mathrm{E}_0^{\mu}(\bar{\mu}'') \equiv \sum_{z \leq \mu} \bar{\mu}''\, \mathrm{P}\{\bar{\mu}''\}$$

and then substitute (10-40) and (10-41) remembering that $\bar{\mu}''$ is a decreasing function of r:

$$\mathrm{E}_0^{\mu}(\bar{\mu}'') = \sum_{r \geq r_\mu} \frac{t+t'}{r+r'-1} f_{nb}\left(r \middle| \frac{t}{t+t'}, r'\right) = \frac{t'}{r'-1} G_{nb}\left(r_\mu \middle| \frac{t}{t+t'}, r'-1\right) \cdot$$

By (10-42) this result is equivalent to (10-43b). ◀

No closed formula for the variance of $\tilde{\bar{\mu}}''$ or for its second moment about the origin can be found, but the latter quantity is approximated by

$$S(J) \equiv \sum_{j=1}^{J} \frac{t'^{j+1}(j-1)!(-1)^{j-1}}{t''^{j-1}(r'-1)(r'-2)\cdots(r'-1-j)} \cdot \tag{10-45a}$$

The sequence $\{S(J)\}$ does not converge, but as long as $J < r' - 1$ it gives values which are alternately above and below the true value of $\mathrm{E}(\tilde{\bar{\mu}}'')^2$, and the error which results from using a finite number of terms is therefore less than the last term included. From (10-45a) we can derive an approximation to the variance which also gives values alternately above and below the true value:

$$\mathrm{V}(\tilde{\bar{\mu}}'') \doteq \breve{\mu}' \frac{t}{t''}\left[\left(1 + \frac{t'}{t}\right) - \frac{r'-1}{r'-3}\frac{t'}{t} + \frac{r'-1}{r'-3}\frac{t'}{t}\sum_{j=3}^{J} \frac{t'^{j-2}(j-1)!(-1)^{j-1}}{t''^{j-2}(r'-4)\cdots(r'-1-j)}\right] \tag{10-45b}$$

where

$$\breve{\mu}' \equiv \mathrm{V}(\tilde{\mu}|r', t') = \frac{t'^2}{(r'-1)^2(r'-2)} \cdot \tag{10-45c}$$

When $r' > 3$, a more suggestive form of the series is obtained by recombining the first two terms:

$$\mathrm{V}(\tilde{\bar{\mu}}'') \doteq \breve{\mu}' \frac{t}{t''}\left[1 - \frac{2}{r'-3}\frac{t'}{t} + \frac{r'-1}{r'-3}\frac{t'}{t}\sum_{j=3}^{J} \frac{t'^{j-2}(j-1)!(-1)^{j-1}}{t''^{j-2}(r'-4)\cdots(r'-1-j)}\right]; \tag{10-45d}$$

but it should be observed that if $r' < 7$ *and* $t < \frac{1}{2}t'(7 - r')/(r' - 4)$ the approximation using a single term of this series *may* be *below* the true value, in which case the approximation using two terms would be worse than the approximation using a single term.

▶ To show that the quantity $S(J)$ defined by (10-45a) has the properties described in the text we shall prove that for any J such that $1 \leq J < r' - 1$,

$$S(J) = \mathrm{E}\left(\tilde{\mu}''^2\left[1 + \frac{J!(-1)^{J-1}}{(\tilde{r} + r' - 2) \cdots (\tilde{r} + r' - 1 - J)}\right]\right) ; \tag{1}$$

since it is immediately apparent that if this is true, then $S(1)$, $S(2)$, $\cdots$, $S(J)$ will be alternately greater and less than $\mathrm{E}(\tilde{\mu}''^2)$.

We start by rewriting the definition (10-45a) of $S(J)$ in the form

$$S(J) = t''^2 \sum_{j=1}^{J} \frac{(j-1)!(-1)^{j-1}}{t''^{j+1}} \cdot \frac{t'^{j+1}}{(r'-1) \cdots (r'-1-j)} . \tag{2}$$

We next observe that we are entitled to write

$$S(J) = t''^2 \sum_{j=1}^{J} \frac{(j-1)!(-1)^{j-1}}{t''^{j+1}} \cdot \mathrm{E}\left[\frac{t''^{j+1}}{(\tilde{r}''-1) \cdots (\tilde{r}''-1-j)}\right] \tag{3}$$

because

$$\mathrm{E}\left[\frac{t''^{j+1}}{(\tilde{r}''-1) \cdots (\tilde{r}''-1-j)}\right] \equiv \sum_{r=0}^{\infty} \frac{(t+t')^{j+1}}{(r+r'-1) \cdots (r+r'-1-j)} f_{nb}\left(r \middle| \frac{t}{t+t'}, r'\right) \tag{4}$$

$$= \frac{t'^{j+1}}{(r'-1) \cdots (r'-1-j)} \sum_{r=0}^{\infty} f_{nb}\left(r \middle| \frac{t}{t+t'}, r'-j-1\right) = \frac{t'^{j+1}}{(r'-1) \cdots (r'-1-j)} .$$

If we now reverse the order of expectation and summation in (3) and then cancel out t''^{j+1}, we obtain

$$S(J) = t''^2\, \mathrm{E}\left[\sum_{j=1}^{J} \frac{(j-1)!(-1)^{j-1}}{(\tilde{r}''-1) \cdots (\tilde{r}''-1-j)}\right] . \tag{5}$$

We can easily verify that the jth term in the summand is equal to

$$-\frac{1}{\tilde{r}''-1} \Delta \frac{(j-1)!(-1)^{j-1}}{(\tilde{r}''-1) \cdots (\tilde{r}''-j)} ,$$

where Δ is the first-difference operator defined by $\Delta a_j = a_{j+1} - a_j$. Then since

$$\Sigma_{j=1}^{J} \Delta a_j = a_{J+1} - a_1 ,$$

we may write

$$S(J) = t''^2\, \mathrm{E}\left(\frac{1}{\tilde{r}''-1}\left[\frac{1}{\tilde{r}''-1} - \frac{J!(-1)^J}{(\tilde{r}''-1) \cdots (\tilde{r}''-J-1)}\right]\right)$$

$$= \mathrm{E}\left(\left[\frac{t''}{\tilde{r}''-1}\right]^2\left[1 + \frac{J!(-1)^{J-1}}{(\tilde{r}''-2) \cdots (\tilde{r}''-J-1)}\right]\right) .$$

The proof of (10-45a) is completed by recalling that

$$\tilde{\mu}'' = \frac{t''}{\tilde{r}''-1} .$$

To derive (10-45bc) from (10-45a) we shall make use of the relation

$$\text{(6)} \qquad V(\tilde{\mu}'') = E(\tilde{\mu}''^2) - [E(\tilde{\mu}'')]^2 \doteq S(J) - [E(\tilde{\mu}'')]^2 \; .$$

Rewriting the series (10-45a) in the form

$$S(J) = S_1 + S_2(J)$$

where

$$\text{(7)} \qquad S_1 = \frac{t'^2}{(r'-1)(r'-2)}\left[1 - \frac{t'}{t''}\frac{1}{r'-3}\right]$$

$$\text{(8)} \qquad S_2(J) = \frac{t'^2}{(r'-1)^2(r'-2)}\frac{t}{t''}\left[\frac{t'}{t}\frac{r'-1}{r'-3}\sum_{j=3}^{J}\frac{t'^{j-2}(j-1)!(-1)^{j-1}}{t''^{j-2}(r'-4)\cdots(r'-1-j)}\right]$$

and making use of (10-43a) for $E(\tilde{\mu}'')$ we can put (6) in the form

$$\text{(9)} \qquad V(\tilde{\mu}'') \doteq S_1 - \frac{t'^2}{(r'-1)^2} + S_2(J) \; .$$

Remembering that $t'' = t' + t$ we can compute

$$\text{(10)} \qquad S_1 - \frac{t'^2}{(r'-1)^2} = \frac{t'^2}{(r'-1)^2(r'-2)}\left[1 - \frac{t'}{t''}\frac{r'-1}{r'-3}\right]$$

$$= \frac{t'^2}{(r'-1)^2(r'-2)}\frac{t}{t''}\left[\frac{t''}{t} - \frac{r'-1}{r'-3}\frac{t'}{t}\right] .$$

The results (10-45bc) are now obtained by substituting (10) and (8) in (9) and then using (10-15c). ◀

Approximations. As t increases, the distribution of $\tilde{\mu}''$ becomes more nearly continuous and approaches the inverted-gamma-1 prior distribution of $\tilde{\mu}$, so that when t is large an inverted-gamma-1 approximation to the distribution of $\tilde{\mu}''$ itself—*not* a Normal approximation—will give good results. It follows from (10-43a) and (10-15) that an inverted-gamma-1 distribution with parameters

$$r^* = \frac{\bar{\mu}'^2}{V(\tilde{\mu}'')} + 2 \; , \qquad t^* = \frac{r^*-1}{r'-1}t' \; , \qquad \text{(10-46)}$$

would have the same mean and variance as the exact distribution of $\tilde{\mu}''$. Once (10-46) has been evaluated approximately by use of (10-45), tail probabilities may be computed by use of (10-14) and linear-loss integrals may be evaluated by use of (10-16).

CHAPTER 11

A. Independent Normal Process, Mean Known

11.1. Prior and Posterior Analysis

11.1.1. Definition of an Independent Normal Process

An Independent Normal process can be defined as a process generating independent random variables $\tilde{x}_1, \cdots, \tilde{x}_i, \cdots$ with identical densities

$$f_N(x|\mu, h) \equiv (2\pi)^{-\frac{1}{2}} e^{-\frac{1}{2}h(x-\mu)^2} h^{\frac{1}{2}} , \qquad \begin{array}{r} -\infty < x < \infty , \\ -\infty < \mu < \infty , \\ h > 0 . \end{array} \tag{11-1}$$

This is the Normal density as defined in (7-63), and therefore the process mean and variance can be obtained from (7-65):

$$\text{E}(\tilde{x}|\mu, h) = \mu , \tag{11-2a}$$
$$\text{V}(\tilde{x}|\mu, h) = 1/h . \tag{11-2b}$$

The parameter h will be called the *precision* of the process. We shall use

$$\sigma \equiv \sqrt{\text{V}(\tilde{x}|\mu, h)} = \sqrt{1/h} \tag{11-3}$$

to denote the process standard deviation.

11.1.2. Likelihood of a Sample When μ Is Known

The likelihood that an Independent Normal process will *generate* ν successive values $x_1, \cdots, x_i, \cdots, x_\nu$ is the product of their individual likelihoods as given by (11-1); this product is obviously

$$(2\pi)^{-\frac{1}{2}\nu} e^{-\frac{1}{2}h\,\Sigma(x_i-\mu)^2} h^{\frac{1}{2}\nu} . \tag{11-4}$$

When the process mean μ is known and only the precision h is unknown, the statistic

$$w \equiv \frac{1}{\nu} \Sigma(x_i - \mu)^2 \tag{11-5}$$

can be computed and will be said to be based on ν *degrees of freedom.* The likelihood (11-4) can be written in terms of this statistic as

$$(2\pi)^{-\frac{1}{2}\nu} e^{-\frac{1}{2}h\nu w} h^{\frac{1}{2}\nu} . \tag{11-6}$$

It is clear from (11-6) that all the information in the sample is conveyed by the statistics w and ν, which are therefore *sufficient* when μ is known and the stopping process is noninformative. Since $(2\pi)^{-\frac{1}{2}\nu}$ in this expression has the same value regardless of the value of the unknown h, it can tell us nothing about h and may therefore be disregarded: the kernel of the likelihood is

$$e^{-\frac{1}{2}h\nu w}\, h^{\frac{1}{2}\nu} \ . \tag{11-7}$$

11.1.3. Conjugate Distribution of $\tilde{h}$

When the mean μ of an Independent Normal process is known but the precision $\tilde{h}$ is treated as a random variable, the most convenient distribution of $\tilde{h}$—the natural conjugate of (11-7)—is the gamma-2 distribution defined by (7-50):

$$f_{\gamma 2}(h|v, \nu) \propto e^{-\frac{1}{2}h\nu v}\, h^{\frac{1}{2}\nu - 1} \ . \tag{11-8}$$

If such a distribution with parameter (v', ν') is assigned to $\tilde{h}$ and if a sample then yields a sufficient statistic (w, ν), the posterior distribution of $\tilde{h}$ will be gamma-2 with parameters

$$v'' = \frac{\nu' v' + \nu w}{\nu' + \nu} \ , \qquad \nu'' = \nu' + \nu \ . \tag{11-9}$$

▶ Multiplying the kernel (11-8) of the prior density by the kernel (11-7) of the likelihood we have

$$\text{D}''(h) \propto e^{-\frac{1}{2}h\nu' v'}\, h^{\frac{1}{2}\nu' - 1} \cdot e^{-\frac{1}{2}h\nu w}\, h^{\frac{1}{2}\nu} = e^{-\frac{1}{2}h\nu'' v''}\, h^{\frac{1}{2}\nu'' - 1} \ ,$$

which is of the form (11-8). ◀

The mean and variance of the gamma-2 distribution (11-8) of $\tilde{h}$ are, by (7-52)

$$\text{E}(\tilde{h}|v, \nu) = \frac{1}{v} \ , \tag{11-10a}$$

$$\text{V}(\tilde{h}|v, \nu) = \frac{1}{\frac{1}{2}\nu v^2} \ . \tag{11-10b}$$

Cumulative probabilities may be obtained from Pearson's tables of the gamma function $I(u, p)$ or from tables of the cumulative Poisson function through the relation (7-51b):

$$F_{\gamma 2}(h|v, \nu) = F_{\gamma^*}(\tfrac{1}{2}\nu v h|\tfrac{1}{2}\nu) = I(hv\sqrt{\tfrac{1}{2}\nu}, \tfrac{1}{2}\nu - 1) = G_P(\tfrac{1}{2}\nu|\tfrac{1}{2}\nu v h) \ . \tag{11-11}$$

Limiting Behavior of the Prior Distribution. As the parameter ν of (11-8) approaches zero while v remains fixed, the mean of the distribution remains fixed at $1/v$, the variance becomes infinite, and the probability becomes more and more concentrated toward $h = 0$.

▶ The behavior of the mean and variance follow immediately from (11-10). To prove that the probability becomes concentrated toward $h = 0$, i.e., to prove that

$$\lim_{\nu \to 0} F_{\gamma 2}(\delta|v, \nu) = 1$$

for any $\delta > 0$ however small, we write

$$F_{\gamma 2}(\delta|v,\nu) = \int_0^\delta \frac{e^{-\frac{1}{2}\nu v h}(\frac{1}{2}\nu v h)^{\frac{1}{2}\nu-1}}{(\frac{1}{2}\nu-1)!}\,\tfrac{1}{2}\nu v\,dh$$

and then substitute

$$u = vh\ , \qquad \delta^* = v\delta\ , \qquad r = \tfrac{1}{2}\nu\ ,$$

to obtain

$$F_{\gamma 2}(\delta|v,\nu) = \int_0^{\delta^*} \frac{e^{-ru}(ru)^{r-1}}{(r-1)!}\,r\,du\ .$$

That this integral approaches 1 as $r \to 0$ was proved at the end of Section 10.1.3. ◀

11.1.4. Conjugate Distribution of $\tilde{\sigma}$

If we assign a gamma-2 distribution to the precision $\tilde{h}$ of an Independent Normal process, then as shown in the proof of (7-58a) we are implicitly assigning an inverted-gamma-2 distribution with parameters $s = \sqrt{v}$ and ν to the process standard deviation $\tilde{\sigma} = \sqrt{1/\tilde{h}}$. This latter distribution is discussed and graphed in Section 7.7.2, but one additional observation is worth making in the present context.

The judgment of a decision maker who must assign a prior distribution to $\tilde{h}$ will often bear much more directly on $\tilde{\sigma}$ than on $\tilde{h}$, and Figure 7.5 may be of service in translating judgments about $\tilde{\sigma}$ into values of the parameters $s = \sqrt{v}$ and ν. Thus if the decision maker feels (1) that there is an "even chance" that $\tilde{\sigma} > 5$ and (2) that there is "one chance in four" that $\tilde{\sigma} > 10$, he can find the corresponding values of the parameters v' and ν' as follows. He first determines by use of the curve for the ratio of the third to the second quartile in Figure 7.5B that the ratio $10/5 = 2$ implies $\nu' = 1$ approximately. He then reads from the curve for the second quartile in Figure 7.5A that when $\nu = 1$ the second quartile is 1.45 times s and computes $s = 5/1.45 = 3.45$ and $v = 3.45^2 \doteq 12$.

Limiting Behavior of the Prior Distribution. As ν approaches 0 with $s = \sqrt{v}$ held fixed, the probability that $\tilde{\sigma}$ lies within any finite interval approaches 0 but the distribution does not become uniform: the ratio of the densities at any two points σ_1 and σ_2 approaches σ_2/σ_1. The mean of the distribution is infinite when $\nu \leq 1$ and the variance is infinite when $\nu \leq 2$.

▶ The behavior of the mean and variance are given directly by (7-59), and the limiting ratio of the densities is easily obtained from the formula for the density (7-57). That the probability of any finite interval approaches 0 follows from the facts (1) that by (7-58a)

$$F_{i\gamma 2}(\sigma|s,\nu) = 1 - F_{\gamma 2}(1/\sigma^2|s^2,\nu)$$

and (2) that as proved at the end of Section 11.1.3

$$\lim_{\nu\to 0} F_{\gamma 2}(\delta|s^2,\nu) = 1$$

for any $\delta > 0$ however small—i.e., for any σ however large. ◀

11.2. Sampling Distributions and Preposterior Analysis With Fixed ν

Throughout Section 11.2 we assume that a sample is to be drawn from an Independent Normal process in such a way that the value of the statistic ν is predetermined, the value of the statistic $\tilde{w}$ being left to chance. We assume further that while the mean μ of the process is known its precision $\tilde{h}$ has a gamma-2 distribution with parameter (v', ν').

11.2.1. Conditional Distribution of $(\tilde{w}|h)$

Given a particular value h of the process precision, the distribution of $\tilde{w}$ is gamma-2 with density

$$D(w|h, \nu) = f_{\gamma 2}(w|h, \nu) \propto e^{-\frac{1}{2}h\nu w} w^{\frac{1}{2}\nu - 1} . \tag{11-12}$$

▶ In the ν-dimensional space of the xs, the random variable $\tilde{w}$ defined by (11-5) has the constant value w on the surface of a hypersphere with center $(\mu, \mu, \cdots, \mu)$ and radius $\sqrt{w}$. Since the area of such a surface is proportional to $(\sqrt{w})^{\nu-1}$ and since an increment dw in w produces an increment proportional to $dw/\sqrt{w}$ in the radius of the sphere, the spherical shell within which the value of $\tilde{w}$ is between w and $w + dw$ is proportional to $w^{\frac{1}{2}\nu-1}\, dw$. Multiplying the density (11-6) by this result and dropping constants we obtain

$$e^{-\frac{1}{2}h\nu w} w^{\frac{1}{2}\nu-1}\, dw ,$$

which apart from the differential element dw is (11-12). ◀

11.2.2. Unconditional Distribution of $\tilde{w}$

If the process precision $\tilde{h}$ is treated as a random variable having the gamma-2 density (11-8) with parameter (v', ν'), the *unconditional* distribution of the statistic $\tilde{w}$ has the inverted-beta-2 density (7-28):

$$D(w|v', \nu'; \nu) = \int_0^\infty f_{\gamma 2}(w|h, \nu)\, f_{\gamma 2}(h|v', \nu')\, dh = f_{i\beta 2}\left(w|\tfrac{1}{2}\nu, \tfrac{1}{2}\nu', \frac{\nu' v'}{\nu}\right) . \tag{11-13}$$

▶ Substituting (7-50) for $f_{\gamma 2}(w)$ and $f_{\gamma 2}(h)$ in the integral (11-13) we obtain, apart from factors constant as regards w and h,

$$D(w) \propto \frac{(\nu w)^{\frac{1}{2}\nu - 1}}{(\nu w + \nu' v')^{\frac{1}{2}(\nu+\nu')}} \int_0^\infty e^{-\frac{1}{2}h(\nu w + \nu' v')} (\tfrac{1}{2}h[\nu w + \nu' v'])^{\frac{1}{2}(\nu+\nu')-1}\, \tfrac{1}{2}(\nu w + \nu' v')\, dh .$$

Comparison with (7-28) shows that the factor preceding the integral is an inverted-beta-2 density with the parameters given in (11-13), while comparison with (7-43) shows that the integral is a constant multiplied by $F_{\gamma 1}(\infty|\frac{1}{2}\nu + \frac{1}{2}\nu', \frac{1}{2}(\nu w + \nu' v')) = 1$. ◀

11.2.3. Distribution of $\tilde{v}''$

The parameter v'' of the posterior distribution of $\tilde{h}$ will, by (11-9), have the value

$$v'' = \frac{\nu' v' + \nu w}{\nu' + \nu} = \frac{\nu' v' + \nu w}{\nu''} . \tag{11-14}$$

This means that when the sample outcome is still unknown and $\tilde{w}$ is therefore a random variable, $\tilde{v}''$ is also a random variable whose distribution can be obtained from the unconditional distribution (11-13) of $\tilde{w}$. Solving (11-14) for w, substituting in the definition (7-28) of $f_{i\beta 2}$, and comparing the result with the definition (7-25) of $f_{i\beta 1}$, we find that the distribution of $\tilde{v}''$ is inverted-beta-1 with density

$$D(v''|v', \nu'; \nu) = f_{i\beta 1}\left(v''|\tfrac{1}{2}\nu', \tfrac{1}{2}\nu'', \frac{\nu' v'}{\nu''}\right) . \tag{11-15}$$

B. Independent Normal Process, Precision Known

11.3. Prior and Posterior Analysis

11.3.1. Likelihood of a Sample When h Is Known

We have already seen in Section 11.1.2 that if the stopping process is non-informative the likelihood of observations $x_1, \cdots, x_i, \cdots, x_n$ on an Independent Normal process is

$$(2\pi)^{-\frac{1}{2}n}\, e^{-\frac{1}{2}h\,\Sigma(x_i-\mu)^2}\, h^{\frac{1}{2}n} \; ; \tag{11-16}$$

the expression is identical to (11-4) except that we are now using n rather than ν to denote the number of observations. If we define the statistic

$$m \equiv \frac{1}{n} \Sigma x_i , \tag{11-17}$$

the likelihood (11-16) can be written

$$(2\pi)^{-\frac{1}{2}n}\, e^{-\frac{1}{2}h\,\Sigma(x_i-m)^2}\, e^{-\frac{1}{2}hn(m-\mu)^2}\, h^{\frac{1}{2}n} . \tag{11-18}$$

When the process precision h is known and only the mean μ is unknown, the only factor in this expression which varies with the unknown parameter is

$$e^{-\frac{1}{2}hn(m-\mu)^2} , \tag{11-19}$$

and this is therefore the kernel of the likelihood. It is clear from (11-19) that the statistic (m, n) is sufficient when h is known and the stopping process is non-informative.

11.3.2. Conjugate Distribution of $\tilde{\mu}$

When the precision h of an Independent Normal process is known but the mean is a random variable $\tilde{\mu}$, the most convenient distribution of $\tilde{\mu}$—the natural conjugate of (11-19)—is the Normal distribution defined by (7-63):

$$f_N(\mu|m, H) \propto e^{-\frac{1}{2}H(\mu-m)^2} .$$

The quantity H in this expression can be thought of as measuring the *precision* of our information on μ, and it will be more instructive to express this measure in units of the *process* precision h. We therefore define the *parameter* (*not* the statistic) n by

$$n \equiv H/h , \tag{11-20}$$

i.e., we say that the information H is equivalent to n observations on the process, and we then write the density of $\tilde{\mu}$ in the form

$$f_N(\mu|m, hn) \propto e^{-\frac{1}{2}hn(\mu-m)^2} . \tag{11-21}$$

If such a distribution with parameter (m', n') has been assigned to $\tilde{\mu}$ and if a sample from the process then yields a sufficient statistic (m, n), the posterior distribution of $\tilde{\mu}$ will be Normal with parameters

$$m'' = \frac{n'm' + nm}{n' + n} , \qquad n'' = n' + n . \tag{11-22}$$

▶ Multiplying the kernel (11-21) of the prior density by the kernel (11-19) of the likelihood and dropping or inserting constants we obtain

$$e^{-\frac{1}{2}hn'(\mu-m')^2-\frac{1}{2}hn(m-\mu)^2} \propto e^{-\frac{1}{2}h([n+n']\mu^2-2[n'm'+nm]\mu)} \propto e^{-\frac{1}{2}h(n+n')\left(\mu-\frac{n'm'+nm}{n'+n}\right)^2} .$$

Apart from constants this is a Normal density like (11-21) but with the parameters specified in (11-22). ◀

Cumulative probabilities under the distribution (11-21) of $\tilde{\mu}$ can be obtained from tables of the unit Normal function F_{N*} by use of the relation (7-64):

$$F_N(\mu|m, hn) = F_{N*}([\mu - m]\sqrt{hn}) . \tag{11-23}$$

The mean, partial expectation, and variance of the distribution are, by (7-65),

$$E(\tilde{\mu}|m, n) \equiv \bar{\mu} = m , \tag{11-24a}$$

$$E^{\mu}_{-\infty}(\tilde{\mu}|m, n) = m\, F_{N*}(u) - f_{N*}(u)/\sqrt{hn} , \qquad u = (\mu - m)\sqrt{hn} , \tag{11-24b}$$

$$V(\tilde{\mu}|m, n) = \frac{1}{hn} . \tag{11-24c}$$

From (11-23) and (11-24b) it follows that the linear-loss integrals are given by

$$L_l(\mu) \equiv \int_{-\infty}^{\mu} (\mu - z)\, f_N(z|m, hn)\, dz = L_{N*}(-u_\mu)/\sqrt{hn} , \tag{11-25a}$$

$$L_r(\mu) \equiv \int_{\mu}^{\infty} (z - \mu)\, f_N(z|m, hn)\, dz = L_{N*}(u_\mu)/\sqrt{hn} , \tag{11-25b}$$

where

$$u_\mu \equiv (\mu - m)\sqrt{hn} , \qquad L_{N*}(u) \equiv f_{N*}(u) - u\, G_{N*}(u) . \tag{11-26}$$

For tables of the function L_{N*}, see Section 7.8.1.

Limiting Behavior of the Prior Distribution. As the parameter n of (11-21) approaches 0 while m is held constant, the mean of the distribution remains fixed at m but the variance becomes infinite and the probability becomes more and more uniformly distributed over the entire interval $-\infty < \mu < \infty$.

▶ The behavior of the mean and variance follow immediately from (11-24). That the distribution approaches a uniform distribution is proved by observing that the ratio of the densities at any two points μ_1 and μ_2 is, by (11-21),

$$e^{-\frac{1}{2}hn([\mu_1 - m]^2 - [\mu_2 - m]^2)} ,$$

and that as $n \to 0$ this quantity approaches 1. ◀

11.4. Sampling Distributions and Preposterior Analysis With Fixed n

Throughout Section 11.4 we assume that a sample is to be taken from an Independent Normal process in such a way that the statistic n is predetermined, the statistic $\tilde{m}$ being left to chance. We assume further that while the precision h of the process is known its mean $\tilde{\mu}$ has a Normal distribution of type (11-21) with parameter (m', n').

11.4.1. *Conditional Distribution of* $(\tilde{m}|\mu)$

Given a particular value μ of the process mean, the conditional distribution of $\tilde{m}$ is Normal with density

$$D(m|\mu; n) = f_N(m|\mu, hn) ; \tag{11-27}$$

the proof will be given in the discussion of (11-38). It follows from (7-65) that

$$E(\tilde{m}|\mu; n) = \mu , \tag{11-28a}$$

$$V(\tilde{m}|\mu; n) = \frac{1}{hn} . \tag{11-28b}$$

11.4.2. *Unconditional Distribution of* $\tilde{m}$

If the process mean $\tilde{\mu}$ is treated as a random variable having a Normal distribution with parameter (m', n'), the unconditional distribution of the statistic $\tilde{m}$ is Normal with density

$$D(m|m', n'; n) = \int_{-\infty}^{\infty} f_N(m|\mu, hn) f_N(\mu|m', hn') d\mu = f_N(m|m', hn_u) \tag{11-29a}$$

where

$$n_u \equiv \frac{n'n}{n' + n} , \qquad \frac{1}{n_u} = \frac{1}{n'} + \frac{1}{n} . \tag{11-29b}$$

By (7-65) the mean and variance of $\tilde{m}$ are

$$E(\tilde{m}|m', n'; n) = m' = E(\tilde{\mu}|m', n') , \tag{11-30a}$$

$$V(\tilde{m}|m', n'; n) = \frac{1}{hn_u} = V(\tilde{\mu}|m', n') + V(\tilde{m}|\mu, n) . \tag{11-30b}$$

▶ To prove (11-29), substitute for $f_N(m)$ and $f_N(\mu)$ in the integrand their formulas (7-63) to obtain, apart from constants,

$$D(m|m', n'; n) \propto \int_{-\infty}^{\infty} e^{-\frac{1}{2}hn(m-\mu)^2 - \frac{1}{2}hn'(\mu - m')^2}\, d\mu \ .$$

Observing that

$$n(m-\mu)^2 + n'(\mu - m')^2 = (n+n')\left(\mu - \frac{n'm' + nm}{n'+n}\right)^2 + \frac{nn'}{n+n'}(m-m')^2$$

and using the definitions (11-22) and (11-29b) of n'', m'', and n_u, we have

$$D(m|m', n'; n) \propto e^{-\frac{1}{2}hn_u(m-m')^2} \int_{-\infty}^{\infty} e^{-\frac{1}{2}hn''(\mu - m'')^2}\, d\mu \ .$$

The integral is a constant times $F_N(\infty|m'', hn'') = 1$, while apart from constants the factor preceding the integral is the right-hand side of (11-29a). ◀

11.4.3. Distribution of $\tilde{m}''$ and $\tilde{\bar{\mu}}''$

If h is known, if the prior distribution of $\tilde{\mu}$ is Normal with parameter (m', n'), and if a sample then yields a sufficient statistic (m, n), formulas (11-24a) and (11-22) show that the parameter m'' and the mean $\bar{\mu}''$ of the posterior distribution of $\tilde{\mu}$ will have the value

$$\bar{\mu}'' = m'' = \frac{n'm' + nm}{n' + n} \ . \tag{11-31}$$

When the $\tilde{m}$ to be obtained by sampling with fixed n is still unknown and $\tilde{\bar{\mu}}'' = \tilde{m}''$ is therefore a random variable, the distribution of $\tilde{\bar{\mu}}'' = \tilde{m}''$ is determined by the unconditional distribution (11-29) of $\tilde{m}$; the density of $\tilde{\bar{\mu}}''$ is

$$D(\bar{\mu}''|m', n'; n) = f_N(\bar{\mu}''|m', hn^*) \tag{11-32a}$$

where

$$n^* \equiv \frac{n' + n}{n} n' \ , \qquad \frac{1}{n^*} = \frac{1}{n'} - \frac{1}{n''} \ . \tag{11-32b}$$

Its mean and variance are

$$E(\tilde{\bar{\mu}}''|m', n'; n) = m' = E(\tilde{\mu}|m', n') \ , \tag{11-33a}$$

$$V(\tilde{\bar{\mu}}''|m', n'; n) = \frac{1}{hn^*} = V(\tilde{\mu}|m', n') - V(\tilde{\mu}|m'', n'') \ . \tag{11-33b}$$

Since this distribution is of the same Normal form as the distribution (11-21) of $\tilde{\mu}$ itself, formulas for cumulative probabilities, partial expectations, and linear-loss integrals can be obtained by simply substituting n^* for n in (11-23) through (11-26).

▶ That the distribution of $\tilde{\bar{\mu}}'' = \tilde{m}''$ is Normal follows from the fact that, by (11-31), $\tilde{\bar{\mu}}'' = \tilde{m}''$ is a linear function of the Normal random variable $\tilde{m}$. That the mean of $\tilde{\bar{\mu}}''$ is equal to the prior mean m' as asserted by (11-33a) follows from (5-27); that the variance of $\tilde{\bar{\mu}}''$ is the prior variance $1/hn'$ less the posterior variance $1/hn''$ as asserted by (11-33b) follows from (5-28); and the parameters of (11-32) are then determined by these values of the moments. ◀

C. *Independent Normal Process, Neither Parameter Known*

11.5. Prior and Posterior Analysis

11.5.1. Likelihood of a Sample When Neither Parameter Is Known

We have already seen in (11-16) that if the stopping process is noninformative the likelihood of observations $x_1, \cdots, x_n$ on an Independent Normal process is

$$(2\pi)^{-\frac{1}{2}n} e^{-\frac{1}{2}h\,\Sigma(x_i-\mu)^2} h^{\frac{1}{2}n} . \tag{11-34}$$

If now we define the statistics

$$m \equiv \frac{1}{n} \Sigma x_i , \tag{11-35a}$$

$$v \equiv \frac{1}{n-1} \Sigma(x_i - m)^2 \qquad (\equiv 0 \text{ if } n = 1) , \tag{11-35b}$$

the likelihood (11-34) can be written

$$(2\pi)^{-\frac{1}{2}n} e^{-\frac{1}{2}h(n-1)v-\frac{1}{2}hn(m-\mu)^2} h^{\frac{1}{2}n} . \tag{11-36}$$

The kernel—i.e. those factors which vary with the unknown parameters μ and h—is

$$e^{-\frac{1}{2}h(n-1)v-\frac{1}{2}hn(m-\mu)^2} h^{\frac{1}{2}n} . \tag{11-37}$$

It is clear from (11-37) that the statistic (m, v, n) is sufficient when the stopping process is noninformative.

11.5.2. Likelihood of the Incomplete Statistics (m, n) and (v, ν)

In many situations a series of observations x may yield useful information on the precision of an Independent Normal process without yielding any information on its mean, as when evidence on the precision of a measuring instrument is obtained by taking a series of measurements of some known standard μ^* before the instrument is used to measure the unknown μ which is of interest in the decision problem at hand. If such a series of measurements is not long enough to determine h "with certainty", the information it contains will be summarized sufficiently by the statistic (w, ν) where ν is the number of observations on μ^* and w is defined by (11-5). Evidence concerning the precision h is also frequently obtainable from a series of observations on some *unknown* quantity μ^* which is itself irrelevant to the decision problem at hand. In this case the sufficient statistic for the observations on μ^* will be (m, v, n) which we have just defined. Although the evidence m on μ^* contained in this triplet tells us nothing about the μ of the decision problem actually at hand, the doublet (v, n) does contain relevant information on h just as (w, ν) contains relevant information on h if μ^* is known. We cannot, however, use the kernel (11-37) of the complete statistic (m, v, n) to exploit the information contained in the incomplete statistic (v, n); what we need is the kernel of the *marginal* likelihood of (v, n), obtained by integrating (11-37) over all sample points $(x_1, \cdots, x_n)$ for which (v, n) has its observed value.

It is this same marginal likelihood of (v, n) which we would require if the value of the component m of the sufficient statistic had been lost; and more for symmetry than for any real application we shall also consider the case where the component v of (m, v, n) is unknown or for some reason irrelevant, so that to exploit the information in the incomplete statistic (m, n) we would require the marginal likelihood of (m, n).

The kernel of the *marginal* likelihood of (m, n) is

$$e^{-\frac{1}{2}hn(m-\mu)^2}\, h^{\frac{1}{2}} \;, \tag{11-38a}$$

while the kernel of the *marginal* likelihood of (v, n) is

$$e^{-\frac{1}{2}h(n-1)v}\, h^{\frac{1}{2}(n-1)} \;. \tag{11-38b}$$

▶ In the n-dimensional space of the xs, construct a radius vector from the origin to the point $(m, m, \cdots, m)$, at this point construct an $(n-1)$-dimensional hyperplane perpendicular to the vector, and on this hyperplane construct a hypercircle with center $(m, m, \cdots, m)$ and radius $\sqrt{(n-1)v}$. The random variable $\tilde{m} = \Sigma\, \tilde{x}_i/n$ then has the constant value m at all points on the hyperplane, and $\tilde{v} = \Sigma(\tilde{x}_i - m)^2/(n-1)$ has the constant value v at all points on the hypercircle. The circumference of the hypercircle is proportional to $(\sqrt{v})^{n-2}$ and an increment dv in v produces an increment proportional to $dv/\sqrt{v}$ in the radius of the hypercircle, so that the area of the solid annulus within which $v \le \tilde{v} \le v + dv$ and $m \le \tilde{m} \le m + dm$ is proportional to $v^{\frac{1}{2}(n-3)}\, dv\, dm$.

Multiplying the density (11-37) by this result we obtain

$$[e^{-\frac{1}{2}hn(m-\mu)^2}\, h^{\frac{1}{2}}\, dm] \cdot [e^{-\frac{1}{2}h(n-1)v}\, v^{\frac{1}{2}(n-1)-1}\, h^{\frac{1}{2}(n-1)}\, dv] \;.$$

Apart from the differential elements and constants involving n but not v, m, h, or μ, this expression is

$$f_N(m|\mu, hn)\, f_{\gamma 2}(v|h, n-1) \;,$$

and this is therefore the exact joint likelihood of the statistics v and m.

Since the integral over $0 \le v < \infty$ of the second factor in the joint likelihood is 1, the marginal likelihood of m is $f_N(m|\mu, hn)$; and since the integral of the first factor over $-\infty < m < \infty$ is 1, the marginal likelihood of v is $f_{\gamma 2}(v|h, n-1)$. Formulas (11-38) result when factors not involving either μ or h are dropped from $f_N(m)$ and $f_{\gamma 2}(v)$. ◀

In order to unify our treatment of cases where only (m, n) or only (v, n) is known and relevant with cases where the entire sufficient statistic (m, v, n) is known and relevant, we would like to express the likelihood of (m, v, n) in such a way that it automatically reduces to (11-38a) when only (m, n) is involved or to (11-38b) when only (v, n) is involved. To do this we define

$$\nu \equiv n - 1 \;, \tag{11-39}$$

which we shall call the number of degrees of freedom in the statistic v, and

$$\delta(n) \equiv \begin{cases} 0 & \text{if} \quad n = 0 \\ 1 & \text{if} \quad n > 0 \end{cases} \;, \tag{11-40}$$

which we shall call the number of degrees of freedom in the statistic m. We then write in place of (11-37)

$$e^{-\frac{1}{2}hn(m-\mu)^2}\, h^{\frac{1}{2}\delta(n)} \cdot e^{-\frac{1}{2}h\nu v}\, h^{\frac{1}{2}\nu} \;. \tag{11-41}$$

If both m and v are known and relevant, this expression is identical to (11-37). If further we adopt the convention that

1. ν (but not n) has the value 0 when v is unknown or irrelevant,
2. n (but not ν) has the value 0 when m is unknown or irrelevant,

then (11-41) reduces to (11-38a) in the first case and to (11-38b) in the second.

11.5.3. *Distribution of* $(\tilde{\mu}, \tilde{h})$

When both parameters of an Independent Normal process are unknown and are to be treated as random variables $\tilde{\mu}$ and $\tilde{h}$, the most convenient joint distribution of the two variables—the natural conjugate of (11-41)—is what we shall call a Normal-gamma distribution, defined by

$$\begin{aligned} f_{N\gamma}(\mu, h|m, v, n, \nu) &= f_N(\mu|m, hn)\, f_{\gamma 2}(h|v, \nu) & -\infty < \mu < \infty\ , \\ &\propto e^{-\frac{1}{2}hn(\mu-m)^2}\, h^{\frac{1}{2}\delta(n)} \cdot e^{-\frac{1}{2}\nu vh}\, h^{\frac{1}{2}\nu-1}\ , & h \geq 0\ , \\ & & v, n, \nu > 0\ , \end{aligned} \tag{11-42}$$

where n is defined by (11-20) and $\delta(n)$ is the delta function defined in (11-40). Since n must be greater than 0 if (11-42) is to represent a *proper* (convergent) density, $\delta(n)$ is here necessarily equal to 1 and the definition of f_N implied by (11-42) is identical to the definition (7-63); the apparently superfluous $\delta(n)$ is added so that the formulas which we shall obtain for the parameters of the posterior distribution will generalize automatically to the limiting case where the prior parameter $n' = 0 = \delta(n')$.

If the prior distribution of $(\tilde{\mu}, \tilde{h})$ is Normal-gamma with parameter (m', n', v', ν') and if a sample then yields a sufficient statistic (m, v, n, ν), the posterior distribution of $(\tilde{\mu}, \tilde{h})$ will be Normal-gamma with parameters

$$m'' = \frac{n'm' + nm}{n' + n}\ , \tag{11-43a}$$

$$n'' = n' + n\ , \tag{11-43b}$$

$$v'' = \frac{[\nu'v' + n'm'^2] + (\nu v + nm^2] - n''m''^2}{[\nu' + \delta(n')] + [\nu + \delta(n)] - \delta(n'')}\ , \tag{11-43c}$$

$$\nu'' = [\nu' + \delta(n')] + [\nu + \delta(n)] - \delta(n'')\ . \tag{11-43d}$$

Observe that when $n' = \nu' = 0$, the parameter (m'', v'', n'', ν'') is equal to the statistic (m, v, n, ν).

▶ Multiplying the prior density (11-42) by the likelihood (11-41) and dropping constants we have

$$e^{-\frac{1}{2}h\nu'v' - \frac{1}{2}hn'(\mu - m')^2}\, h^{\frac{1}{2}\nu' + \frac{1}{2}\delta(n') - 1} \cdot e^{-\frac{1}{2}h\nu v - \frac{1}{2}hn(m-\mu)^2}\, h^{\frac{1}{2}\nu + \frac{1}{2}\delta(n)}\ .$$

We observe first that the posterior exponent of h will be

$$\tfrac{1}{2}\nu' + \tfrac{1}{2}\delta(n') - 1 + \tfrac{1}{2}\nu + \tfrac{1}{2}\delta(n) = \tfrac{1}{2}[\nu' + \delta(n') + \nu + \delta(n) - \delta(n'')] + \tfrac{1}{2}\delta(n'') - 1\ ,$$

which by (11-43d) can be written

$$\tfrac{1}{2}\nu'' + \tfrac{1}{2}\delta(n'') - 1$$

in agreement with the exponent of h in (11-42). We next observe that the posterior exponent of e will be $-\frac{1}{2}hS$ where

$$S \equiv \nu'v' + n'[\mu - m']^2 + \nu v + n[m - \mu]^2$$
$$= \nu'v' + n'm'^2 + \nu v + nm^2 - \frac{(n'm' + nm)^2}{n' + n} + (n' + n)\left[\mu - \frac{n'm' + nm}{n' + n}\right]^2 .$$

Using (11-43ab) our last result can be written

$$S = (\nu'v' + n'm'^2 + \nu v + nm^2 - n''m''^2) + n''(\mu - m'')^2 ;$$

and by (11-43cd) this becomes

$$\nu''v'' + n''(\mu - m'')^2$$

in agreement with the exponent of e in (11-42). ◀

11.5.4. *Marginal Distribution of $\tilde{h}$*

If the joint distribution of the random variable $(\tilde{\mu}, \tilde{h})$ is Normal-gamma as defined by (11-42), the marginal distribution of $\tilde{h}$ is gamma-2 as defined by (7-50):

$$D(h|m, v, n, \nu) = f_{\gamma 2}(h|v, \nu) . \qquad (11\text{-}44)$$

It follows immediately, by (7-51b) and (7-52) that

$$P\{\tilde{h} < h|m, v, n, \nu\} = I(hv\sqrt{\tfrac{1}{2}\nu}, \tfrac{1}{2}\nu - 1) = G_P(\tfrac{1}{2}\nu|\tfrac{1}{2}\nu v h) , \qquad (11\text{-}45)$$

$$E(\tilde{h}|m, v, n, \nu) = \frac{1}{v} , \qquad (11\text{-}46a)$$

$$V(\tilde{h}|m, v, n, \nu) = \frac{1}{\frac{1}{2}\nu v^2} . \qquad (11\text{-}46b)$$

Observe that the distribution (11-44) does not depend on either m or n and is in fact identical to the distribution (11-8) which applies when the process mean μ is known. The marginal distribution of $\tilde{\sigma} = \sqrt{1/\tilde{h}}$ in our present problem is consequently identical to the distribution discussed in Section 11.1.4.

▶ By definition, the marginal density of $\tilde{h}$ is obtained by integrating the joint density (11-42) over $-\infty < \mu < \infty$; and since $f_{\gamma 2}(v|h, \nu)$ does not contain μ this integral is

$$\int_{-\infty}^{\infty} f_N(\mu|m, hn)\, f_{\gamma 2}(h|v, \nu)\, d\mu = f_{\gamma 2}(h|v, \nu) \int_{-\infty}^{\infty} f_N(\mu|m, hn)\, d\mu = f_{\gamma 2}(h|v, \nu) . \quad ◀$$

11.5.5. *Marginal Distribution of $\tilde{\mu}$*

If the joint distribution of $(\tilde{\mu}, \tilde{h})$ is Normal-gamma as defined by (11-42), then as shown by (7-72) the marginal distribution of $\tilde{\mu}$ is the Student distribution defined by (7-69):

$$D(\mu|m, v, n, \nu) = f_S(\mu|m, n/v, \nu) . \qquad (11\text{-}47)$$

Notice that, unlike the marginal distribution of $\tilde{h}$, the marginal distribution of $\tilde{\mu}$ depends on all four parameters of the joint distribution of $(\tilde{\mu}, \tilde{h})$.

Cumulative probabilities under the Student distribution (11-47) are given by (7-70):

$$P\{\tilde{\mu} < \mu|m, v, n, \nu\} = F_{S^*}([\mu - m]\sqrt{n/v}|\nu) \ . \tag{11-48}$$

The mean, variance, and partial expectation are given by (7-71):

$$E(\tilde{\mu}|m, v, n, \nu) \equiv \bar{\mu} = m \ , \qquad \nu > 1 \ , \tag{11-49a}$$

$$V(\tilde{\mu}|m, v, n, \nu) = \frac{v}{n}\frac{\nu}{\nu - 2} \ , \qquad \nu > 2 \ , \tag{11-49b}$$

$$E^{\mu}_{-\infty}(\tilde{\mu}|m, v, n, \nu) = m\, F_{S^*}(t|\nu) - \frac{\nu + t^2}{\nu - 1} f_{S^*}(t|\nu)\sqrt{v/n} \ , \qquad \nu > 1 \ , \tag{11-49c}$$

where

$$t = (\mu - m)\sqrt{n/v} \ . \tag{11-49d}$$

From (11-48) and (11-49cd) it follows that the linear-loss integrals are given by

$$L_l(\mu) \equiv \int_{-\infty}^{\mu} (\mu - z) f_S(z|m, n/v, \nu)\, dz = L_{S^*}(-t|\nu)\sqrt{v/n} \ , \qquad \nu > 1 \tag{11-50a}$$

$$L_r(\mu) \equiv \int_{\mu}^{\infty} (z - \mu) f_S(z|m, n/v, \nu)\, dz = L_{S^*}(t|\nu)\sqrt{v/n} \ , \qquad \nu > 1 \tag{11-50b}$$

where

$$L_{S^*}(t|\nu) \equiv \frac{\nu + t^2}{\nu - 1} f_{S^*}(t|\nu) - t\, G_{S^*}(t|\nu) \ . \tag{11-51}$$

11.5.6. *Limiting Behavior of the Prior Distribution*

As $n \to 0$, both the conditional and the marginal distributions of $\tilde{\mu}$ become increasingly uniform over $-\infty < \mu < \infty$ and their variances become infinite, although their means remain fixed at m. The marginal distribution (11-44) of $\tilde{h}$ is unaffected; the conditional distribution (12-30) of $\tilde{h}$ (derived in Chapter 12 below) approaches a proper gamma-2 distribution with parameters $v\nu/(\nu + 1)$ and $(\nu + 1)$.

As $\nu \to 0$, the conditional distribution of $\tilde{\mu}$ is unaffected. The variance of the marginal distribution of $\tilde{\mu}$ becomes infinite, but the distribution does not become uniform: the ratio of the densities at any two points μ_1 and μ_2 approaches $|\mu_2 - m|/|\mu_1 - m|$. The marginal distribution (11-44) of $\tilde{h}$ becomes more and more concentrated toward $h = 0$ although its mean remains fixed at $1/v$ and its variance becomes infinite; the conditional distribution (12-30) of $\tilde{h}$ approaches a proper gamma-2 distribution with parameters $n(\mu - m)^2$ and 1.

▶ The limiting behavior of the marginal distribution of $\tilde{h}$, which does not contain n as a parameter, has been discussed in Section 11.1.3. The limiting behavior of the Normal conditional distribution of $\tilde{\mu}$, which does not contain ν as a parameter, has been discussed in Section 11.3.2. To determine the limiting behavior of the Student marginal distribution of $\tilde{\mu}$, we observe first that the behavior of the mean and variance follow immediately from (11-49). The question of uniformity is solved by using the formula (7-69) for the density (11-47) to show that the ratio of the densities at any two points μ_1 and μ_2 is

$$\left[\frac{\nu + (\mu_2 - m)^2\, n/v}{\nu + (\mu_1 - m)^2\, n/v}\right]^{\frac{1}{2}\nu + \frac{1}{2}}$$

and observing that this ratio obviously approaches 1 as $n \to 0$ with ν fixed but approaches $|\mu_2 - m|/|\mu_1 - m|$ as $\nu \to 0$ with n fixed. ◀

11.6. Sampling Distributions With Fixed *n*

In Section 11.6 we assume throughout that a sample with predetermined $n > 0$ is to be taken from an Independent Normal process whose parameter $(\tilde{\mu}, \tilde{h})$ has a *proper* Normal-gamma distribution with parameter (m', v', n', ν'). The fact that the distribution is proper implies that $v', n', \nu' > 0$.

11.6.1. Conditional Joint Distribution of $(\tilde{m}, \tilde{v}|\mu, h)$

For a given value of the process parameter (μ, h), the *conditional* joint distribution of the statistic $(\tilde{m}, \tilde{v})$ is the product of the independent densities of $\tilde{m}$ and $\tilde{v}$:

$$D(m, v|\,\mu, h; n, \nu) = f_N(m|\mu, hn)\, f_{\gamma 2}(v|h, \nu) \; . \tag{11-52}$$

The proof was given in the proof of (11-38).

11.6.2. Unconditional Joint Distribution of $(\tilde{m}, \tilde{v})$

If the parameter $(\tilde{\mu}, \tilde{h})$ of the process is treated as having a Normal-gamma density of type (11-42), the *unconditional* joint distribution of the statistic $(\tilde{m}, \tilde{v})$ will have the density

$$D(m, v|m', v', n', \nu'; n, \nu)$$

$$= \int_{-\infty}^{\infty} \int_0^{\infty} f_N(m|\mu, hn)\, f_{\gamma 2}(v|h, \nu)\, f_{N\gamma}(\mu, h|m', v', n', \nu')\, dh\, d\mu$$

$$\propto \frac{(\nu v)^{\frac{1}{2}\nu - 1}}{(\nu' v' + \nu v + n_u[m - m']^2)^{\frac{1}{2}\nu''}} \tag{11-53a}$$

where

$$v', n', \nu' > 0 \; ,$$

$$n_u \equiv \frac{n' n}{n' + n} \; , \qquad \frac{1}{n_u} = \frac{1}{n'} + \frac{1}{n} \; , \tag{11-53b}$$

$$\nu'' = \nu' + \nu + 1 \; .$$

▶ To evaluate the integral (11-53a) we first replace the functions f_N, $f_{\gamma 2}$, and $f_{N\gamma}$ by their definitions (7-63), (7-50), and (11-42), thus obtaining (apart from factors constant as regards m, v, μ, and h)

$$\int_{-\infty}^{\infty} \int_0^{\infty} e^{-\frac{1}{2}hn(m-\mu)^2}\, h^{\frac{1}{2}} \cdot e^{-\frac{1}{2}\nu v h}(vh)^{\frac{1}{2}\nu - 1}\, h \cdot e^{-\frac{1}{2}hn'(\mu - m')^2}\, h^{\frac{1}{2}} \cdot e^{-\frac{1}{2}\nu' v' h}\, h^{\frac{1}{2}\nu' - 1}\, dh\, d\mu \; ,$$

where we have suppressed the $\delta(n')$ of (11-42) because by assumption $n' > 0$ and therefore $\delta(n') = 1$. Using the symbols n'' and m'' defined in (11-43) and the symbol n_u defined in (11-53b) we observe that

$$n(m-\mu)^2 + n'(\mu - m')^2 = n''(\mu - m'')^2 + n_u(m - m')^2 \;;$$

and substituting this result we can write the integral as

$$v^{\frac{1}{2}\nu-1}\int_0^\infty e^{-\frac{1}{2}h(\nu'v'+\nu v+n_u[m-m']^2)}\, h^{\frac{1}{2}(\nu+\nu'+1)-1}\, dh \cdot \int_{-\infty}^{\infty} e^{-\frac{1}{2}hn''(\mu-m'')^2}\, h^{\frac{1}{2}}\, d\mu \;.$$

The second integral is a constant and can be dropped. Substituting in the first integral

$$h = z/A \;, \qquad A = \tfrac{1}{2}(\nu'v' + \nu v + n_u[m - m']^2) \;,$$

we obtain

$$v^{\frac{1}{2}\nu-1}\, A^{-\frac{1}{2}(\nu+\nu'+1)} \int_0^\infty e^{-z}\, z^{\frac{1}{2}(\nu+\nu'+1)-1}\, dz \;.$$

The integral is a constant which can be dropped, leaving

$$\mathrm{D}(m, v) \propto v^{\frac{1}{2}\nu-1}(\nu'v' + \nu v + n_u[m - m']^2)^{-\frac{1}{2}(\nu+\nu'+1)} \;.$$

Because by assumption $\delta(n') = \delta(n) = \delta(n'') = 1$, the final exponent on the right can be written $-\frac{1}{2}\nu''$ in virtue of (11-43d). ◀

11.6.3. Unconditional Distributions of $\tilde{m}$ and $\tilde{v}$

From the unconditional joint distribution (11-53) of $(\tilde{m}, \tilde{v})$ we can obtain two distributions of $\tilde{m}$, both unconditional as regards the parameter (μ, h). The first of these is also unconditional as regards v; its density is

$$\mathrm{D}(m|m', v', n', \nu'; n, \nu) = f_S(m|m', n_u/v', \nu') \;. \tag{11-54}$$

The second is conditional on v even though unconditional as regards (μ, h); its density is

$$\mathrm{D}(m|m', v', n', \nu'; n, \nu; v) = f_S(m|m', n_u/V, \nu' + \nu) \tag{11-55a}$$

where

$$V \equiv \frac{\nu'v' + \nu v}{\nu' + \nu} \;. \tag{11-55b}$$

In the same way the joint distribution of $(\tilde{m}, \tilde{v})$ implies two distributions of $\tilde{v}$. Marginally as regards $\tilde{m}$, the statistic $\tilde{v}$ has the inverted-beta-2 density

$$\mathrm{D}(v|m', v', n', \nu'; n, \nu) = f_{i\beta 2}(v|\tfrac{1}{2}\nu, \tfrac{1}{2}\nu', \nu'v'/\nu) \;; \tag{11-56}$$

while conditional on m, the density of $\tilde{v}$ is

$$\mathrm{D}(v|m', v', n', \nu'; n, \nu; m) = f_{i\beta 2}\left(v|\tfrac{1}{2}\nu, \tfrac{1}{2}[\nu' + 1], \frac{\nu'v' + M}{\nu}\right) \tag{11-57a}$$

where

$$M \equiv n_u(m - m')^2 \;. \tag{11-57b}$$

▶ The marginal density (11-54) of $\tilde{m}$ and the conditional density (11-57) of $(\tilde{v}|m)$ are obtained by writing the kernel (11-53a) of the joint density of $(\tilde{m}, \tilde{v})$ in the form

$$\frac{1}{(\nu'v' + n_u[m - m']^2)^{\frac{1}{2}(\nu''-\nu)}} \times \frac{(\nu v)^{\frac{1}{2}\nu-1}(\nu'v' + n_u[m - m']^2)^{\frac{1}{2}(\nu''-\nu)}}{(\nu v + \nu'v' + n_u[m - m']^2)^{\frac{1}{2}\nu''}}$$

$$\propto f_S(m|m', n_u/v', \nu'' - \nu - 1)\, f_{i\beta 2}\left(v|\tfrac{1}{2}\nu, \tfrac{1}{2}[\nu'' - \nu], \frac{\nu'v' + M}{\nu}\right)$$

where M is defined by (11-57b) and $\nu'' = \nu' + \nu + 1$.

The conditional density (11-55) of $(\tilde{m}|v)$ and the marginal density (11-56) of $\tilde{v}$ are obtained by writing the kernel (11-53a) of the joint density in the form

$$\frac{(\nu v)^{\frac{1}{2}\nu-1}}{(\nu v+\nu' v')^{\frac{1}{2}(\nu''-1)}}\times\frac{(\nu v+\nu' v')^{\frac{1}{2}\nu''}}{(\nu v+\nu' v'+n_u[m-m']^2)^{\frac{1}{2}\nu''}(\nu v+\nu' v')^{\frac{1}{2}}}$$

$$\propto f_{i\beta 2}\left(v|\tfrac{1}{2}\nu,\tfrac{1}{2}[\nu''-1-\nu],\frac{\nu' v'}{\nu}\right)f_S(m|m',n_u/V,\nu''-1)$$

where V is defined by (11-55b). ◀

11.7. Preposterior Analysis With Fixed n

In this section as in the previous one we assume that a sample with predetermined statistic $n > 0$ is to be drawn from an Independent Normal process whose mean $\tilde{\mu}$ and mean precision $\tilde{h}$ are random variables having a proper Normal-gamma distribution with parameter (m', v', n', ν') where $v', n', \nu' > 0$.

11.7.1. Joint Distribution of $(\tilde{m}'', \tilde{v}'')$

By expressing the random variable $(\tilde{m}, \tilde{v})$ in terms of the random variable $(\tilde{m}'', \tilde{v}'')$ we can obtain the joint density of the latter from the joint density (11-53) of the former:

$$D(m'', v''|m', v', n', \nu'; n, \nu) \propto \frac{(\nu'' v''-\nu' v'-n^*[m''-m']^2)^{\frac{1}{2}\nu-1}}{(\nu'' v'')^{\frac{1}{2}\nu''}} \tag{11-58a}$$

where

$$n^* \equiv \frac{n+n'}{n}n', \qquad \frac{1}{n^*}=\frac{1}{n'}-\frac{1}{n''},$$
$$\nu''=\nu'+\nu+1, \tag{11-58b}$$

and the values of the variables m'' and v'' are constrained by

$$\nu'' v'' \geq \nu' v', \qquad n^*(m''-m')^2 \leq \nu'' v''-\nu' v'. \tag{11-58c}$$

▶ The proof of (11-58) proceeds in three steps: we first show that

(1) $$n_u(m-m')^2 = n^*(m''-m')^2\ ;$$

we next show that

(2) $$\nu' v'+\nu v+n_u(m-m')^2=\nu'' v''\ ;$$

and we then substitute these results in (11-53).

To prove (1), we first use the definitions (11-43ab) of m'' and n'' to write

$$n_u(m-m')^2 = n_u\left(\frac{n''m''-n'm'}{n}-m'\right)^2 = n_u\frac{n''^2}{n^2}(m''-m')^2 .$$

From this result and the definitions (11-53b) of n_u and (11-58b) of n^* we have

$$n_u(m-m')^2 = \frac{n'n}{n''}\frac{n''^2}{n^2}(m''-m')^2 = \frac{n'n''}{n}(m''-m')^2 = n^*(m''-m')^2 .$$

To prove (2), we first use the definitions (11-43cd) of ν'' and v'' to write

$$\nu''v'' = (\nu'v' + \nu v) + (n'm'^2 + nm^2 - n''m''^2) \ .$$

The result follows when we use the definitions of n'', m'', and n_u to substitute on the right

$$n'm'^2 + nm^2 - n''m''^2 = n'm'^2 + nm^2 - \frac{1}{n''}(n'm' + nm)^2$$

$$= n'\left(1 - \frac{n'}{n''}\right)m'^2 - 2\frac{n'n}{n''}m'm + n\left(1 - \frac{n}{n''}\right)m^2$$

$$= \frac{n'n}{n''}(m - m')^2 = n_u(m - m')^2 \ .$$

Now using (1) and (2) to replace the variables (m, v) of (11-53) by the variables (m'', v'') we obtain

$$\frac{(\nu''v'' - \nu'v' - n^*[m'' - m']^2)^{\frac{1}{2}\nu - 1}}{(\nu''v'')^{\frac{1}{2}\nu''}} J \ , \tag{3}$$

where J is the Jacobian of the transformation. To evaluate J we write

$$m = n^{-1}(n''m'' - n'm')$$
$$v = \nu^{-1}(\nu''v'' - \nu'v' - n^*[m'' - m']^2)$$

and then compute

$$J \equiv \left|\frac{\partial(m, v)}{\partial(m'', v'')}\right| = \begin{vmatrix} n^{-1}n'' & 0 \\ - & \nu^{-1}\nu'' \end{vmatrix} \ ,$$

thus showing that J is a constant as regards m'' and v''. The proof of (11-58) is completed by observing that it is identical to (3) except for the constant J. ◀

11.7.2. *Distributions of $\tilde{m}''$ and $\tilde{v}''$*

From the joint distribution (11-58) of $(\tilde{m}'', \tilde{v}'')$ we can derive two distributions of $\tilde{m}''$: the distribution of $\tilde{m}''$ conditional on a particular value of $\tilde{v}''$, and the distribution of $\tilde{m}''$ which is marginal as regards $\tilde{v}''$. This latter, *unconditional* distribution of $\tilde{m}''$ has the Student density

$$D(m''|m', v', n', \nu'; n, \nu) = f_S(m''|m', n^*/v', \nu') \tag{11-59}$$

where n^* is defined by (11-58b). The *conditional* distribution of $(\tilde{m}''|v'')$ has the inverted-Student density defined by (8-34):

$$D(m''|m', v', n', \nu'; n, \nu; v'') = f_{iS}^{(1)}(m''|m', n^*/V, \nu) \tag{11-60a}$$

where

$$V \equiv \frac{\nu''v'' - \nu'v'}{\nu} \ . \tag{11-60b}$$

and the variable m'' is constrained by

$$n^*(m'' - m')^2 \leq \nu V \ . \tag{11-60c}$$

In the same way the joint distribution of $(\tilde{m}'', \tilde{v}'')$ implies two distributions of $\tilde{v}''$. *Marginally* as regards $\tilde{m}''$, the density of $\tilde{v}''$ is inverted-beta-1:

$$D(v''|m', v', n', \nu'; n, \nu) = f_{i\beta 1}\left(v''|\tfrac{1}{2}\nu', \tfrac{1}{2}\nu'', \frac{\nu' v'}{\nu''}\right) . \tag{11-61}$$

Conditional on a particular m'', the density is of the same type with different parameters:

$$D(v''|m', v', n', \nu'; n, \nu; m'') = f_{i\beta 1}\left(v''|\tfrac{1}{2}[\nu' + 1], \tfrac{1}{2}\nu'', \frac{\nu' v' + M}{\nu''}\right) \tag{11-62a}$$

where

$$M \equiv n^*(m'' - m')^2 . \tag{11-62b}$$

▶ To obtain the marginal density (11-59) of $\tilde{m}''$ and the conditional density (11-62) of $(\tilde{v}''|m'')$ we first simplify notation by using the definition (11-62b) of M to write the kernel (11-58) of the joint density of $(\tilde{m}'', \tilde{v}'')$ in the form

$$\text{(1)} \qquad \frac{1}{(\nu' v' + M)^{\frac{1}{2}(\nu'' - \nu)}} \times \frac{\left[v'' - \dfrac{\nu' v' + M}{\nu''}\right]^{\frac{1}{2}\nu - 1} \left[\dfrac{\nu' v' + M}{\nu''}\right]^{\frac{1}{2}(\nu'' - \nu)}}{v''^{\frac{1}{2}\nu''}} .$$

Since $\nu'' - \nu = \nu' + 1$ and $\nu = \nu'' - (\nu' + 1)$, this is proportional to

$$f_S(m''|m', n^*/v', \nu')\, f_{i\beta 1}\left(v''|\tfrac{1}{2}[\nu' + 1], \tfrac{1}{2}\nu'', \frac{\nu' v' + M}{\nu''}\right) .$$

To obtain the marginal density (11-61) of $\tilde{v}''$ and the conditional density (11-60) of $(\tilde{m}''|v'')$ we write the kernel (11-58) of the joint density in the form

$$\text{(2)} \qquad \frac{(\nu'' v'' - \nu' v')^{\frac{1}{2}(\nu+1)-1}}{(\nu'' v'')^{\frac{1}{2}\nu''}} \times \frac{(\nu'' v'' - \nu' v' - n^*[m'' - m']^2)^{\frac{1}{2}\nu - 1}}{(\nu'' v'' - \nu' v')^{\frac{1}{2}(\nu+1)-1}} .$$

The first factor is obviously proportional to $f_{i\beta 1}(v''|\tfrac{1}{2}\nu', \tfrac{1}{2}\nu'', \nu' v'/\nu'')$ as defined by (7-25); and if we use the definition (11-60b) of V to write the second factor in the form

$$\frac{[\nu V - n^*(m'' - m')^2]^{\frac{1}{2}\nu - 1}}{(\nu V)^{\frac{1}{2}\nu - \frac{1}{2}}} \propto [\nu - (n^*/V)(m'' - m')^2]^{\frac{1}{2}\nu - 1}(n^*/V)^{\frac{1}{2}}$$

we see that it is proportional to the 1-dimensional ($r = 1$) case of (8-34). ◀

11.7.3. Distribution of $\breve{\tilde{\mu}}''$

The variance of the posterior distribution of $\tilde{\mu}$ will be, by (11-49c),

$$V(\tilde{\mu}|m'', v'', n'', \nu'') \equiv \breve{\mu}'' = \frac{v''}{n''} \frac{\nu''}{\nu'' - 2} , \qquad \nu'' > 2 . \tag{11-63}$$

When $\tilde{v}''$ and therefore $\breve{\tilde{\mu}}''$ are random variables, this formula permits us to obtain the distribution of $\breve{\tilde{\mu}}''$ from the (unconditional) distribution (11-61) of $\tilde{v}''$:

$$P\{\breve{\tilde{\mu}}'' < \breve{\mu}''|m', v', n', \nu'; n, \nu\} = F_{i\beta 1}\left(\breve{\mu}'' \frac{n''[\nu'' - 2]}{\nu''} \middle| \tfrac{1}{2}\nu', \tfrac{1}{2}\nu'', \frac{\nu' v'}{\nu''}\right) . \tag{11-64}$$

The expected value of the kth power of $\breve{\tilde{\mu}}''$ under this distribution is

$$\mathrm{E}(\tilde{\breve{\mu}}''^k|m', v', n', \nu'; n, \nu) = \left[\frac{\nu' v'}{n''(\nu'' - 2)}\right]^k \frac{(\frac{1}{2}\nu'' - 1)!(\frac{1}{2}\nu' - 1 - k)!}{(\frac{1}{2}\nu'' - 1 - k)!(\frac{1}{2}\nu' - 1)!}$$

$$= \breve{\mu}'^k \left[\frac{n'(\nu' - 2)}{n''(\nu'' - 2)}\right]^k \frac{(\frac{1}{2}\nu'' - 1)!(\frac{1}{2}\nu' - 1 - k)!}{(\frac{1}{2}\nu'' - 1 - k)!(\frac{1}{2}\nu' - 1)!} , \qquad k < \tfrac{1}{2}\nu' . \qquad (11\text{-}65)$$

Important special cases are

$$\mathrm{E}(\tilde{\breve{\mu}}''|m', v', n', \nu'; n, \nu) = \frac{v'}{n''} \frac{\nu'}{\nu' - 2} = \frac{n'}{n''} \breve{\mu}', \qquad (11\text{-}66a)$$

$$\mathrm{E}(\sqrt{\tilde{\breve{\mu}}''}|m', v', n', \nu'; n, \nu) \doteq \sqrt{\frac{n'}{n''} \breve{\mu}'} \exp\left[-\frac{3}{8}\left(\frac{1}{\frac{1}{2}\nu' - 1} - \frac{1}{\frac{1}{2}\nu'' - 1}\right)\right] \cdot \qquad (11\text{-}66b)$$

▶ To derive (11-65) we first obtain the kth moment about the origin of $f_{i\beta 1}$ as defined by (7-25):

$$\mu_k' \equiv \int_b^\infty y^k f_{i\beta 1}(y|r, n, b)\, dy = b^k \frac{(n - 1)!(r - k - 1)!}{(n - k - 1)!(r - 1)!} \int_b^\infty f_{i\beta 1}(y|r - k, n - k, b)\, dy$$

$$= b^k \frac{(n - 1)!(r - 1 - k)!}{(n - 1 - k)!(r - 1)!} , \qquad k < r .$$

Substituting herein the parameters of the distribution (11-61) of $\tilde{v}''$ we obtain

$$\mathrm{E}(\tilde{v}''^k) = \left[\frac{\nu' v'}{\nu''}\right]^k \frac{(\frac{1}{2}\nu'' - 1)!(\frac{1}{2}\nu' - 1 - k)!}{(\frac{1}{2}\nu'' - 1 - k)!(\frac{1}{2}\nu' - 1)!} .$$

The first form of (11-65) follows from the fact that by (11-63)

$$\mathrm{E}(\tilde{\breve{\mu}}''^k) = \left[\frac{\nu''}{n''(\nu'' - 2)}\right]^k \mathrm{E}(\tilde{v}''^k) ,$$

and the second form of (11-65) then follows from the fact that by (11-49b)

$$\breve{\mu}' = \frac{v'}{n'} \frac{\nu'}{\nu' - 2} .$$

The derivation of (11-66a) from (11-65) with $k = 1$ is obvious.
To obtain (11-66b) with $k = \frac{1}{2}$ we first substitute

$$x' = \tfrac{1}{2}\nu' - 1 , \qquad x'' = \tfrac{1}{2}\nu'' - 1 ,$$

and write (11-65) in the form

$$\mathrm{E}(\sqrt{\tilde{\breve{\mu}}''}) = \left[\breve{\mu}' \frac{n' x'}{n'' x''}\right]^{\frac{1}{2}} \frac{x''!(x' - \frac{1}{2})!}{(x'' - \frac{1}{2})!\, x'!} .$$

By then using Legendre's formula and Stirling's second approximation

$$(x - \tfrac{1}{2})! = \frac{(2x)!\, \pi^{\frac{1}{2}}}{2^{2x}\, x!} , \qquad x! \doteq (2\pi)^{\frac{1}{2}} x^{x + \frac{1}{2}} \exp\left(-x + \frac{1}{12x}\right) ,$$

we obtain the result (11-66b). (If we use Stirling's first approximation, omitting the term $1/(12x)$, we obtain simply $\sqrt{n'\breve{\mu}'/n''}$.) ◀

11.7.4. *Distribution of $\tilde{\bar{\mu}}''$*

The mean $\bar{\mu}''$ of the posterior distribution of $\tilde{\mu}$ will, by (11-49a), be equal to the parameter m'' of that distribution, so that the prior distribution of $\tilde{\bar{\mu}}''$ is obtained by merely substituting $\bar{\mu}''$ for m'' in the formula (11-59) for the (marginal) density of $\tilde{m}''$:

$$D(\bar{\mu}''|m', v', n', \nu'; n, \nu) = f_S(\bar{\mu}''|m', n^*/v', \nu') \qquad \text{(11-67a)}$$

where

$$n^* \equiv \frac{n + n'}{n} n' \,, \qquad \frac{1}{n^*} = \frac{1}{n'} - \frac{1}{n''} \,. \qquad \text{(11-67b)}$$

It then follows from (7-71) that

$$\begin{aligned} E(\tilde{\bar{\mu}}''|m', v', n', \nu'; n, \nu) &= m' = E(\tilde{\mu}|m', v', n', \nu') \,, && \nu' > 1 \,, \\ V(\tilde{\bar{\mu}}''|m', v', n', \nu'; n, \nu) &= \frac{v'}{n^*} \frac{\nu'}{\nu' - 2} && \nu' > 2 \qquad \text{(11-68)} \\ &= V(\tilde{\mu}|m', v', n', \nu') - E[V(\tilde{\mu}|\tilde{m}'', \tilde{v}'', n'', \nu'')] \,. \end{aligned}$$

Since this distribution of $\tilde{\bar{\mu}}''$ is of the same Student form as the distribution (11-47) of $\tilde{\mu}$ itself, formulas for cumulative probabilities, partial expectations, and linear-loss integrals can be obtained by simply substituting n^* for n in (11-48) through (11-51).

▶ The distribution of $\tilde{\bar{\mu}}''$ can also be derived by observing (1) that the density of $\tilde{\bar{\mu}}''$ for given h is $f_N(\bar{\mu}''|m', hn^*)$ by (11-32), and (2) that the density of $\tilde{h}$ is $f_{\gamma 2}(h|v', \nu')$ by (11-42); the result (11-67) then follows by (7-72). ◀

CHAPTER 12

A. *Independent Multinormal Process, Precision Known*

12.1. Prior and Posterior Analysis

12.1.1. Definition of the Independent Multinormal Process

An r-dimensional Independent Multinormal process can be defined as a process generating independent $r \times 1$ vector random variables $\tilde{x}^{(1)}, \cdots, \tilde{x}^{(j)}, \cdots$ with identical densities

$$f_N^{(r)}(\mathbf{x}|\boldsymbol{\mu}, \mathbf{h}) \equiv (2\pi)^{-\frac{1}{2}r} e^{-\frac{1}{2}(\mathbf{x}-\boldsymbol{\mu})^t \mathbf{h}(\mathbf{x}-\boldsymbol{\mu})} |\mathbf{h}|^{\frac{1}{2}} \qquad \begin{array}{l} -\infty < \mathbf{x} < \infty \; , \\ -\infty < \boldsymbol{\mu} < \infty \; , \\ \mathbf{h} \text{ is PDS} \; . \end{array} \tag{12-1}$$

This is the (multivariate) Normal density defined in (8-17); the process mean and variance are given directly by (8-18):

$$\mathrm{E}(\tilde{\mathbf{x}}|\boldsymbol{\mu}, \mathbf{h}) = \boldsymbol{\mu} \; , \tag{12-2a}$$

$$\mathrm{V}(\tilde{\mathbf{x}}|\boldsymbol{\mu}, \mathbf{h}) = \mathbf{h}^{-1} \; . \tag{12-2b}$$

The parameter $\mathbf{h}$ in (12-1) will be called the *precision* of the process, and we shall find it useful to factor this quantity into two parts: a scalar *mean precision* and a matrix *relative precision.* We do so by defining the mean precision

$$h \equiv |\mathbf{h}|^{1/r} \tag{12-3a}$$

and the relative precision

$$\boldsymbol{\eta} \equiv \mathbf{h}/h = \begin{bmatrix} h_{11}/h & h_{12}/h & \cdots & h_{1r}/h \\ h_{21}/h & h_{22}/h & \cdots & h_{2r}/h \\ \cdots & \cdots & \cdots & \cdots \\ h_{r1}/h & h_{r2}/h & \cdots & h_{rr}/h \end{bmatrix} , \qquad |\boldsymbol{\eta}| = 1 \; . \tag{12-3b}$$

In this notation the definition (12-1) of the Independent Multinormal process becomes

$$f_N^{(r)}(\mathbf{x}|\boldsymbol{\mu}, h\boldsymbol{\eta}) = (2\pi)^{-\frac{1}{2}r} e^{-\frac{1}{2}h(\mathbf{x}-\boldsymbol{\mu})^t \boldsymbol{\eta}(\mathbf{x}-\boldsymbol{\mu})} h^{\frac{1}{2}r} \; , \tag{12-4}$$

while its mean and variance become

$$\mathrm{E}(\tilde{\mathbf{x}}|\boldsymbol{\mu}, h\boldsymbol{\eta}) = \boldsymbol{\mu} \; , \tag{12-5a}$$

$$\mathrm{V}(\tilde{\mathbf{x}}|\boldsymbol{\mu}, h\boldsymbol{\eta}) = (h\boldsymbol{\eta})^{-1} \; . \tag{12-5b}$$

12.1.2. Likelihood of a Sample When $\boldsymbol{\eta}$ Is Known

The likelihood that an Independent Multinormal process will generate n values $\mathbf{x}^{(1)}, \cdots, \mathbf{x}^{(j)}, \cdots, \mathbf{x}^{(n)}$ in that order is the product of the likelihoods of the individual values as given by (12-4):

$$(2\pi)^{-\frac{1}{2}rn} e^{-\frac{1}{2}h\Sigma(x^{(j)}-\mu)^t \eta (x^{(j)}-\mu)} h^{\frac{1}{2}rn} . \tag{12-6}$$

If in addition the stopping process in noninformative in the sense of Section 2.3, this is the likelihood of a sample consisting of the n observations $x^{(1)}, \cdots, x^{(n)}$.

Provided only that the *relative* precision η is known, we can compute all of the following statistics:

$$m = \frac{1}{n} \Sigma x^{(j)} , \tag{12-7a}$$

$$\mathbf{n} = n\eta , \tag{12-7b}$$

$$\nu = r(n-1) , \tag{12-7c}$$

$$v = \frac{1}{\nu} \Sigma(x^{(j)} - m)^t \eta (x^{(j)} - m) \qquad (= 0 \text{ if } \nu = 0) . \tag{12-7d}$$

The statistic $\mathbf{n}$ will be called the *effective sample size.* Dropping the constant $(2\pi)^{-\frac{1}{2}rn}$ from the likelihood (12-6) we can now write its kernel in the form

$$e^{-\frac{1}{2}h\nu v - \frac{1}{2}h(m-\mu)^t \mathbf{n}(m-\mu)} h^{\frac{1}{2}(r+\nu)} . \tag{12-8}$$

► The jth term of the sum in (12-6) can be written

$$\begin{aligned}(x^{(j)} - \mu)^t \eta (x^{(j)} - \mu) &= ([x^{(j)} - m] + [m - \mu])^t \eta ([x^{(j)} - m] + [m - \mu]) \\ &= (x^{(j)} - m)^t \eta (x^{(j)} - m) + (m - \mu)^t \eta (m - \mu) \\ &\quad + (x^{(j)} - m)^t \eta (m - \mu) + (m - \mu)^t \eta (x^{(j)} - m) .\end{aligned}$$

When summed over j the first two terms on the right become respectively νv and $(m - \mu)^t \mathbf{n} (m - \mu)$ while the last two vanish because $\Sigma(x^{(j)} - m) = nm - nm$. ◄

12.1.3. Likelihood of a Sample When Both h and η Are Known

When *both* the relative precision η and the mean precision h are known and the only unknown parameter is the process mean μ, the only factor in (12-8) which varies with an unknown parameter is

$$e^{-\frac{1}{2}h(m-\mu)^t \mathbf{n}(m-\mu)} \tag{12-9}$$

and this is therefore the kernel of the likelihood. The statistic $(m, \mathbf{n})$ is thus sufficient when the process precision $\mathbf{h} = h\eta$ is known and the stopping process is noninformative.

12.1.4. Conjugate Distribution of $\tilde{\mu}$

When the precision $\mathbf{h} = h\eta$ of an Independent Multinormal process is known but the mean $\tilde{\mu}$ is a random variable, the most convenient distribution of $\tilde{\mu}$—the natural conjugate of (12-9)—is the Normal distribution defined by (8-17):

$$f_N(\mu|m, \mathbf{H}) \propto e^{-\frac{1}{2}(\mu - m)^t \mathbf{H}(\mu - m)} . \tag{12-10}$$

The quantity $\mathbf{H}$ in this expression can be thought of as measuring the *precision* of our information on μ, and it will be more instructive to express this measure in units of the *process* mean precision h. We therefore define the *parameter* (*not* the statistic) $\mathbf{n}$ by

$$\mathbf{n} \equiv \mathbf{H}/h \ , \tag{12-11}$$

i.e., we say that the information $\mathbf{H}$ is equivalent to an effective number $\mathbf{n}$ of actual observations on the process, and write the density (12-10) in the form

$$f_N(\boldsymbol{\mu}|\boldsymbol{m}, h\mathbf{n}) \propto e^{-\frac{1}{2}h(\boldsymbol{\mu}-\boldsymbol{m})^t\mathbf{n}(\boldsymbol{\mu}-\boldsymbol{m})} \ . \tag{12-12}$$

The mean and variance of this distribution are given by (8-18):

$$\mathrm{E}(\tilde{\boldsymbol{\mu}}|\boldsymbol{m}, \mathbf{n}) \equiv \bar{\boldsymbol{\mu}} = \boldsymbol{m} \ , \tag{12-13a}$$
$$\mathrm{V}(\tilde{\boldsymbol{\mu}}|\boldsymbol{m}, \mathbf{n}) \equiv \breve{\boldsymbol{\mu}} = (h\mathbf{n})^{-1} \ . \tag{12-13b}$$

If a Normal distribution with parameter $(\boldsymbol{m}', \mathbf{n}')$ is assigned to $\tilde{\boldsymbol{\mu}}$ and if a sample then yields a sufficient statistic $(\boldsymbol{m}, \mathbf{n})$, the posterior distribution of $\tilde{\boldsymbol{\mu}}$ will be Normal with parameters

$$\boldsymbol{m}'' = (\mathbf{n}' + \mathbf{n})^{-1}(\mathbf{n}'\boldsymbol{m}' + \mathbf{n}\,\boldsymbol{m}) \ , \tag{12-14a}$$
$$\mathbf{n}'' = \mathbf{n}' + \mathbf{n} \ . \tag{12-14b}$$

▶ Multiplying the kernel (12-12) of the prior density by the kernel (12-9) of the likelihood we obtain $e^{-\frac{1}{2}hS}$ where

(1) $$S = (\boldsymbol{\mu} - \boldsymbol{m}')^t\mathbf{n}'(\boldsymbol{\mu} - \boldsymbol{m}') + (\boldsymbol{m} - \boldsymbol{\mu})^t\mathbf{n}(\boldsymbol{m} - \boldsymbol{\mu})$$
$$= \boldsymbol{\mu}^t\mathbf{n}'\boldsymbol{\mu} - \boldsymbol{m}'^t\mathbf{n}'\boldsymbol{\mu} - \boldsymbol{\mu}^t\mathbf{n}'\boldsymbol{m}' + \boldsymbol{m}'^t\mathbf{n}'\boldsymbol{m}' + \boldsymbol{m}^t\mathbf{n}\,\boldsymbol{m} - \boldsymbol{\mu}^t\mathbf{n}\,\boldsymbol{m} - \boldsymbol{m}^t\mathbf{n}\,\boldsymbol{\mu} + \boldsymbol{\mu}^t\mathbf{n}\,\boldsymbol{\mu} \ .$$

Remembering that $\mathbf{n}'$ and $\mathbf{n}$ are symmetric we can regroup the terms on the right to obtain

(2) $$S = S_1 + S_2$$

where

(3) $$S_1 = (\boldsymbol{\mu} - [\mathbf{n}' + \mathbf{n}]^{-1}[\mathbf{n}'\boldsymbol{m}' + \mathbf{n}\,\boldsymbol{m}])^t(\mathbf{n}' + \mathbf{n})(\boldsymbol{\mu} - [\mathbf{n}' + \mathbf{n}]^{-1}[\mathbf{n}'\boldsymbol{m}' + \mathbf{n}\,\boldsymbol{m}])$$
$$= (\boldsymbol{\mu} - \boldsymbol{m}'')^t\mathbf{n}''(\boldsymbol{\mu} - \boldsymbol{m}'') \ ,$$

(4) $$S_2 = \boldsymbol{m}'^t\mathbf{n}'\boldsymbol{m}' + \boldsymbol{m}^t\mathbf{n}\,\boldsymbol{m} - (\mathbf{n}'\boldsymbol{m}' + \mathbf{n}\,\boldsymbol{m})^t(\mathbf{n}' + \mathbf{n})^{-1}(\mathbf{n}'\boldsymbol{m}' + \mathbf{n}\,\boldsymbol{m}) \ .$$

1. Since $\boldsymbol{\mu}$ appears only in S_1, the kernel of the conditional distribution of $(\tilde{\boldsymbol{\mu}}|\boldsymbol{m})$ is $e^{-\frac{1}{2}hS_1}$ and the distribution will therefore be nondegenerate Normal with parameters (12-14) provided that $\mathbf{n}''$ is positive definite. Since $\boldsymbol{\eta}$ is positive-definite, $\mathbf{n} = n\boldsymbol{\eta}$ will be positive-definite for any $n > 0$; and therefore $\mathbf{n}'' = \mathbf{n}' + \mathbf{n}$ will be positive-definite for any $n > 0$ even if $\mathbf{n}' = 0$.

Although this completes the proof of (12-14), we anticipate future needs by giving two alternate factorizations of the sum S_2 which plays no role in the distribution of $(\tilde{\boldsymbol{\mu}}|\boldsymbol{m})$.

2. Writing $\mathbf{n}''^{-1}$ in place of $(\mathbf{n}' + \mathbf{n})^{-1}$ in the last term on the right hand side of (4) and then expanding we obtain

$$S_2 = \boldsymbol{m}'^t\mathbf{n}'\boldsymbol{m}' + \boldsymbol{m}^t\mathbf{n}\,\boldsymbol{m} - \boldsymbol{m}'^t\mathbf{n}'\mathbf{n}''^{-1}\mathbf{n}'\boldsymbol{m}'$$
$$- \boldsymbol{m}'^t\mathbf{n}'\mathbf{n}''^{-1}\mathbf{n}\,\boldsymbol{m} - \boldsymbol{m}^t\mathbf{n}\,\mathbf{n}''^{-1}\mathbf{n}'\boldsymbol{m}' - \boldsymbol{m}^t\mathbf{n}\,\mathbf{n}''^{-1}\mathbf{n}\,\boldsymbol{m}$$
$$= \boldsymbol{m}^t(\mathbf{n} - \mathbf{n}\,\mathbf{n}''^{-1}\mathbf{n})\,\boldsymbol{m} - \boldsymbol{m}^t\mathbf{n}\,\mathbf{n}''^{-1}\mathbf{n}'\boldsymbol{m}'$$
$$- \boldsymbol{m}'^t\mathbf{n}'\mathbf{n}''^{-1}\mathbf{n}\,\boldsymbol{m} + \boldsymbol{m}'^t(\mathbf{n}' - \mathbf{n}'\mathbf{n}''^{-1}\mathbf{n}')\,\boldsymbol{m}'$$

The two quantities within parentheses on the right can be reduced:

$$\mathbf{n} - \mathbf{n}\,\mathbf{n}''^{-1}\mathbf{n} = \mathbf{n} - (\mathbf{n}'' - \mathbf{n}')\,\mathbf{n}''^{-1}\mathbf{n} = \mathbf{n}'\mathbf{n}''^{-1}\mathbf{n} \ ,$$
$$\mathbf{n}' - \mathbf{n}'\mathbf{n}''^{-1}\mathbf{n}' = \mathbf{n}' - (\mathbf{n}'' - \mathbf{n})\,\mathbf{n}''^{-1}\mathbf{n}' = \mathbf{n}\,\mathbf{n}''^{-1}\mathbf{n}' \ ,$$

and since

$$\mathbf{n}'\mathbf{n}''^{-1}\mathbf{n} = (\mathbf{n}'' - \mathbf{n})\,\mathbf{n}''^{-1}(\mathbf{n}'' - \mathbf{n}') = \mathbf{n}\,\mathbf{n}''^{-1}\,\mathbf{n}' + \mathbf{n}'' - \mathbf{n}' - \mathbf{n} = \mathbf{n}\,\mathbf{n}''^{-1}\,\mathbf{n}'$$

we may define

$$\mathbf{n}_u \equiv \mathbf{n}'\mathbf{n}''^{-1}\,\mathbf{n} = \mathbf{n}\,\mathbf{n}''^{-1}\,\mathbf{n}'$$

and write

(5) $$S_2 = (m - m')^t\,\mathbf{n}_u(m - m') \ .$$

We shall see later that, when S_2 is written in this form, $e^{-\frac{1}{2}hS_2}$ is the kernel of the marginal or unconditional distribution of $\tilde{m}$.

3. We can express S_2 in terms of m'' rather than m by substituting on the right-hand side of (4)

$$m^t\,\mathbf{n}\,m = (\mathbf{n}^{-1}[\mathbf{n}''m'' - \mathbf{n}'m'])^t\,\mathbf{n}(\mathbf{n}^{-1}[\mathbf{n}''m'' - \mathbf{n}'m']) \ ,$$
$$\mathbf{n}'m' + \mathbf{n}\,m = \mathbf{n}''m'' \ ,$$

thus obtaining

$$S_2 = m''^t(\mathbf{n}''\mathbf{n}^{-1}\,\mathbf{n}'' - \mathbf{n}'')\,m'' - m''^t\,\mathbf{n}''\mathbf{n}^{-1}\,\mathbf{n}'m'$$
$$- m'^t\,\mathbf{n}'\mathbf{n}^{-1}\,\mathbf{n}''m'' + m'^t(\mathbf{n}' + \mathbf{n}'\mathbf{n}^{-1}\,\mathbf{n}')\,m' \ .$$

The two quantities within parentheses can be reduced:

$$\mathbf{n}''\mathbf{n}^{-1}\,\mathbf{n}'' - \mathbf{n}'' = (\mathbf{n}' + \mathbf{n})\,\mathbf{n}^{-1}\,\mathbf{n}'' - \mathbf{n}'' = \mathbf{n}'\mathbf{n}^{-1}\,\mathbf{n}'' \ ,$$
$$\mathbf{n}' + \mathbf{n}'\mathbf{n}^{-1}\,\mathbf{n}' = \mathbf{n}' + (\mathbf{n}'' - \mathbf{n})\,\mathbf{n}^{-1}\,\mathbf{n}' = \mathbf{n}''\mathbf{n}^{-1}\,\mathbf{n}' \ ;$$

and since

$$\mathbf{n}'\mathbf{n}^{-1}\,\mathbf{n}'' = (\mathbf{n}'' - \mathbf{n})\,\mathbf{n}^{-1}(\mathbf{n}' + \mathbf{n}) = \mathbf{n}''\mathbf{n}^{-1}\,\mathbf{n}' + \mathbf{n}'' - \mathbf{n}' - \mathbf{n} = \mathbf{n}''\mathbf{n}^{-1}\,\mathbf{n}'$$

we may define

$$\mathbf{n}^* \equiv \mathbf{n}'\mathbf{n}^{-1}\,\mathbf{n}'' = \mathbf{n}''\mathbf{n}^{-1}\,\mathbf{n}'$$

and write

(6) $$S_2 = (m'' - m')^t\,\mathbf{n}^*(m'' - m') \ .$$

We shall see later that, when S_2 is written in this form, $e^{-\frac{1}{2}hS_2}$ is the kernel of the prior density of $\tilde{m}''$. ◀

12.2. Sampling Distributions With Fixed n

Throughout Section 12.2 we assume that a sample with predetermined $\mathbf{n} = n\eta$ is to be drawn from an r-dimensional Independent Multinormal process whose precision $\mathbf{h} = h\eta$ is known but whose mean $\tilde{\mu}$ is a random variable having a proper Normal distribution with parameter $(m', \mathbf{n}')$, the word proper implying that $\mathbf{n}'$ is PDS.

12.2.1. Conditional Distribution of $(\tilde{m}|\mu)$

Given a particular value μ of the process mean $\tilde{\mu}$, the conditional distribution of $\tilde{m}$ is Normal with density

$$D(m|\mu; \mathbf{n}) = f_N^{(r)}(m|\mu, h\mathbf{n}) \tag{12-15}$$

and with mean and variance

$$\begin{aligned} E(\tilde{m}|\mu; \mathbf{n}) &= \mu \ , \\ V(\tilde{m}|\mu; \mathbf{n}) &= (h\mathbf{n})^{-1} \ . \end{aligned} \tag{12-16}$$

► Since by (12-7a)

$$\tilde{m} = \frac{1}{n} \Sigma\, \tilde{x}^{(j)}$$

is a linear combination of independent Normal random variables, it follows immediately from Corollary 2.2 of Section 8.2.3 that $\tilde{m}$ is Normal, and by this same Corollary in conjunction with (12-5) the mean and variance of $\tilde{m}$ are

$$E(\tilde{m}|\mu, \mathbf{n}) = \frac{1}{n} \Sigma\, E(\tilde{x}^{(j)}|\mu, \boldsymbol{\eta}) = \frac{1}{n} \Sigma\, \mu = \mu \ ,$$

$$V(\tilde{m}|\mu, \mathbf{n}) = \frac{1}{n^2} \Sigma\, V(\tilde{x}^{(j)}|\mu, \boldsymbol{\eta}) = \frac{1}{n^2} \Sigma (h\boldsymbol{\eta})^{-1} = \frac{1}{hn^2} n\boldsymbol{\eta}^{-1} = (hn\boldsymbol{\eta})^{-1} = (h\mathbf{n})^{-1} \ . \quad ◄$$

12.2.2. *Unconditional Distribution of* $\tilde{m}$

When the process mean $\tilde{\mu}$ is treated as a random variable having a Normal distribution with parameter $(m', \mathbf{n}')$, the unconditional distribution of $\tilde{m}$ is Normal with density

$$D(m|m', \mathbf{n}'; \mathbf{n}) \equiv \int_{-\infty}^{\infty} f_N^{(r)}(m|\mu, h\mathbf{n})\, f_N^{(r)}(\mu|m', h\mathbf{n}')\, d\mu$$

$$= f_N^{(r)}(m|m', h\mathbf{n}_u) \tag{12-17a}$$

where

$$\mathbf{n}_u \equiv \mathbf{n}'\mathbf{n}''^{-1}\mathbf{n} = \mathbf{n}\,\mathbf{n}''^{-1}\mathbf{n}' \ , \qquad \mathbf{n}_u^{-1} = \mathbf{n}'^{-1} + \mathbf{n}^{-1} \ . \tag{12-17b}$$

The mean and variance of this distribution are

$$E(\tilde{m}|m', \mathbf{n}'; \mathbf{n}) = m' = E(\tilde{\mu}|m', \mathbf{n}') \ ,$$
$$V(\tilde{m}|m', \mathbf{n}'; \mathbf{n}) = (h\mathbf{n}_u)^{-1} = V(\tilde{\mu}|m', \mathbf{n}') + V(\tilde{m}|\mu, \mathbf{n}) \ . \tag{12-18}$$

► To show that the marginal distribution of $\tilde{m}$ is Normal we need only recall (1) that $(\tilde{m}, \tilde{\mu})$ is Normal because $\tilde{\mu}$ is Normal and $(\tilde{m}|\mu)$ is Normal with linear regression and constant variance, and (2) that any marginal distribution of a joint Normal distribution is Normal.

To find the parameters of the distribution of $\tilde{m}$ we substitute the formula (8-17) for $f_N(m)$ and $f_N(\mu)$ in the integrand of (12-17a), drop factors constant as regards m and μ, and thus find that the kernel of the joint density of $\tilde{\mu}$ and $\tilde{m}$ can be written $e^{-\frac{1}{2}hS^*}$ where

$$S^* = (m - \mu)^t\mathbf{n}(m - \mu) + (\mu - m')^t\mathbf{n}'(\mu - m') \ .$$

Comparison with formula (1) in the proof of (12-14) shows that S^* as defined above is identical to the S of that formula; and we may therefore use the results (3) and (5) obtained in that proof to write

$$S^* = S_1 + S_2$$

where

$$S_1 = (\mu - m'')^t\mathbf{n}''(\mu - m'') \ ,$$
$$S_2 = (m - m')^t\mathbf{n}_u(m - m') \ .$$

Since the integral of $e^{-\frac{1}{2}hS_1}$ over $-\infty < \mu < \infty$ is a constant as regards m, the kernel of the marginal or unconditional density of $\tilde{m}$ is simply $e^{-\frac{1}{2}hS_2}$, in accordance with (12-17).

Alternate proof. (12-17) can also be proved by observing that $\tilde{\mu}$ and $\tilde{\epsilon} = \tilde{m} - \tilde{\mu}$ are independent Normal random variables, from which it follows immediately that

$$\tilde{m} = \tilde{\mu} + \tilde{\epsilon}$$

is Normal with mean and variance

$$\mathrm{E}(\tilde{m}) = \mathrm{E}(\tilde{\mu}) + \mathrm{E}(\tilde{\epsilon}) = \mathrm{E}(\tilde{\mu}) + 0 = m' ,$$
$$\mathrm{V}(\tilde{m}) = \mathrm{V}(\tilde{\mu}) + \mathrm{V}(\tilde{\epsilon}) = \mathrm{V}(\tilde{\mu}) + \mathrm{V}(\tilde{m}|\mu) = (h\mathbf{n}')^{-1} + (h\mathbf{n})^{-1} .$$

Since

$$\mathbf{n}'^{-1} + \mathbf{n}^{-1} = \mathbf{n}^{-1}(\mathbf{n}' + \mathbf{n})\,\mathbf{n}'^{-1} = (\mathbf{n}'\mathbf{n}''^{-1}\mathbf{n})^{-1} \equiv \mathbf{n}_u^{-1} ,$$

the parameters of (12-17) follow from these values of the moments. ◀

12.3. Preposterior Analysis With Fixed n

In this entire section as in the previous one we assume that a sample with predetermined $\mathbf{n}$ is to be drawn from an r-dimensional Independent Multinormal process whose relative precision $\boldsymbol{\eta}$ and mean precision h are both known but whose mean $\tilde{\mu}$ is a random variable having a proper Normal distribution with parameter $(m', \mathbf{n}')$. The fact that the distribution is proper implies that $\mathbf{n}'$ is PDS.

12.3.1. Distribution of $\tilde{m}''$

By (12-14), the parameter m'' of the posterior distribution of $\tilde{\mu}$ will have the value

$$m'' = \mathbf{n}''^{-1}(\mathbf{n}'m' + \mathbf{n}\,m) , \tag{12-19}$$

so that before $\tilde{m}$ is known we can obtain the distribution of $\tilde{m}''$ from the unconditional distribution (12-17) of $\tilde{m}$. The distribution so determined is Normal with density

$$\mathrm{D}(m''|m', \mathbf{n}'; \mathbf{n}) = f_N^{(r)}(m''|m', h\mathbf{n}^*) \tag{12-20a}$$

where

$$\mathbf{n}^* \equiv \mathbf{n}''\mathbf{n}^{-1}\mathbf{n}' = \mathbf{n}'\mathbf{n}^{-1}\mathbf{n}'' , \qquad \mathbf{n}^{*-1} = \mathbf{n}'^{-1} - \mathbf{n}''^{-1} . \tag{12-20b}$$

The mean and variance of the distribution are

$$\begin{aligned} \mathrm{E}(\tilde{m}''|m', \mathbf{n}'; \mathbf{n}) &= m' , \\ \mathrm{V}(\tilde{m}''|m', \mathbf{n}'; \mathbf{n}) &= (h\mathbf{n}^*)^{-1} = (h\mathbf{n}')^{-1} - (h\mathbf{n}'')^{-1} . \end{aligned} \tag{12-21}$$

▶ Since $\tilde{m}''$ as defined by (12-19) is a linear transformation of the Normal random variable $\tilde{m}$, it follows immediately from Corollary 2.1 of Section 8.2.3 that $\tilde{m}''$ is Normal; and by this same Corollary in conjunction with (12-18) the mean and variance of $\tilde{m}''$ are

$$\mathrm{E}(\tilde{m}'') = \mathbf{n}''^{-1}[\mathbf{n}'m' + \mathbf{n}\,\mathrm{E}(\tilde{m})] = \mathbf{n}''^{-1}[\mathbf{n}'m' + \mathbf{n}\,m'] = \mathbf{n}''^{-1}\mathbf{n}''m' = m' ,$$

$$\begin{aligned} \mathrm{V}(\tilde{m}'') &= \mathbf{n}''^{-1}\mathbf{n}\,\mathrm{V}(\tilde{m})\,\mathbf{n}\,\mathbf{n}''^{-1} = \mathbf{n}''^{-1}\mathbf{n}\left(\frac{1}{h}\mathbf{n}^{-1}\mathbf{n}''\mathbf{n}'^{-1}\right)\mathbf{n}\,\mathbf{n}''^{-1} \\ &= \frac{1}{h}\mathbf{n}'^{-1}\mathbf{n}\,\mathbf{n}''^{-1} \equiv (h\mathbf{n}^*)^{-1} . \end{aligned}$$

The parameters of (12-20) are then determined by these values of the moments.

The alternative expressions given in (12-20b) for $\mathbf{n}^*$ and $\mathbf{n}^{*-1}$ are easily derived:

$$\mathbf{n}^* \equiv \mathbf{n}''\mathbf{n}^{-1}\mathbf{n}' = (\mathbf{n}' + \mathbf{n})\,\mathbf{n}^{-1}(\mathbf{n}'' - \mathbf{n}) = \mathbf{n}'\mathbf{n}^{-1}\mathbf{n}'' ;$$
$$\mathbf{n}^{*-1} \equiv \mathbf{n}'^{-1}\mathbf{n}\,\mathbf{n}''^{-1} = \mathbf{n}'^{-1}(\mathbf{n}'' - \mathbf{n}')\,\mathbf{n}''^{-1} = \mathbf{n}'^{-1} - \mathbf{n}''^{-1} .$$

Alternate Proof. (12-20) can also be derived directly from the joint density of $\tilde{\mu}$ and $\tilde{m}$ without first obtaining the marginal density of $\tilde{m}$. We saw in the proof of (12-17) that after we have eliminated μ from this joint density we are left with the kernel $e^{-\frac{1}{2}hS_2}$, and by formula (6) in the proof of (12-14)

$$S_2 = (m'' - m')^t \mathbf{n}^*(m'' - m')$$

in accordance with the exponent called for by (12-20).

Still another derivation of the parameters of (12-20) will be given in the discussion of (12-22). ◀

12.3.2. *Distribution of $\tilde{\bar{\mu}}''$*

The mean $\bar{\mu}''$ of the posterior distribution of $\tilde{\mu}$ will, by (12-13a), be equal to the parameter m'' of that distribution, and therefore the distribution of $\tilde{\bar{\mu}}''$ is given directly by (12-20) with $\bar{\mu}''$ substituted for m''. When, however, we think of this distribution as a distribution of the posterior mean rather than as a distribution of a mere parameter, it becomes instructive to observe that formulas (12-21) for the mean and variance amount to

$$\begin{aligned} \mathrm{E}(\tilde{\bar{\mu}}''|m', \mathbf{n}'; \mathbf{n}) &= \mathrm{E}(\tilde{\mu}|m', \mathbf{n}') \ , \\ \mathrm{V}(\tilde{\bar{\mu}}''|m', \mathbf{n}'; \mathbf{n}) &= \mathrm{V}(\tilde{\mu}|m', \mathbf{n}') - \mathrm{V}(\tilde{\mu}|m'', \mathbf{n}'') \ , \end{aligned} \qquad (12\text{-}22)$$

in accordance with general relations (5-27) and (5-28). These relations constitute, of course, an alternative way of deriving the parameters of the distribution (12-20) of $\tilde{m}''$.

B. Independent Multinormal Process, Relative Precision Known

12.4. Prior and Posterior Analysis

12.4.1. *Likelihood of a Sample When Only $\boldsymbol{\eta}$ Is Known*

We have already seen in (12-8) that if the stopping process is noninformative the kernel of the likelihood of observations $x^{(1)}, \cdots, x^{(j)}, \cdots x^{(n)}$ on an r-dimensional Independent Multinormal process is

$$e^{-\frac{1}{2}hv\nu - \frac{1}{2}h(m-\mu)^t\mathbf{n}(m-\mu)}\, h^{\frac{1}{2}(r+\nu)} \qquad (12\text{-}23)$$

where the statistics m, $\mathbf{n}$, v, and ν have the definitions (12-7). These four statistics are sufficient when both the process mean μ and the mean precision h are unknown and only the relative precision $\boldsymbol{\eta}$ is known.

12.4.2. *Likelihood of the Statistics $(m, \mathbf{n})$ and (v, ν)*

For the same reasons as in the univariate case, Section 11.5.2, we may wish to use the information contained in an incomplete statistic $(m, \mathbf{n})$ or (v, ν), and to do so we need the marginal likelihood of the statistic in question given (μ, h). The kernel of the likelihood of $(m, \mathbf{n})$ is

$$e^{-\frac{1}{2}h(m-\mu)^t\mathbf{n}(m-\mu)}\, h^{\frac{1}{2}r} \ , \qquad (12\text{-}24a)$$

while the kernel of the likelihood of (v, ν) is

$$e^{-\frac{1}{2}h\nu v}\, h^{\frac{1}{2}\nu} \; . \tag{12-24b}$$

▶ To prove formula (12-24) we shall first transform the independent $\tilde{x}^{(j)}$ with correlated components $\tilde{x}_i^{(j)}$ into independent $\tilde{z}^{(j)}$ with independent components $\tilde{z}_i^{(j)}$. Then expressing $\tilde{m}$ and $\tilde{v}$ in terms of the $\tilde{z}_i^{(j)}$ we shall show that their distributions are *independent* and that their kernels are given by (12-24).

Because $\mathbf{h}$ and therefore $\boldsymbol{\eta} = \mathbf{h}/h$ is positive-definite, there exists (by Theorem 7 of Section 8.0.3) a nonsingular matrix $\mathbf{Q}$ such that $\mathbf{Q}^t\,\boldsymbol{\eta}\,\mathbf{Q} = \mathbf{I}$. Defining new variables $z^{(j)}$ by

$$\mathbf{Q}\, z^{(j)} = x^{(j)} - \mu$$

and substituting in the exponent of e in the complete sample likelihood (12-6) we obtain

$$-\tfrac{1}{2}h\, \Sigma(x^{(j)} - \mu)^t\, \boldsymbol{\eta}\, (x^{(j)} - \mu) = -\tfrac{1}{2}h\, \Sigma\, z^{(j)t}\, \mathbf{I}\, z^{(j)} \; ,$$

thus showing that the $\tilde{z}_i^{(j)}$ have *independent* Normal distributions with the same mean 0 and precision h.

Now defining

$$\left.\begin{aligned} \bar{\tilde{z}}_i &\equiv \frac{1}{n}\, \Sigma_{j=1}^n\, \tilde{z}_i^{(j)} \; , \\ \tilde{v}_i &\equiv \frac{1}{n-1}\, \Sigma_{j=1}^n (\tilde{z}_i^{(j)} - \bar{\tilde{z}}_i)^2 \; , \end{aligned}\right\} \qquad i = 1, 2, \cdots, r \; ,$$

we observe (1) that the r pairs $(\bar{\tilde{z}}_i, \tilde{v}_i)$ are mutually independent because the $\tilde{z}_i^{(j)}$ are mutually independent, and (2) that for any i the random variables $\bar{\tilde{z}}_i$ and $\tilde{v}_i$ are independent with joint density

$$D(\bar{z}_i, v_i) = f_N(\bar{z}_i|0, hn)\, f_{\gamma 2}(v_i|h, n-1)$$

by the univariate theory in the proof of (11-38). It follows that the joint density of the r pairs $(\bar{\tilde{z}}_i, \tilde{v}_i)$ is

$$\Pi_{i=1}^r [f_N(\bar{z}_i|0, hn)\, f_{\gamma 2}(v_i|h, n-1) = f_N^{(r)}(\bar{\mathbf{z}}|0, hn\,\mathbf{I})\, \Pi_{i=1}^r f_{\gamma 2}(v_i|h, n-1) \; .$$

The mean $\bar{\tilde{z}}$ is *independent* of the r $\tilde{v}$s.

Formula (12-24a) is now proved by observing that if $\bar{\tilde{z}}$ is Normal with parameter $(0, hn\,\mathbf{I})$, then

$$\tilde{m} \equiv \frac{1}{n}\, \Sigma\, \tilde{x}^{(j)} = \frac{1}{n}\, \Sigma(\mathbf{Q}\, \tilde{z}^{(j)} + \mu) = \mu + \mathbf{Q}\, \bar{\tilde{z}}$$

is Normal with mean μ and variance $\mathbf{Q}\, hn\, \mathbf{I}\, \mathbf{Q}^t = hn\, \boldsymbol{\eta} = h\mathbf{n}$; in other words,

$$D(m) = f_N^{(r)}(m|\mu, h\mathbf{n}) \equiv (2\pi)^{-\frac{1}{2}r}\, e^{-\frac{1}{2}h(m-\mu)^t\mathbf{n}\,(m-\mu)}\, h^{\frac{1}{2}r}|\mathbf{n}|^{\frac{1}{2}} \; .$$

The kernel (12-24a) of the likelihood of m is now obtained by simply dropping from this density all factors constant as regards μ.

To prove formula (12-24b) we write the definition (12-7) of v in the form

$$r(n-1)\, v = \Sigma(x^{(j)} - m)^t\, \boldsymbol{\eta}\, (x^{(j)} - m) \; .$$

Substituting herein

$$x^{(j)} - m = (x^{(j)} - \mu) - (m - \mu) = \mathbf{Q}(z^{(j)} - \bar{z})$$

we obtain

$$r(n-1)v = \Sigma(\mathbf{z}^{(j)} - \bar{\mathbf{z}})^t\,\mathbf{Q}^t\,\boldsymbol{\eta}\,\mathbf{Q}(\mathbf{z}^{(j)} - \bar{\mathbf{z}}) = \Sigma(\mathbf{z}^{(j)} - \bar{\mathbf{z}})^t\,\mathbf{I}(\mathbf{z}^{(j)} - \bar{\mathbf{z}})$$
$$= \Sigma_{j=1}^{n}\,\Sigma_{i=1}^{r}\,(z_i^{(j)} - \bar{z}_i)^2 = \Sigma_{i=1}^{r}\,\Sigma_{j=1}^{n}\,(z_i^{(j)} - \bar{z}_i)^2 \; .$$

Recalling our definition

$$\tilde{v}_i = \frac{1}{n-1}\,\Sigma_{j=1}^{n}\,(\tilde{z}_i^{(j)} - \bar{\tilde{z}}_i)^2$$

and defining

$$\tilde{V} = \Sigma_{i=1}^{r}\,\tilde{v}_i$$

we can express $\tilde{v}$ as

$$\tilde{v} = \frac{\tilde{V}}{r} \; .$$

Because the $\tilde{v}_i$ are independent the density of $\tilde{V} = \Sigma\,\tilde{v}_i$ is the convolution of the r densities of the $\tilde{v}_i$. By (7-53), this convolution is gamma-2 with parameter $(h/r, r[n-1]) = (h/r, \nu)$, so that using the formula (7-50) for $f_{\gamma 2}$ we may write

$$\mathrm{D}(V) = \frac{e^{-\frac{1}{2}\nu(h/r)V}(\frac{1}{2}\nu[h/r]V)^{\frac{1}{2}\nu-1}}{(\frac{1}{2}\nu-1)!}\,\tfrac{1}{2}\nu(h/r) \; .$$

Substituting herein $v = V/r$ and multiplying by the Jacobian $|dV/dv| = r$ we obtain

$$\mathrm{D}(v) = \frac{e^{-\frac{1}{2}h\nu v}(\frac{1}{2}h\nu v)^{\frac{1}{2}\nu-1}}{(\frac{1}{2}\nu-1)!}\,\tfrac{1}{2}h\nu \equiv f_{\gamma 2}(v|h, \nu) \; .$$

The kernel (12-24b) of the likelihood of v is now obtained by simply dropping from this density all factors constant as regards h. ◀

Again for the same reasons as in the univariate case, we wish to express the kernel of the likelihood of the complete sample in such a way that it automatically reduces to (12-24a) when only $(\boldsymbol{m}, \mathbf{n})$ is available and to (12-24b) when only (v, ν) is available. We therefore define

$$p \equiv \mathrm{rank}(\mathbf{n}) = \begin{cases} 0 & \text{if} \quad n = 0 \\ r & \text{if} \quad n > 0 \end{cases} \tag{12-25}$$

and write the kernel (12-23) in the form

$$e^{-\frac{1}{2}h(\boldsymbol{m}-\boldsymbol{\mu})^t\mathbf{n}(\boldsymbol{m}-\boldsymbol{\mu})}\,h^{\frac{1}{2}p} \cdot e^{-\frac{1}{2}h\nu v}\,h^{\frac{1}{2}\nu} \; . \tag{12-26}$$

With the convention that

1. $\nu = 0$ when v is unknown or irrelevant,
2. $\mathbf{n} = \mathbf{0}$ when $\boldsymbol{m}$ is unknown or irrelevant,

the complete kernel (12-26) reduces to (12-24a) in the first case and to (12-24b) in the second.

12.4.3. Conjugate Distribution of $(\tilde{\boldsymbol{\mu}}, \tilde{h})$

When the relative precision $\boldsymbol{\eta}$ of an Independent Multinormal process is known but both the mean $\tilde{\boldsymbol{\mu}}$ and the mean precision $\tilde{h}$ are random variables, the most convenient prior distribution of $(\tilde{\boldsymbol{\mu}}, \tilde{h})$—the natural conjugate of (12-26)—is what we shall call the (multivariate) Normal-gamma distribution, defined by

$$f_{N\gamma}^{(r)}(\boldsymbol{\mu}, h|\boldsymbol{m}, v, \mathbf{n}, \nu) = f_N^{(r)}(\boldsymbol{\mu}|\boldsymbol{m}, h\mathbf{n})\, f_{\gamma 2}(h|v, \nu)$$
$$\propto e^{-\frac{1}{2}h(\boldsymbol{\mu}-\boldsymbol{m})^t\mathbf{n}(\boldsymbol{\mu}-\boldsymbol{m})}\, h^{\frac{1}{2}p} \cdot e^{\frac{1}{2}\nu v h}\, h^{\frac{1}{2}\nu-1} , \qquad (12\text{-}27)$$

where $\mathbf{n}$ expresses the information on $\boldsymbol{\mu}$ in units of the process mean precision h as discussed in Section 12.1.4 and $p \equiv \text{rank}(\mathbf{n})$. Since $\mathbf{n}$ must be PDS and therefore of full rank r if (12-27) is to represent a proper (convergent) density, p will in this case be equal to r and the definition of $f_N^{(r)}$ implied by (12-27) is identical to the definition (8-17); we write p rather than r in (12-27) so that our formulas for the posterior density will generalize automatically to the limiting case where prior information is totally negligible and the decision maker sets $\mathbf{n}' = \mathbf{0}$.

If the prior distribution of $(\tilde{\boldsymbol{\mu}}, \tilde{h})$ is Normal-gamma with parameter $(\boldsymbol{m}', v', \mathbf{n}', \nu')$ where $\mathbf{n}'$ is of rank p' and a sample then yields a statistic $(\boldsymbol{m}, v, \mathbf{n}, \nu)$ where $\mathbf{n}$ is of rank p, the posterior distribution of $(\tilde{\boldsymbol{\mu}}, \tilde{h})$ will be Normal-gamma with parameters

$$\boldsymbol{m}'' = (\mathbf{n}' + \mathbf{n})^{-1}(\mathbf{n}'\boldsymbol{m}' + \mathbf{n}\,\boldsymbol{m}) , \qquad (12\text{-}28a)$$

$$\mathbf{n}'' = \mathbf{n}' + \mathbf{n} , \qquad p'' = \text{rank}(\mathbf{n}'') , \qquad (12\text{-}28b)$$

$$v'' = \frac{[\nu'v' + \boldsymbol{m}'^t\mathbf{n}'\boldsymbol{m}'] + [\nu v + \boldsymbol{m}^t\mathbf{n}\,\boldsymbol{m}] - \boldsymbol{m}''^t\mathbf{n}''\boldsymbol{m}''}{[\nu' + p'] + [\nu + p] - p''} , \qquad (12\text{-}28c)$$

$$\nu'' = [\nu' + p'] + [\nu + p] - p'' . \qquad (12\text{-}28d)$$

▶ Multiplying the kernel of the prior density (12-27) by the kernel of the likelihood (12-26) we have

$$e^{-\frac{1}{2}h\nu'v' - \frac{1}{2}h(\boldsymbol{\mu}-\boldsymbol{m}')^t\mathbf{n}'(\boldsymbol{\mu}-\boldsymbol{m}')}\, h^{\frac{1}{2}\nu'+\frac{1}{2}p'-1}\, e^{-\frac{1}{2}h\nu v - \frac{1}{2}h(\boldsymbol{m}-\boldsymbol{\mu})^t\mathbf{n}(\boldsymbol{m}-\boldsymbol{\mu})}\, h^{\frac{1}{2}\nu+\frac{1}{2}p} .$$

We observe first that the posterior exponent of h will be

$$\tfrac{1}{2}\nu' + \tfrac{1}{2}p' - 1 + \tfrac{1}{2}\nu + \tfrac{1}{2}p = \tfrac{1}{2}[\nu' + p' + \nu + p - p''] + \tfrac{1}{2}p'' - 1 .$$

By (12-28d) this can be written

$$\tfrac{1}{2}p'' + \tfrac{1}{2}\nu'' - 1$$

in agreement with the exponent of h in (12-27). We next observe that the posterior exponent of e will be $-\frac{1}{2}h$ multiplied by

$$\nu'v' + (\boldsymbol{\mu} - \boldsymbol{m}')^t\mathbf{n}'(\boldsymbol{\mu} - \boldsymbol{m}') + \nu v + (\boldsymbol{m} - \boldsymbol{\mu})^t\mathbf{n}(\boldsymbol{m} - \boldsymbol{\mu}) .$$

Using (1) through (4) in the proof of (12-14) this can be written

$$(\boldsymbol{\mu} - \boldsymbol{m}'')^t\mathbf{n}''(\boldsymbol{\mu} - \boldsymbol{m}'') + \nu'v' + \boldsymbol{m}'^t\mathbf{n}'\boldsymbol{m}' + \nu v + \boldsymbol{m}^t\mathbf{n}\,\boldsymbol{m}$$
$$- (\mathbf{n}'\boldsymbol{m}' + \mathbf{n}\,\boldsymbol{m})^t\mathbf{n}''^{-1}(\mathbf{n}'\boldsymbol{m}' + \mathbf{n}\,\boldsymbol{m}) .$$

By (12-28ab) the last term in this expression can be written

$$(\mathbf{n}''\boldsymbol{m}'')^t\mathbf{n}''^{-1}(\mathbf{n}''\boldsymbol{m}'') = \boldsymbol{m}''^t\mathbf{n}''\boldsymbol{m}'' ,$$

and by (12-28cd) the whole expression can therefore be written

$$(\boldsymbol{\mu} - \boldsymbol{m}'')^t\mathbf{n}''(\boldsymbol{\mu} - \boldsymbol{m}'') + \nu''v''$$

in agreement with the exponent of e in (12-27). ◀

12.4.4. *Distributions of* $\tilde{h}$

If the joint distribution of $(\tilde{\mu}, \tilde{h})$ is Normal-gamma as defined by (12-27), the *marginal* distribution of $\tilde{h}$ is gamma-2 as defined by (7-50),

$$D(h|m, v, \mathrm{n}, \nu) = f_{\gamma 2}(h|v, \nu) \ , \tag{12-29}$$

while the *conditional* distribution of $\tilde{h}$ for given μ is of the same type but with different parameters:

$$D(h|\mu; m, v, \mathrm{n}, \nu) = f_{\gamma 2}(h|V, p + \nu) \tag{12-30a}$$

where

$$V \equiv \frac{\nu v + (\mu - m)^t \mathrm{n}(\mu - m)}{p + \nu} \ . \tag{12-30b}$$

Since both these distributions are of exactly the same type as the corresponding distributions of $\tilde{h}$ when $\tilde{\mu}$ is univariate, the discussion of the marginal distribution of $\tilde{h}$ in Section 11.5.4 and the discussion of the implied distribution of $\tilde{\sigma} = \sqrt{1/\tilde{h}}$ in Section 11.1.4 apply here without change.

▶ By definition, the marginal density of $\tilde{h}$ is obtained by integrating the joint density (12-27) over $-\infty < \mu < \infty$, and since $f_{\gamma 2}(v|h, \nu)$ does not contain μ this integral is

$$f_{\gamma 2}(h|v, \nu) \int_{-\infty}^{\infty} f_N^{(r)}(\mu|m, h\mathrm{n}) \, d\mu = f_{\gamma 2}(h|v, \nu) \ .$$

To find the conditional density of $\tilde{h}$ given μ we write the kernel (12-27) of the joint density of $(\tilde{\mu}, \tilde{h})$ in the form

$$e^{-\frac{1}{2}h(\nu v + [\mu - m]^t \mathrm{n} [\mu - m])} \, h^{\frac{1}{2}(p+\nu)-1} \equiv e^{-\frac{1}{2}h(p+\nu)V} \, h^{\frac{1}{2}(p+\nu)-1} \ .$$

Comparison with (7-50) then shows that the density of $(\tilde{h}|\mu)$ is gamma-2 with the parameters specified by (12-30). ◀

12.4.5. *Distributions of* $\tilde{\mu}$

If the joint distribution of $(\tilde{\mu}, \tilde{h})$ is Normal-gamma as defined by (12-27), the *marginal* distribution of $\tilde{\mu}$ is Student as defined by (8-28):

$$D(\mu|m, v, \mathrm{n}, \nu) = f_S^{(r)}(\mu|m, \mathrm{n}/v, \nu) \ . \tag{12-31}$$

It then follows immediately from (8-29) that

$$E(\tilde{\mu}|m, v, \mathrm{n}, \nu) \equiv \bar{\mu} = m \ , \qquad \nu > 1 \ , \tag{12-32a}$$

$$V(\tilde{\mu}|m, v, \mathrm{n}, \nu) \equiv \breve{\mu} = \mathrm{n}^{-1} v \frac{\nu}{\nu - 2} \ , \qquad \nu > 2 \ . \tag{12-32b}$$

The *conditional* density of $\tilde{\mu}$ given h is of course $f_N(\mu|m, h\mathrm{n})$.

▶ By definition, the marginal density of $\tilde{\mu}$ is given by the integral

$$\int_0^{\infty} f_N(\mu|m, h\mathrm{n}) \, f_{\gamma 2}(h|v, \nu) \, dh \ .$$

Substituting herein $h = h^*/v$ and using the definition (7-50) of $f_{\gamma 2}$ we obtain

$$\int_0^\infty f_N(\boldsymbol{\mu}|\boldsymbol{m}, h^*\mathbf{n}/v)\, f_{\gamma 2}(h^*|1, \nu)\, dh^* \;;$$

and (12-31) then follows by (8-28). ◀

12.5. Sampling Distributions With Fixed n

In this section we assume that a sample with predetermined $\mathbf{n} > \mathbf{0}$ and therefore of rank $p = r$ is to be drawn from an r-dimensional Independent Multinormal process whose relative precision $\boldsymbol{\eta}$ is known but whose parameter $(\tilde{\boldsymbol{\mu}}, \tilde{h})$ is a random variable having a proper Normal-gamma distribution with parameter $(\boldsymbol{m}', v', \mathbf{n}', \nu')$. The fact that the distribution is proper implies that $\mathbf{n}'$ is of rank $p = r$ and that $\nu', v' > 0$.

12.5.1. Conditional Joint Distribution of $(\tilde{\boldsymbol{m}}, \tilde{v}|\boldsymbol{\mu}, h)$

The *conditional* joint distribution of the statistic $(\tilde{\boldsymbol{m}}, \tilde{v})$ for a given value of the process parameter $(\boldsymbol{\mu}, h)$ has as its density the product of the independent individual densities of $\tilde{\boldsymbol{m}}$ and $\tilde{v}$:

$$D(\boldsymbol{m}, v|\boldsymbol{\mu}, h, \mathbf{n}, \nu) = f_N^{(r)}(\boldsymbol{m}|\boldsymbol{\mu}, h\mathbf{n})\, f_{\gamma 2}(v|h, \nu) \;. \tag{12-33}$$

The proof was given in the proof of (12-24).

12.5.2. Unconditional Joint Distribution of $(\tilde{\boldsymbol{m}}, \tilde{v})$

The *unconditional* joint distribution of $(\tilde{\boldsymbol{m}}, \tilde{v})$ has the density

$$\begin{aligned} D(\boldsymbol{m}, v|\boldsymbol{m}', v', \mathbf{n}', \nu'; \mathbf{n}, \nu) &= \int_{-\infty}^{\infty}\int_0^\infty f_N^{(r)}(\boldsymbol{m}|\boldsymbol{\mu}, h\mathbf{n})\, f_{\gamma 2}(v|h, \nu)\, f_{N\gamma}^{(r)}(\boldsymbol{\mu}, h|\boldsymbol{m}', v', \mathbf{n}', \nu')\, dh\, d\boldsymbol{\mu} \\ &\propto \frac{(\nu v)^{\frac{1}{2}\nu - 1}}{(\nu' v' + \nu v + [\boldsymbol{m} - \boldsymbol{m}']^t\, \mathbf{n}_u[\boldsymbol{m} - \boldsymbol{m}'])^{\frac{1}{2}\nu''}} \;, \end{aligned} \tag{12-34a}$$

where

$$\begin{aligned} &\mathbf{n}_u \equiv \mathbf{n}'\mathbf{n}''^{-1}\,\mathbf{n} = \mathbf{n}\,\mathbf{n}''^{-1}\,\mathbf{n}' \;, \qquad \mathbf{n}_u^{-1} = \mathbf{n}'^{-1} + \mathbf{n}^{-1} \;, \\ &\nu'' = \nu' + \nu + r \;. \end{aligned} \tag{12-34b}$$

▶ To evaluate the integral (12-34a), we replace $f_{N\gamma}$ by its definition (12-27) and write the integral in the form

$$\int_0^\infty \left[\int_{-\infty}^{\infty} f_N(\boldsymbol{m}|\boldsymbol{\mu}, h\mathbf{n})\, f_N(\boldsymbol{\mu}|\boldsymbol{m}', h\mathbf{n}')\, d\boldsymbol{\mu}\right] f_{\gamma 2}(v|h, \nu)\, f_{\gamma 2}(h|v', \nu')\, dh \;.$$

When the inner integral is evaluated by use of (12-17) this becomes

$$\int_0^\infty f_N(\boldsymbol{m}|\boldsymbol{m}', h\mathbf{n}_u)\, f_{\gamma 2}(v|h, \nu)\, f_{\gamma 2}(h|v', \nu')\, dh \;.$$

Replacing f_N and the $f_{\gamma 2}$'s in this expression by their formulas and dropping constants we have

$$\int_0^\infty e^{-\frac{1}{2}h(\boldsymbol{m}-\boldsymbol{m}')^t\,\mathbf{n}_u(\boldsymbol{m}-\boldsymbol{m}') - \frac{1}{2}h(\nu v + \nu' v')}\, h^{\frac{1}{2}(r+\nu'+\nu)-1}\, v^{\frac{1}{2}\nu - 1}\, dh \;.$$

Substituting herein

$$h = z/A \ , \qquad A = \tfrac{1}{2}(\nu' v' + \nu v + [m - m']^t \mathbf{n}_u [m - m']) \ ,$$

we obtain

$$v^{\frac{1}{2}\nu - 1} A^{-\frac{1}{2}(\nu' + \nu + r)} \int_0^\infty e^{-z} z^{\frac{1}{2}(\nu' + \nu + r) - 1} dz \ .$$

The integral in this expression is simply a constant multiplied by $F_{\gamma^*}(\infty | \frac{1}{2}[\nu' + \nu + r]) = 1$; and since $\nu'' = \nu' + \nu + r$ when $p' = p = p'' = r$, the factors preceding the integral are proportional to (12-34). ◀

12.5.3. *Unconditional Distributions of $\tilde{m}$ and $\tilde{v}$*

From the unconditional joint distribution (12-34) of $(\tilde{m}, \tilde{v})$ we can obtain two distributions of $\tilde{m}$, both unconditional as regards the parameter $(\boldsymbol{\mu}, h)$. The first of these is also unconditional as regards v; its density is

$$D(m | m', v', \mathbf{n}', \nu'; \mathbf{n}, \nu) = f_S^{(r)}(m | m', \mathbf{n}_u / v', \nu') \ . \tag{12-35}$$

The second is conditional on v even though unconditional as regards $(\boldsymbol{\mu}, h)$; its density is

$$D(m | m', v', \mathbf{n}', \nu'; \mathbf{n}, \nu; v) = f_S^{(r)}(m | m', \mathbf{n}_u / V, \nu' + \nu) \tag{12-36a}$$

where

$$V \equiv \frac{\nu' v' + \nu v}{\nu' + \nu} \ . \tag{12-36b}$$

In the same way the joint distribution of $(\tilde{m}, \tilde{v})$ implies two distributions of $\tilde{v}$. Marginally as regards m, the statistic $\tilde{v}$ has the inverted-beta-2 density

$$D(v | m', v', \mathbf{n}', \nu'; \mathbf{n}, \nu) = f_{i\beta 2}(v | \tfrac{1}{2}\nu, \tfrac{1}{2}\nu', \nu' v' / \nu) \ ; \tag{12-37}$$

while conditional on m the density of $\tilde{v}$ is

$$D(v | m', v', \mathbf{n}', \nu'; \mathbf{n}, \nu; m) = f_{i\beta 2}\left(v | \tfrac{1}{2}\nu, \tfrac{1}{2}[\nu' + r], \frac{\nu' v' + M}{\nu}\right) \tag{12-38a}$$

where

$$M \equiv (m - m')^t \mathbf{n}_u (m - m') \ . \tag{12-38b}$$

▶ The marginal density (12-35) of $\tilde{m}$ and the conditional density (12-38) of $(\tilde{v}|m)$ are obtained by writing the kernel (12-34a) of the joint density of $(\tilde{m}, \tilde{v})$ in the form

$$\frac{1}{(\nu' v' + [m - m']^t \mathbf{n}_u [m - m'])^{\frac{1}{2}(\nu'' - \nu)}} \times \frac{(\nu v)^{\frac{1}{2}\nu - 1} (\nu' v' + [m - m']^t \mathbf{n}_u [m - m'])^{\frac{1}{2}(\nu'' - \nu)}}{(\nu v + \nu' v' + [m - m']^t \mathbf{n}_u [m - m'])^{\frac{1}{2}\nu''}}$$

$$\propto f_S^{(r)}(m | m', \mathbf{n}_u / v', \nu'' - \nu - r) \, f_{i\beta 2}\left(v | \tfrac{1}{2}\nu, \tfrac{1}{2}[\nu'' - \nu], \frac{\nu' v' + M}{\nu}\right)$$

where M is defined by (12-38b) and $\nu'' = \nu' + \nu + r$.

The conditional density (12-36) of $(\tilde{m}|v)$ and the marginal density (12-37) of $\tilde{v}$ are obtained by writing the kernel (12-34a) of the joint density in the form

$$\frac{(\nu v)^{\frac{1}{2}\nu-1}}{(\nu v+\nu' v')^{\frac{1}{2}(\nu''-r)}} \times \frac{(\nu v+\nu' v')^{\frac{1}{2}\nu''}}{(\nu v+\nu' v'+[m-m']^t \mathbf{n}_u[m-m'])^{\frac{1}{2}\nu''}(\nu v+\nu' v')^{\frac{1}{2}r}}$$

$$\propto f_{i\beta 2}\left(v|\tfrac{1}{2}\nu, \tfrac{1}{2}[\nu''-r-\nu], \frac{\nu' v'}{\nu}\right) f_S^{(r)}(m|m', \mathbf{n}_u/V, \nu''-r)$$

where V is defined by (12-36b). ◀

12.6. Preposterior Analysis With Fixed n

In this section as in the previous one we assume that a sample with predetermined $\mathbf{n}$ of rank r is to be drawn from an r-dimensional Independent Multinormal process whose relative precision $\boldsymbol{\eta}$ is known but whose mean $\tilde{\boldsymbol{\mu}}$ and mean precision $\tilde{h}$ are random variables having a proper Normal-gamma distribution with parameter $(m', v', \mathbf{n}', \nu')$ where $\mathbf{n}'$ is of rank r and $v', \nu' > 0$.

12.6.1. Joint Distribution of $(\tilde{m}'', \tilde{v}'')$

By expressing the random variable $(\tilde{m}, \tilde{v})$ in terms of the random variable $(\tilde{m}'', \tilde{v}'')$ we can obtain the joint density of the latter from the joint density (12-34) of the former:

$$\mathrm{D}(m'', v''|m', v', \mathbf{n}', \nu'; \mathbf{n}, \nu)$$

$$\propto \frac{(\nu'' v''-\nu' v'-[m''-m']^t \mathbf{n}^*[m''-m'])^{\frac{1}{2}\nu-1}}{(\nu'' v'')^{\frac{1}{2}\nu''}} \qquad (12\text{-}39a)$$

where

$$\mathbf{n}^* \equiv \mathbf{n}''\mathbf{n}^{-1}\mathbf{n}' = \mathbf{n}'\mathbf{n}^{-1}\mathbf{n}'' , \qquad \mathbf{n}^{*-1} = \mathbf{n}'^{-1} - \mathbf{n}''^{-1} ,$$
$$\nu'' = \nu' + \nu + r , \qquad (12\text{-}39b)$$

and the values of the variables m'' and v'' are constrained by

$$\nu'' v'' \geq \nu' v' , \qquad (m''-m')^t \mathbf{n}^*(m''-m') \leq \nu'' v'' - \nu' v' . \qquad (12\text{-}39c)$$

▶ The proof of (12-39) proceeds in three steps: we first show that

$$(1) \qquad (m-m')^t \mathbf{n}_u(m-m') = (m''-m')^t \mathbf{n}^*(m''-m') ;$$

we next show that

$$(2) \qquad \nu' v' + \nu v + (m-m')^t \mathbf{n}_u(m-m') = \nu'' v'' ;$$

and we then substitute these results in (12-34).

Equation (1) follows immediately by comparison of the right-hand sides of (5) and (6) in the proof of (12-14). Comparison of the right-hand sides of (4) and (5) in this same proof then shows that

$$(m-m')^t \mathbf{n}_u(m-m') = m'^t \mathbf{n}' m' + m^t \mathbf{n}\, m - (\mathbf{n}' m' + \mathbf{n}\, m)^t \mathbf{n}''^{-1}(\mathbf{n}' m' + \mathbf{n}\, m)$$
$$= m'^t \mathbf{n}' m' + m^t \mathbf{n}\, m - m''^t \mathbf{n}'' m'' ,$$

and substitution of the left-hand side of this equation in the definition (12-28cd) of $\nu'' v''$ proves equation (2) above. Now using (1) and (2) to replace the variables (m, v) of (12-34) by the variables (m'', v'') we obtain

$$\text{(3)} \qquad \frac{(\nu''v'' - \nu'v' - [m'' - m']^t \mathbf{n}^*[m'' - m'])^{\frac{1}{2}\nu - 1}}{(\nu''v'')^{\frac{1}{2}\nu''}} J ,$$

where J is the Jacobian of the transformation. To evaluate J we write

$$m = \mathbf{n}^{-1}(\mathbf{n}''m'' - \mathbf{n}'m')$$
$$v = \nu^{-1}(\nu''v'' - \nu'v' - [m'' - m']^t \mathbf{n}^*[m'' - m'])$$

and then compute

$$J \equiv \left|\frac{\partial(m, v)}{\partial(m'', v'')}\right| = \begin{vmatrix} \mathbf{n}^{-1}\mathbf{n}'' & 0 \\ - & \nu^{-1}\nu'' \end{vmatrix} ,$$

thus showing that J is a constant as regards m'' and v''. The proof of (12-39) is completed by observing that it is identical to (3) except for the constant J. ◀

12.6.2. *Distributions of $\tilde{m}''$ and $\tilde{v}''$*

From the joint distribution (12-39) of $(\tilde{m}'', \tilde{v}'')$ we can derive two distributions of $\tilde{m}''$: the distribution of $\tilde{m}''$ conditional on a particular value of $\tilde{v}''$, and the distribution of $\tilde{m}''$ which is marginal as regards $\tilde{v}''$. This latter, *unconditional* distribution of $\tilde{m}''$ has the Student density

$$D(m''|m', v', \mathbf{n}', \nu'; \mathbf{n}, \nu) = f_S^{(r)}(m''|m', \mathbf{n}^*/v', \nu') \qquad (12\text{-}40)$$

where $\mathbf{n}^*$ is defined by (12-39b). The *conditional* distribution of $(\tilde{m}''|v'')$ has the inverted-Student density (8-34):

$$D(m''|m', v', \mathbf{n}', \nu'; \mathbf{n}, \nu; v'') = f_{iS}^{(r)}(m''|m', \mathbf{n}^*/V, \nu) \qquad (12\text{-}41a)$$

where

$$V \equiv \frac{\nu''v'' - \nu'v'}{\nu} . \qquad (12\text{-}41b)$$

In the same way the joint distribution of $(\tilde{m}'', \tilde{v}'')$ implies two distributions of $\tilde{v}''$. *Marginally* as regards $\tilde{m}''$, the density of $\tilde{v}''$ is inverted-beta-1:

$$D(v''|m', v', \mathbf{n}', \nu'; \mathbf{n}, \nu) = f_{i\beta 1}\left(v''|\tfrac{1}{2}\nu', \tfrac{1}{2}\nu'', \frac{\nu'v'}{\nu''}\right) . \qquad (12\text{-}42)$$

Conditional on a particular m'', the density is of the same type but with different parameters:

$$D(v''|m', v', \mathbf{n}', \nu'; \mathbf{n}, \nu; m'') = f_{i\beta 1}\left(v''|\tfrac{1}{2}[\nu' + r], \tfrac{1}{2}\nu'', \frac{\nu'v' + M}{\nu''}\right) \qquad (12\text{-}43a)$$

where

$$M \equiv (m'' - m')^t \mathbf{n}^*(m'' - m') . \qquad (12\text{-}43b)$$

These distributions are identical to the corresponding distributions for the case where $\tilde{m}''$ is scalar, i.e., where $r = 1$.

▶ To obtain the marginal density (12-40) of $\tilde{m}''$ and the conditional density (12-43) of $(\tilde{v}''|m'')$ we first simplify notation by using the definition (12-43b) of M to write the kernel (12-39) of the joint density of $(\tilde{m}'', \tilde{v}'')$ in the form

$$(1) \qquad \frac{1}{(\nu'v' + M)^{\frac{1}{2}(\nu''-\nu)}} \times \frac{\left[v'' - \frac{\nu'v' + M}{\nu''}\right]^{\frac{1}{2}\nu-1}\left[\frac{\nu'v' + M}{\nu''}\right]^{\frac{1}{2}(\nu''-\nu)}}{v''^{\frac{1}{2}\nu''}} .$$

Since $\nu'' - \nu = \nu' + r$ and $\nu = \nu'' - [\nu' + r]$, this is proportional to

$$f_S^{(r)}(m''|m', \mathbf{n}^*/v', \nu')\, f_{i\beta1}\left(v''|\tfrac{1}{2}[\nu' + r], \tfrac{1}{2}\nu'', \frac{\nu'v' + M}{\nu''}\right) .$$

To obtain the marginal density (12-42) of $\tilde{v}''$ and the conditional density (12-41) of $(\tilde{m}''|v'')$ we write the kernel (12-39) of the joint density in the form

$$(2) \qquad \frac{(\nu''v'' - \nu'v')^{\frac{1}{2}(\nu+r)-1}}{(\nu''v'')^{\frac{1}{2}\nu''}} \times \frac{(\nu''v'' - \nu'v' - [m'' - m']^t\,\mathbf{n}^*[m'' - m'])^{\frac{1}{2}\nu-1}}{(\nu''v'' - \nu'v')^{\frac{1}{2}(\nu+r)-1}} .$$

The first factor is obviously proportional to $f_{i\beta1}(v''|\frac{1}{2}\nu', \frac{1}{2}\nu'', \nu'v'/\nu'')$; and if we use the definition (12-41b) of V to write the second factor in the form

$$\frac{(\nu V - (m'' - m')^t\,\mathbf{n}^*(m'' - m')]^{\frac{1}{2}\nu-1}}{(\nu V)^{\frac{1}{2}(\nu+r)-1}} \propto [\nu - (m'' - m')^t(\mathbf{n}^*/V)(m'' - m')]^{\frac{1}{2}\nu-1}\,\frac{|\mathbf{n}^*|^{\frac{1}{2}}}{V^{\frac{1}{2}r}}$$

and recall that $|\mathbf{n}^*/V| = |\mathbf{n}^*|/V^r$ we see that this factor is proportional to (12-41a). ◀

12.6.3. Distributions of $\breve{\tilde{\mu}}''$ and $\bar{\tilde{\mu}}''$

The variance $\breve{\mu}''$ of the posterior distribution of $\tilde{\mu}$ will, by (12-32b), be proportional to the parameter v'' of that distribution,

$$\breve{\mu}'' = \mathbf{n}''^{-1}\frac{\nu''}{\nu'' - 2}v'' , \qquad \nu'' > 2 , \qquad (12\text{-}44)$$

so that before the samplé outcome is known the distribution of the random variable $\breve{\tilde{\mu}}''$ can be easily obtained from the (marginal) distribution (12-42) of $\tilde{v}''$. Since this latter distribution is the same as when $\tilde{\mu}$ is univariate, we have by (11-66a) that

$$\begin{aligned} E(\breve{\tilde{\mu}}''|m', v', \mathbf{n}', \nu'; \mathbf{n}, \nu) &= \mathbf{n}''^{-1}\frac{\nu''}{\nu'' - 2}E(\tilde{v}'') = \mathbf{n}''^{-1}\frac{\nu'}{\nu' - 2}v' \\ &= \mathbf{n}''^{-1}\,\mathbf{n}'\breve{\mu}'. \end{aligned} \qquad (12\text{-}45)$$

The mean $\bar{\mu}''$ of the posterior distribution of $\tilde{\mu}$ will, by (12-32a), be equal to the parameter m'' of that distribution, so that the prior distribution of $\bar{\tilde{\mu}}''$ is obtained by merely substituting $\bar{\mu}''$ for m'' in the formula (12-40) for the (marginal) density of $\tilde{m}''$. It then follows by (8-29) that the mean and variance of $\bar{\tilde{\mu}}''$ are

$$E(\bar{\tilde{\mu}}''|m', v', \mathbf{n}', \nu'; \mathbf{n}, \nu) = m' = E(\tilde{\mu}|m', v', \mathbf{n}', \nu') , \qquad \nu' > 1 , \qquad (12\text{-}46a)$$

$$\begin{aligned} V(\bar{\tilde{\mu}}''|m', v', \mathbf{n}', \nu'; \mathbf{n}, \nu) &= \mathbf{n}^{*-1}\frac{\nu'}{\nu' - 2}v' \\ &= V(\tilde{\mu}|m', v', \mathbf{n}', \nu') - E[V(\tilde{\mu}|\tilde{m}'', \tilde{v}'', \mathbf{n}'', \nu'')] , \qquad \nu' > 2 \end{aligned} \qquad (12\text{-}46b)$$

in accordance with the general relations (5-27) and (5-28). These relations of course constitute an alternative method of deriving the parameters of the distribution (12.40) of $\tilde{m}''$.

C. *Interrelated Univariate Normal Processes*

12.7. Introduction

In some situations utilities depend on the joint distribution of some number r of unknown *scalar* quantities $\mu_1, \cdots, \mu_i, \cdots, \mu_r$ each of which is the mean of a distinct *univariate* Normal data-generating process as defined by (11-1). The ith process generates independent random variables $\tilde{x}_i^{(1)}, \cdots, \tilde{x}_i^{(j)}, \cdots$ with identical densities

$$f_N^{(1)}(x_i|\mu_i, h_i) \; , \tag{12-47}$$

and each of the r processes operates independently of all the others. In the general case, the processes are related through a joint prior distribution of their parameters $\tilde{\mu}_1, \cdots, \tilde{\mu}_r, \tilde{h}_1, \cdots, \tilde{h}_r$, but we shall here consider only the special case where the ratios $\tilde{h}_i/\tilde{h}_{i'}$ are known with certainty for all (i, i') even though the actual value of $\tilde{h}_i$ may not be known for any of the processes.

Because the processes operate independently, a different number of observations may be taken on each. The sample from the ith process will yield the statistics

$$n_i\text{: the number of observations on the } i\text{th process} \; , \tag{12-48a}$$

$$m_i = \frac{1}{n_i} \Sigma_j \, x_i^{(j)} \qquad (= 0 \text{ if } n_i = 0) \; , \tag{12-48b}$$

$$\nu_i = n_i - 1 \qquad (= 0 \text{ if } n_i = 0) \; , \tag{12-48c}$$

$$v_i = \frac{1}{\nu_i} \Sigma_j (x_i^{(j)} - m_i)^2 \quad (= 0 \text{ if } \nu_i = 0) \; . \tag{12-48d}$$

The kernel of the joint likelihood of all r samples is the product of the likelihoods of the individual samples as given by (11-37) with $n - 1$ replaced by ν:

$$\Pi[e^{-\frac{1}{2}h_i\nu_i v_i - \frac{1}{2}h_i n_i(m_i - \mu_i)^2} \, h_i^{\frac{1}{2}n_i}] = e^{-\frac{1}{2}\Sigma h_i(\nu_i v_i + n_i[m_i - \mu_i]^2)} \, \Pi \, h_i^{\frac{1}{2}n_i} \; . \tag{12-49}$$

If we now define the *mean precision* of the r processes

$$h \equiv (\Pi \, h_i)^{1/r} \; , \tag{12-50a}$$

the *relative precision* of the ith process

$$\eta_{ii} \equiv h_i/h \; , \tag{12-50b}$$

and the *effective number* of observations on the ith process

$$n_{ii} \equiv n_i \eta_{ii} \; , \tag{12-50c}$$

the kernel can be written (dropping the constant $\Pi \, \eta_{ii}^{\frac{1}{2}n_i}$)

$$e^{-\frac{1}{2}h\Sigma\eta_{ii}\nu_i v_i - \frac{1}{2}h\Sigma n_{ii}(m_i - \mu_i)^2} \, h^{\frac{1}{2}\Sigma n_i} \; ; \tag{12-51}$$

and if we further define

$$\boldsymbol{\mu} \equiv [\mu_1 \cdots \mu_i \cdots \mu_r]^t \tag{12-52a}$$

$$\boldsymbol{m} \equiv [m_1 \cdots m_i \cdots m_r]^t \; , \tag{12-52b}$$

$$\mathbf{n} \equiv \begin{bmatrix} n_{11} & & & & 0 \\ & \cdot & & & \\ & & n_{ii} & & \\ & & & \cdot & \\ 0 & & & & n_{rr} \end{bmatrix} \tag{12-52c}$$

$$p \equiv \text{rank}(\mathbf{n}) = \text{number of processes sampled} \ , \tag{12-52d}$$

$$\nu \equiv \Sigma_{i=1}^{r} \nu_i \ , \tag{12-52e}$$

$$v \equiv \frac{1}{\nu} \Sigma\, \eta_{ii} \nu_i v_i = \frac{1}{\nu} \Sigma_{ij} \frac{h_i}{h} (x_i^{(j)} - m_i)^2 \ , \tag{12-52f}$$

the kernel can be written

$$e^{-\frac{1}{2}h\nu v - \frac{1}{2}h(\mathbf{m}-\boldsymbol{\mu})^t\mathbf{n}(\mathbf{m}-\boldsymbol{\mu})}\, h^{\frac{1}{2}(p+\nu)} \ . \tag{12-53}$$

It is thus apparent that the statistic $(\mathbf{m}, v, \mathbf{n}, \nu)$ is sufficient for the entire set of r samples.

12.8. Analysis When All Processes Are Sampled

If now we at first assume that all r processes are sampled, $p = r$, then the kernel (12-53) of the likelihood of these r samples, one from each of r independent processes, is formally identical to the kernel (12-8) of the likelihood of a single sample from an r-dimensional Multinormal process; and it can also be shown that the conditional distribution of the statistics $(\tilde{\mathbf{m}}, \tilde{v})$ for predetermined values of $(\mathbf{n}, \nu)$ and given values of the parameters $(\boldsymbol{\mu}, h)$ is identical to the corresponding distribution for a sample from a Multinormal process:

$$D(\mathbf{m}, v|\boldsymbol{\mu}, h, \mathbf{n}, \nu) = f_N^{(r)}(\mathbf{m}|\boldsymbol{\mu}, h\mathbf{n})\, f_{\gamma 2}(v|h, \nu) \ . \tag{12-54}$$

It follows that if we assign to $(\tilde{\boldsymbol{\mu}}, \tilde{h})$ a prior distribution of the same type that we assigned in the case of the Multinormal process, the entire analysis of our present problem will be identical to the analysis of the Multinormal process given in Sections 12.1 through 12.6. If the mean precision h of the r processes is known and a Normal prior distribution is assigned to $\tilde{\boldsymbol{\mu}}$,

$$D'(\boldsymbol{\mu}) = f_N^{(r)}(\boldsymbol{\mu}|\mathbf{m}', h\mathbf{n}') \ , \tag{12-55}$$

then the formulas given in Section 12.2 for unconditional distribution of $\tilde{\mathbf{m}}$ and in Section 12.3 for the prior distribution of $\tilde{\mathbf{m}}''$ apply without change. If h is unknown and a Normal-gamma prior distribution is assigned to $(\tilde{\boldsymbol{\mu}}, \tilde{h})$,

$$D'(\boldsymbol{\mu}, h) = f_N^{(r)}(\boldsymbol{\mu}|\mathbf{m}', h\mathbf{n}')\, f_{\gamma 2}(h|v', \nu') \ , \tag{12-56}$$

the formulas in Section 12.5 for the unconditional distributions of $\tilde{\mathbf{m}}$ and $\tilde{v}$ and in Section 12.6 for the prior distributions of $\tilde{\mathbf{m}}''$ and $\tilde{v}''$ apply without change.

▶ To prove (12-54), we start from the fact that, by (11-52), the statistics

$$\tilde{m}_i = \frac{1}{n_i}\Sigma_j\,\tilde{x}_i^{(j)} \qquad \text{and} \qquad \tilde{v}_i = \frac{1}{n_i - 1}\Sigma_j(\tilde{x}_i^{(j)} - \tilde{m}_i)^2$$

of the sample from the *i*th process are independently distributed with densities

$$(1) \qquad f_N(m_i|\mu_i, h_i n_i) \equiv f_N(m_i|\mu_i, hn_{ii}) \propto e^{-\frac{1}{2}hn_{ii}(m_i-\mu_i)^2} ,$$

$$(2) \qquad f_{\gamma 2}(v_i|h_i, \nu_i) \equiv f_{\gamma 2}(v_i|h\eta_{ii}, \nu_i) \propto e^{-\frac{1}{2}h\eta_{ii}\nu_i v_i}(\tfrac{1}{2}h\eta_{ii}\nu_i v_i)^{\frac{1}{2}\nu_i - 1} .$$

Since the processes themselves are also independent, the vector $\tilde{m}$ of the r sample means will be independent of the "pooled" statistic

$$\tilde{v} = \frac{1}{\Sigma\, n_i - r}\Sigma\,\eta_{ii}\nu_i\tilde{v}_i$$

and therefore the joint density of $\tilde{m}$ and $\tilde{v}$ is the product of their individual densities. It follows that (12-54) can be proved by proving that the first factor therein is the density of $\tilde{m}$ and that the second factor is the density of $\tilde{v}$.

To show that $\tilde{m}$ has the density $f_N^{(r)}(m|\mu, h\mathbf{n})$, we need only recall that the independence of the $\tilde{m}_i$ means that their joint density is the product of their individual densities as given by (1) above,

$$D(m) \propto \Pi(e^{-\frac{1}{2}hn_{ii}[m_i-\mu_i]^2}) = e^{-\frac{1}{2}h\Sigma n_{ii}(m_i-\mu_i)^2} = e^{-\frac{1}{2}h(m-\mu)^t\mathbf{n}(m-\mu)} .$$

To show that $\tilde{v}$ has the density $f_{\gamma 2}(v|h, \nu)$, we first substitute

$$w_i = \tfrac{1}{2}h\eta_{ii}\nu_i v_i$$

in (2) to obtain

$$D(w_i) \propto e^{-w_i}\,w_i^{\frac{1}{2}\nu_i - 1} \propto f_{\gamma^*}(w_i|\tfrac{1}{2}\nu_i) .$$

Because the $\tilde{v}_i$ and therefore the $\tilde{w}_i$ are independent, the distribution of

$$\tilde{W} \equiv \Sigma\,\tilde{w}_i$$

has as its density the convolution of the r densities of the individual $\tilde{w}_i$; and by (7-42) this convolution is

$$f_{\gamma^*}(W|\tfrac{1}{2}\Sigma\,\nu_i) \propto e^{-W}\,W^{(\frac{1}{2}\Sigma\nu_i)-1} .$$

Substituting herein

$$\nu \equiv \Sigma\,\nu_i ,$$
$$W \equiv \Sigma\,w_i = \Sigma\,\tfrac{1}{2}h\eta_{ii}\nu_i v_i = \tfrac{1}{2}h\nu v ,$$

we see that

$$D(v) \propto e^{-\frac{1}{2}h\nu v}(\tfrac{1}{2}h\nu v)^{\frac{1}{2}\nu - 1} \propto f_{\gamma 2}(v|h, \nu) .$$ ◀

12.9. Analysis When Only $p < r$ Processes Are Sampled

12.9.1. *Notation*

We next consider the case where observations have been or are to be taken on only $p < r$ of the r processes whose parameters are related a priori by (12-55) or (12-56). To facilitate discussion, we assume without loss of generality that it is the *first* p processes on which observations are taken, partition

$$\mu = \begin{bmatrix}\mu_1\\ \mu_2\end{bmatrix} \qquad \text{where } \begin{matrix}\mu \text{ is } r \times 1 ,\\ \mu_1 \text{ is } p \times 1 ,\end{matrix} \tag{12-57}$$

and partition the parameters of the prior distribution of $(\tilde{\mu}|h)$ accordingly:

$$m' = \begin{bmatrix} m'_1 \\ m'_2 \end{bmatrix} \qquad \text{where} \begin{array}{l} m' \text{ is } r \times 1 \ , \\ m'_1 \text{ is } p \times 1 \ , \end{array}$$

$$\mathbf{n}' = \begin{bmatrix} \mathbf{n}'_{11} & \mathbf{n}'_{12} \\ \mathbf{n}'_{21} & \mathbf{n}'_{22} \end{bmatrix} \qquad \text{where} \begin{array}{l} \mathbf{n}' \text{ is } r \times r \ , \\ \mathbf{n}'_{11} \text{ is } p \times p \ . \end{array} \tag{12-58}$$

The parameters m'' and $\mathbf{n}''$ of the posterior distribution of $\tilde{\mu}$ will of course be partitioned in this same way. To correspond with this notation we define the "natural" statistics

$$m_1 \equiv [m_1 \ \cdots \ m_p]^t \ , \tag{12-59a}$$

$$\mathbf{n}_{11} \equiv \begin{bmatrix} n_{11} & & & 0 \\ & \cdot & & \\ & & \cdot & \\ 0 & & & \cdot \\ & & & n_{pp} \end{bmatrix} \tag{12-59b}$$

$$\nu \equiv \Sigma_{i=1}^p \nu_i = \Sigma\, n_i - p \ , \tag{12-59c}$$

$$v \equiv \frac{1}{\nu} \Sigma\, \eta_{ii}\nu_i v_i \ = \frac{1}{\nu} \Sigma_{ij} \frac{h_i}{h} (x_i^{(j)} - m_i)^2 \ ; \tag{12-59d}$$

the statistics m and $\mathbf{n}$ defined by (12-52bc) are then of the form

$$m \equiv \begin{bmatrix} m_1 \\ 0 \end{bmatrix} \qquad \text{where} \begin{array}{l} m \text{ is } r \times 1 \ , \\ m_1 \text{ is } p \times 1 \ , \end{array}$$

$$\mathbf{n} \equiv \begin{bmatrix} \mathbf{n}_{11} & 0 \\ 0 & 0 \end{bmatrix} \qquad \text{where} \begin{array}{l} \mathbf{n} \text{ is } r \times r \ , \\ \mathbf{n}_{11} \text{ is } p \times p \ , \end{array} \tag{12-60}$$

and will be referred to as "augmented" statistics.

12.9.2. Posterior Analysis

If h is known, if the $r \times 1$ random variable $\tilde{\mu}$ has a Normal distribution with parameter $(m', \mathbf{n}')$, and if samples from p of the r processes yield the augmented statistic $(m, \mathbf{n})$ defined by (12-60), then the posterior distribution of $\tilde{\mu}$ is Normal and its parameters are given by (12-14) without change:

$$\mathbf{n}'' = \mathbf{n}' + \mathbf{n} \ , \tag{12-61a}$$
$$m'' = \mathbf{n}''^{-1}(\mathbf{n}'m' + \mathbf{n}\, m) \ . \tag{12-61b}$$

If h is unknown, if $(\tilde{\mu}, \tilde{h})$ has a Normal-gamma distribution with parameter $(m', v', \mathbf{n}', \nu')$, and if samples from p of the r processes yield a statistic $(m, v, \mathbf{n}, \nu)$ as defined by (12-59) and (12-60), then the posterior distribution of $(\tilde{\mu}, \tilde{h})$ is Normal-gamma with parameters m'' and $\mathbf{n}''$ as given by (12-61) and

$$\nu'' = [\nu' + p'] + [\nu + p] - p'' \ ,$$

$$v'' = \frac{1}{\nu''}([\nu'v' + m'^t\mathbf{n}'m'] + [\nu v + m^t\mathbf{n}\, m] - m''^t\mathbf{n}''m''), \tag{12-62}$$

where $p' \equiv \text{rank}(\mathbf{n}')$ and $p'' \equiv \text{rank}(\mathbf{n}'')$.

▶ The argument which led to (12-53) as the kernel of the joint likelihood of r samples obviously leads to

$$e^{-\frac{1}{2}hv\nu-\frac{1}{2}h(\boldsymbol{m}_1-\boldsymbol{\mu}_1)^t\,\mathbf{n}_{11}(\boldsymbol{m}_1-\boldsymbol{\mu}_1)}\;h^{\frac{1}{2}(p+\nu)}$$

as the likelihood of p samples in terms of the natural statistics $\boldsymbol{m}_1$ and $\mathbf{n}_{11}$. The definitions (12-60) of the augmented statistics $\boldsymbol{m}$ and $\mathbf{n}$ imply, however, that

$$e^{-\frac{1}{2}hv\nu-\frac{1}{2}h(\boldsymbol{m}-\boldsymbol{\mu})^t\,\mathbf{n}(\boldsymbol{m}-\boldsymbol{\mu})}\;h^{\frac{1}{2}(p+\nu)}$$

will have exactly the same value as the previous expression for all $\boldsymbol{\mu}$. Since this latter expression is identical to (12-8) and since the proofs of (12-14) and (12-28) did not depend on any assumption that $\mathbf{n}$ was of full rank, they hold unchanged for (12-61) and (12-62). ◀

12.9.3. Conditional Sampling Distributions With Fixed n

It is obvious that, for given $(\boldsymbol{\mu}, h)$, the conditional distributions of the natural statistics $\tilde{\boldsymbol{m}}_1$ and $\tilde{v}$ as defined by (12-59) depend on $\boldsymbol{\mu}_1$ and h in exactly the same way that $\tilde{\boldsymbol{m}}$ and $\tilde{v}$ as defined by (12-52) depend on $\boldsymbol{\mu}$ and h. There is consequently no need to prove that the joint density of $(\tilde{\boldsymbol{m}}_1, \tilde{v})$ is the product of the independent individual densities of $\tilde{\boldsymbol{m}}_1$ and $\tilde{v}$ and is given by (12-54) modified to read

$$D(\boldsymbol{m}_1, v|\boldsymbol{\mu}_1, h; \mathbf{n}, \nu) = f_N^{(p)}(\boldsymbol{m}_1|\boldsymbol{\mu}_1, h\mathbf{n}_{11})\, f_{\gamma 2}(v|h, \nu) \;. \tag{12-63}$$

12.9.4. Marginal Distribution of $(\tilde{\boldsymbol{\mu}}_1|h)$ *and* $(\tilde{\boldsymbol{\mu}}_1, \tilde{h})$

From (12-63) it is clear that the unconditional distributions of $\tilde{\boldsymbol{m}}_1$ and $\tilde{v}$ will depend on the distribution of $(\tilde{\boldsymbol{\mu}}_1|h)$ if h is known or on the distribution of $(\tilde{\boldsymbol{\mu}}_1, \tilde{h})$ if h is unknown in the same way that the distributions of $\tilde{\boldsymbol{m}}$ and $\tilde{v}$ depended on the distribution of $(\tilde{\boldsymbol{\mu}}|h)$ or $(\tilde{\boldsymbol{\mu}}, \tilde{h})$; and therefore our next step must be to obtain the distributions of $(\tilde{\boldsymbol{\mu}}_1|h)$ and $(\tilde{\boldsymbol{\mu}}_1, \tilde{h})$. In both cases the required distributions are, of course, the distributions *marginal* as regards $\tilde{\boldsymbol{\mu}}_2$.

The *prior* distribution of $(\tilde{\boldsymbol{\mu}}_1|h)$ is given immediately by (8-20): its density is

$$D'(\boldsymbol{\mu}_1|h; \boldsymbol{m}', \mathbf{n}') = f_N^{(p)}(\boldsymbol{\mu}_1|\boldsymbol{m}_1', h\mathbf{n}_m') \tag{12-64a}$$

where

$$\mathbf{n}_m' \equiv \mathbf{n}_{11}' - \mathbf{n}_{12}'\,\mathbf{n}_{22}'^{-1}\,\mathbf{n}_{21}' \;, \qquad \mathbf{n}_m'^{-1} = (\mathbf{n}'^{-1})_{11} \;. \tag{12-64b}$$

Then since the factor $f_{\gamma 2}(h|v', \nu')$ in the prior density (12-56) of $(\tilde{\boldsymbol{\mu}}, \tilde{h})$ is the *marginal* density of $\tilde{h}$, it follows at once that the prior density of $(\tilde{\boldsymbol{\mu}}_1, \tilde{h})$ is

$$D'(\boldsymbol{\mu}_1, h|\boldsymbol{m}', v', \mathbf{n}', \nu') = f_N^{(p)}(\boldsymbol{\mu}_1|\boldsymbol{m}_1', h\mathbf{n}_m')\, f_{\gamma 2}(h|v', \nu') \;. \tag{12-65}$$

The *posterior* distributions of $(\tilde{\boldsymbol{\mu}}_1|h)$ and $(\tilde{\boldsymbol{\mu}}_1, \tilde{h})$ are now easily found. Since the prior distributions are of the same form as the prior distributions of $(\tilde{\boldsymbol{\mu}}|h)$ and $(\tilde{\boldsymbol{\mu}}, \tilde{h})$ and since as was shown in the proof of (12-62) the likelihoods of $\tilde{\boldsymbol{m}}_1$ and $\tilde{v}$ are of the same form as the likelihoods of $\tilde{\boldsymbol{m}}$ and $\tilde{v}$, the posterior distributions of $(\tilde{\boldsymbol{\mu}}_1|h)$ and $(\tilde{\boldsymbol{\mu}}_1, \tilde{h})$ are given directly by (12-14) and (12-28) with r everywhere replaced by p. *Assuming that the prior distributions are proper*, i.e., that $v', \nu' > 0$ and that $\mathbf{n}_m'$ is PDS, we have

$$\mathbf{n}_m'' = \mathbf{n}_m' + \mathbf{n}_{11} \;, \tag{12-66a}$$

$$\boldsymbol{m}_1'' = \mathbf{n}_m''^{-1}(\mathbf{n}_m'\,\boldsymbol{m}_1' + \mathbf{n}_{11}\,\boldsymbol{m}_1) \;, \tag{12-66b}$$

$$\nu'' = \nu' + \nu + p \; , \tag{12-66c}$$

$$v'' = \frac{1}{\nu''}\left([\nu' v' + m_1'^t \, \mathbf{n}_m' \, m_1'] + [\nu v + m_1^t \, \mathbf{n}_{11} \, m_1] - m_1''^t \, \mathbf{n}_m'' \, m_1''\right) . \tag{12-66d}$$

▶ To show that (12-66b) is consistent with the definition of m_1'' in Section 12.9.1 as the first component of $m'' = \mathbf{n}''^{-1}(\mathbf{n}'m' + \mathbf{n}\,m)$, we define

$$\mathbf{u}'' \equiv \mathbf{n}''^{-1} \; ,$$

write (12-61b) in partitioned form as

$$\begin{bmatrix} m_1'' \\ m_2'' \end{bmatrix} = \begin{bmatrix} \mathbf{u}_{11}'' & \mathbf{u}_{12}'' \\ \mathbf{u}_{21}'' & \mathbf{u}_{22}'' \end{bmatrix} \cdot \begin{bmatrix} \mathbf{n}_{11}' \, m_1' + \mathbf{n}_{12}' \, m_2' + \mathbf{n}_{11} \, m_1 \\ \mathbf{n}_{21}' \, m_1' + \mathbf{n}_{22}' \, m_2' \end{bmatrix} ,$$

use (12-61a) and (12-60) to write

$$\begin{bmatrix} \mathbf{u}_{11}'' & \mathbf{u}_{12}'' \\ \mathbf{u}_{21}'' & \mathbf{u}_{22}'' \end{bmatrix} = \begin{bmatrix} \mathbf{n}_{11}' + \mathbf{n}_{11} & \mathbf{n}_{12}' \\ \mathbf{n}_{21}' & \mathbf{n}_{22}' \end{bmatrix}^{-1} ,$$

and use (8-3), (12-64b), and (12-66a) to evaluate

$$\mathbf{u}_{11}'' = (\mathbf{n}_{11}' + \mathbf{n}_{11} - \mathbf{n}_{12}' \, \mathbf{n}_{22}'^{-1} \, \mathbf{n}_{21}')^{-1} \equiv (\mathbf{n}_m' + \mathbf{n}_{11})^{-1} \equiv \mathbf{n}_m''^{-1} \; ,$$
$$\mathbf{u}_{12}'' = -\mathbf{n}_m''^{-1} \, \mathbf{n}_{12}' \, \mathbf{n}_{22}'^{-1} \; ,$$

When these values are substituted in the partitioned formula for m'' given just above, the first row reduces to formula (12-66b) for m_1''. ◀

12.9.5. *Unconditional Distributions of $\tilde{m}_1$ and $\tilde{v}$*

The changes which must be made in the formulas for the unconditional densities of $\tilde{m}$ and $\tilde{v}$ given in Sections 12.2 and 12.5 in order to obtain the corresponding densities of $\tilde{m}_1$ and $\tilde{v}$ are now obvious without proof.

If h is known, the unconditional density of $\tilde{m}_1$ is obtained by modifying (12-17) to read

$$D(m_1|h; m', \mathbf{n}'; \mathbf{n}) = f_N^{(p)}(m_1|m_1', h\mathbf{n}_u) \tag{12-67a}$$

where

$$\mathbf{n}_u \equiv \mathbf{n}_m' \, \mathbf{n}_m''^{-1} \, \mathbf{n}_{11} = \mathbf{n}_{11} \, \mathbf{n}_m''^{-1} \, \mathbf{n}_m' \; , \qquad \mathbf{n}_u^{-1} = \mathbf{n}_m'^{-1} + \mathbf{n}_{11}^{-1} \; . \tag{12-67b}$$

If h is unknown, the unconditional joint density of $(\tilde{m}_1, \tilde{v})$ is obtained by modifying (12-34) to read

$$D(m_1, v|m', v', \mathbf{n}', \nu'; \mathbf{n}, \nu) \propto \frac{(\nu v)^{\frac{1}{2}\nu - 1}}{(\nu' v' + \nu v + [m_1 - m_1']^t \, \mathbf{n}_u [m_1 - m_1'])^{\frac{1}{2}\nu''}} \tag{12-68}$$

where $\mathbf{n}_u$ is defined by (12-67b) and ν'' by (12-66c). The same changes in (12-35) through (12-38) give the densities of $\tilde{m}_1$ and $\tilde{v}$ separately.

12.9.6. *Distributions of $\tilde{m}_1''$ and $\tilde{v}''$*

The distributions of $\tilde{m}_1''$ and $\tilde{v}''$ as defined by (12-66) follow from the distributions of $\tilde{m}''$ and $\tilde{v}''$ given in Sections 12.3 and 12.6 just as obviously as the distributions of $\tilde{m}_1$ and $\tilde{v}$ follows from the distributions of $\tilde{m}$ and $\tilde{v}$.

If h is known, the density of $\tilde{m}_1''$ is obtained by modifying (12-20) to read

$$D(m_1''|h; m', n'; n) = f_N^{(p)}(m_1''|m_1', h n_m^*) \tag{12-69a}$$

where

$$n_m^* \equiv n_m'' \, n_{11}^{-1} \, n_m' = n_m' \, n_{11}^{-1} \, n_m'' \; , \qquad n_m^{*-1} = n_m'^{-1} - n_m''^{-1} \; . \tag{12-69b}$$

If h is unknown, the joint density of $(\tilde{m}_1'', \tilde{v}'')$ is obtained by modifying (12-39) to read

$$D(m_1'', v''|m', v', n', \nu'; n, \nu)$$

$$\propto \frac{(\nu''v'' - \nu'v' - [m_1'' - m_1']^t \, n_m^* [m_1'' - m_1'])^{\frac{1}{2}\nu - 1}}{(\nu''v'')^{\frac{1}{2}\nu''}} \tag{12-70}$$

where n_m^* is defined by (12-69b) and ν'' by (12-66c). The same changes in (12-40) through (12-43) give the densities of $\tilde{m}_1''$ and $\tilde{v}''$ separately.

12.9.7. Preposterior Analysis

The results given in the previous section suffice for preposterior analysis only in the rather special case where utilities depend only on m_1'' and/or v''. More frequently they will also depend on m_2'', and it is here that analysis of our present problem differs sharply from analysis when observations are taken on all r μs. If we rewrite formula (12-61b) considering $\tilde{m}''$ and $\tilde{m}_1$ as random variables we obtain

$$\tilde{m}'' = n''^{-1} \left(n'm' + n \begin{bmatrix} \tilde{m}_1 \\ 0 \end{bmatrix} \right) \, , \tag{12-71}$$

and we see at once that because the only random variable on the right-hand side of this equation is $\tilde{m}_1$ and the distribution of $\tilde{m}_1$ is p-dimensional, the distribution of $\tilde{m}''$ is confined to a p-dimensional subspace within the r-space on which m'' is defined. This means that it would be extremely awkward to try to work with the distribution of the complete $r \times 1$ vector $\tilde{m}''$ expressed in analytic form; we shall do better to express the $(r - p) \times 1$ vector m_2'' as a function of the $p \times 1$ vector m_1'' and work with the p-dimensional distribution of $\tilde{m}_1''$ which we obtained in the previous section.

Although it is quite possible to find the functional relation between m_2'' and m_1'' by using (12-71) to express m_2'' as a function of m_1 and m_1 as a function of m_1'', it will be more instructive to obtain the desired relation by direct consideration of the joint distribution of $\tilde{\mu}_1$ and $\tilde{\mu}_2$ and of the effect of sample information on this distribution. By (8-21) or (8-31), the mean of the prior *conditional* distribution of $\tilde{\mu}_2$, *given* a particular μ_1, is

$$E_{2|1}(\tilde{\mu}_2|\mu_1) = m_2' - n_{22}'^{-1} \, n_{21}'(\mu_1 - m_1') \; , \tag{12-72}$$

where $E_{2|1}$ denotes expectation with respect to $\tilde{\mu}_2$ for fixed μ_1. Since the sample information m_1 bears directly on μ_1 alone, not on μ_2, this information cannot affect the *conditional* distribution of $\tilde{\mu}_2$ given μ_1 and (12-72) holds just as well after the sample has been taken as before. We may therefore take the expectation of both sides of (12-72) with respect to the *posterior* distribution of $\tilde{\mu}_1$ and thus obtain

$$\begin{aligned} m_2'' &= E''(\tilde{\mu}_2) = E_1'' \, E_{2|1}(\tilde{\mu}_2|\tilde{\mu}_1) = E_1''(m_2' - n_{22}'^{-1} \, n_{21}'[\tilde{\mu}_1 - m_1']) \\ &= m_2' - n_{22}'^{-1} \, n_{21}'(m_1'' - m_1') \; . \end{aligned} \tag{12-73}$$

This relation enables us to express as a function of m_1'' any utility originally expressed as a function of m'' and thus to compute expected utility by use of the distribution of $\tilde{m}_1''$ or of $(\tilde{m}_1'', \tilde{v}'')$ obtained in the previous section.

▶ To derive (12-73) algebraically, we start from the partitioned expression for m'' given in the comment on (12-66) and write

$$\begin{aligned} m_2'' &= \mathbf{u}_{21}''(\mathbf{n}_{11}' m_1' + \mathbf{n}_{12}' m_2' + \mathbf{n}_{11} m_1) + \mathbf{u}_{22}''(\mathbf{n}_{21}' m_1' + \mathbf{n}_{22}' m_2') \\ &= (\mathbf{u}_{21}'' \mathbf{n}_{11}' + \mathbf{u}_{22}'' \mathbf{n}_{21}') m_1' + (\mathbf{u}_{21}'' \mathbf{n}_{12}' + \mathbf{u}_{22}'' \mathbf{n}_{22}') m_2' + \mathbf{u}_{21}'' \mathbf{n}_{11} m_1 \ . \end{aligned}$$

Remembering that $\mathbf{u}''\mathbf{n}'' = \mathbf{I}$ and that

$$\mathbf{n}'' = \begin{bmatrix} \mathbf{n}_{11}' + \mathbf{n}_{11} & \mathbf{n}_{12}' \\ \mathbf{n}_{21}' & \mathbf{n}_{22}' \end{bmatrix}$$

and using (12-66a) we can reduce this expression to

$$m_2'' = -\mathbf{u}_{21}'' \mathbf{n}_{11} m_1' + m_2' + \mathbf{u}_{21}'' \mathbf{n}_{11} m_1 = m_2' + \mathbf{u}_{21}''([\mathbf{n}_m' - \mathbf{n}_m''] m_1' + \mathbf{n}_{11} m_1) \ .$$

Transposing the formula for $\mathbf{u}_{12}''$ given in the comment on (12-66) and then using (12-66b) we have

$$\begin{aligned} m_2'' &= m_2' - \mathbf{n}_{22}'^{-1} \mathbf{n}_{21}' \mathbf{n}_m''^{-1}([\mathbf{n}_m' - \mathbf{n}_m''] m_1' + \mathbf{n}_{11} m_1) \\ &= m_2' - \mathbf{n}_{22}'^{-1} \mathbf{n}_{21}'(m_1'' - m_1') \end{aligned}$$

as was to be shown. ◀

CHAPTER 13

Normal Regression Process

13.1. Introduction

13.1.1. *Definition of the Normal Regression Process*

An r-dimensional Normal Regression process can be defined as a process generating independent scalar random variables $\tilde{y}_1, \cdots, \tilde{y}_i, \cdots$ according to the model

$$\tilde{y}_i = \Sigma_{j=1}^{r} x_{ij} \beta_j + \tilde{\epsilon}_i \ , \tag{13-1}$$

where the βs are parameters whose values remain fixed during an entire experiment, the xs are known numbers which in general may vary from one observation to the next although usually (but not necessarily) x_{i1} is a dummy to which the value 1 is assigned on each observation, and the $\tilde{\epsilon}$s are independent random variables with identical Normal densities

$$f_N(\epsilon|0, h) = (2\pi)^{-\frac{1}{2}} e^{-\frac{1}{2}h\epsilon^2} h^{\frac{1}{2}} \ . \tag{13-2}$$

The parameter h will be called the *precision* of the process.

If we define the vectors

$$\begin{aligned} \mathbf{x}_i &\equiv [x_{i1} \quad x_{i2} \quad \cdots \quad x_{ir}]^t \ , \\ \boldsymbol{\beta} &\equiv [\beta_1 \quad \beta_2 \quad \cdots \quad \beta_r]^t \ , \end{aligned} \tag{13-3}$$

the model (13-1) of the ith observation can be written

$$\tilde{y}_i = \mathbf{x}_i^t \boldsymbol{\beta} + \tilde{\epsilon}_i \ . \tag{13-4}$$

It then follows from (13-2) that the density of $\tilde{y}_i$ is

$$f_N(y_i|\mathbf{x}_i^t \boldsymbol{\beta}, h) = (2\pi)^{-\frac{1}{2}} e^{-\frac{1}{2}h(y_i - \mathbf{x}_i^t\boldsymbol{\beta})^2} h^{\frac{1}{2}} \ . \tag{13-5}$$

13.1.2. *Likelihood of a Sample*

The likelihood that a Normal Regression process will generate n values $y_1, \cdots, y_i, \cdots, y_n$ in that order is the product of the likelihood of the individual values as given by (13-5):

$$(2\pi)^{-\frac{1}{2}n} e^{-\frac{1}{2}h\Sigma(y_i - \mathbf{x}_i^t\boldsymbol{\beta})^2} h^{\frac{1}{2}n} \ . \tag{13-6}$$

Provided that the process by which the $\mathbf{x}$s were chosen is noninformative in the sense of Section 2.3, this is the likelihood of the sample described by the n ys and n $\mathbf{x}$s.

The factors in (13-6) which vary with the parameters $\boldsymbol{\beta}$ and h are

$$e^{-\frac{1}{2}h\Sigma(y_i - x_i{}^t\beta)^2}\, h^{\frac{1}{2}n} \,, \tag{13-7}$$

and this can therefore be taken as the kernel of the likelihood. To put this kernel in a form easier to manipulate, we define the vector of all ys observed during an experiment,

$$\boldsymbol{y} \equiv [y_1 \quad \cdots \quad y_i \quad \cdots \quad y_n]^t \tag{13-8a}$$

and the matrix of all the $\boldsymbol{x}$s,

$$\mathbf{X} = \begin{bmatrix} \boldsymbol{x}_1^t \\ \cdots \\ \boldsymbol{x}_i^t \\ \cdots \\ \boldsymbol{x}_n^t \end{bmatrix} = \begin{bmatrix} x_{11} & \cdots & x_{1j} & \cdots & x_{1r} \\ \cdots & \cdots & \cdots & \cdots & \cdots \\ x_{i1} & \cdots & x_{ij} & \cdots & x_{ir} \\ \cdots & \cdots & \cdots & \cdots & \cdots \\ x_{n1} & \cdots & x_{nj} & \cdots & x_{nr} \end{bmatrix} . \tag{13-8b}$$

The kernel (13-7) can then be written

$$e^{-\frac{1}{2}h(\boldsymbol{y} - \mathbf{X}\boldsymbol{\beta})^t(\boldsymbol{y} - \mathbf{X}\boldsymbol{\beta})}\, h^{\frac{1}{2}n} \,. \tag{13-9}$$

If now we let $\boldsymbol{b}$ denote *any* solution of the "normal equations"

$$\mathbf{X}^t\,\mathbf{X}\,\boldsymbol{b} = \mathbf{X}^t\,\boldsymbol{y} \,, \qquad \text{where } \boldsymbol{b} \text{ is } r \times 1 \,, \tag{13-10a}$$

and define

$$\mathbf{n} \equiv \mathbf{X}^t\,\mathbf{X} \,, \qquad \text{so that } \mathbf{n} \text{ is } r \times r \,, \tag{13-10b}$$

$$p \equiv \text{rank}(\mathbf{X}) = \text{rank}(\mathbf{n}) \,, \tag{13-10c}$$

$$\nu \equiv n - p \,, \tag{13-10d}$$

$$v \equiv \frac{1}{\nu}(\boldsymbol{y} - \mathbf{X}\,\boldsymbol{b})^t(\boldsymbol{y} - \mathbf{X}\,\boldsymbol{b}) \,, \tag{13-10e}$$

the kernel (13-9) can be written in the form

$$e^{-\frac{1}{2}h(\boldsymbol{y} - \mathbf{X}\boldsymbol{b})^t(\boldsymbol{y} - \mathbf{X}\boldsymbol{b})}\, h^{\frac{1}{2}\nu}\, e^{-\frac{1}{2}h(\mathbf{X}\boldsymbol{b} - \mathbf{X}\boldsymbol{\beta})^t(\mathbf{X}\boldsymbol{b} - \mathbf{X}\boldsymbol{\beta})}\, h^{\frac{1}{2}p} \tag{13-11a}$$

or equivalently

$$e^{-\frac{1}{2}h\nu v}\, h^{\frac{1}{2}\nu}\, e^{-\frac{1}{2}h(\boldsymbol{b} - \boldsymbol{\beta})^t\mathbf{n}(\boldsymbol{b} - \boldsymbol{\beta})}\, h^{\frac{1}{2}p} \,. \tag{13-11b}$$

If the $n \times r$ matrix $\mathbf{X}$ is of rank r, then $\mathbf{X}^t\,\mathbf{X} = \mathbf{n}$ has an inverse and $\boldsymbol{b}$ is uniquely determined by

$$\boldsymbol{b} = \mathbf{n}^{-1}\,\mathbf{X}^t\,\boldsymbol{y} \,. \tag{13-12}$$

If $\text{rank}(\mathbf{X}) = p < r$, there will be an infinity of $\boldsymbol{b}$s satisfying (13-10a); we can assign arbitrary values to $r - p$ components of $\boldsymbol{b}$ and then solve for the remaining p components in terms of these. In either case, however, the $n \times 1$ vector $\mathbf{X}\,\boldsymbol{b}$ defined by (13-10a) is unique and there is no ambiguity about the sample likelihood as given by either version of (13-11).

▶ To prove (13-11), observe that the exponent of e in (13-9) can be written as $-\frac{1}{2}h$ multiplied by

$$\begin{aligned}(\boldsymbol{y} - \mathbf{X}\,\boldsymbol{b} + \mathbf{X}\,\boldsymbol{b} - \mathbf{X}\,\boldsymbol{\beta})^t(\boldsymbol{y} - \mathbf{X}\,\boldsymbol{b} + \mathbf{X}\,\boldsymbol{b} - \mathbf{X}\,\boldsymbol{\beta}) \\ = (\boldsymbol{y} - \mathbf{X}\,\boldsymbol{b})^t(\boldsymbol{y} - \mathbf{X}\,\boldsymbol{b}) + (\mathbf{X}\,\boldsymbol{b} - \mathbf{X}\,\boldsymbol{\beta})^t(\mathbf{X}\,\boldsymbol{b} - \mathbf{X}\,\boldsymbol{\beta}) \\ + (\boldsymbol{y} - \mathbf{X}\,\boldsymbol{b})^t\,\mathbf{X}(\boldsymbol{b} - \boldsymbol{\beta}) + (\boldsymbol{b} - \boldsymbol{\beta})^t\,\mathbf{X}^t(\boldsymbol{y} - \mathbf{X}\,\boldsymbol{b}) \,,\end{aligned}$$

and the last two terms on the right vanish because

$$\mathbf{X}^t(y - \mathbf{X}\,b) = [(y - \mathbf{X}\,b)^t\,\mathbf{X}]^t = 0$$

by (13-10a).

That the "normal equations" (13-10a) always have a solution and that $\mathbf{X}\,b$ is unique even if b is not can easily be seen by thinking of y as a point in Euclidean n-space $R^{(n)}$. If $\mathbf{X}$ is of rank p, the columns of $\mathbf{X}$ span a p-dimensional subspace $R^{(p)}$ within $R^{(n)}$, and (13-10a) defines $\mathbf{X}\,b$ as the projection of y on $R^{(p)}$ by requiring that $(y - \mathbf{X}\,b)$ be orthogonal to the columns of $\mathbf{X}$. This projection necessarily exists and is unique even though its representation $\mathbf{X}\,b$ as a linear combination b of all r columns of $\mathbf{X}$ will not be unique unless these columns are *just* sufficient in number to span $R^{(p)}$, i.e., unless $p = r$. ◀

13.1.3. Analogy With the Multinormal Process

Comparison of the kernel (13-11b) of a sample from the Normal Regression process with the kernel (12-8) of a sample from the Independent Multinormal process shows that the two are formally identical when $\mathbf{X}$ and therefore $\mathbf{n}$ are of full rank r. It follows that in this case the entire analysis of the Regression problem will be *formally* identical to the analysis in Chapters 12A and 12B of the Multinormal process; but there is nevertheless an important *practical* difference between the two problems: computation of the Multinormal statistic m is a real aid to data processing, whereas computation of the Regression statistic b by solution of the normal equations (13-10a) is too laborious to be worth while. We shall therefore present most of our results both in terms of b, to show the analogy between the two problems, and in terms of y, for practical applications.

When $\mathbf{X}$ and $\mathbf{n}$ are of rank $p < r$ and b is not unique, there is a very close resemblance between our present problem and the problem of using the Multinormal model to represent r distinct univariate Normal processes on only $p < r$ of which observations are actually taken. The statistic $\mathbf{n}$ defined for this latter problem in (12-60) is of rank p just like the statistic $\mathbf{n}$ in our present problem, and the statistic m there defined contains p "real" and $r - p$ arbitrary elements just like the statistic b of our present problem; and it will turn out that these partly arbitrary statistics behave in our present problem just as they did in the previous problem. For simple posterior analysis, they can be used exactly like ordinary statistics; but their sampling distributions require special treatment as do the preposterior distributions based on these sampling distributions.

A. Normal Regression Process, Precision Known

13.2. Prior and Posterior Analysis

13.2.1. Likelihood of a Sample When h Is Known

When the process precision h is known, the only factor in the complete sample kernel (13-11) which varies with unknown parameters is

$$e^{-\frac{1}{2}h(\mathbf{X}b - \mathbf{X}\beta)^t(\mathbf{X}b - \mathbf{X}\beta)} \tag{13-13a}$$

or equivalently

$$e^{-\frac{1}{2}h(b-\beta)^t\mathbf{n}(b-\beta)} \tag{13-13b}$$

Either of these expressions can therefore be taken as the kernel when h is known: the unique statistic $(\mathbf{X}, \mathbf{X}\, b)$ or *any* statistic $(b, \mathbf{n})$ satisfying (13-10a) is sufficient when h is known, whether or not b itself is unique.

13.2.2. Distribution of $\tilde{\beta}$

When the process precision h is known but the regression coefficient $\tilde{\beta}$ is a random variable, the most convenient distribution of $\tilde{\beta}$—the natural conjugate of the sample kernel (13-13b)—is a Normal distribution as defined by (8-17):

$$f_N(\beta|b, h\mathbf{n}) \propto e^{-\frac{1}{2}h(\beta - b)^t \mathbf{n}(\beta - b)} , \qquad (13\text{-}14)$$

where $\mathbf{n}$ expresses the information on β in units of the *process* precision h; cf. Section 12.1.4. The mean and variance of this distribution are, by (8-18):

$$\begin{aligned} E(\tilde{\beta}|b, \mathbf{n}) &= b \ , \\ V(\tilde{\beta}|b, \mathbf{n}) &= (h\mathbf{n})^{-1} \ . \end{aligned} \qquad (13\text{-}15)$$

If a Normal prior distribution with parameters $(b', \mathbf{n}')$ is assigned to $\tilde{\beta}$ and if a sample then yields data $(\mathbf{X}, y)$ or a sufficient statistic $(b, \mathbf{n})$, the posterior distribution of $\tilde{\beta}$ will be Normal with parameters

$$\mathbf{n}'' = \mathbf{n}' + \mathbf{X}^t\, \mathbf{X} = \mathbf{n}' + \mathbf{n} \ , \qquad (13\text{-}16a)$$
$$b'' = \mathbf{n}''^{-1}(\mathbf{n}'b' + \mathbf{X}^t\, y) = \mathbf{n}''^{-1}(\mathbf{n}'b' + \mathbf{n}\, b) \ . \qquad (13\text{-}16b)$$

▶ To prove (13-16) we can start from the sample kernel in the form (13-13b) and then use identically the same argument used to prove (12-14) with β substituted for μ and b for m; notice that this proof does not require the existence of $\mathbf{n}^{-1}$ or the uniqueness of b.

Alternatively, we can prove (13-16) without even formal reference to b by using the likelihood in the form (13-9). Multiplying the kernel of this likelihood by the kernel of the prior density (13-14) we obtain $e^{-\frac{1}{2}hT}$ where

(1) $$T = (y - \mathbf{X}\,\beta)^t(y - \mathbf{X}\,\beta) + (\beta - b')^t\,\mathbf{n}'(\beta - b') \ .$$

Regrouping terms we can write

(2) $$T = T_1 + T_2$$

where

(3) $$\begin{aligned} T_1 &= (\beta - [\mathbf{n}' + \mathbf{X}^t\,\mathbf{X}]^{-1}[\mathbf{n}'b' + \mathbf{X}^t\,y])^t(\mathbf{n}' + \mathbf{X}^t\,\mathbf{X})(\beta - [\mathbf{n}' + \mathbf{X}^t\,\mathbf{X}]^{-1}[\mathbf{n}'b' + \mathbf{X}^t\,y]) \\ &= (\beta - b'')^t\,\mathbf{n}''(\beta - b'') \ , \end{aligned}$$

(4) $$\begin{aligned} T_2 &= b'^t\,\mathbf{n}'b' + y^t\,y - (\mathbf{n}'b' + \mathbf{X}^t\,y)^t(\mathbf{n}' + \mathbf{X}^t\,\mathbf{X})^{-1}(\mathbf{n}'b' + \mathbf{X}^t\,y) \\ &= b'^t\,\mathbf{n}'b' + y^t\,y - b''^t\,\mathbf{n}''b'' \ . \end{aligned}$$

1. Since β appears only in T_1, the kernel of the conditional distribution of $(\tilde{\beta}|y)$ is $e^{-\frac{1}{2}hT_1}$ and the distribution will therefore be nondegenerate Normal with parameters (13-16) provided that $\mathbf{n}'' = (\mathbf{n}' + \mathbf{X}^t\,\mathbf{X})$ is positive-definite. Since $\mathbf{X}^t\,\mathbf{X}$ is positive-semidefinite by its form, $\mathbf{n}''$ will be positive-definite if *either* (a) $\mathbf{n}'$ is positive-definite, i.e., if the prior distribution is proper, *or* (b) $\mathbf{X}$ is of rank r, i.e., if the observations y suffice to establish a distribution of $\tilde{\beta}$ even though the prior parameter $\mathbf{n}' = \mathbf{0}$. The matrix $\mathbf{n}''$ may of course be positive-definite even though *neither* $\mathbf{n}'$ nor $\mathbf{X}^t\,\mathbf{X}$ is positive-definite; but this can happen only if $\mathbf{n}'$ is of rank less than r but greater than 0, and such prior distributions will rarely occur in practice.

2. Although this completes the proof of (13-16), we anticipate future needs by giving

an alternate factorization of the sum T_2 which plays no role in the distribution of $(\tilde{\boldsymbol\beta}|y)$. Writing $\mathbf{n}''^{-1}$ in place of $(\mathbf{n}' + \mathbf{X}^t\,\mathbf{X})^{-1}$ in the last term of the middle member of (4) and then expanding we obtain

$$\begin{aligned} T_2 &= b'^t\,\mathbf{n}'b' + y^t\,y - b'^t\,\mathbf{n}'\mathbf{n}''^{-1}\,\mathbf{n}'b' \\ &\quad - b'^t\,\mathbf{n}'\mathbf{n}''^{-1}\,\mathbf{X}^t\,y - y^t\,\mathbf{X}\,\mathbf{n}''^{-1}\,\mathbf{n}'b' - y^t\,\mathbf{X}\,\mathbf{n}''^{-1}\,\mathbf{X}^t\,y \\ &= y^t(\mathbf{I} - \mathbf{X}\,\mathbf{n}''^{-1}\,\mathbf{X}^t)\,y - y^t\,\mathbf{X}\,\mathbf{n}''^{-1}\,\mathbf{n}'b' \\ &\quad - b'^t\,\mathbf{n}'\mathbf{n}''^{-1}\,\mathbf{X}^t\,y + b'^t(\mathbf{n}' - \mathbf{n}'\mathbf{n}''^{-1}\,\mathbf{n}')\,b' \ . \end{aligned}$$

Substituting $(\mathbf{n}'' - \mathbf{X}^t\,\mathbf{X})$ for every $\mathbf{n}'$ on the right-hand side of this result we reduce it to

$$\begin{aligned} T_2 &= y^t(\mathbf{I} - \mathbf{X}\,\mathbf{n}''^{-1}\,\mathbf{X}^t)\,y - y^t\,\mathbf{X}(\mathbf{I} - \mathbf{n}''^{-1}\,\mathbf{X}^t\,\mathbf{X})\,b' \\ &\quad - b'^t(\mathbf{I} - \mathbf{X}^t\,\mathbf{X}\,\mathbf{n}''^{-1})\,\mathbf{X}^t\,y + b'^t(\mathbf{X}^t\,\mathbf{X} - \mathbf{X}^t\,\mathbf{X}\,\mathbf{n}''^{-1}\,\mathbf{X}^t\,\mathbf{X})\,b' \\ &= (y - \mathbf{X}\,b')^t(\mathbf{I} - \mathbf{X}\,\mathbf{n}''^{-1}\,\mathbf{X}^t)(y - \mathbf{X}\,\mathbf{b}') \ . \end{aligned} \tag{5}$$

We shall see later that, when T_2 is written in this form, $e^{-\frac{1}{2}hT_2}$ is the kernel of the marginal or unconditional distribution of $\tilde{y}$. ◀

13.3. Sampling Distributions With Fixed X

In this section we assume that a sample with predetermined $\mathbf{X}$ and therefore with predetermined $\mathbf{n} = \mathbf{X}^t\,\mathbf{X}$ is to be drawn from an r-dimensional Normal Regression process whose precision h is known but whose regression coefficient $\tilde{\boldsymbol\beta}$ is a random variable having a proper Normal distribution with parameter $(b', \mathbf{n}')$. The fact that the distribution is proper implies that $\mathbf{n}'$ is PDS and thus of rank r.

13.3.1. Conditional Distribution of $(\tilde{y}|\boldsymbol\beta)$

Because the individual $\tilde{y}$s are independent by the definition of the Normal regression process, the joint density of n $\tilde{y}$s for a given value of the process parameter $\boldsymbol\beta$ is the product of the individual densities as given by (13-5), and the definitions (13-8) of y and $\mathbf{X}$ make it clear that this product can be written

$$\mathrm{D}(y|\boldsymbol\beta;\mathbf{X}) = f_N^{(n)}(y|\mathbf{X}\,\boldsymbol\beta, h\mathbf{I}) \ . \tag{13-17}$$

The mean and variance of the distribution are

$$\begin{aligned} \mathrm{E}(\tilde{y}|\boldsymbol\beta;\mathbf{X}) &= \mathbf{X}\,\boldsymbol\beta \ , \\ \mathrm{V}(\tilde{y}|\boldsymbol\beta;\mathbf{X}) &= (h\mathbf{I})^{-1} \ . \end{aligned} \tag{13-18}$$

These results hold whether $\mathbf{X}$ is of rank r or of rank $p < r$.

13.3.2. Unconditional Distribution of $\tilde{y}$

The *unconditional* density of $\tilde{y}$—i.e., the density given $\mathbf{X}$ but not $\boldsymbol\beta$—is

$$\begin{aligned} \mathrm{D}(y|b', \mathbf{n}';\mathbf{X}) &\equiv \int_{-\infty}^{\infty} f_N^{(n)}(y|\mathbf{X}\,\boldsymbol\beta, h\mathbf{I})\, f_N^{(r)}(\boldsymbol\beta|b', h\mathbf{n}')\, d\boldsymbol\beta \\ &= f_N^{(n)}(y|\mathbf{X}\,b', h\mathbf{n}_y) \end{aligned} \tag{13-19a}$$

where

$$\mathbf{n}_y \equiv \mathbf{I} - \mathbf{X}\,\mathbf{n}''^{-1}\,\mathbf{X}^t \ , \qquad \mathbf{n}_y^{-1} = \mathbf{X}\,\mathbf{n}'^{-1}\,\mathbf{X}^t + \mathbf{I} \ . \tag{13-19b}$$

The mean and variance of this distribution are

$$\begin{aligned} E(\tilde{y}|b', n'; X) &= X\, b' , \\ V(\tilde{y}|b', n'; X) &= (h n_y)^{-1} . \end{aligned} \tag{13-20}$$

Again these results apply whether $\mathbf{X}$ is of rank r or of rank $p < r$.

▶ To show that the marginal distribution of $\tilde{y}$ is Normal we need only recall (1) that the joint distribution of $(\tilde{y}, \tilde{\beta})$ is Normal because the distributions of $\tilde{\beta}$ and of $(\tilde{y}|\beta)$ are Normal, and (2) that any marginal distribution of a joint Normal distribution is Normal.

To find the parameters of the distribution of $\tilde{y}$ we substitute the formula (8-17) for $f_N(y)$ and $f_N(\beta)$ in the integrand of (13-19a), drop factors constant as regards y and β, and thus find that the kernel of the joint density of $\tilde{\beta}$ and $\tilde{y}$ can be written $e^{-\frac{1}{2}hT^*}$ where

$$T^* = (y - X\beta)^t(y - X\beta) + (\beta - b')^t n'(\beta - b') .$$

Comparison with formula (1) in the proof of (13-16) shows that T^* as defined above is identical to the T of that formula; and we may therefore use the results (3) and (5) obtained in that proof to write

$$T^* = T_1 + T_2$$

where

$$\begin{aligned} T_1 &= (\beta - b'')^t n''(\beta - b'') , \\ T_2 &= (y - X\, b')^t(I - X\, n''^{-1} X^t)(y - X\, b') . \end{aligned}$$

Since the integral of $e^{-\frac{1}{2}hT_1}$ over $-\infty < \beta < \infty$ is a constant as regards y, the kernel of the marginal or unconditional density of $\tilde{y}$ is simply $e^{-\frac{1}{2}hT_2}$, in accordance with (13-19).

To prove that the formula for n_y^{-1} in (13-19b) follows from the definition there given of n_y itself, i.e., to prove that

$$I - X\, n''^{-1} X^t = (X\, n'^{-1} X^t + I)^{-1} ,$$

we multiply both sides by the nonsingular matrix $(X\, n'^{-1} X^t + I)$ and obtain

$$(I - X\, n''^{-1} X^t)(X\, n'^{-1} X^t + I) = I ,$$

implying that

$$X\, n'^{-1} X^t = X(n''^{-1} + n''^{-1} X^t X\, n'^{-1})\, X^t .$$

The equation is now proved by substituting $(n'' - n')$ for $X^t X$ inside the parentheses.

Alternate proof. (13-19) can also be proved by making use of the facts (1) that

$$\tilde{y} = X\tilde{\beta} + \tilde{\epsilon}$$

is by definition a sum of independent Normal random variables, (2) that $\tilde{\epsilon}$ has density $f_N(\epsilon|0, hI)$, and (3) that the density of $X\tilde{\beta}$ can be found from the fact that $\tilde{\beta}$ has density $f_N(\beta|b', hn')$. From (1) it follows immediately that $\tilde{y}$ is Normal and from (2) and (3) that its mean and variance are

$$\begin{aligned} E(\tilde{y}) &= E(X\tilde{\beta}) + E(\tilde{\epsilon}) = X\, E(\tilde{\beta}) + 0 = X\, b' , \\ V(\tilde{y}) &= V(X\tilde{\beta}) + V(\tilde{\epsilon}) = X\, V(\tilde{\beta})\, X^t + (hI)^{-1} \\ &= X(hn')^{-1} X^t + (hI)^{-1} = (hn_y)^{-1} . \end{aligned}$$

The parameters of (13-19) are then determined by these values of the moments. ◀

13.3.3. Distributions of $\tilde{b}$ When $\mathbf{X}$ Is of Rank r

When $\mathbf{X}$ is of rank r and the statistic b is unique, the *conditional* distribution of $\tilde{b}$ for a given value of the process parameter β is given by

$$D(b|\beta; \mathbf{X}) = f_N^{(r)}(b|\beta, h\mathbf{n}) \tag{13-21}$$

where $\mathbf{n} = \mathbf{X}^t\mathbf{X}$. The *unconditional* distribution of $\tilde{b}$, given $\mathbf{X}$ but not β, is

$$D(b|b', \mathbf{n}'; \mathbf{X}) = f_N^{(r)}(b|b', h\mathbf{n}_u) \tag{13-22a}$$

where

$$\mathbf{n}_u^{-1} \equiv \mathbf{n}'^{-1} + \mathbf{n}^{-1}, \qquad \mathbf{n}_u = \mathbf{n}'\mathbf{n}''^{-1}\mathbf{n} = \mathbf{n}\,\mathbf{n}''^{-1}\mathbf{n}' . \tag{13-22b}$$

For the corresponding formulas for the case where $\mathbf{X}$ is of rank $p < r$, see (13-51) and (13-56).

▶ To prove (13-21) we first observe that when $\mathbf{X}$ is of rank r the statistic b has the definition (13-12)

$$\tilde{b} = \mathbf{n}^{-1}\mathbf{X}^t\tilde{y}$$

and is thus a linear combination of Normal random variables. Now $\mathbf{n}^{-1}\mathbf{X}^t$ is an $r \times r$ matrix of rank r and r cannot exceed the number n of the rows of $\mathbf{X}$; and since the dimensionality of y is n, it follows by (8-26) in conjunction with (13-18) that $\tilde{b}$ is Normal with mean and variance

$$\begin{aligned} E(\tilde{b}|\beta, \mathbf{n}) &= \mathbf{n}^{-1}\mathbf{X}^t E(\tilde{y}|\beta) = \mathbf{n}^{-1}\mathbf{X}^t\mathbf{X}\beta = \beta , \\ V(\tilde{b}|\beta, \mathbf{n}) &= \mathbf{n}^{-1}\mathbf{X}^t V(\tilde{y}|\beta)\mathbf{X}\mathbf{n}^{-1} = \mathbf{n}^{-1}\mathbf{X}^t(h\mathbf{I})^{-1}\mathbf{X}\mathbf{n}^{-1} \\ &= h^{-1}\mathbf{n}^{-1}\mathbf{n}\,\mathbf{n}^{-1} = (h\mathbf{n})^{-1} . \end{aligned}$$

To prove (13-22) we use (8-26) in conjunction with (13-20):

$$\begin{aligned} E(\tilde{b}|b', \mathbf{n}'; \mathbf{n}) &= \mathbf{n}^{-1}\mathbf{X}^t(\mathbf{X}\,b') = b' , \\ V(\tilde{b}|b', \mathbf{n}'; \mathbf{n}) &= \mathbf{n}^{-1}\mathbf{X}^t\frac{1}{h}(\mathbf{I} + \mathbf{X}\mathbf{n}'^{-1}\mathbf{X}^t)\mathbf{X}\mathbf{n}^{-1} = \frac{1}{h}\mathbf{n}^{-1}(\mathbf{n} + \mathbf{n}\,\mathbf{n}'^{-1}\mathbf{n})\,\mathbf{n}^{-1} \\ &= \frac{1}{h}(\mathbf{n}^{-1} + \mathbf{n}'^{-1}) \equiv (h\mathbf{n}_u)^{-1} . \end{aligned}$$

Alternatively we could argue that $\tilde{\beta}$ and

$$\tilde{e} \equiv \tilde{b} - \tilde{\beta} = \mathbf{n}^{-1}\mathbf{X}^t(\mathbf{X}\tilde{\beta} + \tilde{\epsilon}) - \tilde{\beta} = \mathbf{n}^{-1}\mathbf{X}^t\tilde{\epsilon}$$

are independent Normal random variables and therefore that

$$\tilde{b} = \tilde{\beta} + \tilde{e}$$

is Normal with mean and variance

$$\begin{aligned} E(\tilde{b}) &= E(\tilde{\beta}) + E(\tilde{e}) , \\ V(\tilde{b}) &= V(\tilde{\beta}) + V(\tilde{e}) . \end{aligned}$$

The mean and variance of $\tilde{\beta}$ are given by (13-15) as

$$E(\tilde{\beta}) = b' , \qquad V(\tilde{\beta}) = (h\mathbf{n}')^{-1} ;$$

and from the fact that $\tilde{\epsilon}$ has density $f_N(\epsilon|0, h\mathbf{I})$ we can compute

$$\begin{aligned} E(\tilde{e}) &= \mathbf{n}^{-1}\mathbf{X}^t E(\tilde{\epsilon}) = \mathbf{n}^{-1}\mathbf{X}^t 0 = 0 , \\ V(\tilde{e}) &= \mathbf{n}^{-1}\mathbf{X}^t V(\tilde{\epsilon})\mathbf{X}\mathbf{n}^{-1} = \mathbf{n}^{-1}\mathbf{X}^t(h\mathbf{I})^{-1}\mathbf{X}\mathbf{n}^{-1} = (h\mathbf{n})^{-1} . \end{aligned}$$

Substituting these results in the previous formulas for the mean and variance of $\tilde{b}$ we obtain

$$E(\tilde{b}) = b' , \qquad V(\tilde{b}) = \frac{1}{h}(\mathbf{n}'^{-1} + \mathbf{n}^{-1}) \equiv (h\mathbf{n}_u)^{-1} .$$

From the definition

$$\mathbf{n}_u^{-1} \equiv \mathbf{n}'^{-1} + \mathbf{n}^{-1} \; ,$$

the alternative formulas in (13-22b) are easily derived with the aid of (8-1):

$$\begin{aligned} \mathbf{n}_u = (\mathbf{n}'^{-1} + \mathbf{n}^{-1})^{-1} &= \mathbf{n}'(\mathbf{n}' + \mathbf{n})^{-1}\mathbf{n} = \mathbf{n}'\,\mathbf{n}''^{-1}\mathbf{n} \\ &= \mathbf{n}(\mathbf{n}' + \mathbf{n})^{-1}\mathbf{n}' = \mathbf{n}\,\mathbf{n}''^{-1}\mathbf{n}' \end{aligned}$$ ◀

13.4. Preposterior Analysis With Fixed X of Rank r

In this section we assume as in the previous section that a sample with predetermined $\mathbf{X}$ is to be drawn from an r-dimensional Normal Regression process whose precision h is known but whose regression coefficient $\tilde{\boldsymbol{\beta}}$ is a random variable having a Normal distribution with parameter $(\boldsymbol{b}', \mathbf{n}')$ where $\mathbf{n}'$ is PDS and therefore of rank r. We also assume in this section that $\mathbf{X}$ and therefore $\mathbf{n} = \mathbf{X}^t\,\mathbf{X}$ are of rank r, and our results will *not* apply without change if this assumption is violated.

13.4.1. Distribution of $\tilde{\boldsymbol{b}}''$

The parameter $\boldsymbol{b}''$ of the posterior distribution of $\tilde{\boldsymbol{\beta}}$ will, by (13-16), have the value

$$\boldsymbol{b}'' = \mathbf{n}''^{-1}\,\mathbf{n}'\boldsymbol{b}' + \mathbf{n}''^{-1}\,\mathbf{X}^t\,\boldsymbol{y} \; , \tag{13-23}$$

so that before $\tilde{\boldsymbol{y}}$ is known we can obtain the distribution of $\tilde{\boldsymbol{b}}''$ from the unconditional distribution (13-19) of $\tilde{\boldsymbol{y}}$. When $\mathbf{X}$ and therefore $\mathbf{n}''^{-1}\,\mathbf{X}^t$ are of rank r, the distribution so determined is Normal with density

$$\mathrm{D}(\boldsymbol{b}''|\boldsymbol{b}', \mathbf{n}'; \mathbf{X}) = f_N^{(r)}(\boldsymbol{b}''|\boldsymbol{b}', h\mathbf{n}^*) \tag{13-24a}$$

where

$$\mathbf{n}^* \equiv \mathbf{n}''\mathbf{n}^{-1}\,\mathbf{n}' = \mathbf{n}'\mathbf{n}^{-1}\,\mathbf{n}'' \; , \qquad \mathbf{n}^{*-1} = \mathbf{n}'^{-1} - \mathbf{n}''^{-1} \; . \tag{13-24b}$$

For the case where $\mathbf{X}$ is of rank $p < r$, see (13-60).

▶ Since $\tilde{\boldsymbol{b}}''$ as defined by (13-23) is an $r \times 1$ linear combination of $n \geq r$ Normal random variables, it follows from (8-26) in conjunction with (13-20) that if the $r \times n$ matrix $\mathbf{n}''^{-1}\,\mathbf{X}^t$ is of rank r, then $\tilde{\boldsymbol{b}}''$ is Normal with mean and variance

$$\mathrm{E}(\tilde{\boldsymbol{b}}'') = \mathbf{n}''^{-1}(\mathbf{n}'\boldsymbol{b}' + \mathbf{X}^t\mathrm{E}[\tilde{\boldsymbol{y}}]) = \mathbf{n}''^{-1}(\mathbf{n}'\boldsymbol{b}' + \mathbf{X}^t\mathbf{X}\,\boldsymbol{b}') = \boldsymbol{b}' \; ,$$

$$\begin{aligned} \mathrm{V}(\tilde{\boldsymbol{b}}'') &= \mathbf{n}''^{-1}\,\mathbf{X}^t\,\mathrm{V}(\tilde{\boldsymbol{y}})\,\mathbf{X}\,\mathbf{n}''^{-1} = \mathbf{n}''^{-1}\,\mathbf{X}^t\,\frac{1}{h}\,(\mathbf{X}\,\mathbf{n}'^{-1}\,\mathbf{X}^t + \mathbf{I})\,\mathbf{X}\,\mathbf{n}''^{-1} \\ &= \frac{1}{h}\mathbf{n}''^{-1}(\mathbf{n}\,\mathbf{n}'^{-1}\mathbf{n} + \mathbf{n})\,\mathbf{n}''^{-1} = \frac{1}{h}\mathbf{n}''^{-1}([\mathbf{n}'' - \mathbf{n}']\,\mathbf{n}'^{-1}\mathbf{n} + \mathbf{n})\,\mathbf{n}''^{-1} \\ &= \frac{1}{h}\mathbf{n}'^{-1}\,\mathbf{n}\,\mathbf{n}''^{-1} \; . \end{aligned}$$

Because $\mathbf{n}$ is of full rank, we may define

$$\mathbf{n}^* \equiv \mathbf{n}''\,\mathbf{n}^{-1}\,\mathbf{n}'$$

and write our last result in the form

$$V(\bar{\tilde{b}}'') = (h\mathbf{n}^*)^{-1} .$$

For the derivation of the alternative formulas for $\mathbf{n}^*$, see the proof of (12-20). ◀

13.4.2. Distribution of $\bar{\tilde{\beta}}''$

The mean of the posterior distribution of $\tilde{\beta}$ will, by (13-15), be equal to the parameter b'' of that distribution; and the prior distribution of $\bar{\tilde{\beta}}''$ can therefore be obtained by simply substituting $\tilde{\beta}''$ for b'' in (13-24). The mean and variance of the distribution of $\bar{\tilde{\beta}}''$ thus determined are, by (8-18),

$$\begin{aligned} E(\bar{\tilde{\beta}}''|b', \mathbf{n}'; \mathbf{X}) &= b' = E(\tilde{\beta}|b', \mathbf{n}') , \\ V(\bar{\tilde{\beta}}''|b', \mathbf{n}'; \mathbf{X}) &= (h\mathbf{n}^*)^{-1} = V(\tilde{\beta}|b', \mathbf{n}') - V(\tilde{\beta}|b'', \mathbf{n}'') , \end{aligned} \tag{13-25}$$

in accordance with the general relations (5-27) and (5-28). These relations constitute, of course, an alternative way of deriving the parameters of (13-24).

B. Normal Regression Process, Precision Unknown

13.5. Prior and Posterior Analysis

13.5.1. Likelihood of a Sample When Neither β Nor h Is Known

We have already seen in (13-9) and (13-11) that the kernel of the likelihood of observations $y_1, \cdots, y_n$ on a Normal Regression process is

$$e^{-\frac{1}{2}h(y-\mathbf{X}\beta)^t(y-\mathbf{X}\beta)}\, h^{\frac{1}{2}n} \tag{13-26}$$

$$= e^{-\frac{1}{2}h(y-\mathbf{X}b)^t(y-\mathbf{X}b)}\, h^{\frac{1}{2}\nu}\, e^{-\frac{1}{2}h(\mathbf{X}b-\mathbf{X}\beta)^t(\mathbf{X}b-\mathbf{X}\beta)}\, h^{\frac{1}{2}p} \tag{13-27}$$

$$= e^{-\frac{1}{2}h\nu v}\, h^{\frac{1}{2}\nu}\, e^{-\frac{1}{2}h(b-\beta)^t\mathbf{n}(b-\beta)}\, h^{\frac{1}{2}p} \tag{13-28}$$

and that these expressions have a unique value whether or not b itself is unique. The statistic $(b, v, \mathbf{n}, \nu)$ is sufficient; $p \equiv \text{rank}(\mathbf{X}) = \text{rank}(\mathbf{n})$ is simply an auxiliary abbreviation.

For exactly the same reasons as in the Multinormal or univariate Normal problem, we may wish to use an incomplete statistic $(b, \mathbf{n})$ or (v, ν) when the complete sufficient statistic $(b, v, \mathbf{n}, \nu)$ is unavailable or irrelevant—cf. Section 12.4.2. To use $(b, \mathbf{n})$ we must substitute for the likelihood of the complete sufficient statistic $(b, v, \mathbf{n}, \nu)$ the marginal likelihood of $(b, \mathbf{n})$, the kernel of which is

$$e^{-\frac{1}{2}h(b-\beta)^t\mathbf{n}(b-\beta)}\, h^{\frac{1}{2}p} = e^{-\frac{1}{2}h(\mathbf{X}b-\mathbf{X}\beta)^t(\mathbf{X}b-\mathbf{X}\beta)}\, h^{\frac{1}{2}p} ; \tag{13-29a}$$

while to use (v, ν) we must use *its* marginal likelihood, the kernel of which is

$$e^{-\frac{1}{2}h\nu v}\, h^{\frac{1}{2}\nu} = e^{-\frac{1}{2}h(y-\mathbf{X}b)^t(y-\mathbf{X}b)}\, h^{\frac{1}{2}\nu} . \tag{13-29b}$$

With the convention that

1. $\nu = 0$ when v is unknown or irrelevant,
2. $\mathbf{n} = \mathbf{0}$ when b is unknown or irrelevant,

the complete kernel (13-27) or (13-28) reduces to (13-29a) in the former case, to (13-29b) in the latter.

▶ To show that the formulas (13-29) give the kernels of the *marginal* likelihoods of $(\boldsymbol{b}, \mathbf{n})$ and (v, ν) we shall prove the stronger proposition that the likelihoods of $(\boldsymbol{b}, \mathbf{n})$ and (v, ν) are *independent* with kernels given by (13-29). To do so we return to the geometry introduced in the proof of (13-11), regard $\boldsymbol{y}$ as a point in Euclidean n-space, and decompose this space into the p-dimensional subspace $R^{(p)}$ spanned by the columns of $\mathbf{X}$ and the orthogonal subspace $R^{(\nu)}$ where $\nu = n - p$. We now observe that because $(\boldsymbol{y} - \mathbf{X}\,\boldsymbol{b})$ is orthogonal to $R^{(p)}$ while $\mathbf{X}\,\boldsymbol{b}$ and $\mathbf{X}\,\boldsymbol{\beta}$ lie wholly within $R^{(p)}$, we may regard $(\mathbf{X}\,\boldsymbol{b} - \mathbf{X}\,\boldsymbol{\beta})$ and $(\boldsymbol{y} - \mathbf{X}\,\boldsymbol{b})$ as the orthogonal components of $(\boldsymbol{y} - \mathbf{X}\,\boldsymbol{\beta})$ in $R^{(p)}$ and $R^{(\nu)}$ respectively. Since $(\tilde{\boldsymbol{y}} - \mathbf{X}\,\boldsymbol{\beta})$ is spherical Normal with mean $\boldsymbol{0}$ and precision h, its orthogonal components are spherical Normal with mean $\boldsymbol{0}$ and precision h, and the right-hand sides of (13-29) are the kernels of these two Normal densities in p-space and ν-space respectively.

This means that (13-29a) is the kernel of the total likelihood of all $\boldsymbol{y}$ in $R^{(n)}$ corresponding to a particular $\mathbf{X}\,\boldsymbol{b}$ in $R^{(p)}$ and that (13-29b) is the kernel of the total likelihood of all $\boldsymbol{y}$ in $R^{(n)}$ corresponding to a particular $(\boldsymbol{y} - \mathbf{X}\,\boldsymbol{b})$ in $R^{(\nu)}$. Since there are many $(\boldsymbol{y} - \mathbf{X}\,\boldsymbol{b})$ in $R^{(\nu)}$ which yield the same value of $(\boldsymbol{y} - \mathbf{X}\,\boldsymbol{b})^t(\boldsymbol{y} - \mathbf{X}\,\boldsymbol{b}) = \nu v$, we would have to integrate (13-29b) over all these points in order to compute the actual marginal likelihood of (v, ν); but since this integral would be simply (13-29b) multiplied by a volume element which does not involve $\boldsymbol{\beta}$ or h, the *kernel* of the marginal likelihood of (v, ν) is simply (13-29b). The same kind of argument shows that we can take (13-29a) as the kernel of the marginal likelihood of $(\boldsymbol{b}, \mathbf{n})$ without even asking whether or not $R^{(p)}$ contains more than one $\mathbf{X}\,\boldsymbol{b}$ corresponding to a given $(\boldsymbol{b}, \mathbf{n})$. ◀

13.5.2. Distribution of $(\tilde{\boldsymbol{\beta}}, \tilde{h})$

The most convenient distribution of $(\tilde{\boldsymbol{\beta}}, \tilde{h})$—the natural conjugate of (13-28)—is a Normal-gamma distribution formally identical to (12-27):

$$f_{N\gamma}^{(r)}(\boldsymbol{\beta}, h|\boldsymbol{b}, v, \mathbf{n}, \nu) \propto e^{-\frac{1}{2}h(\boldsymbol{\beta}-\boldsymbol{b})^t\mathbf{n}(\boldsymbol{\beta}-\boldsymbol{b})}\, h^{\frac{1}{2}p}\, e^{-\frac{1}{2}h\nu v}\, h^{\frac{1}{2}\nu-1}\ , \tag{13-30}$$

where $\mathbf{n}$ measures the information on $\boldsymbol{\beta}$ in units of the *process* precision h and the abbreviation p denotes rank$(\mathbf{n})$. For (13-30) to represent a proper distribution, we must have $p = r$; cf. Section 12.4.3.

If the prior distribution of $(\tilde{\boldsymbol{\beta}}, \tilde{h})$ is Normal-gamma with parameter $(\boldsymbol{b}', v', \mathbf{n}', \nu')$ where rank$(\mathbf{n}') = p'$, and if a sample then yields data $(\boldsymbol{y}, \mathbf{X})$ where $\mathbf{X}$ is of rank p or a sufficient statistic $(\boldsymbol{b}, v, \mathbf{n}, \nu)$ where $\mathbf{n}$ is of rank p, the posterior distribution of $(\tilde{\boldsymbol{\beta}}, \tilde{h})$ will be Normal-gamma with parameters

$$\mathbf{n}'' = \mathbf{n}' + \mathbf{X}^t\,\mathbf{X} = \mathbf{n}' + \mathbf{n}\ , \tag{13-31a}$$

$$\boldsymbol{b}'' = \mathbf{n}''^{-1}(\mathbf{n}'\boldsymbol{b}' + \mathbf{X}^t\,\boldsymbol{y}) = \mathbf{n}''^{-1}(\mathbf{n}'\boldsymbol{b}' + \mathbf{n}\,\boldsymbol{b})\ , \tag{13-31b}$$

$$p'' = \text{rank}(\mathbf{n}'')\ , \tag{13-31c}$$

$$\nu'' = [\nu' + p'] + [\nu + p] - p''\ , \tag{13-31d}$$

$$\begin{aligned} v'' &= \frac{1}{\nu''}\,[(\nu'v' + \boldsymbol{b}'^t\,\mathbf{n}'\boldsymbol{b}') + \boldsymbol{y}^t\,\boldsymbol{y} - \boldsymbol{b}''^t\,\mathbf{n}''\boldsymbol{b}''] \\ &= \frac{1}{\nu''}\,[(\nu'v' + \boldsymbol{b}'^t\,\mathbf{n}'\boldsymbol{b}') + (\nu v + \boldsymbol{b}^t\,\mathbf{n}\,\boldsymbol{b}) - \boldsymbol{b}''^t\,\mathbf{n}''\boldsymbol{b}'']\ . \end{aligned} \tag{13-31e}$$

These results apply regardless of the rank of either $\mathbf{n}'$ or $\mathbf{X}$.

▶ To prove the formulas expressed in terms of y and $\mathbf{X}$, we multiply the sample kernel in the form (13-26) by the prior kernel (13-30), thus obtaining

$$e^{-\frac{1}{2}h(T+\nu'v')}\, h^{\frac{1}{2}[\nu'+p'+n]-1}$$

where, as we saw in the proof of (13-16),

$$T = (\beta - b'')^t\, \mathbf{n}''(\beta - b'') + b'^t\, \mathbf{n}' b' + y^t\, y - b''^t\, \mathbf{n}'' b'' \ .$$

The definitions (13-31) now allow us to write

$$T + \nu' v' = (\beta - b'')^t\, \mathbf{n}''(\beta - b'') + \nu'' v''$$

in agreement with the exponent of e in (13-30); and since $n = p + \nu$ by the definition (13-10d) of ν they also allow us to write

$$\nu' + p + n = \nu' + p' + \nu + p = \nu'' + p''$$

in agreement with the exponent of h in (13-30).

To prove formulas (13-31ab) expressed in terms of b and $\mathbf{n}$, we need only refer to the proof of (13-16). To prove (13-31e) in terms of b and $\mathbf{n}$, we write

$$\begin{aligned} y^t\, y &= ([y - \mathbf{X}\, b] + \mathbf{X}\, b)^t([y - \mathbf{X}\, b] + \mathbf{X}\, b) \\ &= (y - \mathbf{X}\, b)^t(y - \mathbf{X}\, b) + (\mathbf{X}\, b)^t(\mathbf{X}\, b) + (y - \mathbf{X}\, b)^t\, \mathbf{X}\, b + (\mathbf{X}\, b)^t(y - \mathbf{X}\, b) \ . \end{aligned}$$

The first term on the right is νv by the definition (13-10e), the second term is $b^t\, \mathbf{n}\, b$ by (13-10b), and the last two terms vanish because

$$\mathbf{X}^t(y - \mathbf{X}\, b) = \mathbf{X}^t\, y - \mathbf{X}^t\, \mathbf{X}\, b = 0$$

by (13-10a). We may therefore write

$$y^t\, y = \nu v + b^t\, \mathbf{n}\, b \ ,$$

and substitution of this result in the first version of (13-31e) produces the second. ◀

13.5.3. *Marginal and Conditional Distributions of $\tilde{h}$*

If the joint distribution of $(\tilde{\beta}, \tilde{h})$ is Normal-gamma as defined by (13-30) and the parameter ν of this distribution is greater than 0, the discussion in Section 12.4.4 of the analogous distribution (12-27) shows that the *marginal* distribution of $\tilde{h}$ is gamma-2 with density

$$\mathrm{D}(h|b, v, \mathbf{n}, \nu) = f_{\gamma 2}(h|v, \nu) \tag{13-32}$$

while the *conditional* distribution of $\tilde{h}$ given β is gamma-2 with density

$$\mathrm{D}(h|b, v, \mathbf{n}, \nu; \beta) = f_{\gamma 2}(h|V, \nu + p) \tag{13-33a}$$

where

$$V \equiv \frac{\nu v + (\beta - b)^t\, \mathbf{n}(\beta - b)}{\nu + p} \ . \tag{13-33b}$$

13.5.4. *Marginal and Conditional Distributions of $\tilde{\beta}$*

If $(\tilde{\beta}, \tilde{h})$ is Normal-gamma and the parameter $\mathbf{n}$ of its distribution is of rank r, the discussion in Section 12.4.5 shows that the *marginal* distribution of $\tilde{\beta}$ is Student with density

$$\mathrm{D}(\boldsymbol{\beta}|\boldsymbol{b}, v, \mathbf{n}, \nu) = f_S^{(r)}(\boldsymbol{\beta}|\boldsymbol{b}, \mathbf{n}/v, \nu) \tag{13-34}$$

and therefore, by (8-29), with mean and variance

$$\mathrm{E}(\tilde{\boldsymbol{\beta}}|\boldsymbol{b}, v, \mathbf{n}, \nu) = \boldsymbol{b} , \tag{13-35a}$$

$$\mathrm{V}(\tilde{\boldsymbol{\beta}}|\boldsymbol{b}, v, \mathbf{n}, \nu) = \mathbf{n}^{-1} v \frac{\nu}{\nu - 2} \cdot \tag{13-35b}$$

The *conditional* distribution of $\tilde{\boldsymbol{\beta}}$ given h is of course Normal with density $f_N^{(r)}(\boldsymbol{\beta}|\boldsymbol{b}, h\mathbf{n})$.

13.6. Sampling Distributions With Fixed X

In this section we assume that a sample with predetermined $\mathbf{X}$ and therefore with predetermined $\mathbf{n} = \mathbf{X}^t\,\mathbf{X}$ is to be drawn from an r-dimensional Normal Regression process whose parameter $(\tilde{\boldsymbol{\beta}}, \tilde{h})$ is a random variable having a proper Normal-gamma distribution with parameter $(\boldsymbol{b}', v', \mathbf{n}', \nu')$. The fact that the distribution is proper implies that $v', \nu' > 0$ and that $\mathbf{n}'$ is PDS and thus of rank r.

13.6.1. Unconditional Distribution of $\tilde{\boldsymbol{y}}$

The *conditional* distribution of $\tilde{\boldsymbol{y}}$ for a particular value of the process parameter $(\boldsymbol{\beta}, h)$ was given in (13-17). The *unconditional* distribution of $\tilde{\boldsymbol{y}}$ is Student with density

$$\mathrm{D}(\boldsymbol{y}|\boldsymbol{b}', v', \mathbf{n}', \nu'; \mathbf{X}) = \int_{-\infty}^{\infty} \int_0^{\infty} f_N^{(n)}(\boldsymbol{y}|\mathbf{X}\,\boldsymbol{\beta}, h\mathbf{I})\, f_{N\gamma}^{(r)}(\boldsymbol{\beta}, h|\boldsymbol{b}', v', \mathbf{n}', \nu')\, dh\, d\boldsymbol{\beta}$$

$$= f_S^{(n)}(\boldsymbol{y}|\mathbf{X}\,\boldsymbol{b}', \mathbf{n}_y/v', \nu') \tag{13-36a}$$

where as defined in (13-19b)

$$\mathbf{n}_y \equiv \mathbf{I} - \mathbf{X}\,\mathbf{n}''^{-1}\,\mathbf{X}^t , \qquad \mathbf{n}_y^{-1} = \mathbf{X}\,\mathbf{n}'^{-1}\,\mathbf{X}^t + \mathbf{I} . \tag{13-36b}$$

The mean and variance of this distribution are

$$\begin{aligned} \mathrm{E}(\tilde{\boldsymbol{y}}|\boldsymbol{b}', v', \mathbf{n}', \nu'; \mathbf{X}) &= \mathbf{X}\,\boldsymbol{b}' , \\ \mathrm{V}(\tilde{\boldsymbol{y}}|\boldsymbol{b}', v', \mathbf{n}', \nu'; \mathbf{X}) &= \mathbf{n}_y^{-1} v' \frac{\nu}{\nu - 2} \cdot \end{aligned} \tag{13-37}$$

These results hold whether $\mathbf{X}$ is of rank r or of rank $p < r$.

► To prove (13-36) we recall that by (13-19) the conditional density of $\tilde{\boldsymbol{y}}$ given h but not $\boldsymbol{\beta}$ is $f_N(\boldsymbol{y}|\mathbf{X}\boldsymbol{b}', h\mathbf{n}_y)$ and that by (13-32) the marginal density of $\tilde{h}$ is $f_{\gamma 2}(h|v', \nu')$. The conclusion then follows by the argument used to prove (12-31). ◄

13.6.2. Conditional Joint Distribution of $(\tilde{\boldsymbol{b}}, \tilde{v}|\boldsymbol{\beta}, h)$

Provided that $\mathbf{X}$ and therefore $\mathbf{n}$ are of rank r, the conditional distribution of $(\tilde{\boldsymbol{b}}, \tilde{v})$ for given $(\boldsymbol{\beta}, h)$ as well as $(\mathbf{n}, \nu)$ is the product of the independent densities of $(\tilde{\boldsymbol{b}}|\boldsymbol{\beta}, h)$ and $(\tilde{v}|h)$:

$$\mathrm{D}(\boldsymbol{b}, v|\boldsymbol{\beta}, h; \mathbf{n}, \nu) = f_N^{(r)}(\boldsymbol{b}|\boldsymbol{\beta}, h\mathbf{n})\, f_{\gamma 2}(v|h, \nu) . \tag{13-38}$$

For the case where $\mathbf{X}$ is of rank $p < r$, see (13-51).

▶ To prove (13-38) we return to the proof of (13-29), where it was shown that the densities of $\mathbf{X}\,\tilde{b}$ and $(\tilde{y} - \mathbf{X}\,\tilde{b})$ were independent and respectively proportional to

(1) $$e^{-\frac{1}{2}h(\mathbf{X}b - \mathbf{X}\beta)^t(\mathbf{X}b - \mathbf{X}\beta)}\, h^{\frac{1}{2}p}$$

defined on $R^{(p)}$ and

(2) $$e^{-\frac{1}{2}h(y - \mathbf{X}b)^t(y - \mathbf{X}b)}\, h^{\frac{1}{2}\nu}$$

defined on $R^{(\nu)}$. Writing (1) in the form

$$e^{-\frac{1}{2}h(b-\beta)\mathbf{X}^t\mathbf{X}(b-\beta)}\, h^{\frac{1}{2}p} ,$$

we see immediately that if $\mathbf{X}^t\,\mathbf{X} = \mathbf{n}$ is of full rank r and therefore positive-definite, the $r \times 1$ vector $\tilde{b}$ has density $f_N^{(r)}(b|\beta, h\mathbf{n})$. Next interpreting the quantity

$$(y - \mathbf{X}\,b)^t(y - \mathbf{X}\,b) = \nu v$$

in (2) as the square of the length of the ν-dimensional vector $(y - \mathbf{X}\,b)$, we see that the random variable $\tilde{v}$ is constant on the surface of a ν-dimensional hypersphere of radius proportional to $\sqrt{v}$. Since the volume of such a hypersphere is proportional to $(\sqrt{v})^\nu = v^{\frac{1}{2}\nu}$, the volume of the spherical shell within which $v \le \tilde{v} \le v + dv$ will be proportional to $d(v^{\frac{1}{2}\nu}) \propto v^{\frac{1}{2}\nu-1}\,dv$. Substituting νv in (2) and multiplying by $v^{\frac{1}{2}\nu-1}$ we see that the density of $\tilde{v}$ is proportional to

$$e^{-\frac{1}{2}h\nu v}\, h^{\frac{1}{2}\nu}\, v^{\frac{1}{2}\nu-1} ,$$

which apart from constants is $f_{\gamma 2}(v|h, \nu)$. Finally, since the distributions of $\mathbf{X}\,\tilde{b}$ and $(\tilde{y} - \mathbf{X}\,\tilde{b})$ are independent, the distributions of $\tilde{b}$ and $\tilde{v}$ are independent, from which it follows that the joint density of $(\tilde{b}, \tilde{v})$ is simply the product of the individual densities, thus completing the proof of (13-38). ◀

13.6.3. *Unconditional Distributions of $\tilde{b}$ and $\tilde{v}$*

Provided that $\mathbf{X}$ and therefore $\mathbf{n}$ are of rank r, the joint distribution of $(\tilde{b}, \tilde{v})$ unconditional on (β, h) has the density

$$\text{D}(b, v|b', v', \mathbf{n}', \nu'; \mathbf{n}, \nu) \propto \frac{(\nu v)^{\frac{1}{2}\nu-1}}{(\nu' v' + \nu v + [b - b']^t\, \mathbf{n}_u[b - b'])^{\frac{1}{2}\nu''}} \tag{13-39a}$$

where

$$\mathbf{n}_u \equiv \mathbf{n}'\mathbf{n}''^{-1}\,\mathbf{n} = \mathbf{n}\,\mathbf{n}''^{-1}\,\mathbf{n}' , \qquad \mathbf{n}_u^{-1} = \mathbf{n}'^{-1} + \mathbf{n}^{-1} ,$$
$$\nu'' = \nu' + \nu + r . \tag{13-39b}$$

This joint density can also be written as either (1) the product of the marginal density of $\tilde{b}$ and the conditional density of $\tilde{v}$ given b,

$$f_S^{(r)}(b|b', \mathbf{n}_u/v', \nu')\, f_{i\beta 2}\left(v|\tfrac{1}{2}\nu, \tfrac{1}{2}[\nu' + r], \frac{\nu' v' + B}{\nu}\right) \tag{13-40a}$$

where

$$B \equiv (b - b')^t\, \mathbf{n}_u(b - b') , \tag{13-40b}$$

or (2) the product of the marginal density of $\tilde{v}$ and the conditional density of $\tilde{b}$ given v,

$$f_{i\beta 2}(v|\tfrac{1}{2}\nu, \tfrac{1}{2}\nu', \nu' v'/\nu)\, f_S^{(r)}(b|b', \mathbf{n}_u/V, \nu' + \nu) \tag{13-41a}$$

where

$$V \equiv \frac{\nu' v' + \nu v}{\nu' + \nu} \cdot \tag{13-41b}$$

For the case where $\mathbf{X}$ is of rank $p < r$, see Section 13.9.3.

▶ The proofs of (13-39) through (13-41) are identical to the proofs of the corresponding formulas (12-34) through (12-38) with $\boldsymbol{\beta}$ and $\boldsymbol{b}$ substituted for $\boldsymbol{\mu}$ and $\boldsymbol{m}$. ◀

13.7. Preposterior Analysis With Fixed X of Rank *r*

In this section as in the previous one we assume that a sample with predetermined $\mathbf{X}$ and therefore with predetermined $\mathbf{n}$ and ν is to be drawn from an r-dimensional Normal Regression process whose parameter $(\tilde{\boldsymbol{b}}, \tilde{h})$ is a random variable having a proper Normal-gamma distribution with parameter $(\boldsymbol{b}', v', \mathbf{n}', \nu')$ where $v', \nu' > 0$ and $\text{rank}(\mathbf{n}') = r$. We also assume in this section that $\mathbf{X}$ and therefore $\mathbf{n} = \mathbf{X}^t\mathbf{X}$ are of rank r, and our results will *not* apply without change if this assumption is violated.

13.7.1. Joint Distribution of $(\tilde{\boldsymbol{b}}'', \tilde{v}'')$

By using (13-31) to express the random variable $(\tilde{\boldsymbol{b}}, \tilde{v})$ in terms of the random variable $(\tilde{\boldsymbol{b}}'', \tilde{v}'')$ we can obtain the joint density of the latter from the unconditional joint density (13-39) of the former:

$$D(\boldsymbol{b}'', v''|\boldsymbol{b}', v', \mathbf{n}', \nu'; \mathbf{n}, \nu) \propto \frac{(\nu''v'' - \nu'v' - [\boldsymbol{b}'' - \boldsymbol{b}']^t\,\mathbf{n}^*[\boldsymbol{b}'' - \boldsymbol{b}'])^{\frac{1}{2}\nu - 1}}{(\nu''v'')^{\frac{1}{2}\nu''}} \tag{13-42a}$$

where

$$\mathbf{n}^* \equiv \mathbf{n}''\mathbf{n}^{-1}\mathbf{n}' = \mathbf{n}'\mathbf{n}^{-1}\mathbf{n}'' , \qquad \mathbf{n}^{*-1} = \mathbf{n}'^{-1} - \mathbf{n}''^{-1} ,$$
$$\nu'' = \nu' + \nu + r , \tag{13-42b}$$

and the variables $\boldsymbol{b}''$ and v'' are constrained by

$$\nu''v'' \geq \nu'v' ,$$
$$(\boldsymbol{b}'' - \boldsymbol{b}')^t\,\mathbf{n}^*(\boldsymbol{b}'' - \boldsymbol{b}') \leq \nu''v'' - \nu'v' . \tag{13-42c}$$

▶ The proof of (13-42) proceeds in three steps: we first show that

(1) $$(\boldsymbol{b} - \boldsymbol{b}')^t\,\mathbf{n}_u(\boldsymbol{b} - \boldsymbol{b}') = (\boldsymbol{b}'' - \boldsymbol{b}')^t\,\mathbf{n}^*(\boldsymbol{b}'' - \boldsymbol{b}') ;$$

we next show that

(2) $$\nu'v' + \nu v + (\boldsymbol{b} - \boldsymbol{b}')^t\,\mathbf{n}_u(\boldsymbol{b} - \boldsymbol{b}') = \nu''v'' ;$$

and we then substitute these results in (13-39).

To prove equation (1) we first use the definitions (13-31ab) of $\boldsymbol{b}''$ and $\mathbf{n}''$ to write

$$\boldsymbol{b} - \boldsymbol{b}' = \mathbf{n}^{-1}(\mathbf{n}''\boldsymbol{b}'' - \mathbf{n}'\boldsymbol{b}') - \boldsymbol{b}' = \mathbf{n}^{-1}\mathbf{n}''\boldsymbol{b}'' - \mathbf{n}^{-1}(\mathbf{n}'' - \mathbf{n})\,\boldsymbol{b}' - \boldsymbol{b}'$$
$$= \mathbf{n}^{-1}\mathbf{n}''(\boldsymbol{b}'' - \boldsymbol{b}') .$$

Substituting this result and the definition (13-39b) of $\mathbf{n}_u$ in the left-hand side of (1) we then obtain

$$(b'' - b')^t \mathbf{n}''\mathbf{n}^{-1}(\mathbf{n}'\mathbf{n}''^{-1}\mathbf{n})\,\mathbf{n}^{-1}\mathbf{n}''(b'' - b') = (b'' - b')^t\mathbf{n}''\,\mathbf{n}^{-1}\mathbf{n}'(b'' - b') \;;$$

and by the definition (13-42b) of $\mathbf{n}^*$ this is the right-hand side of (1).

To prove equation (2), we first expand

$$(b - b')^t\mathbf{n}_u(b - b') = b^t\,\mathbf{n}_u\,b + b'^t\mathbf{n}_u\,b' - b^t\,\mathbf{n}_u\,b' - b'^t\mathbf{n}_u\,b \;.$$

We then derive alternate formulas for $\mathbf{n}_u$ from formulas (13-39b)

$$\begin{aligned}\mathbf{n}_u &\equiv \mathbf{n}'\mathbf{n}''^{-1}\mathbf{n} = (\mathbf{n}'' - \mathbf{n})\,\mathbf{n}''^{-1}\mathbf{n} = \mathbf{n} - \mathbf{n}\,\mathbf{n}''^{-1}\mathbf{n}\\ &= \mathbf{n}\,\mathbf{n}''^{-1}\mathbf{n}' = (\mathbf{n}'' - \mathbf{n}')\,\mathbf{n}''^{-1}\mathbf{n}' = \mathbf{n}' - \mathbf{n}'\mathbf{n}''^{-1}\mathbf{n}'\end{aligned}$$

and substitute in the previous equation to obtain

$$\begin{aligned}(b - b')^t\mathbf{n}_u(b - b') &= b^t(\mathbf{n} - \mathbf{n}\,\mathbf{n}''^{-1}\mathbf{n})\,b + b'^t(\mathbf{n}' - \mathbf{n}'\mathbf{n}''^{-1}\mathbf{n}')\,b'\\ &\quad - b^t(\mathbf{n}\,\mathbf{n}''^{-1}\mathbf{n}')\,b' - b'^t(\mathbf{n}'\mathbf{n}''^{-1}\mathbf{n})\,b\\ &= b^t\mathbf{n}\,b + b'^t\mathbf{n}'b' - (\mathbf{n}'b' + \mathbf{n}\,b)^t\mathbf{n}''^{-1}(\mathbf{n}'b' + \mathbf{n}\,b)\\ &= b^t\mathbf{n}\,b + b'^t\mathbf{n}'b' - b''^t\mathbf{n}''b'' \;.\end{aligned}$$

Equation (2) is proved by substituting this result in the left-hand member and comparing the result with the second formula for $\nu''v''$ in (13-31e).

We can now use (1) and (2) to obtain (13-42) from (13-39); for the details, see the proof of (12-39). ◀

13.7.2. *Distributions of* $\tilde{b}''$ *and* $\tilde{v}''$

The marginal densities of $\tilde{b}''$ and $\tilde{v}''$ determined by the joint density (13-42) are respectively

$$\mathrm{D}(b''|b', v', \mathbf{n}', \nu'; \mathbf{n}, \nu) = f_S^{(r)}(b''|b', \mathbf{n}^*/v', \nu') \;, \tag{13-43}$$

where $\mathbf{n}^*$ is defined by (13-42b), and

$$\mathrm{D}(v''|b', v', \mathbf{n}', \nu'; \mathbf{n}, \nu) = f_{i\beta 1}(v''|\tfrac{1}{2}\nu', \tfrac{1}{2}\nu'', \nu'v'/\nu'') \;. \tag{13-44}$$

The proofs are identical to the proofs of (12-40) and (12-42) with b substituted for m; and the conditional distributions of $(\tilde{b}''|v'')$ and $(\tilde{v}''|b'')$ can be obtained from (12-41) and (12-43) by the same translation.

13.7.3. *Distributions of* $\bar{\tilde{\beta}}''$ *and* $\breve{\tilde{\beta}}''$

The mean $\bar{\beta}''$ of the posterior distribution of $\tilde{\beta}$ will, by (13-35a), be equal to the parameter b'' of that distribution, and therefore the prior distribution of $\bar{\tilde{\beta}}''$ is given directly by (13-43) with $\bar{\beta}''$ substituted for b''. The variance $\breve{\beta}''$ of the posterior distribution of $\tilde{\beta}$ will, by (13-35b), be in simple proportion to the parameter v'' of that distribution, and therefore the prior distribution of $\breve{\tilde{\beta}}''$ can be obtained by obvious modifications of (13-44).

Since these distributions are formally identical to the distributions of the quantities $\bar{\tilde{\mu}}''$ and $\breve{\tilde{\mu}}''$ which were discussed in Section 12.6.3, we refer the reader to that section for further comment.

C. Normal Regression Process, $\mathbf{X}^t\,\mathbf{X}$ *Singular*

13.8. Introduction

When the rank p of the matrix $\mathbf{X}$ is less than the number r of parameters $\boldsymbol{\beta}$, the statistic $\boldsymbol{b}$ is not completely determinate. As seen in Section 13.1.2, $r - p$ elements may be given arbitrary values after which the p remaining elements will be determined by these assignments *and* the observations $\boldsymbol{y}$. Although this arbitrariness in the definition of $\boldsymbol{b}$ makes no difference whatever in simple posterior analysis, as we saw in Sections 13.2 and 13.5, it obviously means that the sampling distributions of Sections 13.3 and 13.6 will not apply. What is much more important, it means that the preposterior analysis in Sections 13.4 and 13.7 will not apply, since if the distribution of $\tilde{\boldsymbol{b}}$ is p-dimensional, the distribution of

$$\tilde{\boldsymbol{b}}'' = \mathbf{n}''^{-1}(\mathbf{n}'\boldsymbol{b}' + \mathbf{n}\,\tilde{\boldsymbol{b}})$$

will necessarily be confined to a p-dimensional subspace within the r-space on which $\boldsymbol{b}''$ is defined.

The problem of obtaining these degenerate distributions in usable form is very much like the problem which arose in Section 12.9 when we wished to analyze the relations among the means of r univariate processes only p of which were actually to be sampled, since in both cases the statistic $\boldsymbol{m}$ or $\boldsymbol{b}$ has $r - p$ arbitrary elements and the statistic $\mathbf{n}$ is of rank p. The present problem has one additional complication, however: the conditional distribution of the observations $\boldsymbol{y}$ will in general depend on all r components of the parameter $\boldsymbol{\beta}$, whereas in Section 12.9 our task was simplified by the fact that the conditional distribution of the observations depended on only those p components of $\boldsymbol{\mu}$ which represented the means of the processes actually sampled.

13.8.1. Definitions and Notation

Recalling from the comment on (13-12) in Section 13.1.2 that our difficulties are due to the fact that the unique projection of the sample point $\boldsymbol{y}$ on the p-dimensional column space of $\mathbf{X}$ can be expressed as an infinity of different combinations $\boldsymbol{b}$ of the $r > p$ nonindependent columns of $\mathbf{X}$, we shall select some one set of p columns which *are* independent and express the projection of $\boldsymbol{y}$ as a *unique* combination $\boldsymbol{b}^*$ of these columns alone.

To simplify notation we shall assume without loss of generality that the *first* p columns of $\mathbf{X}$ are linearly independent and take them as our chosen set. We then partition

$$\mathbf{X} = [\mathbf{X}_1 \quad \mathbf{X}_2] \qquad \text{where} \qquad \begin{matrix}\mathbf{X} \text{ is } n \times r\ , \\ \mathbf{X}_1 \text{ is } n \times p\ ,\end{matrix} \tag{13-45}$$

$$\mathbf{n} = \mathbf{X}^t\,\mathbf{X} = \begin{bmatrix}\mathbf{X}_1^t\,\mathbf{X}_1 & \mathbf{X}_1^t\,\mathbf{X}_2 \\ \mathbf{X}_2^t\,\mathbf{X}_1 & \mathbf{X}_2^t\,\mathbf{X}_2\end{bmatrix} = \begin{bmatrix}\mathbf{n}_{11} & \mathbf{n}_{12} \\ \mathbf{n}_{21} & \mathbf{n}_{22}\end{bmatrix}, \tag{13-46}$$

and digress slightly from our immediate objective to point out that if we define

$$\mathbf{Q} \equiv \mathbf{n}_{11}^{-1}\,\mathbf{n}_{12} \;, \tag{13-47}$$

we can write

$$\mathbf{X}_2 = \mathbf{X}_1\,\mathbf{Q} \;. \tag{13-48}$$

▶ It is shown in Section 8.0.4 that the projection of any vector $\boldsymbol{v}$ on the column space of $\mathbf{X}_1$ is $[\mathbf{X}_1(\mathbf{X}_1^t\,\mathbf{X}_1)^{-1}\,\mathbf{X}_1^t]\,\boldsymbol{v}$, and accordingly the kth column of $\mathbf{X}_1\mathbf{Q} = [\mathbf{X}_1(\mathbf{X}_1^t\,\mathbf{X}_1)^{-1}\,\mathbf{X}_1^t]\,\mathbf{X}_2$ is the projection of the kth column of $\mathbf{X}_2$ on the column space of $\mathbf{X}_1$. Then since the fact that every column of $\mathbf{X}_2$ is by hypothesis a linear combination of the columns of $\mathbf{X}_1$ means that every column of $\mathbf{X}_2$ lies wholly within the column space of $\mathbf{X}_1$, the projection of any column of $\mathbf{X}_2$ is simply the column itself: the kth column of $\mathbf{X}_1\mathbf{Q}$ *is* the kth column of $\mathbf{X}_2$. ◀

We are now ready to find the unique p-tuple $\boldsymbol{b}^*$ which will express the projection of $\boldsymbol{y}$ on the column space of $\mathbf{X}$ as a linear combination of the p independent columns of $\mathbf{X}_1$. To have this property, $\boldsymbol{b}^*$ must satisfy the revised normal equations (cf. 13-10a).

$$\mathbf{X}_1^t\,\mathbf{X}_1\,\boldsymbol{b}^* = \mathbf{X}_1^t\,\boldsymbol{y} \;; \tag{13-49a}$$

and since $\mathbf{X}_1$ and therefore $\mathbf{X}_1^t\,\mathbf{X}_1 = \mathbf{n}_{11}$ are of full rank, these equations have the unique solution

$$\boldsymbol{b}^* = \mathbf{n}_{11}^{-1}\,\mathbf{X}_1^t\,\boldsymbol{y} \;. \tag{13-49b}$$

If we also define

$$\boldsymbol{b} = \begin{bmatrix}\boldsymbol{b}^*\\ \boldsymbol{0}\end{bmatrix} \qquad \text{where} \qquad \begin{array}{l}\boldsymbol{b} \text{ is } r \times 1 \;,\\ \boldsymbol{b}^* \text{ is } p \times 1 \;,\end{array} \tag{13-50}$$

the fact that $\mathbf{X}\,\boldsymbol{b} = \mathbf{X}_1\,\boldsymbol{b}^*$ makes it obvious that the projection of $\boldsymbol{y}$ on the column space of $\mathbf{X}_1$ or $\mathbf{X}$ can also be expressed as $\mathbf{X}\,\boldsymbol{b}$ and therefore that $\boldsymbol{b}$ as thus defined satisfies the original normal equations (13-10a).

13.9. Distributions of $\tilde{\boldsymbol{b}}^*$ and $\tilde{v}$

13.9.1. Conditional Distributions of $\tilde{\boldsymbol{b}}^*$ and $\tilde{v}$

For given $(\boldsymbol{\beta}, h)$ as well as $\mathbf{X}$, the joint density of the statistics $\tilde{\boldsymbol{b}}^*$ and $\tilde{v}$ is the product of their independent individual densities,

$$\mathrm{D}(\boldsymbol{b}^*, v|\boldsymbol{\beta}, h; \mathbf{X}) = f_N^{(p)}(\boldsymbol{b}^*|\boldsymbol{\beta}^*, h\mathbf{n}_{11})\,f_{\gamma 2}(v|h, \nu) \tag{13-51a}$$

where ν is defined by (13-10d) and

$$\boldsymbol{\beta}^* \equiv [\mathbf{I} \quad \mathbf{Q}]\,\boldsymbol{\beta} = \boldsymbol{\beta}_1 + \mathbf{Q}\,\boldsymbol{\beta}_2 \;, \tag{13-51b}$$

$\boldsymbol{\beta}_1$ and $\boldsymbol{\beta}_2$ being respectively the first p and last $(r - p)$ elements of the $r \times 1$ vector $\boldsymbol{\beta}$.

▶ By the definition (13-50) of $\boldsymbol{b}$,

$$\mathbf{X}\,\boldsymbol{b} = [\mathbf{X}_1 \quad \mathbf{X}_2]\begin{bmatrix}\boldsymbol{b}^*\\ \boldsymbol{0}\end{bmatrix} = \mathbf{X}_1\,\boldsymbol{b}^* \;,$$

and by (13-48) and (13-51b),

$$\mathbf{X}\,\boldsymbol{\beta} = \mathbf{X}_1[\mathbf{I} \quad \mathbf{Q}]\,\boldsymbol{\beta} = \mathbf{X}_1\,\boldsymbol{\beta}^* \ .$$

The proof of (13-51) is now identical to the proof of (13-38) with $(\mathbf{X}\,\boldsymbol{b} - \mathbf{X}\,\boldsymbol{\beta})$ everywhere replaced by $(\mathbf{X}_1\,\boldsymbol{b}^* - \mathbf{X}_1\,\boldsymbol{\beta}^*)$ and with $\mathbf{n} = \mathbf{X}^t\,\mathbf{X}$ everywhere replaced by $\mathbf{n}_{11} = \mathbf{X}_1^t\,\mathbf{X}_1$. ◀

13.9.2. *Distributions of* $(\tilde{\boldsymbol{\beta}}^*|h)$ *and* $(\tilde{\boldsymbol{\beta}}^*, \tilde{h})$

From (13-51) it is clear that the unconditional distributions of $\tilde{\boldsymbol{b}}^*$ and $\tilde{v}$ will depend on the distribution of $(\tilde{\boldsymbol{\beta}}^*|h)$ if h is known or on the distribution of $(\tilde{\boldsymbol{\beta}}^*, \tilde{h})$ if h is unknown in the same way that the distributions of $\tilde{\boldsymbol{b}}$ and $\tilde{v}$ depended on the distributions of $(\tilde{\boldsymbol{\beta}}|h)$ or $(\tilde{\boldsymbol{\beta}}, \tilde{h})$. The *prior* distribution of $(\tilde{\boldsymbol{\beta}}^*|h)$ is given immediately by the definition (13-51b) of $\tilde{\boldsymbol{\beta}}^*$ in conjunction with (8-26):

$$D(\boldsymbol{\beta}^*|h;\boldsymbol{b}',\mathbf{n}') = f_N^{(p)}(\boldsymbol{\beta}^*|\boldsymbol{b}^{*\prime}, h\mathbf{n}_m') \tag{13-52}$$

where

$$\begin{aligned} \boldsymbol{b}^{*\prime} &= [\mathbf{I} \quad \mathbf{Q}]\,\boldsymbol{b}' = \boldsymbol{b}_1' + \mathbf{Q}\,\boldsymbol{b}_2' \ , \\ \mathbf{n}_m'^{-1} &= [\mathbf{I} \quad \mathbf{Q}]\,\mathbf{n}'^{-1}\,[\mathbf{I} \quad \mathbf{Q}]^t \ . \end{aligned} \tag{13-53}$$

Then since by (13-32) the prior density of $\tilde{h}$ marginally as regards $\tilde{\boldsymbol{\beta}}$ is $f_{\gamma 2}(h|v', \nu')$, it follows at once that the prior density of $(\tilde{\boldsymbol{\beta}}^*, \tilde{h})$ is

$$D(\boldsymbol{\beta}^*, h|\boldsymbol{b}', v', \mathbf{n}', \nu') = f_N^{(p)}(\boldsymbol{\beta}^*|\boldsymbol{b}^{*\prime}, h\mathbf{n}_m')\,f_{\gamma 2}(h|v', \nu') \ . \tag{13-54}$$

The *posterior* distributions of $(\tilde{\boldsymbol{\beta}}^*|h)$ and $(\tilde{\boldsymbol{\beta}}^*, \tilde{h})$ are now given by formulas exactly analogous to those which apply when $\mathbf{X}$ is of rank r. *Assuming that the prior distributions are proper*, i.e. that $v', \nu' > 0$ and that $\mathbf{n}_m'$ is PDS, we have

$$\mathbf{n}_m'' = \mathbf{n}_m' + \mathbf{n}_{11} \ , \tag{13-55a}$$

$$\boldsymbol{b}^{*\prime\prime} = \mathbf{n}_m''^{-1}(\mathbf{n}_m'\,\boldsymbol{b}^{*\prime} + \mathbf{n}_{11}\,\boldsymbol{b}^*) \ , \tag{13-55b}$$

$$\nu'' = \nu' + \nu + p \ , \tag{13-55c}$$

$$v'' = \frac{1}{\nu}\,([\nu'v' + \boldsymbol{b}^{*\prime t}\,\mathbf{n}_m'\,\boldsymbol{b}^{*\prime}] + [\nu v + \boldsymbol{b}^{*t}\,\mathbf{n}_{11}\,\boldsymbol{b}^*] - \boldsymbol{b}^{*\prime\prime t}\,\mathbf{n}_m''\,\boldsymbol{b}^{*\prime\prime} \ . \tag{13-55d}$$

13.9.3. *Unconditional Distributions of* $\tilde{\boldsymbol{b}}^*$ *and* $\tilde{v}$

We can now obtain the unconditional densities of $\tilde{\boldsymbol{b}}^*$ and $\tilde{v}$ by obvious modifications in the formulas given in Sections 13.3.3 and 13.6.3 for the unconditional densities of $\tilde{\boldsymbol{b}}$ and $\tilde{v}$.

If h is known, the density of $\tilde{\boldsymbol{b}}^*$ is obtained by modifying (13-22) to read

$$D(\boldsymbol{b}^*|\boldsymbol{b}', \mathbf{n}'; \mathbf{X}) = f_N^{(p)}(\boldsymbol{b}^*|\boldsymbol{b}^{*\prime}, h\mathbf{n}_u) \tag{13-56a}$$

where

$$\mathbf{n}_u = \mathbf{n}_m'\,\mathbf{n}_m''^{-1}\,\mathbf{n}_{11} = \mathbf{n}_{11}\,\mathbf{n}_m''^{-1}\,\mathbf{n}_m' \ , \qquad \mathbf{n}_u^{-1} = \mathbf{n}_m'^{-1} + \mathbf{n}_{11}^{-1} \ . \tag{13-56b}$$

If h is unknown, the joint density of $(\tilde{\boldsymbol{b}}^*, \tilde{v})$ is obtained by modifying formula (13-39) to read

$$D(\boldsymbol{b}^*, v|\boldsymbol{b}', v', \mathbf{n}', \nu'; \mathbf{X}) \propto \frac{(\nu v)^{\frac{1}{2}\nu - 1}}{(\nu'v' + \nu v + [\boldsymbol{b}^* - \boldsymbol{b}^{*\prime}]^t\,\mathbf{n}_u[\boldsymbol{b}^* - \boldsymbol{b}^{*\prime}])^{\frac{1}{2}\nu''}} \tag{13-57}$$

where $\mathbf{n}_u$ is defined by (13-56b) and ν'' by (13-55c). The same changes in (13-40) give the marginal density of $\tilde{b}^*$ separately,

$$D(b^*|b', v', \mathbf{n}', \nu'; \mathbf{X}) = f_S^{(p)}(b^*|b^{*\prime}, \mathbf{n}_u/v', \nu') \;, \tag{13-58}$$

while formula (13-41) for the marginal density of $\tilde{v}$ applies with no changes at all.

13.10. Preposterior Analysis

The problem of preposterior analysis when only p of r univariate processes were to be sampled was handled in Sections 12.9.6 and 12.9.7 by first obtaining the distributions of the posterior parameters $\tilde{m}_1''$ and $(\tilde{m}_1'', \tilde{v}'')$ and then showing that $\tilde{m}_2''$ could be expressed as a function of $\tilde{m}_1''$. In our present problem such an approach is not convenient because on the one hand we cannot be sure that $\tilde{b}_1''$ will have a nondegenerate distribution while on the other hand the posterior utilities are not naturally expressed in terms of the well-behaved quantity $\tilde{b}^{*\prime\prime}$ and therefore there is no advantage to be gained by obtaining its prior distribution. We shall therefore attack our present problem by direct use of the unconditional distribution of the statistic $\tilde{b}^*$ or of the joint unconditional distribution of the *statistic* $\tilde{b}^*$ and the posterior *parameter* $\tilde{v}''$.

13.10.1. *Utilities Dependent on $\tilde{b}''$ alone*

Whether or not the process parameter h is known, formulas (13-16b) and (13-31b) show that before the sample outcome is known the random variable $\tilde{b}''$ is related to the random variable $\tilde{b}$ by

$$\tilde{b}'' = \mathbf{n}''^{-1}(\mathbf{n}'b' + \mathbf{n}\,\tilde{b}) \;, \tag{13-59}$$

and by the definition (13-50) of $\tilde{b}$ for our present problem this may be written

$$\tilde{b}'' = \mathbf{n}''^{-1}\left(\mathbf{n}'b' + \begin{bmatrix}\mathbf{n}_{11}\\ \mathbf{n}_{21}\end{bmatrix}\tilde{b}^*\right) \cdot \tag{13-60}$$

By use of this formula utilities originally expressed in terms of $\tilde{b}''$ can be reexpressed in terms of $\tilde{b}^*$ and expected utilities can then be computed by use of the distribution (13-56) if h is known or by use of (13-58) if h is unknown.

13.10.2. *Utilities Dependent on $(\tilde{b}'', \tilde{v}'')$ Jointly*

If h is unknown and utilities depend on both $\tilde{b}''$ and $\tilde{v}''$, expected utilities can be computed by first using (13-60) to replace $\tilde{b}''$ by $\tilde{b}^*$ and then using the joint distribution of $(\tilde{b}^*, \tilde{v}'')$, whose density is

$$D(b^*, v''|b', v', \mathbf{n}', \nu'; \mathbf{X}) \propto \frac{(\nu''v'' - \nu'v' - [b^* - b^{*\prime}]^t\,\mathbf{n}_u[b^* - b^{*\prime}])^{\frac{1}{2}\nu-1}}{(\nu''v'')^{\frac{1}{2}\nu''}} \tag{13-61}$$

where $\mathbf{n}_u$ is defined by (13-56b) and ν'' by (13-55c).

▶ The joint distribution of $(\tilde{b}^{*\prime\prime}, \tilde{v}'')$ can obviously be obtained by substituting $b^{*\prime\prime}$, $b^{*\prime}$, and $\mathbf{n}_m^*$ for b'', b', and $\mathbf{n}^*$ in (13-42), thus obtaining

$$\frac{(\nu''v'' - \nu'v' - [b^{*\prime\prime} - b^{*\prime}]^t\, \mathbf{n}_m^*[b^{*\prime\prime} - b^{*\prime}]^{\frac{1}{2}\nu - 1}}{(\nu''v'')^{\frac{1}{2}\nu''}} \;;$$

and equation (1) in the proof of (13-42) then shows that we may replace $b^{*\prime\prime}$ and $\mathbf{n}_m^*$ in this result by b^* and $\mathbf{n}_u$. ◀

13.10.3. Distribution of $\tilde{v}''$

The marginal distribution of $\tilde{v}''$ is the same whether or not $\mathbf{X}$ is of rank r; it is given by (13-44) as

$$D(\tilde{v}''|b', v', \mathbf{n}', \nu'; \mathbf{X}) = f_{i\beta 1}(v''|\tfrac{1}{2}\nu', \tfrac{1}{2}\nu'', \nu'v'/\nu'') \;. \tag{13-62}$$

▶ Formula (13-62) follows immediately from the fact that the joint distribution of $(\tilde{b}^{*\prime\prime}, \tilde{v}'')$ when $\mathbf{X}$ is of rank p is exactly analogous to the joint distribution of $(\tilde{b}'', \tilde{v}'')$ when $\mathbf{X}$ is of rank r. ◀

Table I

STANDARDIZED STUDENT DENSITY FUNCTION

$f_{S*}(t|n)$

t	$n=1$	$n=2$	$n=3$	$n=4$	$n=5$	$n=6$	$n=7$	$n=8$
·05	·3175162	·3528915	·3669407	·3744146	·3790377	·3821751	·3844419	·3861557
·15	·3113056	·3476700	·3621007	·3697771	·3745279	·3777513	·3800812	·3818427
·25	·2995859	·3376051	·3527034	·3607428	·3657201	·3690998	·3715433	·3733915
·35	·2835724	·3233913	·3392790	·3477591	·3530188	·3565942	·3591812	·3611391
·45	·2647068	·3059329	·3225401	·3314454	·3369826	·3407543	·3434890	·3455574
·55	·2443839	·2862204	·3033028	·3125209	·3182744	·3222033	·3250551	·3272186
·65	·2237680	·2652188	·2824067	·2917505	·2976083	·3016202	·3045385	·3067564
·75	·2037184	·2437833	·2606467	·2698820	·2756981	·2796934	·2826058	·2848232
·85	·1847953	·2226122	·2387219	·2476003	·2532132	·2570792	·2599031	·2620557
·95	·1673114	·2022282	·2172082	·2254950	·2307468	·2343702	·2370201	·2390419
1·05	·1513960	·1829920	·1965463	·2040439	·2087951	·2120731	·2144703	·2162992
1·15	·1370549	·1650211	·1770484	·1836082	·1877489	·1906975	·1926761	·1942595
1·25	·1242185	·1487194	·1589119	·1644395	·1678943	·1703535	·1719653	·1732634
1·35	·1127759	·1338070	·1422386	·1466924	·1494216	·1512570	·1525725	·1535635
1·45	·1025979	·1203447	·1270562	·1304403	·1324372	·1337391	·1346482	·1353157
1·55	·0935518	·1082557	·1133371	·1156930	·1169791	·1177592	·1182687	·1186200
1·65	·0855098	·0974411	·1010161	·1024127	·1030315	·1033193	·1034502	·1035013
1·75	·0783532	·0877915	·0900033	·0905284	·0905388	·0903768	·0901612	·0899370
1·85	·0719751	·0791955	·0801962	·0799476	·0794183	·0788576	·0783350	·0778669
1·95	·0662801	·0715447	·0714866	·0705662	·0695706	·0686667	·0678807	·0672040
2·05	·0611840	·0647366	·0637669	·0622752	·0608872	·0596969	·0586927	·0578440
2·15	·0566136	·0586769	·0569334	·0549665	·0532574	·0518360	·0506578	·0496734
2·25	·0525047	·0532797	·0508888	·0485357	·0465721	·0449716	·0436610	·0425755
2·35	·0488018	·0484682	·0455437	·0428848	·0407274	·0389952	·0375906	·0364352
2·45	·0454566	·0441735	·0408165	·0379236	·0356262	·0338045	·0323397	·0311425
2·55	·0424272	·0403348	·0366342	·0335699	·0311794	·0293046	·0278091	·0265941
2·65	·0396771	·0368986	·0329314	·0297496	·0273061	·0254094	·0239079	·0226954
2·75	·0371749	·0338176	·0296507	·0263968	·0239340	·0220412	·0205541	·0193607
2·85	·0348928	·0310505	·0267397	·0234531	·0209988	·0191308	·0176746	·0165133
2·95	·0328070	·0285609	·0241549	·0208669	·0184435	·0166174	·0152044	·0140852
3·05	·0308964	·0263171	·0218562	·0185929	·0162184	·0144463	·0130866	·0120168
3·15	·0291426	·0242912	·0198093	·0165916	·0142797	·0125715	·0112716	·0102562
3·25	·0275295	·0224588	·0179838	·0148283	·0125894	·0109518	·0097162	·0087582
3·35	·0260429	·0207984	·0163535	·0132729	·0111145	·0095519	·0083832	·0074840
3·45	·0246704	·0192913	·0148951	·0118992	·0098263	·0083412	·0072405	·0064002
3·55	·0234008	·0179209	·0135886	·0106843	·0087000	·0072933	·0062603	·0054782
3·65	·0222245	·0166727	·0124162	·0096085	·0077142	·0063856	·0054192	·0046936
3·75	·0211326	·0155339	·0113626	·0086544	·0068502	·0055986	·0046968	·0040256
3·85	·0201175	·0144932	·0104142	·0078071	·0060921	·0049155	·0040759	·0034566
3·95	·0191724	·0135406	·0095592	·0070535	·0054261	·0043218	·0035417	·0029714
4·05	·0182910	·0126675	·0087871	·0063822	·0048401	·0038053	·0030817	·0025575
4·15	·0174680	·0118655	·0080890	·0057835	·0043238	·0033554	·0026851	·0022041
4·25	·0166982	·0111281	·0074566	·0052486	·0038683	·0029630	·0023129	·0019020
4·35	·0159774	·0104491	·0068831	·0047701	·0034660	·0026203	·0020471	·0016435
4·45	·0153015	·0098229	·0063621	·0043413	·0031100	·0023206	·0017912	·0014220
4·55	·0146655	·0092447	·0058881	·0039567	·0027946	·0020582	·0015695	·0012321
4·65	·0140705	·0087099	·0054563	·0036110	·0025148	·0018282	·0013773	·0010690
4·75	·0135092	·0082147	·0050624	·0033000	·0022661	·0016261	·0012103	·0009288
4·85	·0129803	·0077556	·0047025	·0030197	·0020449	·0014484	·0010650	·0008081
4.95	·0124815	·0073294	·0043734	·0027668	·0018479	·0012920	·0009386	·0007041
5·05	·0120106	·0069333	·0040719	·0025382	·0016720	·0011540	·0008283	·0006144
5·15	·0115654	·0065647	·0037954	·0023315	·0015149	·0010322	·0007320	·0005368
5·25	·0111443	·0062214	·0035415	·0021444	·0013743	·0009246	·0006478	·0004697
5·35	·0107455	·0059012	·0033080	·0019742	·0012484	·0008292	·0005740	·0004115
5·45	·0103675	·0056022	·0030931	·0018198	·0011355	·0007447	·0005094	·0003611
5·55	·0100090	·0053228	·0028951	·0016794	·0010340	·0006697	·0004527	·0003173
5·65	·0096685	·0050614	·0027124	·0015515	·0009427	·0006030	·0004028	·0002791
5·75	·0093449	·0048165	·0025436	·0014350	·0008605	·0005436	·0003588	·0002459
5·85	·0090371	·0045871	·0023875	·0013286	·0007864	·0004907	·0003201	·0002169
5·95	·0087442	·0043717	·0022431	·0012313	·0007195	·0004431	·0002859	·0001916
6.05	·0084651	·0041694	·0021091	·0011424	·0006590	·0004012	·0002557	·0001695
6.15	·0081991	·0039793	·0019851	·0010609	·0006043	·0003635	·0002290	·0001501
6.25	·0079453	·0038032	·0018696	·0009861	·0005547	·0003297	·0002053	·0001331
6.35	·0077031	·0036320	·0017625	·0009175	·0005897	·0002993	·0001843	·0001182
6.45	·0074716	·0034732	·0016628	·0008545	·0004688	·0002721	·0001657	·0001051
6.55	·0072504	·0033235	·0015699	·0007965	·0004317	·0002476	·0001491	·0000935
6.65	·0070387	·0031821	·0014834	·0007431	·0003979	·0002256	·0001343	·0000834
6.75	·0068362	·0030486	·0014027	·0006939	·0003671	·0002057	·0001211	·0000744
6.85	·0066422	·0029224	·0013273	·0006485	·0003390	·0001878	·0001093	·0000664
6.95	·0064562	·0028030	·0012568	·0006066	·0003133	·0001716	·0000988	·0000594
7.05	·0062780	·0026899	·0011910	·0005678	·0002899	·0001570	·0000894	·0000532
7.15	·0061070	·0025827	·0011293	·0005319	·0002684	·0001437	·0000810	·0000477
7.25	·0059428	·0024812	·0010715	·0004987	·0002488	·0001317	·0000734	·0000428

Table I (*continued*)

STANDARDIZED STUDENT DENSITY FUNCTION

$f_{S^*}(t|n)$

t	$n=9$	$n=10$	$n=12$	$n=15$	$n=20$	$n=24$	$n=30$	$n=60$	$n=\infty$
·05	·3874964	·3885738	·3901977	·3918304	·3934709	·3942957	·3951216	·3967788	·3984239
·15	·3832204	·3843283	·3860103	·3876765	·3893644	·3902118	·3910612	·3927671	·3944793
·25	·3748377	·3760004	·3777534	·3795165	·3812902	·3821805	·3830735	·3848650	·3866681
·35	·3626709	·3639048	·3657646	·3676365	·3695203	·3704665	·3714158	·3733221	·3752403
·45	·3471796	·3484852	·3504582	·3524416	·3544118	·3554475	·3564528	·3584842	·3605270
·55	·3289158	·3302831	·3323488	·3344322	·3365338	·3375913	·3386531	·3407889	·3429439
·65	·3085018	·3099026	·3120272	·3141730	·3163407	·3174325	·3185299	·3207413	·3229724
·75	·2865667	·2879743	·2901058	·2922619	·2944132	·2955434	·2966494	·2988817	·3011374
·85	·2637504	·2651199	·2671960	·2692868	·2714289	·2725040	·2735865	·2757715	·2779849
·95	·2406352	·2419227	·2438770	·2458779	·2478664	·2488803	·2499020	·2519658	·2540591
1·05	·2177405	·2189049	·2206729	·2224645	·2242811	·2251985	·2261226	·2279900	·2298821
1·15	·1955051	·1965110	·1980358	·1995786	·2011396	·2019277	·2027201	·2043188	·2059363
1·25	·1742808	·1750999	·1763366	·1775824	·1788376	·1794676	·1801004	·1813711	·1826491
1·35	·1543282	·1549419	·1558612	·1567771	·1576887	·1581428	·1585949	·1594938	·1603833
1·45	·1358248	·1362251	·1368126	·1373823	·1379317	·1381980	·1384583	·1389596	·1394306
1·55	·1188730	·1190610	·1193177	·1195417	·1197251	·1198021	·1198680	·1199648	·1200090
1·65	·1035100	·1034955	·1034358	·1033255	·1031596	·1030540	·1029317	·1026364	·1022649
1·75	·0897216	·0895212	·0891703	·0887533	·0882633	·0879890	·0876937	·0870349	·0862773
1·85	·0774531	·0770880	·0764795	·0757912	·0750151	·0745916	·0741422	·0731630	·0720649
1·95	·0666207	·0661153	·0652880	·0643699	·0633522	·0628030	·0622249	·0609769	·0595947
2·05	·0571219	·0565021	·0554966	·0543933	·0531821	·0525331	·0518527	·0503940	·0487920
2·15	·0488427	·0481339	·0469921	·0457481	·0443927	·0436700	·0429157	·0413056	·0395500
2·25	·0416649	·0408919	·0396528	·0383113	·0368590	·0360882	·0352862	·0335835	·0317397
2·35	·0354712	·0346562	·0333562	·0319569	·0304512	·0296561	·0288314	·0270897	·0252182
2·45	·0301485	·0293115	·0279826	·0265608	·0250403	·0242415	·0234160	·0216830	·0198374
2·55	·0255903	·0247484	·0234182	·0220037	·0205015	·0197167	·0189030	·0172249	·0154493
2·65	·0216986	·0208662	·0195574	·0181750	·0167179	·0159613	·0151862	·0135826	·0119122
2·75	·0183846	·0175729	·0163039	·0149730	·0135817	·0128643	·0121333	·0106338	·0090936
2·85	·0155685	·0147867	·0135712	·0123063	·0109959	·0103255	·0096462	·0082671	·0068728
2·95	·0131799	·0124343	·0112824	·0100938	·0088744	·0082559	·0076334	·0063834	·0051426
3·05	·0111566	·0104519	·0093703	·0082642	·0071417	·0065776	·0060139	·0048964	·0038098
3·15	·0094448	·0087837	·0077762	·0067557	·0057322	·0052231	·0047184	·0037316	·0027943
3·25	·0079976	·0073816	·0034496	·0055154	·0045900	·0041349	·0036876	·0028262	·0020290
3·35	·0067750	·0062043	·0053474	·0044980	·0036677	·0032642	·0028714	·0021276	·0014587
3·45	·0057424	·0052162	·0044327	·0036649	·0029251	·0025702	·0022282	·0015923	·0010383
3·55	·0048704	·0043874	·0036744	·0029841	·0023290	·0020190	·0017236	·0011849	·0007317
3·65	·0041340	·0036924	·0030462	·0024286	·0018516	·0015827	·0013293	·0008770	·0005105
3·75	·0035120	·0031095	·0025261	·0019758	·0014703	·0012383	·0010223	·0006456	·0003526
3·85	·0029864	·0026207	·0020956	·0016071	·0011662	·0009671	·0007843	·0004729	·0002411
3·95	·0025421	·0022106	·0017393	·0013072	·0009242	·0007542	·0006003	·0003446	·0001633
4·05	·0021662	·0018665	·0014446	·0010634	·0007319	·0005874	·0004584	·0002500	·0001094
4·15	·0018480	·0015775	·0012006	·0008652	·0005792	·0004569	·0003494	·0001805	·0000726
4·25	·0015785	·0013347	·0009986	·0007042	·0004582	·0003551	·0002659	·0001299	·0000477
4·35	·0013499	·0011305	·0008313	·0005734	·0003624	·0002757	·0002020	·0000929	·0000310
4·45	·0011559	·0009587	·0006927	·0004672	·0002866	·0002139	·0001532	·0000663	·0000200
4·55	·0009911	·0008140	·0005778	·0003809	·0002266	·0001659	·0001161	·0000470	·0000127
4·65	·0008509	·0006920	·0004825	·0003107	·0001792	·0001286	·0000878	·0000333	·0000080
4·75	·0007315	·0005890	·0004033	·0002537	·0001418	·0000997	·0000664	·0000235	·0000050
4·85	·0006297	·0005020	·0003375	·0002073	·0001122	·0000772	·0000502	·0000165	·0000031
4·95	·0005419	·0004284	·0002828	·0001695	·0000888	·0000598	·0000379	·0000116	·0000019
5·05	·0004686	·0003661	·0002372	·0001388	·0000703	·0000464	·0000286	·0000081	·0000012
5·15	·0004051	·0003133	·0001993	·0001138	·0000557	·0000359	·0000215	·0000056	·0000007
5·25	·0003507	·0002685	·0001676	·0000931	·0000441	·0000278	·0000162	·0000039	·0000004
5·35	·0003040	·0002304	·0001411	·0000767	·0000350	·0000216	·0000122	·0000027	·0000002
5·45	·0002639	·0001980	·0001190	·0000631	·0000278	·0000167	·0000092	·0000019	·0000001
5·55	·0002294	·0001703	·0001005	·0000519	·0000221	·0000130	·0000069	·0000013	·0000001
5·65	·0001997	·0001468	·0000849	·0000428	·0000176	·0000101	·0000052	·0000009	
5·75	·0001740	·0001266	·0000719	·0000353	·0000140	·0000078	·0000039	·0000007	
5·85	·0001519	·0001094	·0000610	·0000292	·0000112	·0000061	·0000030	·0000004	
5·95	·0001328	·0000947	·0000517	·0000241	·0000089	·0000047	·0000022	·0000003	
6·05	·0001162	·0000820	·0000440	·0000200	·0000071	·0000037	·0000017	·0000002	
6·15	·0001018	·0000711	·0000374	·0000166	·0000057	·0000029	·0000013	·0000001	
6·25	·0000893	·0000618	·0000319	·0000138	·0000045	·0000022	·0000010	·0000001	
6·35	·0000785	·0000538	·0000272	·0000115	·0000036	·0000018	·0000007	·0000001	
6·45	·0000690	·0000468	·0000233	·0000095	·0000029	·0000014	·0000006		
6·55	·0000608	·0000408	·0000199	·0000080	·0000023	·0000011	·0000004		
6·65	·0000537	·0000357	·0000171	·0000067	·0000019	·0000008	·0000003		
6·75	·0000474	·0000312	·0000146	·0000056	·0000015	·0000007	·0000002		
6·85	·0000419	·0000273	·0000126	·0000047	·0000012	·0000005	·0000002		
6·95	·0000371	·0000239	·0000108	·0000039	·0000010	·0000004	·0000001		
7·05	·0000329	·0000210	·0000093	·0000033	·0000008	·0000003	·0000001		
7·15	·0000292	·0000184	·0000080	·0000028	·0000006	·0000003	·0000001		
7·25	·0000259	·0000162	·0000069	·0000023	·0000005	·0000002	·0000001		

Table II

UNIT-NORMAL LINEAR-LOSS INTEGRAL

$$L_{N^*}(u) \equiv \int_u^\infty (t - u)(2\pi)^{-\frac{1}{2}} e^{-\frac{1}{2}t^2}\, dt = f_{N^*}(u) - u\, G_{N^*}(u)$$

$$L_{N^*}(-u) = u + L_{N^*}(u)$$

Examples: $L_{N^*}(2.47) = .002199$; $L_{N^*}(-2.47) = 2.472199$

u	decimal prefix ↓	.00	.01	.02	.03	.04	.05	.06	.07	.08	.09	.10	tenths of the mean tabular difference (negative) 1	2	3	4	5
.0	.	3989	3940	3890	3841	3793	3744	3697	3649	3602	3556	3509	5	10	14	19	24
.1	.	3509	3464	3418	3373	3328	3284	3240	3197	3154	3111	3069	4	9	13	18	22
.2	.	3069	3027	2986	2944	2904	2863	2824	2784	2745	2706	2668	4	8	12	16	20
.3	.	2668	2630	2592	2555	2518	2481	2445	2409	2374	2339	2304	4	7	11	15	18
.4	.	2304	2270	2236	2203	2169	2137	2104	2072	2040	2009	1978	3	7	10	13	16
.5	.	1978	1947	1917	1887	1857	1828	1799	1771	1742	1714	1687	3	6	9	12	15
.6	.	1687	1659	1633	1606	1580	1554	1528	1503	1478	1453	1429	3	5	8	10	13
.7	.	1429	1405	1381	1358	1334	1312	1289	1267	1245	1223	1202	2	5	7	9	11
.8	.	1202	1181	1160	1140	1120	1100	1080	1061	1042	1023	1004	2	4	6	8	10
.9	.	1004	0986	0968	0950	0933	0916	0899	0882	0865	0849	0833	2	3	5	7	9
1.0	.0	8332	8174	8019	7866	7716	7568	7422	7279	7138	6999	6862	15	29	44	59	74
1.1	.0	6862	6727	6595	6465	6336	6210	6086	5964	5844	5726	5610	13	25	38	50	63
1.2	.0	5610	5496	5384	5274	5165	5059	4954	4851	4750	4650	4553	11	21	32	42	53
1.3	.0	4553	4457	4363	4270	4179	4090	4002	3916	3831	3748	3667	9	18	27	35	44
1.4	.0	3667	3587	3508	3431	3356	3281	3208	3137	3067	2998	2931	7	15	22	29	37
1.5	.0	2931	2865	2800	2736	2674	2612	2552	2494	2436	2380	2324	6	12	18	24	30
1.6	.0	2324	2270	2217	2165	2114	2064	2015	1967	1920	1874	1829	5	10	15	20	25
1.7	.0	1829	1785	1742	1699	1658	1617	1578	1539	1501	1464	1428	4	8	12	16	20
1.8	.0	1428	1392	1357	1323	1290	1257	1226	1195	1164	1134	1105	3	6	10	13	16
1.9	.0	1105	1077	1049	1022	0996	0970	0945	0920	0896	0872	0849	3	5	8	10	13
2.0	$.0^2$	8491	8266	8046	7832	7623	7418	7219	7024	6835	6649	6468	20	40	61	81	101
2.1	$.0^2$	6468	6292	6120	5952	5788	5628	5472	5320	5172	5028	4887	16	32	47	63	79
2.2	$.0^2$	4887	4750	4616	4486	4358	4235	4114	3996	3882	3770	3662	12	24	37	49	61
2.3	$.0^2$	3662	3556	3453	3352	3255	3159	3067	2977	2889	2804	2720	9	19	28	38	47
2.4	$.0^2$	2720	2640	2561	2484	2410	2337	2267	2199	2132	2067	2004	7	14	21	29	36
2.5	$.0^2$	2004	1943	1883	1826	1769	1715	1662	1610	1560	1511	1464	5	11	16	22	27
2.6	$.0^2$	1464	1418	1373	1330	1288	1247	1207	1169	1132	1095	1060	4	8	12	16	20
2.7	$.0^2$	1060	1026	0993	0961	0929	0899	0870	0841	0814	0787	0761	3	6	9	12	15
2.8	$.0^3$	7611	7359	7115	6879	6650	6428	6213	6004	5802	5606	5417	22	44	66	88	110
2.9	$.0^3$	5417	5233	5055	4883	4716	4555	4398	4247	4101	3959	3822	16	32	48	64	80
3.0	$.0^3$	3822	3689	3560	3436	3316	3199						12	25	37	50	62
							3199	3087	2978	2873	2771	2673	11	21	32	42	53
3.1	$.0^3$	2673	2577	2485	2396	2311	2227						9	18	27	36	45
							2227	2147	2070	1995	1922	1852	8	15	23	30	38
3.2	$.0^3$	1852	1785	1720	1657	1596	1537						6	13	19	25	32
							1537	1480	1426	1373	1322	1273	5	11	16	21	26
3.3	$.0^3$	1273	1225	1179	1135	1093	1051						4	9	13	18	22
							1051	1012	0973	0937	0901	0867	4	7	11	15	18
3.4	$.0^4$	8666	8335	8016	7709	7413	7127						31	62	92	123	154
							7127	6852	6587	6331	6085	5848	26	51	77	102	128
3.5	$.0^4$	5848	5620	5400	5188	4984	4788						21	42	64	85	106
							4788	4599	4417	4242	4073	3911	18	35	53	70	88
3.6	$.0^4$	3911	3755	3605	3460	3321	3188						14	29	43	58	72
							3188	3059	2935	2816	2702	2592	12	24	36	48	60
3.7	$.0^4$	2592	2486	2385	2287	2193	2103						10	20	29	39	49
							2103	2016	1933	1853	1776	1702	8	16	24	32	40
3.8	$.0^4$	1702	1632	1563	1498	1435	1375						7	13	20	26	33
							1375	1317	1262	1208	1157	1108	5	11	16	21	27
3.9	$.0^4$	1108	1061	1016	0972	0931	0891						4	9	13	17	22
							0891	0853	0816	0781	0747	0715	4	7	11	14	18

Interpolation by *mean* tabular difference:
$L_{N^*}(2.473) = .002199 - .000021 = .002178.$
Error never exceeds .3% of true value.

Interpolation by *exact* tabular difference:
$L_{N^*}(2.473) = .002199 - (.3)(.000067) = .002179.$
Error never exceeds .03% of true value.

9 780471 383499